Renewable Energy

Renewable Energy

POWER FOR A SUSTAINABLE FUTURE

THIRD EDITION

Edited by Godfrey Boyle

OXFORD

UNIVERSITY PRESS

Published by Oxford University Press, Great Clarendon Street, Oxford OX2 6DP in association with
The Open University, Walton Hall, Milton Keynes MK7 6AA

Oxford University Press is a department of the University of Oxford.
It furthers the University's objective of excellence in research, scholarship,
and education by publishing worldwide. Oxford is a registered trade mark of
Oxford University Press in the UK and in certain other countries

First published 1996. Second edition 2004. This edition 2012.

Edited and designed by The Open University

Typeset by SR Nova Pvt. Ltd, Bangalore, India

Printed and bound in the United Kingdom by Latimer Trend and Company Ltd, Plymouth

This book forms part of Open University teaching materials. Details of these and other Open University modules can be obtained
from the Student Registration and Enquiry Service, The Open University, PO Box 197, Milton Keynes MK7 6BJ, United Kingdom
(tel. +44 (0)845 300 60 90, email general-enquiries@open.ac.uk).

www.open.ac.uk

British Library Cataloguing in Publication Data available on request

Library of Congress Cataloguing in Publication Data available on request

ISBN 978 0 19 954533 9

3.1

Preface

How can the planet's future energy demands be supplied, cleanly, safely and sustainably?

At current consumption levels, our existing energy systems have many deleterious effects on human health and the natural environment. In particular, carbon dioxide and other greenhouse gases released by fossil fuel burning threaten to cause unprecedented changes in the Earth's climate, with mainly adverse consequences. Moreover, by the middle of this century, if current trends continue, the world's population is likely to increase to some nine billion: in the absence of major improvements in energy efficiency, world primary energy demand could increase by around 50%, presenting even greater challenges.

There is now a broad consensus that the world must shift to low- or zero-carbon energy sources if the impacts of climate change are to be mitigated. The renewable energy sources described in this book are essentially carbon-free or low-carbon and appear to be generally more sustainable than fossil or nuclear fuels, although many technologies are still under development and, at the time of writing, the costs of some are high.

This third edition of *Renewable Energy* reflects the remarkable progress that has been made in the field since the publication of the first edition in 1996: renewable energy sources are rapidly entering the mainstream. With that progress has come greatly increased recognition among professionals, politicians and the public that renewables could provide a major proportion of the world's energy needs by the middle of the twenty-first century, given adequate investment in research, development and deployment.

However, if that potential is to be realized, the world will need many more professional people with a thorough knowledge of renewable energy systems, their underlying physical and technological principles, their economics, their environmental impact and how they can be integrated into the world's energy systems.

Renewable Energy is intended to provide a basis for this knowledge. It is aimed at students and staff in universities, and at professionals, policymakers and members of the public interested in creating a sustainable energy future.

This book has been designed both to to be used on its own and also to work alongside a companion volume, *Energy Systems and Sustainability*, which considers current energy demand and usage, and describes traditional energy sources, outlining the economics and consequences (environmental and otherwise) of using coal, oil, gas and nuclear power.

We hope that *Renewable Energy* will contribute to an improved understanding of renewable energy as a key potential solution to the sustainability problems of our present energy systems. We also hope that it conveys something of the enthusiasm we feel for this complex, fascinating and increasingly important subject.

Godfrey Boyle
Professor of Renewable Energy, The Open University

New to this edition
This edition has been updated throughout to reflect new data, advances in technology, new developments around the world and the integration of renewables into existing energy systems. It demonstrates the increased appreciation of the economic, social and environmental impacts of renewables. The appendix on unit conversions and costing analysis has been expanded to form two appendices. A summary at the end of each chapter consolidates the information contained.

About the authors

Godfrey Boyle is Professor of Renewable Energy at The Open University

Bob Everett is a Lecturer in Renewable Technology at The Open University

Dick Morris is a retired Senior Lecturer in Systems at The Open University

Jonathan Scurlock is renewable energy advisor to the UK National Farmers' Union and a visiting research fellow at The Open University

Janet Ramage is a visiting lecturer at The Open University

David Elliott is Professor of Technology Policy at The Open University

Derek Taylor is an architect and visiting lecturer at The Open University

Les Duckers is Principal Lecturer in Environmental Management at Coventry University

John Garnish is a consultant and former director of Geothermal programmes at the European Commission

Geoff Brown. The late Geoff Brown was Professor of Earth Sciences at The Open University

Gary Alexander is a retired Senior Lecturer in Telematics at The Open University

Contents

Chapter 1

Introducing renewable energy

By Godfrey Boyle, Bob Everett and Gary Alexander

1.1 Introduction

Renewable energy sources, derived principally from the enormous power of the Sun's radiation, are at once the most ancient and the most modern forms of energy used by humanity.

Solar power, both in the form of direct solar radiation and in indirect forms such as bioenergy, water or wind power, was the energy source upon which early human societies were based. When our ancestors first used fire, they were harnessing the power of photosynthesis, the solar-driven process by which plants are created from water and atmospheric carbon dioxide. Societies went on to develop ways of harnessing the movements of water and wind, both caused by solar heating of the oceans and atmosphere, to grind corn, irrigate crops and propel ships. As civilizations became more sophisticated, architects began to design buildings to take advantage of the Sun's energy by enhancing their natural use of its heat and light, so reducing the need for artificial sources of warmth and illumination.

Technologies for harnessing the power of Sun, firewood, water and wind continued to improve right up to the early years of the industrial revolution. However, by then the advantages of coal, the first of the fossil fuels to be exploited on a large scale, had become apparent. These highly-concentrated energy sources soon displaced wood, wind and water in the homes, industries and transport systems of the industrial nations. Today the fossil fuel trio of coal, oil and natural gas provides over 80% of the world's energy.

Concerns about the adverse environmental and social consequences of fossil fuel use, such as air pollution or mining accidents, and about the finite nature of supplies, have been voiced intermittently for several centuries. But it was not until the 1970s, with the steep price rises of the 'oil crisis' and the advent of the environmental movement, that humanity began to take more seriously the prospect of fossil fuels 'running out', and the possibility that their continued use could be destabilizing the planet's natural ecosystems and the global climate (see Section 1.3 below).

The development of nuclear energy following Second World War raised hopes of a cheap, plentiful and clean alternative to fossil fuels. However, nuclear power development has stalled in some countries in recent years, due to increasing concerns about safety, cost, waste disposal and weapons proliferation, although in other countries nuclear expansion is continuing.

Continuing concerns about the 'sustainability' of both fossil and nuclear fuel use have been a major catalyst of renewed interest in the renewable energy sources in recent decades. Ideally, a **sustainable energy source** is one that:

■ is not substantially depleted by continued use

■ does not entail significant pollutant emissions or other environmental problems

■ does not involve the perpetuation of substantial health hazards or social injustices.

In practice, only a few energy sources come close to this ideal, but as this and subsequent chapters will show, the 'renewables' (see Section 1.4 for an explicit definition) appear generally more sustainable than fossil or nuclear fuels: they are essentially inexhaustible and their use usually entails fewer health hazards and much lower emissions of greenhouse gases or other pollutants.

Before going on to introduce the renewables in more detail, it is first useful to review some basic energy concepts that may be unfamiliar to readers who do not have a scientific background. For a more detailed discussion of basic energy concepts, see, for example, *Energy Systems and Sustainability* (Everett et al., 2012).

Force, energy and power

The word energy is derived from the Greek *en* (in) and *ergon* (work). The scientific concept of **energy** (broadly defined as 'the capacity to do work') serves to reveal the common features in processes as diverse as burning fuels, propelling machines or charging batteries. These and other processes can be described in terms of diverse **forms of energy**, such as *thermal* energy (heat), *chemical* energy (in fuels or batteries), *kinetic* energy (in moving substances), *electrical* energy, *gravitational* potential energy, and various others.

In the main, this book uses the international SI system of units. The conversion factors between these and other units commonly used in the field of energy can be found in Appendix A.

The scientific world agreed on a single set of units, the SI system (Système International d'Unités) in 1960. There are seven basic units, of which the three which are relevant here are the metre (m), the kilogram (kg) and the second (s). The units for many other quantities are derived from the basic units. For some of the derived units, such as metres per second (m s^{-1}), the unit for speed, the base units are obvious. Others have been given specific names, such as the:

■ newton (N) for force

■ the joule (J) for energy

■ the watt (W) for power.

Large quantities are specified using multipliers (see Table 1.1 for some examples). Thus, a kilowatt (written as 1 kW) is a thousand watts.

Note that, unless otherwise stated, in this book the multiplier M and the terms billion and trillion are as defined in Table 1.1 (Appendix A has a fuller list of multiplier prefixes and more details about international variations in definition).

Table 1.1 Multiplier prefixes

Symbol	Prefix	Multiply by	... as a power of ten
k	kilo-	one thousand	10^3
M	mega-	one million	10^6
G	giga-	one billion (one thousand million)	10^9
T	tera-	one trillion (one million million)	10^{12}
P	peta-	one quadrillion (one billion million)	10^{15}
E	exa-	one quintillion (one billion billion)	10^{18}

In order to change the motion of any object, a force is needed, and the formal SI unit for force, the **newton (N)**, is defined as that force which will accelerate a mass of one kilogram (kg) at a rate of one metre per second per second (m s^{-2}). Expressed more generally:

force (N) = mass (kg) $\times$ acceleration (m s^{-2}).

Thus the derived unit, the newton, is equivalent to kg m s^{-2}.

In the real world, force is often needed to move an object even at a steady speed, but this is because there are opposing forces such as friction to be overcome.

Whenever a force is accelerating something or moving it against an opposing force, it must be providing energy. The unit of energy, the **joule (J)**, is defined as the energy supplied by a force of one newton in causing movement through a distance of 1 metre. In general:

energy (J) = force (N) $\times$ distance (m).

So a joule is dimensionally equivalent to one newton metre (N m).

The terms *energy* and *power* are often used informally as though they were synonymous (e.g. wind energy/wind power), but in scientific discussion it is important to distinguish them. **Power** is the *rate* at which energy is being converted from one form to another, or transferred from one place to another. Its unit is the **watt (W)**, and one watt is defined as one joule per second (hence a watt is equivalent to one J s^{-1}). A 100 watt incandescent light bulb, for example, is converting one hundred joules of electrical energy into light (and 'waste' heat) each second. In popular speech, 'power' is often taken to denote *electricity,* but scientifically it applies to any situation where energy is transferred or converted. Occasionally (for example in Chapter 9) a power rating maybe specifically defined as MW$_e$ or MW$_t$ where the subscripts e and t refer to electrical and thermal energy respectively.

In practice, it is often convenient to measure energy in terms of the power used over a given time period. If the power of an electric heater is 1 kW, and it runs for an hour, we say that it has consumed one **kilowatt-hour (kWh)** of energy. As 1 kilowatt is 1000 watts, from the definition of the watt this is 1000 joules per second. There are 3600 seconds in an hour, so:

1 kWh = 3600 $\times$ 1000 = 3.6 $\times 10^6$ joules (3.6 MJ).

Energy is also often measured simply in terms of quantities of fuel, and national energy statistics often use the units 'tonnes of coal equivalent' (tce), 'tonnes of oil equivalent' (toe) or even 'barrels of oil equivalent' (boe). The most common units and their conversion factors are listed in Appendix A.

Energy conservation: the First Law of Thermodynamics

The renewable energy technologies described in this book transform one form of energy into another (the final form in many cases being electricity). In any such transformation of energy, the total quantity of energy remains unchanged. This principle, that energy is always conserved, is expressed by the **First Law of Thermodynamics**. So if the electrical energy *output* of a power station, for example, is less than the energy content of the fuel *input,* then some of the energy must have been converted to another form (usually waste heat).

If the total quantity of energy is always the same, how can we talk of *consuming* it? Strictly speaking, we don't: we just convert it from one form into other forms. We consume *fuels*, which are sources of readily available energy. We may burn fuel in a vehicle engine, converting its stored chemical energy into heat and then into the kinetic energy of the moving vehicle. By using a wind turbine we can extract kinetic energy from moving air and convert it into electrical energy, which can in turn be used to heat the filament of an incandescent lamp causing it to radiate light energy.

Forms of energy

At the most basic level, the diversity of energy forms can be reduced to four:

- kinetic
- gravitational
- electrical
- nuclear.

Kinetic energy

The **kinetic energy** possessed by any moving object is equal to half the mass (m) of the object times the square of its velocity (v), i.e.:

$$\text{kinetic energy} = \tfrac{1}{2}mv^2$$

where energy is in joules (J), mass in kilograms (kg) and velocity in metres per second (m s^{-1}).

Less obviously, the kinetic energy *within* a material determines its temperature. All matter consists of atoms, or combinations of atoms called molecules. In a gas, such as the air that surrounds us, these move freely. In a solid or a liquid, they form a more or less loosely linked network in which every particle is constantly vibrating. **Thermal energy**, or heat, is the name given to the energy associated with this rapid random motion. The higher the temperature of a body, the faster its molecules are moving. In the temperature scale that is most natural to scientific theory, the Kelvin (K) scale, zero corresponds to zero molecular motion. In the more commonly used Celsius scale of temperature (written as °C), the size of one degree is the same as 1 kelvin, but zero corresponds to the freezing point of water and 100 °C to the boiling point of water at atmospheric pressure. The two scales are therefore related by a simple formula:

temperature (K) = temperature (°C) + 273.

Gravitational energy

A second fundamental form of energy is **gravitational energy**. On Earth, an input of energy is required to lift an object because the gravitational pull of the Earth opposes that movement. If an object, such as an apple, is lifted above your head, the input energy is stored in a form called **gravitational potential energy** (often just 'potential energy' or 'gravitational energy'). That this stored energy exists is obvious if you release the apple and observe the subsequent conversion to kinetic energy. The gravitational force pulling an object towards the Earth is called the *weight* of the object, and is equal to its *mass, m*, multiplied by the acceleration due to gravity, g (which is 9.81 m s^{-2}, although for rough calculations needing less than 2% precision a value of 10 m s^{-2} is often used). Note that although everyday language may treat mass and weight as the same, science does not. The potential energy (in joules) stored in raising an object of mass m (in kilograms) to a height H (in metres) is given by the following equation (see Figure 1.1):

potential energy = force × distance = weight × height = $m \times g \times H$.

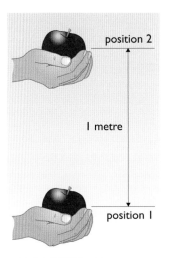

Figure 1.1 The amount of energy required to raise a 100 g apple vertically through 1 m is approximately one joule (1 J)

Electrical energy

Gravity is not the only force influencing the objects around us. On a scale far too small for the eye to see, electrical forces hold together the atoms and molecules of all materials; gravity is an insignificant force at the molecular level. The **electrical energy** associated with these forces is the third of the basic forms. Every atom can be considered to consist of a cloud of electrically charged particles, electrons, moving incessantly around a central nucleus. When atoms bond with other atoms to form molecules, the distribution of electrons is changed, often with dramatic effect. Thus **chemical energy**, viewed at the atomic level, can be considered to be a form of electrical energy. When a fuel is burned, the energy liberated (the chemical energy) is converted into heat energy. Essentially, the electrical energy released as the electrons are rearranged (that is, the net release of energy from the breaking and forming of bonds) is converted to the kinetic energy of the molecules of the combustion products.

A more familiar form of electrical energy is that carried by **electric currents** – organized flows of electrons in a material, usually a metal. In metals, one or two electrons from each atom can become detached and move freely through the lattice structure of the material. These 'free electrons' allow metals to carry electrical currents.

To maintain a steady current of electrons requires a constant input of energy because the electrons continually lose energy in collisions with the metal lattice (which is why wires get warm when they carry electric currents). Voltage (in volts) is a measure of the electrical 'potential difference' between two points in an electrical circuit, analogous to height in the measurement of gravitational potential energy (see above). The power (in watts) delivered by an electrical supply, or used by an appliance, is given by multiplying the voltage (in volts) by the current (measured in amperes, or 'amps'):

power (W) = voltage (V) × current (I).

In a typical power station, the input fuel is burned and used to produce high-pressure steam, which drives a rotating turbine. This in turn drives an electrical generator, which operates on a principle discovered by Michael Faraday in 1832: a voltage is induced in a coil of wire that spins in a magnetic field. Connecting the coil to an electric circuit will then allow a current to flow. The electrical energy can in turn be transformed into heat, light, motion or whatever, depending upon what is connected to the circuit. Electricity is often used in this way, as an *intermediary* form of energy: it allows energy released from one source to be converted to another quite different form, usually at some distance from the source.

Another form of electrical energy is that carried by electromagnetic radiation. More properly called **electromagnetic energy**, this is the form in which, for example, solar energy reaches the Earth. Electromagnetic energy is radiated in greater or lesser amounts by every object. It travels as a wave that can carry energy through empty space. The length of the wave (its wavelength) characterizes its form, which includes X-rays, ultraviolet and infrared radiation, visible light, radio waves and microwaves.

Nuclear energy

The fourth and final basic form of energy, bound up in the central nuclei of atoms, is called **nuclear energy**. The technology for releasing it was developed during the Second World War for military purposes, and subsequently in a more controlled version for the commercial production of electricity. Nuclear power stations operate on much the same principles as fossil fuel plants, except that the furnace in which the fuel burns is replaced by a nuclear reactor in which atoms of uranium are split apart in a 'fission' process that generates large amounts of heat.

The energy source of the Sun is also of nuclear origin. Here the process is not nuclear fission but nuclear *fusion*, in which hydrogen atoms fuse to form helium atoms – such enormous numbers of these reactions take place that massive amounts of solar radiation are generated in the process. Attempts to imitate the Sun by creating power-producing nuclear fusion reactors have been the subject of many decades of research and development effort but have yet to come to fruition.

Conversion, efficiencies and capacity factors

When energy is converted from one form to another, the useful output is never as much as the input. The ratio of the useful output to the required input (usually expressed as a percentage) is called the **efficiency** of the process:

percentage efficiency = (energy output/energy input) × 100.

This efficiency can be as high as 90% in a water turbine or well-run electric motor, around 35–40% in a coal-fired power station (if the 'waste' heat is not put to use), and as low as 10–20% in a typical internal combustion engine. Some inefficiencies can be avoided by good design, but others are inherent in the nature of the type of energy conversion.

In the systems mentioned above, the difference between the high and low conversion efficiencies is because the latter are **heat engines** involving the conversion of heat into mechanical or electrical energy. Heat, as already indicated, is the kinetic energy of randomly moving molecules, an essentially chaotic form of energy. No machine can convert this chaos completely into the ordered state associated with mechanical or electrical energy. This is the essential message of the *Second Law of Thermodynamics*: that there is necessarily a limit to the efficiency of any heat engine. Some energy must always be lost to the external environment, usually as low-temperature heat. (Box 2.4 of Chapter 2 looks at the efficiency of heat engines in more detail).

When considering the economics of a power plant, rather than just its efficiency, it is useful to have a measure of its productivity in practice.

One measure of this is the plant's **capacity factor (CF)**: its actual output over a given period of time divided by the maximum possible output. The units for the output quantities can be kWh, MWh, GWh, etc., and the result can be expressed as either a fraction or a percentage.

There are 8760 hours in a year (365 days × 24 hrs/day = 8760 hours), so a 1 MW plant running constantly at its full rated capacity for one year would generate 8760 MWh of output, and would have an annual capacity factor of 1, or 100%.

A 1 MW wind turbine might, in practice, typically produce 3000 MWh of electricity in a year (because the wind doesn't always blow at the full rated speed for which the turbine is designed) – in such a case its annual capacity factor would be:

$$(3000/8760) = 0.342 = 34.2\%$$

The period to which a capacity factor relates is not always a year – weekly or monthly capacity factors are often quoted.

The terms 'plant factor' and 'load factor' are also sometimes used as synonyms for 'capacity factor' in the context of power systems.

1.2 Present-day energy use

World energy supplies

The energy used by a final consumer is usually the end result of a series of energy conversions. For example, energy from burning coal may be converted in a power station to electricity, which is then distributed to households and used in immersion heaters to heat water in domestic hot water tanks. The energy released when the coal is burned is called the **primary energy** required for that use. The amount of electricity reaching the consumer, after conversion losses in the power station and transmission losses in the electricity grid, is the **delivered energy**. After further losses in the tank and pipes, a final quantity, called the **useful energy**, comes out of the hot tap.

World total annual consumption of all forms of primary energy increased more than tenfold during the twentieth century, and by the year 2009 had reached an estimated 502 EJ (exajoules), or some 12 000 million tonnes of oil equivalent (Mtoe) (Figure 1.2). As the figure reveals, fossil fuels provided more than four fifths of the total. The world population in 2009 was some 6.8 billion, so the annual average energy consumption per person was about 74 GJ (gigajoules), equivalent to the energy content of approximately 5.5 litres of oil per day for every man, woman and child.

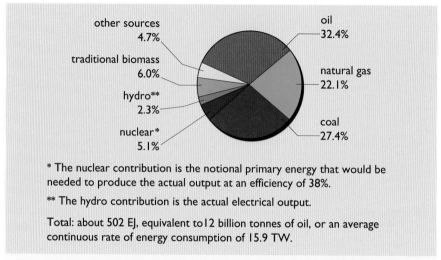

* The nuclear contribution is the notional primary energy that would be needed to produce the actual output at an efficiency of 38%.

** The hydro contribution is the actual electrical output.

Total: about 502 EJ, equivalent to12 billion tonnes of oil, or an average continuous rate of energy consumption of 15.9 TW.

Figure 1.2 Percentage contributions to world primary energy consumption in 2009. 'Other sources' are 'new' biomass, solar and geothermal energy, and energy from wind, wave, tide and wastes (sources: authors' estimates based on BP, 2010; IEA, 2009; WWEA, 2010)

But these figures conceal major differences. The average North American consumes more than 250 GJ per year, most people in Europe use roughly half this amount, and many of those in the poorer countries of the world less than one fifth – much of it in the form of local 'biofuels' (see Chapter 4).

How much do renewables contribute to world energy supplies? As Figure 1.2 shows, traditional biomass, hydro power and a range of other renewable sources contributed an estimated 13% of world primary energy in 2009. Figure 1.3 gives a more detailed breakdown, for the year 2008.

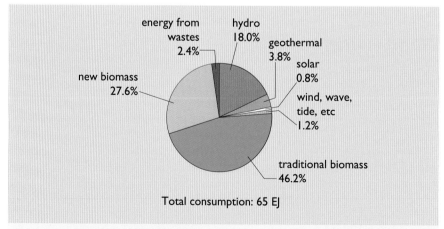

Figure 1.3 Chart showing percentage breakdown of individual renewable energy sources' contributions to world primary energy supplies in 2008 (sources: authors' estimates based on IEA, 2009; BP, 2010)

The largest contribution is an estimated value of 30 EJ from 'traditional biomass' (wood, straw, dung, etc.) mainly used in developing countries. Since most of this isn't traded, it often doesn't enter into national economic statistics and its true magnitude is only known approximately. The next largest category is 'new biomass'. This includes wood and other crops specifically grown for energy purposes, biogas, and biofuels such as ethanol and biodiesel. This is a commodity that is likely to be traded and so its magnitude is more certain. Hydro power is the next largest category, supplying over 2% of the world's primary energy. 'New biomass', together with energy from wastes, geothermal energy, solar energy and energy from wind, wave and tidal power make up the 'other sources' shown in Figure 1.2.

In practice, many electricity-generating fossil fuelled and renewable energy technologies produce large amounts of unused 'waste' heat. Renewable energy proportions based on *primary* energy may thus give a misleading picture. Proportions of renewable energy in national (and global) statistics are now often quoted in terms of *gross final energy consumption* (see Box 1.1).

BOX 1.1 Primary energy, delivered energy and gross final energy

Figure 1.2 showed the estimated total global *primary energy* consumption for 2009. About 25–30% of this was turned into waste heat in power stations, most of which is dumped uselessly into seas, lakes or the sky via cooling towers. Only a small percentage of this waste heat is put to good use in district heating schemes. This wastage occurs in fossil fuelled power stations, nuclear power plants and renewable power plants fuelled by wood or landfill gas. For example a landfill gas plant may consume 4 kWh of (primary) gas to produce 1 kWh of output electricity. However, other technologies such as wind, PV and hydro power can generate useful electricity directly with minimal losses.

When comparing technologies or compiling national statistics, those based on *delivered energy*, i.e. the fuel, useful heat or electricity actually received by an end user, probably give a better representation of the overall picture.

The 2009 European Union Renewable Energy Directive (CEC, 2009) sets out requirements for expressing future national renewable energy contributions in terms of the **gross final energy consumption**. This is basically defined as the *delivered energy* to the end users but with two additional small contributions: firstly, the losses in transmission of electricity and heat (in district heating schemes); and secondly the electricity and heat consumed in energy industries (such as in oil refineries or within power stations). However, it does *not* include the very large waste heat losses in electricity generation that feature in primary energy figures.

The overall effect of using statistics based on gross final energy is to give more prominence to hydro, wind and PV technologies and less to low efficiency electricity generation technologies based on fossil fuels, nuclear power or renewables where the waste heat is not put to good use. It also focuses attention on the need to improve electricity generation efficiencies.

How long will the world's fossil fuel reserves last? At current consumption rates, it is estimated that world coal reserves could last for about 120 years, oil for approximately 45 years and natural gas for around 60 years (BP, 2010). However, in the more immediate future there are likely to be serious constraints on the *rate* at which fossil fuels can be produced, particularly oil. Existing oilfields have a limited life; once exhausted they have to be replaced with new ones. In order just to maintain the world's oil production at its current level, a large number of new oil fields will have to be developed. Even more challenging is the need for new fields to be continuously *discovered* (Figure 1.4). According to the International Energy Agency, the 'easy oil' has been largely used up. What remains is likely to be more expensive and in difficult areas such as the Arctic or in deep offshore wells (IEA, 2010a).

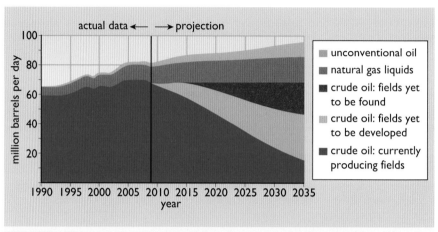

Figure 1.4 An International Energy Agency chart indicating the challenges involved in maintaining current levels of conventional oil production (source: IEA, 2010a)

According to a study of global oil depletion by the UK Energy Research Centre:

> a peak of conventional oil production before 2030 appears likely and there is a significant risk of a peak before 2020.
>
> (Sorrell et al., 2009)

So although large worldwide reserves of oil will remain, the overall production of *conventional* oil seems likely to 'plateau' or even decline. This has serious implications for the UK, where oil production peaked in 1999 and gas production peaked in 2000.

Energy use in the UK

In the UK, as in most countries, energy demand is categorized in official statistics into four main sectors: *domestic, commercial and institutional (or services), industrial* and *transport*.

As Figure 1.5 shows, almost one third of UK primary energy is lost in the process of conversion and delivery – most of it in the form of 'waste' heat from power stations. These losses are greater than the country's demand for space and water heating energy. Even when energy has been delivered to customers in the various sectors, it is often used very wastefully. The UK energy system is described in more detail in Chapter 10 of this book.

In the UK, the contribution of renewables to *primary* energy supply in 2009 was quite small: some 3.3% (see Figures 1.5 and 1.6(a)). The percentage

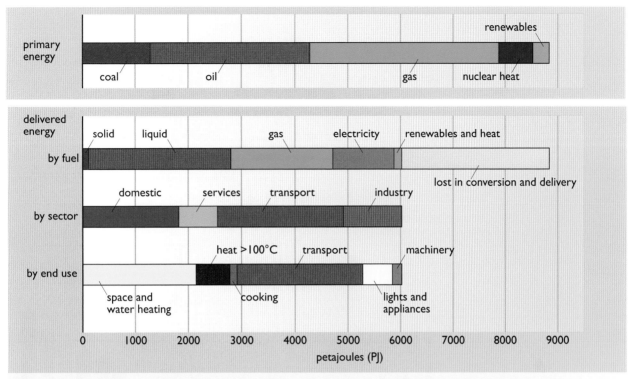

Figure 1.5 UK primary and delivered energy use, 2009 (sources: DECC, 2010a, DECC, 2010b). Note: in the second bar, 'electricity' includes renewable electricity; 'renewables and heat' includes biofuels for transport and heat from CHP plants

contribution of renewables to electricity supplies was somewhat larger, however (see Figure 1.6(b)). Some 6.7% of UK electricity came from renewable and waste sources, mainly in the form of wind power, with smaller contributions from biomass, waste and landfill gas combustion and hydro power (DECC, 2010b). The UK Government aims to increase the proportion of primary energy from renewables to 15% by 2020 (see Section 1.5).

1.3 Fossil fuels and climate change

Society's current use of fossil and nuclear fuels has many adverse consequences. These include air pollution, acid rain, the depletion of natural resources and the dangers of nuclear radiation. This brief introduction concentrates on one of these problems: global climate change caused by emissions of greenhouse gases from fossil fuel combustion.

The surface temperature of the Earth establishes itself at an equilibrium level where the incoming energy from the Sun balances the outgoing infrared energy re-radiated from the surface back into space (see Chapter 2, Figure 2.5). If the Earth had no atmosphere its surface temperature would be −18 °C; but its atmosphere, which includes 'greenhouse gases' – principally, water vapour, carbon dioxide and methane – acts like the panes of a greenhouse, allowing solar radiation (which lies in the range from ultra-violet to short wave infrared) to enter but inhibiting the outflow of long-wave infrared radiation. The natural 'greenhouse effect' that these

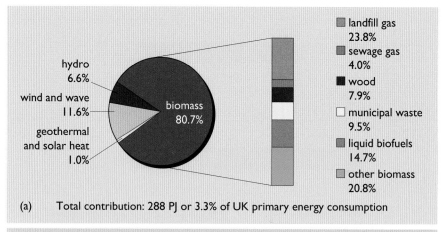

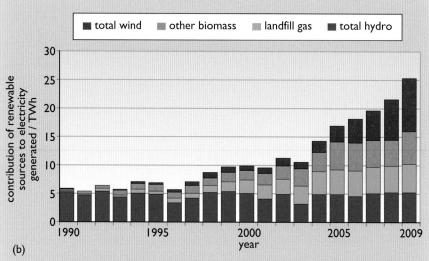

Figure 1.6 (a) Primary energy contributions from renewable energy in the UK, 2009. The total, 6875 Mtoe, is equivalent to 288 PJ. The main contributors were wind, biomass in various forms, and hydro power (b) Growth in electricity generation from renewable sources in the UK 1990–2009. In 2009 renewables contributed 6.7% of UK electricity (source: DECC 2010b)

gases cause is essential in maintaining the Earth's surface temperature at a level suitable for life at around 15 °C.

Since the Industrial Revolution, however, human activities have been adding extra greenhouse gases to the atmosphere. The principal contributor to these increased emissions is carbon dioxide from the combustion of fossil fuels. Humanity's rate of emission of CO_2 from these fuels has increased enormously since 1950 (see Figure 1.7).

There have also been significant additional contributions from emissions of methane.

Scientists estimate (IPCC, 2007a) that these 'anthropogenic' (human-induced) emissions caused a rise in the Earth's global mean surface temperature of approximately 0.7 °C between 1950 and 2005 (Figure 1.8). If emissions are not curbed, the Intergovernmental Panel on Climate Change (IPCC) estimates

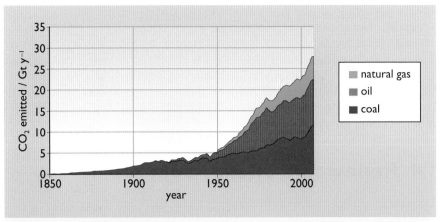

Figure 1.7 CO_2 emissions from the burning of fossil fuels 1850–2009 (sources: Boden et al., 2010; IEA, 2010b; BP, 2010)

that the Earth's surface temperature could rise by between 1.4 and 5.8 °C (depending on the assumptions made) by the end of the twenty-first century. Such rises would probably be associated with an increased frequency of climatic extremes, such as floods or droughts, and serious disruptions to agriculture and natural ecosystems. The thermal expansion of the world's oceans could mean that sea levels would rise by around 0.5 m by the end of the century, which could inundate some low-lying areas. Beyond 2100, or perhaps before, much greater sea level rises could occur if major Antarctic ice sheets were to melt.

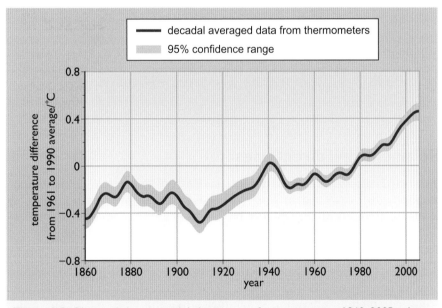

Figure 1.8 Observed changes in global average surface temperature 1860–2005, relative to corresponding averages for the period 1961–1990. The term '95% confidence range' indicates that there is only a one in 20 chance of a measurement lying outside this range (source: IPCC, 2007b)

The threat of global climate change, mainly caused by carbon dioxide emissions from fossil fuel combustion, is one of the main reasons why there is a growing consensus on the need to reduce such emissions. In order to ensure that global mean temperature rises do not exceed 2 °C above pre-industrial levels by 2050, studies show that global carbon emissions will need to be reduced by approximately 80% by that date. This implies that global CO_2 emissions need to peak almost immediately and then fall sharply over the course of the rest of this century (Allen et al., 2009). Emission reductions on this scale will inevitably involve a switch to low- or zero-carbon energy sources such as renewables.

1.4 Renewable energy sources

Renewable energy can be defined as:

> *energy obtained from the continuous or repetitive currents of energy recurring in the natural environment*
>
> (Twidell and Weir, 1986)

or as

> *energy flows which are replenished at the same rate as they are 'used'*
>
> (Sorensen, 2000)

From Figure 1.9, which summarizes the origins and magnitudes of the Earth's renewable energy sources, it is clear that their principal source is solar radiation. Approximately 30% of the 5.4 million EJ per year arriving at the Earth is reflected back into space. The remaining 70% is, in principle, available for use on Earth, and amounts to approximately 3.8 million EJ, more than 8200 times the rate of consumption of fossil and nuclear fuels in 2009, some 462 EJ. (If biofuels and hydro power are included, total world energy consumption was 502 EJ, as stated earlier.) Two non-solar, renewable energy sources are also shown on the figure: the motion of the ocean tides and geothermal heat from the Earth's interior, which manifests itself in convection in volcanoes and hot springs, and in conduction in rocks.

Solar energy: direct uses

Solar radiation can be converted into useful energy *directly*, using various technologies. Absorbed in solar 'collectors', it can provide hot water or space heating. Buildings can also be designed with 'passive solar' features that enhance the contribution of solar energy to their space heating and lighting requirements.

Solar energy can also be concentrated by mirrors to provide high-temperature heat for generating electricity. Such 'solar thermal-electric' power stations are in commercial operation in countries like the USA and Spain. *Solar thermal energy* conversion is described in Chapter 2.

Solar radiation can also be converted directly into electricity using photovoltaic (PV) modules, normally mounted on the roofs or facades of buildings. At the time of writing, electricity from photovoltaics is more expensive than that from conventional sources, but prices are falling fast and the industry is expanding rapidly. *Solar photovoltaics* is described in Chapter 3.

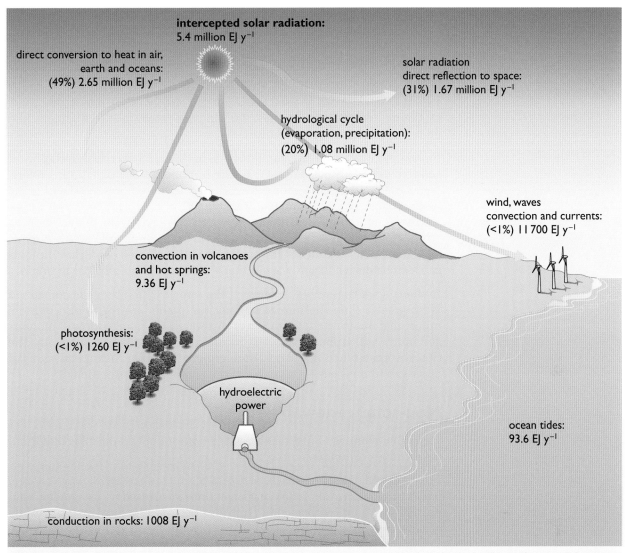

Figure 1.9 The various forms of renewable energy depend primarily on incoming solar radiation, which totals some 5.4 million EJ per year

Solar energy: indirect uses

Solar radiation can be converted to useful energy *indirectly*, via other energy forms. A large fraction of the radiation reaching the Earth's surface is absorbed by the oceans, warming them and adding water vapour to the air. The water vapour condenses as rain to feed rivers, into which dams and turbines can be located to extract the energy of the flowing water.

Hydropower, described in Chapter 5, has steadily grown during the twentieth century, and at the time of writing provides about a sixth of the world's electricity.

Sunlight falls in a more perpendicular direction in tropical regions and more obliquely at high latitudes, heating the tropics to a greater degree than the polar regions. The result is a massive heat flow towards the poles, carried by currents in the oceans and the atmosphere. The energy in such currents can be harnessed, for example by wind turbines. *Wind power*, described in Chapter 7, has developed on a large scale only in the past few decades, but is one of the fastest-growing of the 'new' renewable sources of electricity.

Where winds blow over long stretches of ocean, they create waves, and a variety of devices can be used to extract that energy. *Wave power*, described in Chapter 8, is attracting new funding for research, development and demonstration in several countries.

Bioenergy, discussed in Chapter 4, is another indirect manifestation of solar energy. Through photosynthesis in plants, solar radiation converts water and atmospheric carbon dioxide into carbohydrates, which form the basis of more complex molecules. Biomass, in the form of wood or other 'biofuels', is a major world energy source, especially in the developing world. Gaseous and liquid fuels derived from biological sources make significant contributions to the energy supplies of some countries. Biofuels can also be derived from wastes, many of which are biological in origin.

Biofuels are a renewable resource if the rate at which they are consumed is no greater than the rate at which new plants are re-grown – which, unfortunately, is often not the case. Although the combustion of biofuels generates atmospheric CO_2 emissions, these should be offset by the CO_2 absorbed when the plants were growing, but significant emissions of other greenhouse gases can result if the combustion is inefficient.

Non-solar renewables

Two other sources of renewable energy do not depend on solar radiation: *tidal* and *geothermal* energy.

Tidal energy, discussed in Chapter 6, is often confused with wave energy, but its origins are quite different. Ocean tides are caused by the gravitational pull of the Moon (with a small contribution from the Sun) on the world's oceans, causing a regular rise and fall in water levels as the Earth rotates. The power of the tides can be harnessed by building a low dam or 'barrage' behind which the rising waters are captured and then allowed to flow back through electricity-generating turbines.

It is also possible to harness the power of strong underwater currents, which are mainly tidal in origin. Various devices for exploiting this energy source, such as marine current turbines (rather like underwater wind turbines) are at the demonstration stage.

Heat from within the Earth is the source of geothermal energy, discussed in Chapter 9. The high temperature of the interior was originally caused by gravitational contraction of the planet as it was formed, but has since been enhanced by the heat from the decay of radioactive materials deep within the Earth.

In some places where hot rocks are very near to the surface, water is heated in underground aquifers. These have been used for centuries to provide hot water or steam. In some countries, geothermal steam is used to produce electricity and, in others, hot water from geothermal wells is used for heating.

If steam or hot water is extracted at a greater rate than heat is replenished from surrounding rocks, a geothermal site will cool down and new holes will have to be drilled nearby. When operated in this way, geothermal energy is not strictly renewable. However, it is possible to operate in a renewable mode by keeping the rate of extraction below the rate of renewal.

1.5 Renewable energy in a sustainable future

Renewable energy sources are already providing a significant proportion of the world's primary energy; Chapter 10 *Integrating renewable energy* will describe a number of long-term energy studies suggesting that renewables are likely to be providing a much greater proportion of the world's energy by the second half of the twenty-first century. However, this introductory chapter concludes with a very brief look at the prospects for renewable energy in the European Union (EU) as a whole, and in the United Kingdom in particular, in the coming decades.

The EU in its '20:20:20' Directive, passed in 2009, set a target for Europe to achieve by 2020 a 20% reduction in carbon emissions combined with a 20% contribution to gross final energy consumption from renewable sources. Within this overall target, individual member states have been given different specific targets suited to their climates and circumstances: the target for the UK is to produce 15% of primary energy from renewables by 2020 (European Union, 2011).

EU member countries have been directed to produce National Renewable Energy Action Plans showing how they propose to achieve their 20:20:20 targets. The UK Government's Action Plan (DECC, 2010c) concludes that delivering 15% of UK gross final energy consumption by 2020 is likely to involve renewables supplying approximately:

- 30% of electricity demand, including 2% from small-scale sources
- 12% of heat demand
- 10% of transport demand.

To ensure that these targets are met, the UK government has put in place a number of support measures which include the following.

- A Renewables Obligation (RO): an obligation on large-scale electricity suppliers to source a significant proportion of their supplies from renewable sources.
- A 'Clean Energy Cash-Back' scheme (similar to the 'feed-in tariff' schemes operating in Germany and some other countries), under which premium prices are paid to small-scale generators of renewable electricity.
- A Renewable Heat Incentive (RHI), offering incentives to encourage the use of heat-producing renewables.
- A Renewable Transport Fuel Obligation (RTFO), in which road fuel suppliers are obliged to blend in a proportion of biofuels.
- A Green Investment Bank to channel capital towards renewable energy and energy efficiency projects.

Looking a little further ahead, to 2030, the UK Committee on Climate Change in its 2011 report on Renewable Energy (CCC, 2011) envisaged four scenarios in which the renewable energy contribution would rise to:

- between 35% and 65% of electricity supplies
- between 35% and 50% of UK heat demand
- between 11% and 25% of transport energy needs.

The precise values achieved will depend on the policies pursued and their effectiveness. The total contribution of renewables to gross final consumption could be between 28% and 46% (Figure 1.10).

1.6 **Summary**

The analysis on which Figure 1.10 is based, together with other similar analyses, suggests that the prospects for renewable energy in the coming decades look bright. Chapter 10 examines in more detail the future prospects for renewables in the UK, the EU and the World as a whole.

In the chapters that follow, each of the principal renewable energy sources are examined in turn: in each case their physical principles, the main technologies involved, their costs and environmental impact, the size of the potential resource and their future prospects are discussed. To start, we discuss the renewable source that is the basis of most of the others: solar energy.

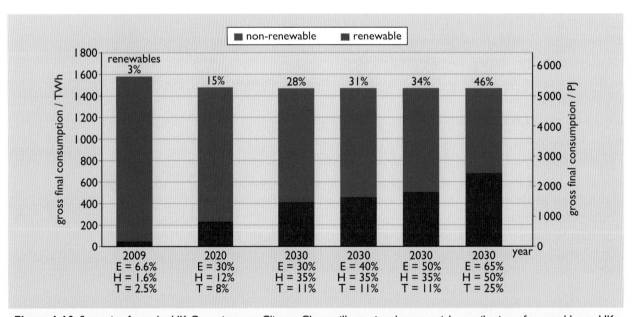

Figure 1.10 Scenarios from the UK Committee on Climate Change illustrating the potential contribution of renewables to UK heat (H), electricity (E) and transport energy (T), and to overall gross final energy consumption, by 2020 and 2030, compared with the contributions in 2009 (adapted from CCC, 2011). Note: 'gross final consumption' shown above is approximately equivalent to 'delivered energy', excluding losses in conversion and delivery, as shown in the last three bars of Figure 1.5 above

References

Allen, M., Frame, D., Frieler, K., Hare, W., Huntingford, C., Jones, C., Knutti, R., Lowe, J., Meinshausen, M., Meinshausen, N. and Raper, S. (2009) 'The exit strategy', *Nature Reports Climate Change*, Issue 5 [online], doi:10.1038/climate.2009.38 http://www.nature.com/climate/2009/0905/full/climate.2009.38.html (accessed 3 October 2011).

Boden, T. A., Marland, G. and R. J. Andres (2010) *Global, Regional, and National Fossil-Fuel CO_2 Emissions* [online], Carbon Dioxide Information Analysis Center, Oak Ridge National Laboratory, U.S. Department of Energy, Oak Ridge, TN., http://cdiac.ornl.gov/trends/emis/tre_glob.html (accessed 3 October 2011).

BP (2010) *BP Statistical Review of World Energy 2010*, London, The British Petroleum Company; available at http://www.bp.com (accessed 3 October 2011).

CCC (2011) *The Renewable Energy Review*, London, The Committee on Climate Change, available at http://hmccc.s3.amazonaws.com/Renewables%20Review/The%20renewable%20energy%20review_Printout.pdf (accessed 3 October 2011).

CEC (2009) *Directive 2009/28/EC on the promotion of energy from renewable sources*, Commission of the European Communities, available at http://eur-lex.europa.eu (accessed 20 September 2011).

DECC (2010a) *Digest of United Kingdom Energy Statistics 2010 (DUKES)*, Department of Energy and Climate Change; available at http://www.decc.gov.uk/en/content/cms/statistics/publications/dukes/dukes.aspx (accessed 3 October 2011).

DECC (2010b) *Energy in Brief 2010*, Department of Energy and Climate Change; available at http://www.decc.gov.uk/en/content/cms/statistics/publications/brief/brief.aspx (accessed 3 October 2011).

DECC (2010c) *National Renewable Energy Action Plan for the United Kingdom*, Department of Energy and Climate Change; available at http://www.decc.gov.uk/en/content/cms/meeting_energy/renewable_ener/uk_action_plan/uk_action_plan.aspx (accessed 3 October 2011).

European Union (2011) *Renewable Energy Targets by 2020*, available at: http://ec.europa.eu/energy/renewables/targets_en.htm (accessed 1 November 2011).

Everett, B., Boyle, G. A., Peake S. and Ramage, J. (eds) (2012) *Energy Systems and Sustainability: Power for a Sustainable Future* (2nd edn), Oxford, Oxford University Press/Milton Keynes, The Open University.

IEA (2009) *Renewables and Waste in World in 2008* [online], http://www.iea.org/stats/renewdata.asp?COUNTRY_CODE=29 (accessed 3 October 2011).

IEA (2010a) *World Energy Outlook 2010*, Paris, International Energy Agency.

IEA (2010b) *CO$_2$ Emissions from Fuel Combustion (2010 edition): Highlights* [online]: Paris, International Energy Agency; http://www.iea.org/co2highlights/ (accessed 3 October 2010).

IPCC (2007a) *Climate Change 2007: The Physical Scientific Basis*, Cambridge University Press. http://www.ipcc.ch/publications_and_data/publications_and_data_reports.shtml (accessed 23 October 2011).

IPCC (2007b) *Climate Change 2007: Synthesis Report. Contribution of Working Groups I, II and III to the Fourth Assessment Report of the Intergovernmental Panel on Climate Change*, Geneva, IPCC.

Sorensen, B. (2000) *Renewable Energy* (2nd edn), London, Academic Press.

Sorrell, S., Speirs, J., Bentley, R., Brandt, A. and Miller, R. (2009) *Global Oil Depletion: An assessment of the evidence for a near-term peak in global oil production*, Report for United Kingdom Energy Research Centre; available at http://www.ukerc.ac.uk/support/Global%20Oil%20Depletion (accessed 3 October 2011).

Twidell, J. and Weir, A. (1986) *Renewable Energy Resources*, London, E. and F. N. Spon.

WWEA (2010) *World Wind Energy Report 2010,* WWEA, http://www.wwindea.org (accessed 21 November 2011).

Chapter 2

Solar thermal energy

By Bob Everett

2.1 Introduction

As we saw in Chapter 1, the Sun is the ultimate source of most of our renewable energy supplies. Since there is a long history of the Sun being regarded as a deity, the direct use of solar radiation has a deep appeal to engineer and architect alike.

In this chapter, we look at some of the methods employed to gather solar thermal or heat energy. Solar photovoltaic (PV) energy, the direct conversion of the Sun's rays to electricity, is dealt with in Chapter 3. Solar thermal collection methods are many and varied, so we can only give the briefest introduction and supply points to further reading for those interested in studying the subject in greater depth.

What sorts of system can be used to collect solar thermal energy?

Most systems for low-temperature solar heating depend on the use of glazing, in particular its ability to transmit visible light but block infrared radiation. High-temperature solar collection is more likely to employ mirrors to concentrate the Sun's radiation. In practice, solar energy systems of both types can take a wide range of forms. These include:

Active solar heating. This always involves a discrete **solar collector**, usually mounted on the roof of a building, to gather solar radiation. Mostly, collectors are quite simple and the heat will be at low temperature (under 100 °C) and used for domestic hot water or swimming pool heating.

Passive solar heating. This term has two slightly different meanings.

- In the 'narrow' sense, it means the absorption of solar energy directly into a building to reduce the energy required for heating the habitable spaces (i.e. what is called **space heating**). Passive solar heating systems mostly use air to circulate the collected energy, usually without pumps or fans – indeed the 'collector' is often an integral part of the building.

- In the 'broad' sense, it means the whole process of integrated low-energy building design, effectively to reduce the heat demand to the point where small passive solar and other 'free heat' gains make a significant contribution in winter. A large solar contribution to a large heat load may look impressive, but what really counts is to minimize the total fossil fuel consumption and thus achieve the minimum cost. This is a key feature of *superinsulated* or *Passivhaus* building design.

Daylighting. This means making the best use of natural daylight, through both careful building design and the use of controls to switch off artificial lighting when there is sufficient natural light available.

Solar thermal engines. These are an extension of active solar heating, usually using more complex collectors to produce temperatures high enough to drive steam turbines to produce electric power. They can come in a number of different types, but most of the world's solar thermally generated electricity is produced in California in multi-megawatt plants using large parabolic mirrors.

This chapter also briefly describes *heat pumps*. These draw heat energy from the outside environment which, in the 2009 EU Renewable Energy Directive (CEC, 2009), is classified as renewable energy. The heat gains from air-source heat pumps may sometimes be categorized as 'solar energy' although they do not *directly* depend on the availability of solar radiation. The gains from ground-source heat pumps are categorized in the EU Directive as 'geothermal energy', a topic described further in Chapter 9.

It must be stressed at the outset that making the best use of solar energy requires a careful understanding of the climate of any particular location. Indeed, many of our present energy problems stem from attempts to produce buildings inappropriate to the local climate. This can mean that the economics of solar technologies commonly used in southern Europe may be disappointing when transferred, for example, to northern Scotland.

However, most of the methods described in this chapter have been well tried and tested over the past century. Even the most spectacular of modern solar thermal-electric power stations are just uprated versions of inventive systems built at the beginning of the 20th century. The skill of using solar thermal energy, in all its forms, perhaps lies in producing systems that are cheap enough to compete with 'conventional' systems based on fossil fuels at current prices.

2.2 The rooftop solar water heater

For most people, 'solar heating' means the rooftop solar water heater. In Europe by 2010 there were almost 35 million square metres of solar collectors installed, 13 million m² in Germany but only about 570 000 m² in the UK (ESTIF, 2011). Most use simple flat plate collectors. There are two basic forms of system: pumped or thermosyphon.

The pumped solar water heater

This is the form most common in northern Europe, normally roof-mounted (Figure 2.1). A typical flat plate pumped system consists of three elements as shown in Figure 2.2.

(1) A collector panel, typically of 3–5 square metres in area, tilted to face the Sun and mounted on the normal pitched roof of a house, as in Figure 2.1. This panel itself normally consists of three components (see Figure 2.3). The main absorber might be a steel plate bonded to copper or steel tubing through which water circulates. The plate is sprayed with a special black paint or coated with a selective surface to maximize the solar absorption. It is normally covered with a single sheet of glass or plastic and the whole assembly is insulated on the back to cut heat losses.

Figure 2.1 Solar panels mounted on roof (photo courtesy of Arcon)

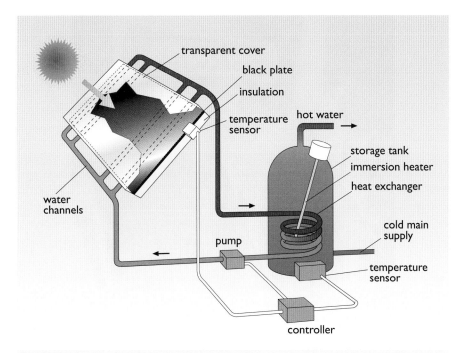

Figure 2.2 Pumped active solar water heater

Figure 2.3 Components of a solar panel

(2) A storage tank, typically of around 200 litres capacity, which often doubles as the normal domestic hot water cylinder. This usually contains an electric immersion heater for winter use. The tank is insulated all round, typically with 50 mm of glass fibre or polyurethane foam. The hot water from the panel circulates through a heat exchanger at the bottom of the tank.

(3) A pumped circulation system to transfer the heat from the panel to the store. Sensors detect when the collector is becoming hot and switch on an electric circulating pump. Since in northern Europe the collector has to be able to survive freezing temperatures, the circulating water contains an antifreeze. Non-toxic propylene glycol is often used (instead of the poisonous ethylene glycol commonly used in car engines).

In the UK, field trial results suggest that such a system can typically provide over 1100 kWh y^{-1} or about 40% of a household's hot water (EST, 2011a).

The thermosyphon solar water heater

In frost-free climates where it is safe to mount the storage tank outdoors, a simpler **thermosyphon** arrangement can be used, as shown in Figure 2.4.

This design dispenses with the circulation pump. It relies on the natural convection of hot water rising from the collector panel to carry heat up to the storage tank, which must be installed above the collector. There is no need for a heat exchanger as the required domestic hot water circulates directly through the panel.

Normally the storage tank also contains an electric immersion heater for top-up and use on cloudy days. Mediterranean systems are usually designed

Figure 2.4 A typical Mediterranean thermosyphon solar water heater – the insulated storage tank is at the top

to be free-standing for mounting on buildings with flat roofs. Given the higher levels of solar radiation in these countries, they are usually sold with only around 2 m² of collector area.

2.3 The nature and availability of solar radiation

The wavelengths of solar radiation

The Sun is an enormous nuclear fusion reactor, which converts hydrogen into helium at the rate of 4 million tonnes per second. It radiates energy by virtue of its high surface temperature, approximately 6000 °C. Of this radiation, approximately one-third of that incident on Earth is simply reflected back. The rest is absorbed and eventually retransmitted to deep space as long-wave infrared radiation. On average the Earth re-radiates just as much energy as it receives and sits in a stable energy balance at a temperature suitable for life.

We perceive solar radiation as white light. In fact it spreads over a wide spectrum of wavelengths, from 'short-wave' infrared (longer than red light) to ultraviolet (shorter than violet). The pattern of wavelength distribution is critically determined by the temperature of the surface of the Sun (see Figure 2.5).

The Earth, which has an average atmospheric temperature of −20 °C and a surface temperature of 15 °C, radiates energy as long-wave infrared to deep space, the temperature of which is only a few degrees above the absolute zero value, −273 °C. We tend to forget this outgoing radiation, but its effects can be observed on a clear night when a ground frost can occur as heat radiates first to the cold upper atmosphere and then out into space.

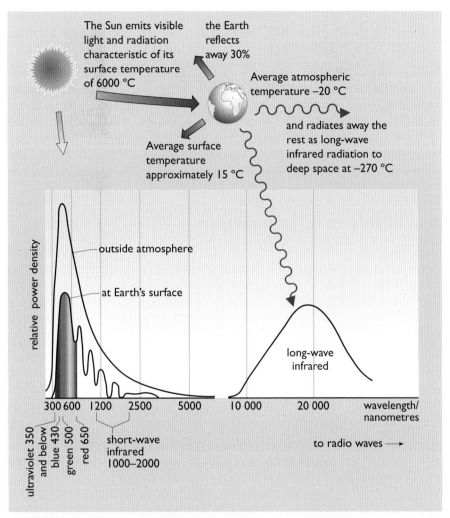

The Sun emits visible light and radiation characteristic of its surface temperature of 6000 °C

the Earth reflects away 30%

Average atmospheric temperature −20 °C

and radiates away the rest as long-wave infrared radiation to deep space at −270 °C

Average surface temperature approximately 15 °C

relative power density

outside atmosphere

at Earth's surface

long-wave infrared

300 600   1200   2500   5000   10 000   20 000   wavelength/ nanometres

ultraviolet 350 and below

blue 430

green 500

red 650

short-wave infrared 1000–2000

to radio waves ⟶

Figure 2.5 Radiation of energy to and from the Earth

As we shall see, most of the art of low-temperature solar energy collection depends on our ability to use glass and surfaces with selective properties that allow solar radiation to pass through but block the re-radiation of long-wave infrared. The gathering of solar energy for high-temperature applications, such as driving steam engines, mainly involves concentrating solar energy using complex mirrors.

Direct and diffuse radiation

When the Sun's rays hit the atmosphere, more or less of the light is scattered, depending on the cloud cover. A proportion of this scattered light comes to Earth as **diffuse radiation**. On the ground this appears to come from all over the sky. Some of it we see as the blue colour of a clear sky, but most is the 'white' light scattered from clouds.

What we normally call 'sunshine', that portion of light that appears to come straight from the Sun, is known as **direct radiation**. On a clear day, this can approach a power density of 1 kilowatt per square metre (1 kW m^{-2}), known as '1 sun' for solar collector testing purposes. Generally in northern Europe and in urban locations in southern Europe, practical peak power densities are around 900–1000 watts per square metre.

In northern Europe, on average over the year approximately 50% of the radiation is diffuse and 50% is direct. In southern Europe, where solar radiation levels are higher, most of the extra contribution is in direct radiation, especially in summer. Both diffuse and direct radiation are useful for most solar thermal applications, but only direct radiation can be focused to generate very high temperatures. On the other hand it is the diffuse radiation that provides most of the 'daylight' in buildings, particularly in north-facing rooms.

Availability of solar radiation

Interest in solar energy has prompted the accurate measurement and mapping of solar energy resources over the globe. This is normally done using **solarimeters** (see Figure 2.6). These contain carefully calibrated thermoelectric elements fitted under a glass cover, which is open to the whole vault of the sky. A voltage proportional to the total incident light energy is produced and then recorded electronically.

Figure 2.6 A solarimeter (also known as a pyranometer)

Most solarimeter measurements are recorded simply as **total energy incident on the horizontal surface**. More detailed measurements separate the direct and diffuse radiation. These can be mathematically recombined to calculate the radiation on tilted and vertical surfaces.

As we might expect, annual total solar radiation on a horizontal surface is highest near the equator, over 2000 kilowatt-hours per square metre per year ($kWh\ m^{-2}\ y^{-1}$), and especially high in sunny desert areas. These are more favoured than northern Europe, which typically only receives about $1000\ kWh\ m^{-2}\ y^{-1}$. Many experimental projects, such as solar thermal power stations, have been built in areas like southern France or Spain, where radiation levels are around $1500\ kWh\ m^{-2}\ y^{-1}$, or the southern USA, where levels can reach $2500\ kWh\ m^{-2}\ y^{-1}$.

It is obvious that in Europe summers are sunnier than winters, but what does that mean in energy terms?

On average in July, the solar radiation on a horizontal surface in northern Europe (e.g. Ireland, UK, Denmark and northern Germany) is between 4.5 and $5\ kWh\ m^{-2}$ per day (see Figure 2.7). Five kilowatt-hours represents about half of the daily average energy consumption for water heating for an average UK household. At 2009 UK domestic fuel prices, this amount of heat would cost approximately 25p if it was obtained from a high-efficiency gas boiler, or slightly more using off-peak electricity. In southern Europe (Spain, Italy and Greece), July solar radiation levels are higher, between 6 and $7.5\ kWh\ m^{-2}$ per day.

In winter, however, the amount of solar radiation is far lower. In January on average in northern Europe it can be only one-tenth of its July value, around $0.5\ kWh\ m^{-2}$ per day (see Figure 2.8), yet in southern Europe there may still be appreciable amounts, 1.5 to $2\ kWh\ m^{-2}$ per day.

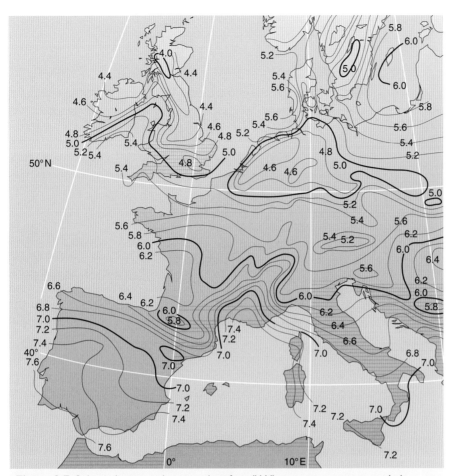

Figure 2.7 Solar radiation on horizontal surface (kWh per square metre per day), Europe, July (source: CEC, 1994)

The implications of this are that in northern Europe we need to look for applications that require energy mainly in the summer. In southern Europe, there may be enough radiation in the winter to consider year-round applications.

Tilt and orientation

So far, we have talked about solar radiation on the horizontal surface. To collect as much radiation as possible, a surface should face south (assuming it is in the northern hemisphere) and must be tilted towards the Sun. How much it should be tilted is dependent on the latitude and at what time of year most solar collection is required.

If the tilt angle between a surface and the horizontal is equal to the latitude, it will be perpendicular to the Sun's rays at midday in March and September (see Figure 2.9).

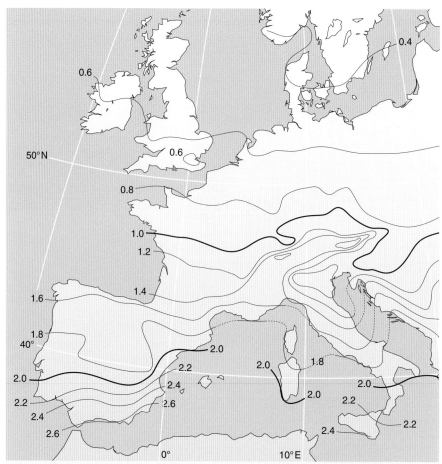

Figure 2.8 Solar radiation on horizontal surface (kWh per square metre per day), Europe, January (source: CEC, 1994)

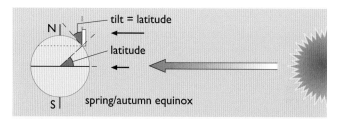

Figure 2.9 A surface tilted at the latitude angle will be perpendicular to the Sun's rays at mid-day on the spring or autumn equinox

There is also the difference between summer and winter to consider. Box 2.1 explains why countries at high latitudes (such as the UK) receive more solar energy in summer than winter.

To maximize solar collection in summer (when there is most radiation to be had), the tilt angle should be less than the latitude. To maximize solar collection in winter (when more solar radiation may be needed) the tilt angle should be greater than the latitude angle (see Figure 2.12).

BOX 2.1 Solar radiation and the seasons

If the energy output of the Sun is constant, why does the UK receive more radiation in summer than in winter?

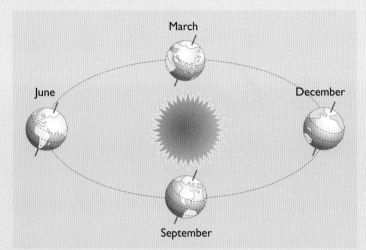

Figure 2.10 The Earth revolves around the Sun with its axis tilted at an angle of 23.5° to the plane of rotation

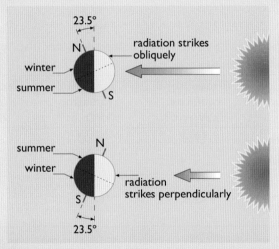

Figure 2.11 The tilt of the Earth's axis creates summer and winter

The Earth circles the Sun with its polar axis tilted towards the plane of rotation (Figure 2.10). In June, the North Pole is tilted towards the Sun. The Sun's rays thus strike the northern hemisphere more perpendicularly and the Sun appears higher in the sky (Figure 2.11). In December the North Pole is tilted away from the Sun and its rays strike more obliquely, giving a lower energy density on the ground (i.e. fewer kilowatt-hours reach each square metre of ground per day).

Another important factor is that the lower the Sun in the sky, the further its rays have to pass through the atmosphere, giving them more opportunity to be scattered back into space. When the Sun is at 60° to the vertical its peak energy density will have fallen to one-quarter of that when it is vertically overhead. This topic will be revisited in Chapter 3.

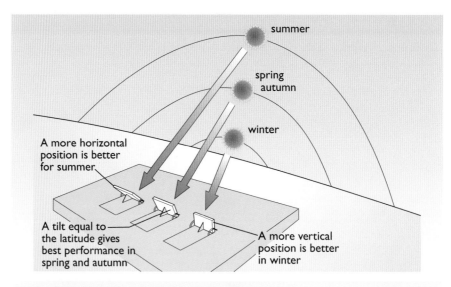

Figure 2.12 Optimizing the tilt for different seasons

Fortunately the effects of tilt and orientation are not particularly critical. Table 2.1 gives totals of energy incident on various tilted surfaces for Kew, near London.

Table 2.1 Effect of tilting a south-facing collection surface (data for Kew, near London, latitude 52° N)

Tilt /°	Annual total radiation /kWh m^{-2}	June total radiation /kWh m^{-2}	December total radiation /kWh m^{-2}
0 – Horizontal	944	153	16
30	1068	153	25
45	1053	143	29
60	990	126	30
90 – Vertical	745	82	29

Source: Achard and Gicquel, 1986

Similarly, the effects of orientation away from south are relatively small. For most solar heating applications, collectors can be faced anywhere from south-east to south-west. This relative flexibility means that a large proportion of existing buildings have roof orientations suitable for solar energy systems. This conclusion applies to both solar thermal and solar photovoltaic (PV) systems.

2.4 The magic of glass

Most low-temperature solar collection is dependent on the properties of one rather curious substance – glass. It is hard to imagine a world without glazed windows. They have been around since the time of the Romans, who invented a process for making plate glass, although the ability to make large sheets of glass was lost in the aptly named 'Dark Ages' and did not reappear until the 17th century.

Transparency

Glass is transparent to visible light and short-wave infrared radiation but has the added advantage of being opaque to long-wave infrared re-radiated from a solar collector or building behind it (see Figure 2.13).

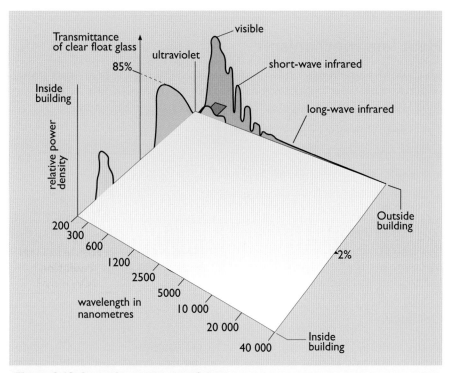

Figure 2.13 Spectral transmittance of glass

Over the past few decades, enormous effort has been put into improving the performance of glazing, both to increase its transparency to visible radiation, and to prevent heat escaping through it.

Manufacturers strive to make glass as transparent as possible, i.e. they try to maximize its **transmittance**, the fraction of incident light that passes through it. They usually do this by minimizing the iron content of the glass. Certain plastics that have optical properties similar to glass can be used instead, although normally they must be protected from the damaging effects of ultraviolet light.

Table 2.2 shows the optical properties of commonly used glazing materials. They share the property of a high solar transmittance (close to 1.0), but the long-wave infrared transmittance is very low by comparison.

Table 2.2 Optical properties of commonly used glazing materials

Material	Thickness /mm	Solar transmittance	Long-wave infrared transmittance
Float glass (normal window glass)	3.9	0.83	0.02
Low-iron glass	3.2	0.90	0.02
Perspex	3.1	0.82	0.02
Polyvinyl fluoride (tedlar)	0.1	0.92	0.22
Polyester (mylar)	0.1	0.87	0.18

Heat loss mechanisms

Much development work has also gone into reducing the heat loss through windows and solar collector glazing. Heat energy will flow through any substance where the temperature on the two sides is different.

The *rate* of this energy flow depends on:

- the temperature difference between the two sides
- the total area available for the flow
- the insulating qualities of the material.

It is obvious that more heat is lost through a large window than a small one, and on a cold day than a warm one. In order to understand how this heat loss occurs, and how it can be minimized, we need to look at the three mechanisms that are involved in the transmission of heat: conduction, convection and radiation.

Conduction

In any material, heat energy will flow by **conduction** from hotter to colder regions. The rate of flow will depend on the first two factors listed above, and on the **thermal conductivity** of the material.

Generally, *metals* have very high thermal conductivities and can transmit large amounts of heat for small temperature differences. Where the frames of glazing systems are made of metal, they should include an insulated thermal break to minimize heat loss.

Insulators require a large temperature differential to conduct only a small amount of heat. Still air is a very good insulator. Most practical forms of insulation rely on very small pockets of air, trapped for example between the panes of glazing, as bubbles in a plastic medium, or between the fibres of mineral wool.

Convection

A warmed fluid, such as air, will expand as it warms, becoming less dense and rising as a result, creating a fluid flow known as **convection**. This is one of the principal modes of heat transfer through windows and out to the environment (see Figure 2.14). It occurs between the air and the glass

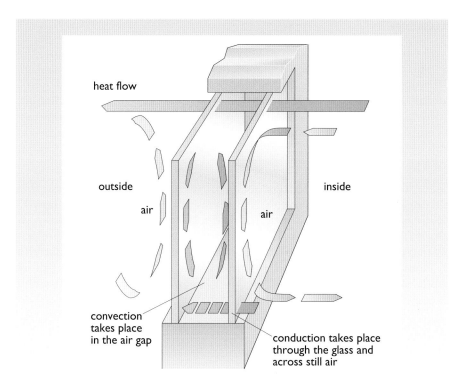

heat flow

outside

inside

air

air

convection takes place in the air gap

conduction takes place through the glass and across still air

Figure 2.14 How heat escapes from a double-glazed window. The air space is normally about 16 mm wide. If it is too narrow, convection will be difficult but conduction will be easy because there is only a small thickness of air to conduct across. If it is too wide, convection currents can easily circulate. In addition there is infrared radiation across the air space which can be reduced by using low emissivity coatings

on the inside and outside surfaces, and, in double glazing, in the air space between the panes. The convection effects can be reduced by filling double glazing with heavier, less mobile gas molecules, most commonly argon, though carbon dioxide or krypton can be used.

Convection can also be reduced by limiting the space available for gas movement. This is the principle used in the insulation materials mentioned above.

Various forms of **transparent insulation** have been developed that use a transparent plastic medium containing bubbles of trapped insulating gas. These materials could eventually revolutionize the concept of windows (and walls), but at present the materials are expensive, not robust and need protection from the rigours of weather and ultraviolet light.

Alternatively double glazing can be evacuated. Convection currents cannot flow in a vacuum. However, a very high vacuum is required and it needs to last for the whole life of the window, 50 years or more. Also the window will need internal structural spacers to stop it collapsing inwards under the air pressure on the outside. These spacers conduct heat across the gap, slightly reducing the overall performance.

A simpler way to reduce the convection effects is to insert extra panes of glass or of transparent plastic film between the other two, turning double glazing into triple or quadruple glazing.

Radiation

Heat energy can be **radiated**, in the same manner as it is radiated from the Sun to the Earth. The quantity of radiation is dependent on the temperature difference between the radiating body and its surroundings. The roof of a building, for example, will radiate heat (i.e. long-wave infrared radiation) away to the atmosphere. It also depends on a quality of the surface known as **emissivity**. Most materials used in buildings have high emissivities of approximately 0.9, that is, they radiate 90% of the theoretical maximum for a given temperature.

Other surfaces can be produced that have low emissivities. This means that although they may be hot, they will radiate little heat outwards. '**Low-E' coatings** are now normally used inside double glazing to cut radiated heat losses from the inner pane to the outer one across the air gap. There are two basic types. 'Hard coat' uses a thin layer of tin oxide, giving an emissivity of about 0.15. 'Soft coat' uses very thin layers of optically transparent silver sandwiched between layers of metal oxide and gives a better emissivity of about 0.05.

Window *U*-value

Conduction, convection and radiation all contribute to the complex process of heat loss through a wall, window, roof, etc. In practice, the actual performance of any particular building element is usually specified by a **U-value**, defined so that:

heat flow rate per square metre = *U*-value × temperature difference.

The units in which *U*-values are expressed are thus watts per square metre per kelvin (W m^{-2} K^{-1}). As pointed out in Chapter 1 temperatures can be measured in degrees Celsius (°C) or kelvins (K). The size of a degree is the same on both scales, so temperature differences are identical in °C and K and *U*-values will often be seen written in units of W m^{-2} °C^{-1}.

The lower the *U*-value, the better the insulation performance. Table 2.3 gives typical *U*-values of various types of window glazing system (the precise values will depend on construction details, particularly the details of the frames). By way of comparison: 10 cm of opaque fibreglass insulation has a *U*-value of 0.35 W m^{-2} K^{-1}. Box 2.2 gives a sample energy calculation.

Table 2.3 Indicative *U*-values for windows with wood or PVC-U frames

Glazing type	W m^{-2} K^{-1}
Single glazing	4.8
Double glazing (normal glass, air filled)	2.7
Double glazing (hard coat low-e, emissivity = 0.15, air filled)	2.0
Double glazing (hard coat low-e, emissivity = 0.2, argon filled)	2.0
Double glazing (soft coat low-e, emissivity = 0.05, argon filled)	1.7
Triple glazing (soft coat low-e, emissivity = 0.05, argon filled)	1.3

Source: BRE, 2005

BOX 2.2 *U-value and heat loss*

What is the rate of heat loss through a large single-glazed window with an area of 2 m², on a day when the outdoor and indoor temperatures are 5 °C and 20 °C respectively?

Table 2.3 shows that the *U*-value for this window is 4.8 W m⁻² K⁻¹, so the loss rate is

$$2 \times 4.8 \times (20 - 5) = 144 \text{ W}$$

Note that, if the temperature difference remained the same throughout 24 hours, the total loss would be almost 3.5 kWh. If this window was replaced with the best of the glazing types shown in Table 2.3, this loss would be reduced to under 1 kWh. Although windows are an important route by which solar energy can be collected, their role in cutting unnecessary heat losses is just as important.

2.5 Low-temperature solar energy applications

We have seen how solar radiation can produce low-temperature heat. Just how useful is this?

As we saw in Chapter 1, Figure 1.5, in the UK about a third of all the end-use of fuel is for low-temperature space and water heating. About 80% of delivered energy use in the domestic sector is in this form (see Figure 2.15).

Although simple solar systems are in principle ideal for supplying this heat, there are other potential competitors. These include:

- district heating fed by waste heat from existing conventional power stations or from industrial processes
- small-scale combined heat and power generation plant
- heat pumps (see Box 2.3).

All of these merit further development, and unlike solar heating, most have the advantage of being able to run all year round.

Swimming pools are another potential application. They do not use a significant proportion of Europe's total energy consumption, since there are not very many of them, but individually they can be enormous energy users. A large, indoor leisure pool in northern Europe can use 1 kW of power for every square metre of pool area continuously throughout the year. This kind of establishment is a prime candidate for the technologies listed above.

Outdoor pools, usually unheated, are rather different. Here the aim is to make the water a little more attractive when people come to use them, which is usually on sunny, warm days. This is ideal solar heating territory.

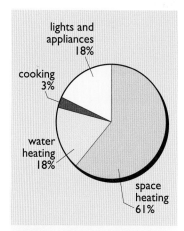

Figure 2.15 Breakdown of UK domestic sector energy use 2009 (DECC, 2011a)

BOX 2.3 **Heat pumps**

A heat pump is essentially very similar to a refrigerator, except that in the UK it is likely to be primarily used for heating rather than cooling. Figure 2.16 shows the key elements of the most common type.

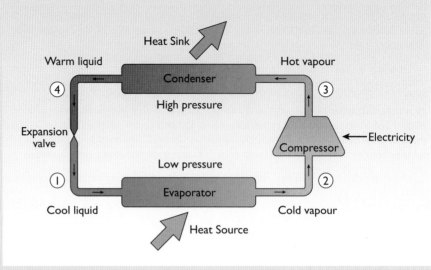

Figure 2.16 Schematic diagram of a heat pump (source: EST, 2010)

Figure 2.17 Fan-coil unit for an air-source heat pump

Figure 2.18 Evaporator pipes laid in a trench for a ground-source heat pump

The heat pumping process is made possible by the use of a special **refrigerant** liquid that boils at low temperature (typically about −15 °C at atmospheric pressure, higher at higher pressures). At point (1) in Figure 2.16 this starts as a cool liquid at a low pressure. In order to convert a liquid to a vapour, it must be given energy – the so-called **latent heat of evaporation**. The refrigerant absorbs heat in a heat exchanger called the **evaporator** and vaporizes (point (2)). The vapour then enters an electrically driven compressor that raises both its pressure and temperature (point (3)). The hot vapour then enters another heat exchanger, the **condenser** where it condenses to a warm liquid and gives up its latent heat of evaporation (point (4)). Finally it is forced through a fine **expansion valve** or throttle where it loses pressure, vaporizing and dropping in temperature. It then repeats the cycle.

Overall, heat is 'pumped' from a low temperature in the evaporator to a higher temperature in the condenser. In a domestic refrigerator, the heat is absorbed in an evaporator inside the refrigerated compartment, thus lowering its temperature, and pumped to a condenser on the back of the refrigerator, where the heat is released, warming one's kitchen in the process.

In buildings a heat pump may be used for heating or cooling (more commonly known as air conditioning). When used for heating the evaporator is located somewhere in the external environment. An **air-source heat pump** is likely to have a *fan coil unit* such as that shown in Figure 2.17. A **ground-source heat pump** uses pipes buried in the soil. These may be laid in a shallow trench, as in Figure 2.18 or in a deep vertical borehole which may be 10 metres or more deep. Such systems are described in Chapter 9.

Heat is then pumped from the outside environment to a condenser inside the building, normally connected to a central heating system. The temperature of the heat is sufficient to be useful for heating purposes. Energy (usually electrical) is, of course, required to operate the compressor. The ratio of the heat output to the electrical input is known as the **coefficient of performance** (COP). For systems installed in the UK this typically has values of 2–3 (EST, 2010). In order to function in mid-winter, the evaporator has to be able to absorb heat from the external environment even though the external temperature may be very low (down to −5 °C in the UK). This requirement potentially limits the performance of air-source heat pumps. Burying the evaporator coil in the ground (or in a lake or river) provides a more stable temperature environment in extreme winter conditions and can result in higher COPs.

In an air-source heat pump, the heat that is drawn from the external environment is taken immediately from the outside air, cooling it in the process. In a ground-source heat pump, the same process takes place, but by cooling the ground (by only a degree or two) so that heat flows down into it from the air over a large area and a long time period.

Under the 2009 EU Renewable Energy Directive (CEC, 2009) heat *gains* from heat pumps, i.e.

the difference between the heat output and the electricity input, are classified as renewable energy. Those from air-source heat pumps are termed *aerothermal*; those from rivers or lakes are *hydrothermal* and those from 'energy stored in the form of heat beneath the surface of solid earth' are classified as *geothermal*.

This system of classifications may be a source of confusion. Heat gains from heat pumps are beginning to feature in national renewable energy statistics and they may be classified together with solar gains although they do not *directly* depend on solar radiation for their performance. Ground-source heat pumps, particularly those using vertical borehole heat exchangers as described in Chapter 9, are likely to be classified as 'geothermal heat pumps'.

Domestic water heating

Domestic water heating is perhaps the best overall potential application for active solar heating in Europe. It is a demand that continues all year round and still needs to be satisfied in the summer when there is plenty of sunshine. In the UK in 2009 it accounted for approximately 5% of the total national delivered energy use. A typical UK household uses approximately 13 kWh per day of delivered energy for this purpose (DECC, 2011a). In practice, much of this can be simply lost as waste heat. Uninsulated hot water cylinders and unlagged pipework are common causes of such losses and even solar water heaters can suffer from this failing.

Incoming mains water is usually at a temperature close to that of the ground at about 1 metre depth, approximately 12 °C in the UK, varying only slightly over the year, and it has to be heated up to 60 °C. In many books it is suggested that temperatures as low as 45 °C are adequate, but recent concerns over Legionnaires' Disease, caused by *Legionella pneumophila* bacteria multiplying in warm water, have highlighted the need for a higher temperature.

Domestic water heating is usually done in one of three ways:

- By electricity, with an immersion heater in a hot-water storage cylinder.
- Again using a storage cylinder, but with a heat exchanger coil inside connected to a central heating boiler (usually gas-fired) or possibly to a district heating supply system.
- By an 'instantaneous' heater, usually powered by gas or electricity.

In the UK, natural gas is the dominant fuel for domestic heating. As will be described later in Section 2.10, obtaining heat by burning natural gas directly involves less CO_2 production than using electricity generated from fossil fuels. Where solar heat can usefully substitute for heat produced by burning natural gas, every kilowatt-hour will save on the emission of about 0.2 kg CO_2. Where it substitutes for UK-generated electricity the figure is about 0.5 kg CO_2.

However, in many sunnier countries the majority of homes may use electric water heating throughout the year. In countries such as Greece, the electricity generation mix has a higher proportion of coal use than the UK and the emission savings may be closer to 0.7 kg CO_2 per kWh of heat produced (IEA, 2011). Also, in such countries, the national electricity demand is likely to peak in the summer, with ever-increasing demands for refrigeration and air-conditioning, rather than in the winter as in the UK. Thus every solar water heater installed saves not only on fuel, but, equally importantly, on building new power plants. Put another way, where solar heat can be substituted for fossil-fuelled electricity used for low-temperature heating purposes, this can be considered as beneficial as building a PV or solar thermal-electric power plant to generate an equivalent amount of extra electricity.

Domestic space heating

Space heating involves warming the interior spaces of buildings to internal temperatures of approximately 20 °C. In the UK, it consumes almost 20% of the country's delivered energy, yet with an appropriate heating system it can in principle be carried out with water at only 45 °C. It is an activity that only occurs over the **heating season**. For normal UK buildings, this extends from about mid-September to April, although, as we shall see later in the section on passive solar heating, this can vary considerably with location and level of insulation.

However, there is a fundamental problem that for this application in the UK, as in much of northern Europe, the availability of solar radiation is completely out of phase with the overall demand for heat (see Figure 2.19). Although the total amount of solar radiation over a whole year on a particular site may far exceed the total building heating needs, the amount available during the heating season may be quite small.

Even with south-facing vertical surfaces, the amount of radiation intercepted in the UK over the winter is relatively small. In London, for example, over a typical six-month winter period of October to March, 1 m² of south-facing vertical surface will only intercept 250 kWh of solar radiation.

It is important to emphasize that the suitability of solar energy for space heating is dependent on the local climate. Textbooks may show quite grandiose solar buildings, but these may only be appropriate in particular locations, often places that are both cold and sunny in winter.

In summer, the UK has similar temperatures, and receives a similar amount of solar radiation, to other European countries on the same latitude.

In winter, the picture is different. The UK has relatively mild winters. However, the winter solar radiation remains largely dependent on latitude alone. As can be seen in Figure 2.20, average January temperatures in London are virtually identical to those in the south of France (follow the 5 °C contour).

Why then do northern Europeans go south for the winter? Because it is sunnier. As we saw from Figure 2.8, the south of France receives three times as much solar radiation on the horizontal surface in mid-winter as does London.

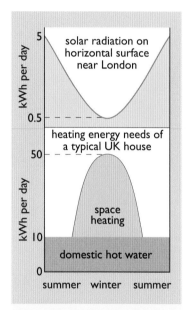

Figure 2.19 The availability of solar radiation is out of phase with space heating demand in the UK

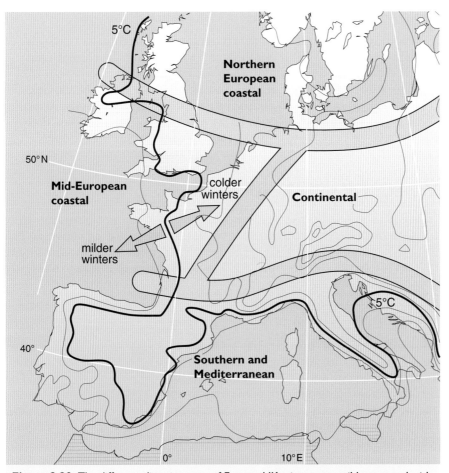

Figure 2.20 The different climatic zones of Europe. UK winters are mild compared with much of the rest of Europe. The 5 °C contour is for average January temperatures.

Broadly speaking, the climate of western Europe can be split into four regions (Figure 2.20):

(1) Northern European coastal zone: cold winters with little solar radiation in mid-winter; mild summers.

(2) Mid-European coastal zone: cool winters with modest amounts of solar radiation; mild summers.

(3) Continental zone: very cold winters with modest amounts of solar radiation; hot summers.

(4) Southern and Mediterranean zone: mild winters with high solar radiation; hot sunny summers.

It is no coincidence that many solar experimental projects have been built on the boundaries of regions 3 and 4, in the Pyrenees and the area around the Alps. This kind of climate is also typical of Colorado in the central USA, another area where solar-heated houses abound.

The broad view of passive solar heating is really about the subtle influence of climate on building design. Without this appreciation, it is all too easy to design buildings that are inappropriate to their surroundings.

As the Roman architect Vitruvius said in the first century BC:

> We must begin by taking note of the countries and climates in which homes are to be built if our designs for them are to be correct. One type of house seems appropriate for Egypt, another for Spain … one still different for Rome, and so on with lands of varying characteristics. This is because one part of the Earth is directly under the Sun's course, another is far away from it … It is obvious that designs for homes ought to conform to the diversities of climate.
>
> (Cited in Butti and Perlin, 1980)

Varieties of solar heating system

In practice, the categories 'active' and 'passive' are not clear cut: they blend into each other, with a whole range of possibilities in between. The following examples illustrate the range of solar heating systems available in addition to the roof-mounted solar water heaters that we have already considered.

Swimming pool heating

For swimming pool heating, the solar system can be extremely simple. Pool water is pumped through a large area of collector, usually unglazed. Typically, the collector will be about half the area of the pool itself. The best results are achieved with pools that do not have other forms of heating and are consequently at relatively low temperatures (under 20 °C). The aim may not necessarily be to save energy as much as to make the pool temperature more acceptable to bathers.

Conservatory (or 'sunspace')

A conservatory or greenhouse on the south side of a building can be thought of as a kind of habitable solar collector (see Figure 2.21(a)). Air is the heat transfer fluid, carrying energy into the building behind. The energy store is the building itself, especially the wall at the back of the conservatory.

Trombe wall

With a Trombe wall (named after its French inventor, Félix Trombe), the conservatory is replaced by a thin air space in front of a storage wall (see Figure 2.21(b)). This is a solar collector with the storage immediately behind. Solar radiation warms the store and is radiated into the house in an even fashion from its inner side. In addition, on sunny days, air is circulated through the air space into the house behind. At night and on cold days, the air flow is cut off.

This concept can take many forms. Small collector panels can be built directly on to the existing walls of buildings. In the extreme, the air path can be omitted, and walls simply covered with 'transparent insulation'.

Direct gain

Direct gain is the simplest and most common of all passive solar heating systems (see Figure 2.21(c)). All glazed buildings make use of this to some degree. The Sun's rays simply penetrate the windows and are absorbed

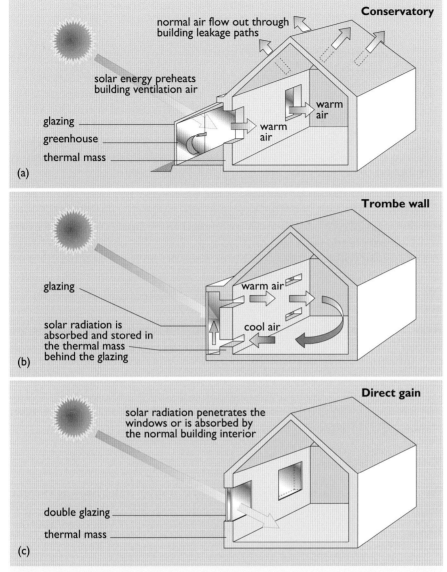

Figures 2.21 Different types of passive solar heating system: (a) conservatory; (b) Trombe wall; (c) direct gain

into the interior. If the building is 'thermally massive' enough, i.e. built of heavy materials such as concrete, and the heating system responsive, the gains are likely to be useful. If the building is too 'thermally lightweight', such as one of timber frame construction, it may overheat on sunny days and the occupants will perceive the effect as a nuisance.

2.6 Active solar heating

History

A solar water heater could be made simply by placing a tank of water behind a normal window. Indeed, many of the first systems produced in the USA in the 1890s were little more than this.

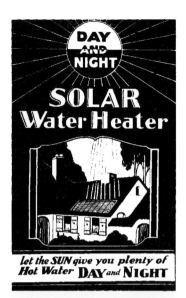

Figure 2.22 An advertisement for Bailey's thermosyphon solar water heaters, circa 1915

The thermosyphon solar water heater as we know it was patented in 1909 by William J. Bailey in California. Since the system had an insulated tank, which could keep water hot overnight, Bailey called his business the 'Day and Night' Solar Water Heater Company (Figure 2.22). He sold approximately 4000 systems before the local discovery of cheap natural gas in the 1920s virtually closed his business.

In Florida, the solar water heating business flourished until the 1940s. Eighty per cent of new homes built in Miami between 1935 and 1941 had solar systems. Possibly as many as 60 000 were sold over this period in this area alone. Yet by 1950 the US solar heating industry had completely succumbed to cheap fossil fuel (Butti and Perlin, 1980).

It was not until the oil price rises of the 1970s that the commercial solar collector reappeared. Low fuel prices in the 1990s discouraged their use, but they since have been heavily promoted in countries such as Austria and Germany and, more recently, China. Over the period 2005–10 the growth rate of collector sales was approximately 16% per annum. By 2010 an estimated total of 280 million m² (i.e. 280 square kilometres) of solar collectors had been installed worldwide, over 60% of them in China (Weiss and Mauthner, 2011).

As will be explained in Chapter 3, solar photovoltaic (PV) panels are normally specified by their peak electrical output expressed in kW_p under full sunshine. For statistical comparability solar thermal collectors are now often quoted in terms of their peak *thermal* output, where 1 m² of collector is taken to have a peak thermal output of 0.7 kW_{th} (ESTIF, 2011). The estimated total world installed capacity, given the installed collector areas above, is thus 196 GW_{th}.

Solar collectors

We have already considered the basic form of a solar water heating system, but what about the choices involved in selecting the components?

Just as solar heating systems can have several variants, so can solar collectors. Figure 2.23 illustrates the most common types for low-temperature use.

Unglazed panels (Figure 2.23(a)) are most suitable for swimming pool heating, where it is only necessary for the water temperature to rise by a few degrees above ambient air temperature, so heat losses are relatively unimportant.

Flat plate air collectors (Figure 2.23(b)) are not so common as water collectors and are mainly used for applications such as crop drying.

Glazed flat plate water collectors (Figure 2.23(c)) are, outside of China, the mainstay of domestic solar water heating. Usually they are only single glazed but may have an additional second glazing layer, sometimes of plastic. The more elaborate the glazing system, the higher the temperature difference that can be sustained between the absorber and the external air.

The absorber plate usually has a very black surface that absorbs nearly all of the incident solar radiation, i.e. it has a high **absorptivity**. Most normal black paints still reflect approximately 10% of the incident radiation (a white surface, by way of comparison, might reflect back 70–80%). Some panels use a **selective surface** that has both high absorptivity in

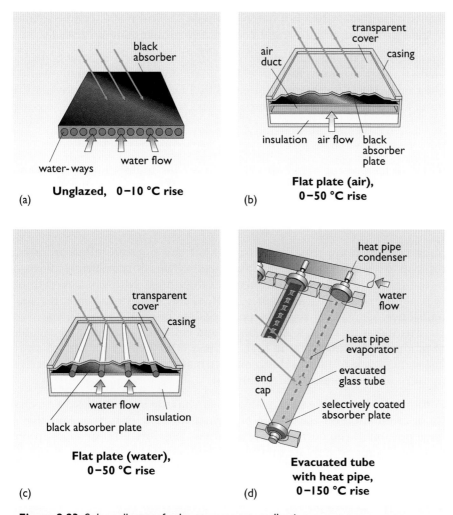

Figure 2.23 Solar collectors for low-temperature collection

the visible region and a low emissivity in the long-wave infrared to cut heat losses.

Many designs of absorber plate have been tried with success, including specially made pressed aluminium panels and small-bore copper pipes soldered to thick copper or steel sheet. Generally, an absorber plate must have high thermal conductivity, to transfer the collected energy to the water with minimum temperature loss.

Evacuated tube collectors. The example shown in Figure 2.23(d) takes the form of a set of modular tubes superficially similar to fluorescent lamps. The absorber plate is a metal strip down the centre of each tube. Convective heat losses are suppressed by a vacuum in the tube. The absorber plate uses a special 'heat pipe' to carry the collected energy to the water, which circulates along a header pipe at the top of the array. A **heat pipe** is a device that takes advantage of the thermal properties of a boiling fluid to carry large amounts of heat. A hollow tube is filled with a liquid at a pressure chosen so that it can be made to boil at the 'hot' end, but the vapour will condense at the 'cold' end. The tube in effect has a thermal conductivity many times greater than if it had been made of solid metal, and is capable of transferring large amounts of heat for a small temperature rise.

An alternative design, commonly used in China, uses a narrow collector blade through which water circulates contained within a double-walled glass tube. Heat losses are suppressed by having a vacuum between the inner and outer walls of the tube.

Evacuated collectors are generally more expensive than flat plate ones, but they have a lower heat loss, allowing a better performance in winter.

Other *concentrating collector* designs used for high-temperature steam raising and electricity generation are described later, in Section 2.9.

Robustness, mounting and orientation

Solar collectors are usually roof-mounted and once installed are often difficult to reach for maintenance and repairs. They must be firmly attached to the roof in a leak-proof manner and then must withstand everything that nature can throw at them – frost, wind, acid rain, sea spray and hailstones. They also have to be proof against internal corrosion and very large temperature swings. A double-glazed collector is potentially capable of producing boiling water in high summer if the heat is not carried away fast enough. It is quite an achievement to make something that can survive up to 20 or more years of this treatment.

Fortunately, as we have seen, panels do not have to be installed to a precise tilt or orientation for acceptable performance. This, in turn, means that a large portion of the current building stock, possibly 50% or more in the UK, could support a solar collector.

Active solar space heating and interseasonal storage

So far, we have looked in detail at domestic solar water heaters with only a few square metres of collector. It might be tempting to think that if a larger collector together with a much larger storage tank were fitted, solar energy could supply the annual low-temperature space heating needs of a building. However, as pointed out earlier, solar radiation is least available in mid-winter when it would be most needed for space heating.

One possibility is to increase the size of the storage tank so that summer sun could be saved right through to the winter. This is known as **interseasonal storage**. However, the difficulties of this should not be underestimated. The volume of hot water storage needed to supply the heating needs of a house may be almost the same size as the house itself. Also such a storage tank might need insulation half a metre thick to retain most of its heat from summer to winter. In order to reduce the ratio of surface area to volume, and hence heat loss, it pays to make the storage tank very large. This essentially limits the technology to large buildings or communal schemes.

Although experimental active solar space heating systems have been built, in practice it has generally proved simpler and more economical to save a kilowatt-hour of space heating energy through better insulation than to *supply* an extra one from active solar heating.

Solar district heating

There are considerable economies of scale for large projects where solar collectors can be purchased and erected in bulk. Since the 1980s there has been a steady stream of construction of large arrays of solar collectors for district heating systems in mainland Europe, mainly in Denmark, Sweden and Germany. The arrays can be very large. The 18 000 m² array shown in Figure 2.24 has a 12 100 m³ heat store and supplies 30% of the annual heat requirement for the district heating system supplying 1600 households. This represents a collector area of 11 m² per household. The array is scheduled to be increased by a further 15 000 m² with a further 75 000 m³ of storage. This should increase the share of heat production for the district heating system to 55% (Sunmark, 2011).

Figure 2.24 An 18 000 m² array of collectors feeding a district heating system at Marstal in Denmark

2.7 Passive solar heating

History

All glazed buildings are already to some extent passively solar heated – effectively they are live-in solar collectors. The art of making the best use of this dates back to the Romans, who put glass to good use in their favourite communal meeting place, the bath house. Window openings 2 m wide and 3 m high have been found at Pompeii.

After the fall of the Roman Empire, the ability to make really large sheets of glass vanished for over a millennium. It was not until the end of the 17th century that the plate glass process reappeared in France, allowing sheets of 2 m² or more to be made.

Even so, cities of the 18th and 19th centuries were overcrowded and the houses ill-lit. It was not until the late 19th century that pioneering urban planners set out to design better conditions. They became obsessed with the medical benefits of sunlight after it was discovered that ultraviolet light killed bacteria. Sunshine and fresh air became the watchwords of 'new towns' in the UK like Port Sunlight near Liverpool, built to accommodate the workers of a soap factory.

The planners then did not realize that ultraviolet light does not penetrate windows, but the tradition of allowing access for plenty of sunlight continues, reinforced by findings that exposure to bright light in winter is essential to maintain human hormone balances. Without it, people are likely to develop mid-winter depression.

Given the UK's past plentiful supply of coal, there was little interest in using solar energy to cut fuel bills. The construction of the Wallasey School building in Cheshire in 1961, inspired by earlier US and French buildings, was thus something of a novelty (see Figures 2.25 and 2.26).

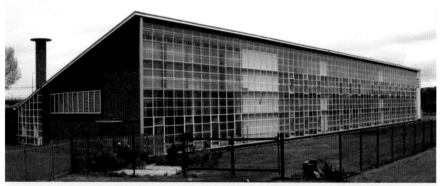

Figure 2.25 Wallasey School, Cheshire, UK – built in 1961

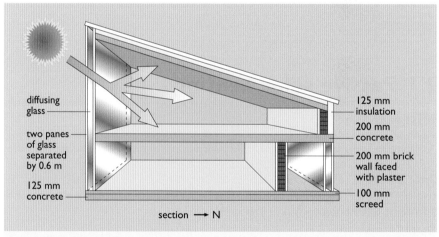

Figure 2.26 Wallasey School – section

Direct gain buildings as solar collectors

The Wallasey School building is a classic direct gain design. It has the essential features required for passive solar heating:

(1) a large area of south-facing glazing to capture the sunlight

(2) thermally heavyweight construction (dense concrete or brickwork). This stores the thermal energy through the day and into the night

(3) thick insulation on the outside of the structure to retain the heat.

After its construction, the oil-fired heating system originally installed was found to be unnecessary and was for a time removed, leaving the building totally heated by a mixture of solar energy, heat from incandescent lights and the body heat of the students.

Passive solar heating versus superinsulation

Although the Wallasey School building is one style of low-energy building, there are others. The Wates house, built at Machynlleth in Wales in 1975 (Figure 2.27) was one of the first 'superinsulated' buildings in the UK. It features 450 mm of wall insulation and small quadruple-glazed windows. This was a radical design, given that at this time normal UK houses were built with single glazing, no wall insulation, and new building regulations requiring a mere 25 mm of loft insulation were only just being introduced.

Figure 2.27 The superinsulated Wates House at Machynlleth

Situated low in a mountain valley, the Wates house is certainly not well-placed for passive solar heating. In fact, it was intended to be heated and lit by electricity from a wind turbine.

Which of the two design approaches – passive solar or superinsulation – is better? There are no easy answers to this question. The art of design for passive solar heating is to understand the energy flows in a building and make the most of them. There need to be sufficient solar gains to meet a substantial proportion of the winter heating needs. These can be reduced by good levels of insulation. Solar energy is also needed to provide adequate lighting, but not so much in summer that there is overheating.

Window energy balance

We can think of a south-facing window as a kind of passive solar heating element. Solar radiation enters during the day, and, if the building's internal temperature is higher than that outside, heat will be conducted, convected and radiated back out.

The question is whether more heat flows in than out, so that the window provides a net energy benefit. The answer depends on several things:

(1) the building's average internal temperature
(2) the average external temperature
(3) the available solar radiation
(4) the transmittance characteristics of the window, its orientation and shading
(5) the *U*-value of the window, which is, in turn, dependent on whether it is single or double glazed (or even better insulated).

Figure 2.28 shows the average monthly 'energy balance' of a south-facing window in the vicinity of London for a building with an average internal temperature of 18 °C. In the dull, cold months of December and January, a single-glazed window is a net energy loser and a double-glazed one only just breaks even. However, in the autumn and spring months, November and March, a double-glazed window becomes a positive contributor to space heating needs. Its performance can be further improved by insulating it at night.

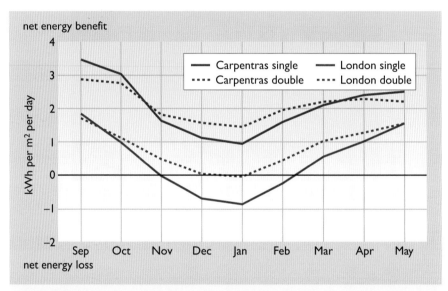

Figure 2.28 Window energy balance: London and Carpentras, in the south of France

We can compare this with a similar energy balance for Carpentras, near Avignon, in the south of France (also shown in Figure 2.28). Although the mid-winter months there are almost as cold as in London, they are far sunnier, being at a lower latitude. The incoming solar radiation is far greater than the heat flowing out, even in mid-winter, and the energy balance is markedly positive.

The heating season and free heat gains

To return to the UK, we need to consider how best use can be made of the solar energy available.

With a long heating season, a south-facing double-glazed window is a good thing. It can perhaps supply extra heat during October and November, March and April. On the other hand, with a very short heating season confined to the dullest months, say just December and January, it is not really much use at all.

How long is the heating season?

In order to answer this question, we must consider the rest of the building, its insulation standards and its so-called 'free' heat gains.

In a typical house, to keep the inside warmer than the outside air temperature, it is necessary to inject heat. The greater the temperature difference between the inside and the outside, the more heat needs to be supplied. In summer it may not be necessary to supply any heat at all, but in mid-winter large amounts will be needed. The total amount of heat that needs to be supplied over the year can be called the **gross heating demand**.

This will have to be supplied from three sources:

(1) 'free heat gains', which are those energy contributions to the space heating load of the building from the normal activities that take place in it: the body heat of people, and heat from cooking, washing, lighting and appliances. Taken individually, these are quite small. In total, they can make a significant contribution to the total heating needs. In a typical UK house, this can amount to 15 kWh per day

(2) passive solar gains, mainly through the windows

(3) fossil fuel energy, from the normal heating system.

Let us now consider, for example, the monthly average gross heat demand of a poorly insulated 1970s UK house (similar houses will be found right across northern and central Europe). As shown in Figure 2.29, this will

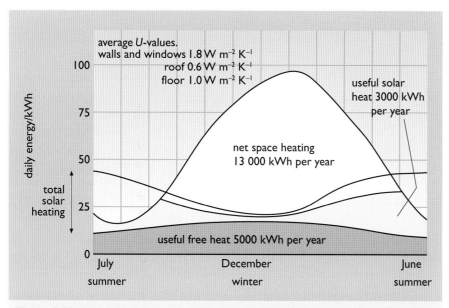

Figure 2.29 Contribution to the net space heating demand in a typical poorly insulated UK house of the 1970s

be higher in the cold mid-winter months than in the warmer spring and autumn. In summer, when the outside air temperature is high, this heating requirement almost drops to zero.

As shown in Figure 2.29, for this particular house, over the whole year, out of a total gross heating demand of 21 000 kWh, 5000 kWh come from free heat gains and 3000 kWh from solar gains (we have assumed to count free heat gains before solar gains).

Put another way, a perfectly ordinary house is already 14% passive solar heated. The **net heating demand**, to be supplied by the normal fossil fuel heating system, is simply the outstanding heat requirement, namely 13 000 kWh. This will have to be supplied from mid-September to the end of May.

It is possible to cut the house heat demand by putting in cavity wall and loft insulation and double instead of single glazing. This will reduce the gross heating demand and, as shown in Figure 2.30, allow the free heat gains and normal solar gains to maintain the internal temperature of the house for a longer period of the year. The insulation levels shown are approximately those of the 2002 UK Building Regulations for new housing. They have since been tightened further.

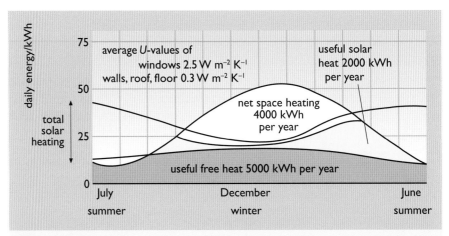

Figure 2.30 Contribution to net space heating demand in a house of normal design but reasonably well insulated

The heating season will then be reduced to between October and the end of April. Out of a total gross heat demand of 11 000 kWh, 5000 kWh still come from free heat gains, but as a result of the improved insulation, it will be possible to utilize only 2000 kWh of solar gains. Finally, 4000 kWh will remain to be supplied from the normal heating system.

By insulating the house, 9000 kWh per year will have been saved in fossil fuel heating, but the solar contribution (using our slightly arbitrary accounting system) will have fallen from 3000 to 2000 kWh.

It might be thought that improvements in the efficiency of domestic appliances and lighting would be leading to reduced levels of free heat gains from their use. UK statistics do show that in 2009 homes used less energy for cooking than they did in 1970, and that energy use for lighting and refrigerators has been falling since the late 1990s. However this has been more than offset by increased electricity use for consumer electronics, including televisions and home computers (DECC, 2011a). All of these unintentionally provide heating energy as well.

There are two ways in which the space heating demand could be cut further.

(1) By providing extra insulation. If the house was superinsulated, using insulation of 200 mm or greater thickness, the space heating load might disappear almost completely, leaving just a small need on the coldest, dullest days. Solar gains might not be essential.

(2) By providing appropriate glazing to ensure that the best use is made of the mid-winter sun.

Which of these methods is chosen will depend on the local climate and the relative expense of insulation materials and glazing. Per square metre, insulated wall tends to be a lot cheaper than good quality insulated glazing. The desired aesthetics of the building and the need for natural daylight inside will dictate whether or not it is easier to collect an extra 100 kWh of solar energy or to save 100 kWh with extra insulation.

Passivhaus design

In Germany the idea of superinsulation has been promoted in the form of the **PassivHaus standard**, developed during the 1990s. The house is 'passive' in the sense that it does not need a conventional large heating system (BRE, 2008). This form of design has been used in over 30 000 buildings across the world to date. It involves using thick insulation, good quality windows and airtight construction to reduce the space heating demand of a building to a low level. It can then be heated mainly by solar gains and heat from appliances and the occupants themselves.

Although the *Passivhaus* approach has mainly been applied to new buildings, even the thermal performance of existing buildings can be radically improved if they are adequately insulated. In the late 1990s an estate of apartment blocks originally built in the 1950s in Ludwigshafen in south-west Germany (Figure 2.31) was given a thorough thermal modernization including:

▪ At least 200 mm thickness of foam insulation on the roof and in the walls (see Figure 2.32)

▪ Triple-glazed windows with argon filling and low-emissivity coatings on the glass to cut the heat loss

▪ Mechanical ventilation with heat recovery (MVHR); this uses heat recovered from outgoing air to preheat incoming fresh air

▪ A fuel cell based combined heat and power (CHP) unit (see Chapter 10 for details of fuel cells).

Monitoring showed that the net space heating energy use (i.e. that to be supplied by the heating system) fell by a factor of seven from 210 kWh

Figure 2.31 The Brunck Estate in Ludwigshafen, Germany

Figure 2.32 Laying blocks of foam insulation on the roof

per square metre of floor area per year to only 30 kWh m² y⁻¹. 30 kWh is equivalent to the energy content of 3 litres of heating oil – hence the project name, the 3 Litre House (Luwoge, 2008).

General passive solar heating techniques

There are some basic general guidelines for optimizing the use of passive solar heating in buildings.

(1) They should be well-insulated to keep down the overall heat losses.

(2) They should have a responsive, efficient heating system.

(3) They should face south (anywhere from south-east to south-west is fine). The glazing should be concentrated on the south side, as should the main living rooms, with little-used rooms, such as bathrooms, on the north.

(4) They should avoid overshading by other buildings in order to benefit from the essential mid-winter sun.

(5) They should be 'thermally massive' to avoid overheating in summer.

These guidelines were used, broadly in the order above, to design some low-energy, passive solar-heated houses on the Pennyland estate in Milton Keynes in central England in the late 1970s. The design steps (see Figure 2.33) were carefully costed and the energy effects evaluated by computer model.

The resulting houses had a form that was somewhere between the Wallasey School building and the Wates house. The houses faced south, there was not too much glazing, but not too little, and the main living rooms were concentrated on the south side (see Figures 2.34–2.36).

An entire estate of these houses was built and the final product carefully monitored. At the end of the exercise it was found that the steps 1–5 listed

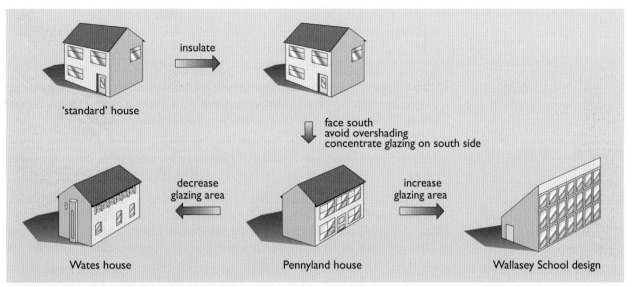

Figure 2.33 Design steps in low-energy housing

Figure 2.34 Passive solar housing at Pennyland – south elevation – the main living rooms have large windows and face south

Figure 2.35 Passive solar housing at Pennyland – the north side has smaller windows

Figure 2.36 Pennyland floor plans

above produced houses that used only half as much gas for low-temperature heating as 'normal' houses built in the preceding year. The extra cost was 2.5% of the total construction cost and the payback time was four years.

Here we come back to the difference between the 'broad' and 'narrow' definitions of passive solar heating. In its broad sense, it encompasses all the energy-saving ideas (1–5 above) put into these houses. In its narrow sense, it covers only the points that are rigidly solar based (3–5).

In this project, insulation and efficient heating saved the vast bulk of the energy, but approximately 500 kWh per year of useful space heating energy came from applying points 3–5 (Chapman et al., 1985).

Put another way, this 500 kWh is the difference in energy consumption between a solar and a non-solar house of the same insulation standard. We can call this figure the **marginal passive solar gain**. As we saw in Figures 2.29 and 2.30, even non-solar houses have some solar gains. What we are doing is trying to maximize them.

It is rather difficult to calculate the extra cost involved in producing marginal passive solar gains. After all, the passive solar 'heater' is an integral part of the building, not a bolt-on extra. Careful costing studies of different building designs and layouts have shown that modest marginal solar gains can be had at minimal extra cost.

Essentially, in its narrow sense, passive solar heating is largely free, being simply the result of good practice. In the 'wider' sense of integrated low-energy house design, the energy savings have to be balanced against the cost of a whole host of energy conservation measures, some of which involve glazing and perhaps have a solar element, and others which do not.

The balance between insulation savings and passive solar gains is also highly dependent on the local climate. In practice the most ambitious passive solar buildings are built in climates that have high levels of sunshine during cold winters.

Conservatories, greenhouses and atria

Direct gain design is really for new buildings: it cannot do much for existing ones. However, for many old buildings, conservatories or greenhouses could be added on to the south sides, just as they can be incorporated into new buildings (Figure 2.37).

Add-on conservatories and greenhouses are expensive and cannot normally be justified on energy savings alone. Rather, they are built as extra areas of unheated habitable space. A strong word of caution is necessary here. A conservatory only saves energy if it is not heated like other areas of the house. There is a danger that it will be looked on as just another room and equipped with radiators connected to the central heating system. One house built like this can easily negate the energy savings of ten others with unheated conservatories.

The costs can be reduced for new buildings if they are integrated into the design. The Hockerton housing estate in Nottinghamshire in the UK, completed in 1998 (see Figure 2.38), combines a 'Wallasey school' type design with a full-width conservatory (EST, 2003). Not only do the walls

Figure 2.37 Conservatory on a Victorian terraced house

Figure 2.38 These low energy houses at Hockerton in Northamptonshire, completed in 1998, feature a full width conservatory, thick insulation and earth sheltering

and roof have 300 mm of insulation, but the rear of the houses is also built up with earth, giving extra thermal mass and protection from the worst winter weather. This is known as **earth sheltering**.

Glazed atria are also becoming increasingly common. At their simplest, they are just glazed-over light wells in the centre of office buildings. At the other extreme, entire shopping streets can be given a glazed roof, creating an unheated but well-lit circulation space. Again a strong word of caution is necessary. 'Heated' shops may have wide-open doors, or even entire frontages, onto the 'unheated' circulation space, giving rise to unnecessary heat losses.

Avoiding overshading

One important aspect of design for passive solar heating is to make sure that the mid-winter sun can penetrate to the main living spaces without being obstructed by other buildings. This will require careful spacing of the buildings.

There are many design aids to doing this, but a useful tool is the **sunpath diagram** (see Figure 2.39). For a given latitude, this shows the apparent path of the Sun through the sky as seen from the ground.

In practice, the contours of surrounding trees and buildings can be plotted on it to see at what times of day during which months the Sun will be obscured. The Pennyland houses were laid out so that the midday sun in mid-December just appeared over the roof-tops of the houses immediately to the south.

However, we need to ask whether it is advisable to cut down all offending overshading trees in the area to let the Sun through. To obtain maximum benefit from passive solar heating in its broad sense, it is necessary to follow another guideline:

Houses should be sheltered from strong winter winds.

Computer modelling suggests that, in houses such as those built at Pennyland, sheltering can produce energy savings of the same order of magnitude as marginal passive solar gains, approximately 500 kWh per year per house.

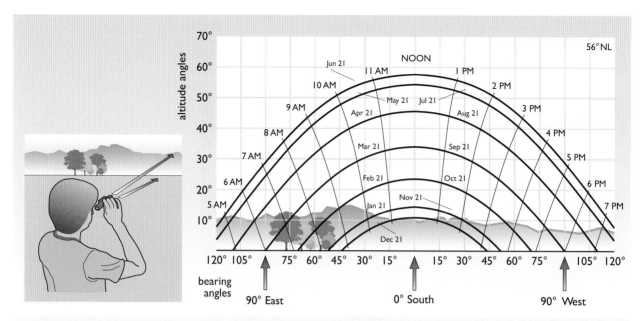

Figure 2.39 Plotting the skyline on a sunpath diagram can give important information about overshading. The sunpath diagram shown is for 56° N, which is approximately the latitude of Glasgow or Edinburgh

Where do the winter winds come from? This is immensely dependent on the local micro-climate of the site. In large parts of the UK, the prevailing wind is from the south-west. It would thus be ideal if every house could have a big row of trees on its south-west side. But is it possible to provide shelter from the wind without blocking out the winter sun? This is where housing layout becomes an art. Every site is different and needs solutions appropriate to it.

2.8 Daylighting

As well as providing heat, the Sun provides daylight. This is a commodity that we all take for granted. Replacing it with artificial light was, before the middle of the 20th century, very expensive (a topic discussed in Everett et al., 2012). Large mirrors were used in narrow London city streets to gather valuable daylight (Figure 2.40). With the coming of cheap electricity and efficient fluorescent lighting, daylight has been neglected and most modern office buildings are designed to rely heavily on electric light.

Houses are traditionally well-designed to make use of natural daylight. Indeed, most of those that were not have long ago been designated slums and duly demolished. In the UK in 2010, domestic lighting accounted for only 2.6% of the sector's delivered energy use.

In some commercial offices, however, lighting can account for up to 30% of the delivered energy use. Modern factory units and hypermarket buildings are built with barely any indows. Modern 'deep-plan' office buildings, such as those at Canary Wharf in London (Figure 2.41), have plenty of windows on the outside but there are many offices, central corridors and stairwells on the inside that require continuous lighting, even when the Sun is shining brightly outside.

Figure 2.40 Mirrors used to catch valuable daylight in narrow London streets before the Second World War

Although in winter the heat from lights can usefully contribute to space heating energy, in summer (when there is most light available) it can cause overheating, especially in well-insulated buildings. Making the best use of natural light saves both on energy and on the need for air conditioning.

Daylighting is a combination of energy conservation and passive solar design. It aims to make the most of the natural daylight that is available. Many of the design details will be found in the better quality 19th century buildings. Traditional techniques include:

- shallow-plan design, allowing daylight to penetrate all rooms and corridors
- light wells in the centre of buildings
- roof lights
- tall windows, which allow light to penetrate deep inside rooms
- the use of task lighting directly over the workplace, rather than lighting the whole building interior.

Figure 2.41 Modern deep-plan office buildings, such as those at Canary Wharf in London, require continuous artificial lighting in the centre, which may create overheating in summer

Other experimental techniques include the use of steerable mirrors to direct light into light wells, and the use of optical fibres and light ducts.

When artificial light has to be used, it is important to make sure that it is used efficiently and is turned off as soon as natural lighting is available. Control systems can be installed that reduce artificial lighting levels when photoelectric cells detect sufficient natural light. Payback times on these energy conservation techniques can be very short and savings of 50% or more are feasible.

In designing new buildings, there is a conflict between lighting design and thermal design. Deep-plan office buildings have a smaller surface area per unit volume than shallow-plan ones. They will need less heating in winter. As with all architecture, there are seldom any simple answers and compromises usually have to be made.

2.9 Solar thermal engines and electricity generation

So far, we have considered only low-temperature applications for solar energy. If the Sun's rays are concentrated using mirrors, high enough temperatures can be generated to boil water to drive steam engines. These can produce mechanical work for water pumping or, more commonly nowadays, for driving an electric generator. This solar thermal-electric generation is known as **concentrating solar power (CSP)**.

The systems used have a long history and many modern plants differ little from the prototypes built 100 years ago. Indeed, if cheap oil and gas had not appeared in the 1920s, solar engines might have developed to be commonplace in sunny countries.

Concentrating solar collectors

Legend has it that in 212 BC Archimedes used the reflective power of the polished bronze shields of Greek warriors to set fire to Roman ships besieging the fortress of Syracuse. Although long derided as myth, Greek navy experiments in 1973 showed that 60 men each armed with a mirror 1 m by 1.5 m could indeed ignite a wooden boat at 50 m.

If each mirror perfectly reflected all its incident direct beam radiation squarely on to the same target location as the other 59 mirrors, the system could be said to have a **concentration ratio** of 60. Given an incident direct beam intensity of, say, 800 W m^{-2}, the target would receive 48 kW m^{-2}, roughly equivalent to the power density of a boiling ring on an electric cooker.

This use of steerable mirrors forms the basis of the modern 'power tower' systems described later.

The most common method of concentrating solar energy is to use a parabolic mirror. All rays of light that enter parallel to the axis of a mirror formed in this particular shape will be reflected to one point, the focus (Figure 2.42(a)). However, if the rays enter slightly off-axis, they will not pass through this point. It is therefore essential that the mirror tracks the Sun.

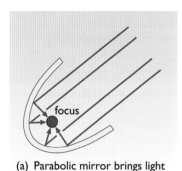

(a) Parabolic mirror brings light to a precise focus in centre

In the *line focus or trough collector* the Sun's rays are focused onto a pipe running down the centre of a trough (Figure 2.42(b)). The pipe is likely to carry a high temperature heat transfer fluid such as a mineral oil. Such systems are mainly used for generating steam for electricity generation. The trough can be pivoted to track the Sun up and down (i.e. in **elevation**) or east to west. A line focus collector can be oriented with its axis in either a horizontal or a vertical plane.

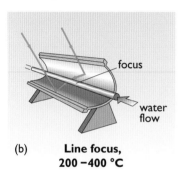

(b)   **Line focus, 200 – 400 °C**

In the *point focus or dish collector*, the Sun's image is concentrated on a steam boiler or a Stirling engine in the centre of the mirror. For optimum performance, the axis must be pointed directly at the Sun at all times, so it needs to track the Sun both in elevation and in **azimuth** (that is, side to side).

Most mirrors are assembled from sheets of curved or flat glass fixed to a framework.

There are trade-offs between the complexity of design of a concentrating system and its concentration ratio. A well-built and well-aimed parabolic dish collector can achieve a concentration ratio of over 1000. A line focus parabolic trough collector may achieve a concentration ratio of 50, but this is adequate for most power plant systems. The ratio required depends on the desired target temperature.

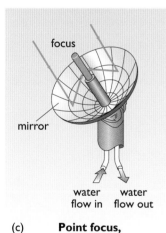

(c)   **Point focus, >1000 °C**

Figure 2.42 Parabolic mirrors for high-temperature applications (a) principles of focusing (b) a line focus or trough collector (c) a point focus or dish collector

Unless the incident solar energy is carried away by some means, the target, be it a boat or a boiler, will settle at an equilibrium temperature where the incoming radiation balances heat losses to the surrounding air. The latter will be mainly by convection and re-radiation of infrared energy and will be dependent on the surface area of the target and its exposure to wind. A line focus parabolic trough collector can produce a temperature of 200–400 °C. A dish point focus system can produce a temperature of over 1500 °C.

What is important to appreciate is that no concentrating collector can deliver in total any more energy than falls on it, but what it does receive is all concentrated into one small area.

The first solar engine age

The process of converting the concentrated power of the Sun into useful mechanical work started in the 19th century. When, in the 1860s, France lacked a supply of cheap coal, Augustin Mouchot, a mathematics professor from Tours, had the answer: solar-powered steam engines. In the 1870s and 1880s, Mouchot and his assistant, Abel Pifre, produced a series of machines ranging from the solar printing press shown in Figure 2.43 to solar wine stills, solar cookers and even solar engines driving refrigerators.

Figure 2.43 Abel Pifre's solar-powered printing press

Their basic collector design was a parabolic concentrating collector with a steam boiler mounted at the focus. Steam pipes ran down to a reciprocating engine (like a steam railway engine) on the ground.

Although these systems were widely acclaimed, they suffered from the fundamental low-power density of solar radiation and low overall efficiency.

In order to understand some of the problems of engines powered by solar energy, it is necessary to consider the Second Law of Thermodynamics (see Box 2.4).

The early French solar steam engines were not capable of producing steam at really high temperatures and as a result their thermal efficiencies were poor. It required a machine that occupied 40 m^2 of land just to drive a one-half horsepower engine (less than the power of a modern domestic vacuum cleaner!)

By the 1890s, it was clear that this was not going to compete with the new supplies of coal in France, which were appearing as a result of increased investment in mines and railways.

BOX 2.4 Heat engines and Carnot efficiency

The steam engine is familiar enough. It works by boiling water to produce a high-pressure vapour. This then goes to an 'expander', which extracts energy and from which low-pressure vapour is exhausted. The expander can be a reciprocating engine or a turbine. Such systems are known as **heat engines**.

All heat engines are subject to fundamental limits on their efficiency set by the **Second Law of Thermodynamics** (a topic discussed in more detail in Everett et al., 2012). They all produce work by taking in heat at a high temperature, T_{in}, and rejecting it at a lower one, T_{out}. In the ideal case, the maximum efficiency they could be expected to achieve is given by:

$$\text{maximum efficiency} = 1 - \frac{T_{out}}{T_{in}}$$

where T_{in} and T_{out} are expressed in the Kelvin temperature scale (or degrees Celsius plus 273). This ideal efficiency is known as the **Carnot efficiency**, after the 19th century French scientist Sadi Carnot.

For example, in a modern CSP plant well-designed parabolic trough collectors might produce steam at 350 °C. This would be fed to a steam turbine and low-temperature heat would be rejected in cooling towers at 30 °C. The theoretical efficiency of the system would therefore be:

$$1 - (30 + 273)/(350 + 273) = 0.51$$

i.e. 51%.

Its practical efficiency is more likely to be about 25%, due to various losses.

Systems that use turbines are often referred to as **Rankine cycles**, after another pioneer of thermodynamics, William Rankine.

Normally, to boil water, its temperature must be raised to at least 100 °C. This may be difficult to achieve with simple non-concentrating solar collectors or with other sources of heat. It would be more convenient to work with a fluid with a lower boiling point. In order to do this, a 'closed cycle' system must be adopted, with a condenser that changes the exhaust vapour back to a liquid and allows it to be returned to the boiler.

Systems have been developed that use stable organic chemicals with suitably low boiling points, similar to the refrigerants used in heat pumps (see Box 2.3). One that uses an organic fluid and a turbine is known as an **organic Rankine cycle (ORC)**. These are commonly used with low-temperature solar engines such as ocean thermal energy conversion (OTEC) systems, described later, and some types of geothermal plant as described in Chapter 9.

These low-temperature systems are likely to have poor efficiencies. For example, the theoretical Carnot efficiency of a heat engine that was fed with relatively low-temperature vapour at 85 °C, say from a flat plate solar collector, and exhausted heat at 35 °C would be only 14%.

At the beginning of the 20th century, in the USA, an entrepreneur named Frank Shuman applied the principle again, this time with large parabolic trough collectors. He realized that the best potential would be in really sunny climates. After building a number of prototypes, he raised enough financial backing for a large project at Meadi in Egypt. This used five parabolic trough collectors, each 80 m long and 4 m wide. At the focus, a finned cast iron pipe carried away steam to an engine.

In 1913, his system, producing 55 horsepower, was demonstrated to a number of VIPs, including the British government's Lord Kitchener. The payback time would have been only four years since the alternative fuel in Egypt at the time was coal, which had to be imported from the UK.

By 1914, Shuman was talking of building 20 000 square miles of collector in the Sahara, which would 'in perpetuity produce the 270 million horsepower required to equal all the fuel mined in 1909' (see Butti and Perlin, 1980, for the full story). Then came the First World War and immediately afterwards the era of cheap oil. Interest in solar steam engines collapsed and lay dormant for virtually half a century.

The new solar age

Solar engines revived with the coming of the space age. When, in 1945, a UK scientist and writer, Arthur C. Clarke, described a possible future 'geostationary satellite', which would broadcast television to the world, it was to be powered by a solar steam engine. In fact, by the time such satellites materialized, 25 years later, photovoltaics (see Chapter 3) had been developed as a reliable source of electricity.

Elsewhere, space rockets, guided missiles and nuclear reactors needed facilities where components could be tested at high temperatures without contamination from the burning of fuel needed to achieve them. The French solved this problem in 1969 by building an eight-storey-high parabolic mirror at Odeillo in the Pyrenees. This faced north towards a large field of **heliostats**: steerable flat mirrors, which, like those held by Archimedes' warriors, track the Sun. This huge mirror could produce temperatures of 3800 °C at its focus, but only in an area of 50 cm^2.

Power towers

In the early 1980s, the first serious, large, experimental solar thermal electricity generation schemes were built to make use of high temperatures. These are now known as concentrating solar power (CSP) plants. Several were of the 'power tower' type, using a large array of heliostats on the ground which focus the Sun's rays onto a central receiver at the top of a tower (Figure 2.44). This is a chamber where either steam can be produced directly, or a heat transfer fluid such as mineral oil or molten salt can be raised to a high temperature to be pumped away to generate steam at ground level. The steam is then used to drive a turbine to generate electricity.

A 10 megawatt (MW) plant, *Solar One*, was built at Barstow in California in 1981. Initially the Barstow plant used high-temperature synthetic oils to carry away the heat to a steam boiler. In 1995, it was rebuilt as Solar Two, and between 1996 and 1999 it operated using a molten salt at over 500 °C (this involved a mixture of sodium nitrate and potassium nitrate that has a melting point of over 200 °C). Its design included heat storage, allowing it to produce electricity potentially on a 24-hour basis.

More recently this technology has been taken up near Seville in southern Spain, where two plants have been commissioned, the 11 MW *Planta Solar* 10 (PS10) in 2007 and the adjacent *Planta Solar* 20 (PS20) in 2009 (Figure 2.45). These have limited heat storage and can use natural gas

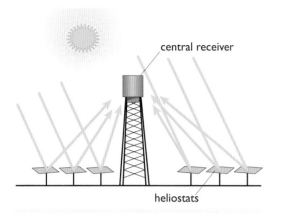

central receiver

heliostats

Figure 2.44 The central receiver on a power tower is heated by a large array of steerable heliostat mirrors on the ground

Figure 2.45 The PS10 and PS20 power tower plants near Seville in Spain

as back-up. The newer *Gemasolar* 20 MW plant also near Seville and commissioned in 2011 has increased thermal storage using molten salt.

Parabolic trough concentrating collector systems

Until very recently most of the world's electricity produced by solar thermal generation came from several large solar power stations developed by Luz International in the Mojave desert in California. Between 1984 and 1990, Luz constructed nine Solar Electricity Generating Systems (SEGS) of between 13 and 80 MW rating and totalling 354 MW. These are essentially massively uprated versions of Shuman's 1913 design, using large fields of parabolic trough collectors (see Figures 2.46 and 2.47). Each successive project has concentrated on increasing economies of scale in purchasing mirror glass and the use of commercially available steam turbines. The last 80 MW SEGS to be built (SEGS IX) has 484 000 m² of collector area.

Figure 2.46 SEGS solar collector field at Kramer Junction in southern California

Figure 2.47 SEGS solar collector field – aerial view

The collectors heat synthetic oil to 390 °C, which can then produce high-temperature steam via a heat exchanger. In recent years the five plants at Kramer Junction (SEGS III to VII) have recorded annual plant efficiencies of 14% and peak efficiencies of up to 21.5% (Solarpaces, 2010). This is competitive with commercially available PV systems.

The SEGS plants were intended to compete with fossil fuel generated electricity to feed the peak afternoon air-conditioning demands in California, and for several years this objective was successfully achieved. In 1992, reductions in the price of gas, to which the price paid for electricity from the plant was tied, brought financial difficulties for the Luz company and the construction of new plants ceased. However those already built have continued to operate reliably and cheaply and this is now regarded as a 'mature technology'. The construction of large solar power plants resumed in the USA in 2007 with the commissioning of *Nevada Solar One*, a 64 MW parabolic trough plant outside Boulder City, Nevada.

Concerns of global warming have meant that the state of California has set a target of having 33% of its electricity from renewable sources by 2020. This has led to proposals for over 4 GW of concentrating solar projects, most of them large parabolic trough systems. Some of the proposed plants are solar-fossil fuel hybrids where the steam turbine is powered by the Sun during the day, by stored heat in the evening and by natural gas at night. There are good thermodynamic reasons for combined gas and solar operation, since it allows the steam turbine to be run at its maximum operating temperature and thermal efficiency.

Fresnel mirror systems

Fresnel mirror systems are a halfway house between power tower and parabolic trough designs. A raised linear collector is heated by steerable strips of flat mirror. Although this has limited temperature-raising performance it only requires ground-mounted flat mirrors and could potentially be cheaper than parabolic trough systems. A prototype 5 MW plant started operation in 2008 near Bakersfield in California.

Parabolic dish concentrator systems

Instead of conveying the solar heat from the collector down to a separate engine, an alternative approach is to put the engine itself at the focus of a mirror. This has been tried both with small steam engines and with Stirling engines.

Stirling engines are described in Everett et al., 2012, and have a long history (they were invented in 1816). Although steam engines have fundamental difficulties when operating with input temperatures above 700 °C, Stirling engines, given the right materials, can be made to operate at temperatures of up to 1000 °C, with consequent higher efficiencies. Current experimental solar systems using these have managed very high overall conversion efficiencies, approaching 30% on average over the day.

A pilot scheme of 60 dishes each driving a 25 kW Stirling engine was constructed in Arizona in 2010 (Figure 2.48) though it is not clear whether

Figure 2.48 An array of 60 parabolic dishes each with a 25 kW Stirling engine constructed at Maricopa in Arizona in 2010

or not further proposed multi-megawatt projects requiring many hundreds of dishes will go ahead.

Low-temperature systems

The systems above rely on producing high temperatures. As described in Box 2.4 this is essential to maximize generation efficiency. It also minimizes the land area required for a given power output. However, other low-temperature systems have been tried and remain of interest.

Solar ponds

Solar ponds use a large, salty lake as a kind of flat plate collector. If the lake has the right gradient of salt concentration (salty water at the bottom and fresh water at the top) and the water is clear enough, solar energy is absorbed at the bottom of the pond.

The hot, salty water cannot rise, because it is heavier than the fresh water on top. The upper layers of water effectively act as an insulating blanket and the temperature at the bottom of the pond can reach 90 °C. This is a high enough temperature to run an organic Rankine cycle (ORC) engine. However, the thermodynamic limitations of the relatively low temperatures mean low solar-to-electricity conversion efficiencies, typically less than 2%. Nevertheless, a system of 5 MW peak electrical output, fed from a lake of over 20 hectares, was demonstrated in Israel in the 1980s. The large thermal mass of the pond acts as a heat store, and electricity generation can go on day or night, as required. Their best location is in the large areas of the world where natural flat salt deserts occur.

In practice, the system has disadvantages. Large amounts of fresh water are required to maintain the salt gradient. These can be hard to find in the solar pond's natural location, the desert. Indeed, the best use for solar ponds may be to generate heat for water desalination plants, creating enough fresh water to maintain themselves and also supply drinking water.

Ocean thermal energy conversion (OTEC)

Ocean thermal energy conversion essentially uses the sea as a solar collector. It exploits the small temperature difference between the warm surface of the sea and the cold water at the bottom (Figure 2.49). In deep tropical waters, 1000 m deep or more, this can amount to 20 °C, which is sufficient to drive an ORC engine.

Although the efficiency is likely to be low and the ORC system used needs to be finely tuned to boil at just the right temperature, there is an extremely large amount of water available. The technology is of considerable interest on islands that have to import fuel for electricity generation.

Initial experiments made on a ship in the Caribbean in the 1930s were only marginally successful. Water had to be pumped from a great depth to obtain a significant temperature difference, and the whole system barely produced more energy than it used in pumping. In the 1970s a more successful 50 kW OTEC device was tested in Hawaii by the US Lockheed Martin company. In 2009 the US Navy awarded the company a US$9 million contract to continue the research in conjunction with the University of Hawaii.

The engineering difficulties are enormous. An OTEC station producing 10 MW of electricity would need to pump nearly 500 cubic metres per second of both warm and cold water through its heat exchangers, whilst remaining moored in sea 1000 metres deep.

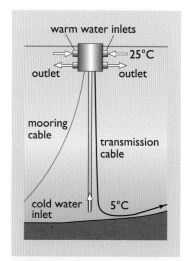

Figure 2.49 Ocean Thermal Energy Converter (OTEC) floating platform

Solar updraft tower devices

These exploit warm air produced in a very large greenhouse. The air is allowed to rise through a tall chimney. Solar updraft towers are sometimes referred to as 'solar chimneys', but this term may also be used to refer to devices using rising warm air to ventilate buildings.

The updraft is used to turn an air turbine at the base of the chimney, driving a generator to produce electricity. This sounds simple enough. What is not so simple is the scale of construction. A 50 kW prototype built in Manzanares in Spain in 1981 (Figure 2.50) and operational until 1989 used a greenhouse collector 240 m in diameter feeding warm air to a chimney 195 m high.

There are large economies of scale to be had. In a very sunny region of the world, a cost-optimum plant might have an output of 100 MW using a collector 3.6 km in diameter and feeding a chimney 950 m tall (Schlaich, 1995). However, because such design only produces warm air (a 35 °C temperature rise has been assumed in the calculations) the overall generation efficiency would be low, around 1.3%. Such a system would thus require considerably more land area than one of equivalent output using high-temperature concentrating collectors. However the ground beneath the collector provides a limited amount of energy storage, extending the output by a couple of hours into the evening.

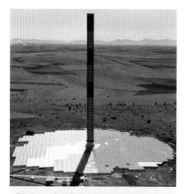

Figure 2.50 This prototype 50 kW solar updraft tower at Manzanares in Spain operated between 1981 and 1989

A 200 kW prototype has been constructed at Jinshawan in Inner Mongolia in China and there are proposals for multi-megawatt plants in a number of countries.

2.10 Economics, potential and environmental impact

Although the motto of Bailey's 'Day and Night' Solar Water Heater company was 'Solar energy, like salvation, is free' the company still had to charge its customers money for the collection equipment. Fortunately, the assessment of the economics of solar thermal systems doesn't pose any great problems. Most solar thermal heating systems can be regarded in the same manner as conventional heating plant or building components and simple notions of 'payback times' often give an adequate assessment (see Appendix B). Solar thermal electricity plants have similar ratings and life expectancies to gas or diesel power stations. The difference is that they have no fuel costs and reasonably low operating and maintenance costs.

Domestic active solar water heating

In terms of collector area, domestic water heating for residential buildings makes up the bulk of the world solar thermal market. In 2010 the UK had an estimated '**solar park**' (i.e. total installed area) of approximately 570 000 m^2 of glazed and evacuated tube collectors (ESTIF, 2011). This represents an almost fourfold increase since 2001. However the figure contrasts with the 13 million m^2 installed in Germany.

It is likely that 50% of existing UK dwellings are suitable to have a solar water heater fitted. If this potential were to be fully taken up by 2025, this would require the deployment of approximately 12 million systems (allowing for a rise in the UK housing stock), or about 50 million m^2 of collectors. This could save an estimated 34.6 PJ (9.6 TWh) per year of oil and gas and 2.4 TWh per year of electricity, resulting in reductions of about 1% of current UK CO_2 emissions (DTI, 1999).

The same DTI study suggested that there were only about 100 000 swimming pools in the UK, and even if this number were to double by 2025, the total potential energy saving from the use of solar heating would only amount to an extra 7% over and above that for domestic water heating.

At the time of writing (2011), sales of solar water heaters in the UK are low and prices are high. According to the Energy Saving Trust (EST), the typical cost of an installed solar water heater in the UK is £4800. They estimate that, based on the results of a recent field trial (EST, 2011a), typical savings from a well-installed and properly used system are £55 per year when replacing gas heating and £80 per year when replacing electric immersion heating. They also estimate the typical carbon savings at around 230 kg CO_2 y^{-1} when replacing gas and 510 kg CO_2 y^{-1} when replacing electric immersion heating (EST, 2011b).

This gives a payback time of 87 years against gas and 60 years against electricity. This might not seem encouraging. However, the savings in

CO_2 emissions can be given a financial value, a 'carbon price' reflecting the reduced damage to the environment (a topic discussed a little further in Chapter 10, Section 10.3. Taking a value of £40 per tonne of CO_2 improves the payback time to 75 years against gas and 48 years against electricity.

As pointed out earlier, solar collectors are in competition with other energy saving devices such as heat pumps (although these do require considerable amounts of electricity for their operation). The UK Committee on Climate Change (CCC) estimates that the levelized cost of heat from solar thermal panels for domestic use is nearly 27 p kWh^{-1} (see Appendix B for an explanation of 'levelized cost'). This compares with values of 6.8–10.5 p kWh^{-1} for heat from a conventional gas boiler (including the costs of the boiler itself) and 10.9–18.6 p kWh^{-1} for heat from heat pumps (CCC, 2011).

Looking further afield, the economics are more promising. In central and southern Europe solar thermal heat may be cost competitive with both gas and electricity (ESTTP, 2009). In southern Europe there is more sunshine, so a system may produce twice as much energy per square metre of collector as in the UK. It is perhaps not surprising that Greece has a high level of solar water heater ownership and that sales have been running at 150 000–200 000 $m^2 \ y^{-1}$.

But it is somewhat surprising that Greece was overtaken in 1999 in terms of collector area per head of population by Austria. Between 1995 and 1998 the rate of installation there was running at around 200 000 $m^2 \ y^{-1}$ and this rose to about 350 000 $m^2 \ y^{-1}$ in 2008 and 2009 (ESTIF, 2011). Much of the initial popularity of solar water heating was attributable to the success of 'do-it-yourself' schemes.

In Europe and its associated (EU-27) states sales of solar collectors more than tripled between 2000 and 2010, the main countries being Germany, Austria, France, Greece, Italy and Spain.

Looking further afield, China has almost 60% of the world's installed capacity (over 100 GW of peak thermal capacity in 2009). Over 90% of these are thermosyphon systems and use evacuated tubes (see Figure 2.51) (Weiss and Mauthner, 2011). Volume manufacture and competition between many suppliers means that collector prices are typically a third of those for systems outside China (IPCC, 2011).

Figure 2.51 A typical Chinese solar water heater using evacuated tube collectors

As for environmental impact, that of solar water heating schemes in the UK is likely to be small. The materials used are those of everyday building and plumbing. Pumped solar collectors can be installed to be visually almost indistinguishable from normal roof lights, with storage tanks hidden inside the roof space. Elsewhere, the use of free-standing thermosyphon systems on flat roofs can be highly visually intrusive. It is not so much the collector that is the problem but the storage tank above it. (Bailey's 'Day and Night' Solar Water Heater Company also had to face these problems. It offered to disguise the storage tank as a chimney.)

The situation is perhaps a little different for the kind of large district heating array shown in Figure 2.24 which obviously takes up a significant amount of urban ground area that could be used for other purposes.

Passive solar heating and daylighting

In its narrow sense of producing an increase in the amount of solar energy directly used in providing useful space heating, passive solar heating is highly economic, indeed possibly free. It should be borne in mind that to some extent passive solar heating and the use of daylight are already features of normal buildings, reducing the UK's energy demand by an estimated 0.5 EJ (DECC, 2011b). This is a figure that does not appear in national renewable energy statistics.

Buildings specifically designed to use passive solar heating have been generally well-received by their occupants and are of interest to architects. However, the potential is limited by the low rate of replacement of the building stock.

In it broader, *Passivhaus,* sense of reducing overall heating energy demand, including retrofit projects, the potential is enormous. Yet, perversely, the superinsulation of buildings that contain large numbers of heat-producing appliances is likely to reduce the heating season to the dull mid-winter months, making the use of solar energy less important.

Designing buildings to take advantage of daylighting involves both energy conservation and passive solar heating considerations. In warmer countries, daylighting may be far more important, since cutting down on summer electricity use for lighting can also save on air conditioning costs.

Designing and laying out buildings to make the best use of sunlight has been part of the architectural tradition for centuries. It is generally seen as environmentally beneficial and has already shaped many towns and cities. For example, when in 1904 the city council of Boston, Massachusetts, USA, was faced with proposals for a 100 m high skyscraper, it commissioned an analysis of the shading of other buildings that this would cause. It was not pleased with the results and imposed strict limits on building heights.

However, a word of caution is necessary. In the UK, the tradition of new town development has been based partly on Victorian notions of the health aspects of 'light and air' in contrast to the overcrowded squalor of existing cities. This has been beneficial in terms of better penetration of solar energy into buildings. On the other hand, the encouragement of low building densities has led to vast tracts of sprawling suburbs and the consumption of enormous quantities of energy in transportation.

Solar thermal engines and electricity generation

As the original pioneers realized, it pays to build solar thermal power systems in really sunny places. In order to generate the high temperatures necessary for thermodynamically efficient operation, the local climate has to have plenty of direct solar radiation – diffuse radiation will not do. Although there was a surge of interest in solar thermal electricity in the 1980s, low fossil fuel prices around the world dampened interest during the 1990s – in contrast to continued enthusiasm for photovoltaics. It is only since about 2005 that construction of new solar thermal power plants has revived and there are now several gigawatts of plant under construction around the world.

Currently, in sunny desert locations, solar thermal electricity, particularly that from large parabolic trough plants, is considered competitive with large-scale PV power. The technology also has potential for thermal heat storage and integration with conventional fossil-fuelled steam power plants. In the USA, the cost of electricity from parabolic trough plants without heat storage is estimated to be around 17 to 20 US cents per kWh (10.6 to 12.5 pence per kWh) (Solarpaces, 2010). The bulk of this cost consists of repayments on the initial capital investment. Operation and maintenance costs are a modest 3 cents per kWh (2 pence per kWh) (Greenpeace/ESTELA/Solarpaces, 2009).

If six hours of heat storage is included then the price rises to 20 to 30 US cents per kWh (12.5 to 18.8 p per kWh) (IPCC, 2011). Although this technology could not compete with cheap natural gas in the USA in the 1990s, its economics (and low CO_2 emissions) are sufficiently good to support a resurgence of construction in the USA and in countries such as Spain.

The overall potential for such systems is enormous. Back in 1914, Frank Shuman was talking of building 20 000 square miles (50 000 km^2) of collector in the Sahara desert. With modern concentrating solar power plant 1 km^2 of land is enough to generate as much as 100–130 GWh of electricity per year (Greenpeace/ESTELA/Solarpaces, 2009). Ignoring the relative availability over the year, 3000 km^2 would thus be adequate to supply all of the UK's electricity and 35 000 km^2 (about 12% of the area of the state of Nevada) sufficient to supply all of the USA's electricity.

Sunny deserts, within striking distance of large urban electricity demands, are needed. In California, the Mojave desert is ideal. In Europe, central and southern Spain and other southern Mediterranean countries are possibilities, with proposals for grid links to North Africa (see Chapter 10, Section 10.8).

The problem would be to find a suitable way of conveying the output to the loads. Although long-distance electricity transmission can be used, the manufacture and distribution of hydrogen might be another possibility. The possibilities of a future 'hydrogen economy' are discussed in Chapter 10.

The environmental consequences of solar thermal power stations are somewhat mixed. A major problem is the sheer quantity of land required. Although, typically, the collectors only take up one-third of the land area, it may be physically difficult to use it for anything else. This is unlike wind farms where the turbines are very widely spaced and crops can grow underneath. CSP plants also require a certain amount of water for washing dust off the mirrors. This may be hard to come by in desert locations.

Solar ponds and solar thermal updraft projects need even larger areas of flat land than CSP plants because of their low thermodynamic efficiency.

The environmental consequences of OTEC systems may be mixed. On the one hand, it is claimed that the vast amounts of water being pumped circulate nutrients and can increase the amount of fish life. On the other, dissolved carbon dioxide can be released from the deep sea water, thereby negating some of the benefits of renewable energy generation. Only further experiments will resolve these issues.

2.11 **Summary**

Solar energy is a resource that is there for the taking. All that is needed is to produce the necessary hardware.

This chapter started by introducing the most basic 'active solar' technology: the rooftop solar water heater. Although this has limited potential in the UK (not the sunniest of countries) it is widely used in Mediterranean countries and there has been a phenomenal growth in its use in recent years in China.

Section 2.3 described the nature of solar radiation starting with the important distinction between 'light' (i.e. short-wave radiation) and 'heat' (long-wave radiation). Direct radiation, the unobstructed rays of the Sun, is important for concentrating collectors, and diffuse radiation is important for natural lighting. The total amount of solar radiation available over a year varies from country to country and is higher in those at a lower latitude (i.e. closer to the equator). Optimizing the amount of radiation falling on a flat plate collector requires an understanding of the appropriate tilt, which will depend on the latitude of the site.

Section 2.4 turned to the properties of glass and other plastic glazing materials, particularly their ability to transmit light but block the re-radiation of long-wave infrared radiation (heat). It described the three mechanisms of heat loss: conduction, convection and radiation in the context of a double glazed window. It introduced the U-value as a measure of heat loss and described ways in which window U-values could be reduced.

Section 2.5 described some possible low-temperature applications for solar energy, particularly domestic solar water heating and space heating. It pointed out that the need for winter space heating is highly dependent on the local climate. It also introduced the 'heat pump', which although not directly a 'solar' technology, does draw heat from the outside environment and is a rival for the production of low-temperature heat. The section ended with a brief description of three 'passive solar' technologies where the solar collector is an integral part of a building: the conservatory, the Trombe wall and direct gain collection.

Section 2.6 looked at 'active solar heating' in more detail, describing four basic collector types and their applications.

Section 2.7 described the complex topic of passive solar heating in detail, particularly 'direct gain' design, looking at a south-facing window as a solar collector and the actual heating energy needs of house. These in turn relate to its insulation level and the role of 'free heat gains' from people, lights and appliances. It discussed the choice between 'solar' and 'superinsulated' design, the latter being recently expressed in the form of *Passivhaus* projects. It briefly described other passive solar technologies: conservatories, greenhouses and atria, and the use of sunpath diagrams as a tool for avoiding overshading.

Section 2.8 briefly described the use of daylighting in avoiding excessive use of artificial light.

Section 2.9 turned to the use of solar thermal energy to drive engines and particularly to generate electricity, in what is now known as concentrating solar power (CSP). It described the basic forms of concentrating collector: the line focus or parabolic trough, and the point focus or dish collector. These were used in large CSP plants in California in the 1980s and there has been a resurgence of construction since 2005. It also introduced the 'power tower' design, where steerable flat mirrors or *heliostats* are used to concentrate the Sun's rays on a central receiver. These form the basis of new CSP plants constructed in Spain. Box 2.4 described the importance of achieving high temperatures in order to get a high overall engine efficiency. It also introduced the organic Rankine cycle (ORC) used in other low-temperature systems, and also briefly described solar ponds, ocean thermal energy conversion (OTEC) and solar updraft towers.

Finally Section 2.10 looked at the economics, potential and environmental impact of solar thermal systems. Although heat pumps may be more economic in the UK, solar water heating is highly viable in central and southern Europe. Mass manufacturing of solar water heaters in China has led to low collector costs and been responsible for the extraordinary growth in collector sales there over the past decade. The economics of concentrating solar power is sufficiently good to support proposals for several gigawatts of new construction in the USA alone.

Globally, the future outlook for solar heating and concentrating solar power generation looks very good. Other 'solar' technologies such as heat pumps and *Passivhaus* energy conservation technology offer considerable potential for reducing global CO_2 emissions.

References

Achard, P. and Gicquel, R. (eds) (1986) *European Passive Solar Handbook* (preliminary edition), Commission of the European Communities, DG XII.

BRE (2005) *SAP 2005 – The Government's Standard Assessment Procedure for Energy Rating of Dwellings*, Watford: Building Research Establishment, [online] available from: http://www.projects.bre.co.uk/SAP2005 (accessed 13 September 2011).

BRE (2008) *Passivhaus Primer: Introduction, An aid to understanding the key principles of the Passivhaus Standard*, Watford: Building Research Establishment, [online] available from: http://www.passivhaus.org.uk (accessed 20 September 2011).

Butti, K. and Perlin, J. (1980) *A Golden Thread: 2500 Years of Solar Architecture and Technology*, London: Marion Boyars.

CCC (2011) *The Renewable Energy Review*, London: Committee on Climate Change, [online] available from: http://theccc.org.uk (accessed 23 September 2011).

CEC (1994) *'Solar Radiation', European Solar Radiation Atlas Volume 1*: Report EUR 9344, Directorate for General Science.

CEC (2009) *Directive 2009/28/EC on the Promotion of Energy from Renewable Sources*, Commission of the European Communities, [online] available from: http://eur-lex.europa.eu (accessed 20 September 2011).

Chapman, J., Lowe, R. and Everett, R. (1985). *The Pennyland Project*, Energy Research Group, Milton Keynes, UK: Open University, [online] available from: http://oro.open.ac.uk/19860/ (accessed 20 September 2011).

DECC (2011a) *Energy Consumption in the United Kingdom: data tables*, Department of Energy and Climate Change, [online] available from: http://www.decc.gov.uk (accessed 18 September 2011).

DECC (2011b) *Digest of UK Energy Statistics (DUKES): Chapter 7 – Renewable Energy Sources*, [online] available from: http://www.decc.gov.uk (accessed 2 February 2012).

DTI (1999) *New and Renewable Energy: Prospects in the UK for the 21st Century – Supporting Analysis*, ETSU-R122, Department of Trade and Industry.

EST (2003) *The Hockerton Housing Project*, Energy Efficiency Best Practice in Housing New Practice Profile 119, Energy Saving Trust, [online] available from: http://www.est.org.uk (accessed 13 September 2011).

EST (2010) *Getting Warmer: a Field Trial of Heat Pumps*, Energy Saving Trust, [online] available from: http://www.energysavingtrust.org.uk (accessed 13 September 2011).

EST (2011a) *Here Comes the Sun: a Field Trial of Solar Water Heating Systems*, Energy Saving Trust, [online] available from: http://www.energysavingtrust.org.uk (accessed 13 September 2011).

EST (2011b) *Solar Water Heating*, Energy Saving Trust website, [online] available from: http://www.energysavingtrust.org.uk (accessed 23 September 2011).

ESTIF (2011) *Solar Thermal Markets in Europe*, European Solar Thermal Industry Federation, [online] available from: http://www.estif.org (accessed 19 September 2011).

ESTTP (2009) *Solar Heating and Cooling for a Sustainable Energy Future in Europe*, European Solar Thermal Technology Platform, Brussels, [online] available from: http://esttp.org (accessed 24 September 2011).

Everett, B., Boyle, G. A., Peake, S. and Ramage, J. (eds) (2012) *Energy Systems and Sustainability: Power for a Sustainable Future* (2nd edn), Oxford, Oxford University Press/Milton Keynes, The Open University.

Greenpeace/ESTELA/Solarpaces (2009) *Concentrating Solar Power: Global Outlook 2009*, Greenpeace/European Solar Thermal Electricity Association/ IEA Solarpaces, [online] available from: http://www.greenpeace.org (accessed 23 September 2011).

IEA (2011) *CO_2 emissions from fuel combustion: highlights: data tables*, International Energy Agency, Paris, [online] available from: http://www.iea.org (accessed 14 October 2011).

IPCC (2011) *Special Report on Renewable Energy*, Intergovernmental Panel on Climate Change, [online] available from: http://www.ipcc.ch (accessed 24 September 2011).

Luwoge (2008) *Das 3-Liter-Haus*, [online] available from: http://www.luwoge.de (accessed 20 September 2011).

REN21 (2011) *Renewables 2011 Global Status Report*, Renewable Energy Policy Network for the 21st Century, Paris, [online] available from: http://www.ren21.net (accessed 14 October 2011).

Solarpaces (2010) *Annual Report, IEA Solar Power and Chemical Energy Systems,* [online] available from: http://www.solarpaces.org (accessed 22 September 2011).

Sunmark (2011) *Marstal district heating*, [online] available from: http://www.sunmark.com (accessed 16 October 2011).

Schlaich, J. (1995) *The Solar Chimney – Electricity from the Sun*, Edition Axel Menges, Stuttgart.

Weiss, W. and Mauthner, F. (2011) *Solar Heat Worldwide*, IEA Solar Heating and Cooling Programme, AEE INTEC, Austria, [online] available from: http://www.iea-shc.org (accessed 24 September 2011).

Further reading

Baker, N. V., Franchiotti, A. and Steemers, K. A. (1993) *Daylighting in Architecture – A European Reference Book, Luxembourg: European Commission Handbook EUR 15006 EN.*

Duffie, J. A. and Beckman, W. A. (2006) *Solar Engineering of Thermal Processes*, 3rd edition, New York: John Wiley. [A classic textbook on the physics and engineering of solar thermal energy systems.]

Goulding, J. R., Lewis, J. O. and Steemers, T. C. (eds.) (1992) *Energy in Architecture – The European Passive Solar Handbook*, Batsford. [A mine of technical information, highly recommended.]

Mazria, E. (1979) *Passive Solar Energy Book*, Rodale Press. [A classic beautifully illustrated introduction to design for passive solar heating.]

Stine, W. B. and Geyer, M. (2001) *Power from the Sun*, available at http://www.powerfromthesun.net (accessed 22 September 2011). [A thorough online book covering most aspects of solar design with online solar calculators.]

Chapter 3

Solar photovoltaics

By Godfrey Boyle

3.1 Introduction

In Chapter 2 we saw how solar energy can be used to generate electricity by producing high-temperature heat to power an engine, which then produces mechanical work to drive an electrical generator. This chapter is concerned with a more direct method of generating electricity from solar radiation, namely **photovoltaics**: the conversion of solar energy *directly* into electricity in a solid-state device.

If one were asked to design the ideal energy conversion system, it would be difficult to devise something better than the solar photovoltaic (PV) cell. In it we have a device which harnesses an energy source that is by far the most abundant of those available on the planet: as earlier chapters have emphasized, the net solar power input to the Earth is more than 8000 times humanity's current rate of use of fossil and nuclear fuels.

The PV cell itself is, in its most common form, made almost entirely from silicon, the second most abundant element in the Earth's crust. It has no moving parts and can therefore in principle, if not yet in practice, operate for an indefinite period without wearing out. Furthermore its output is electricity, probably the most useful of all energy forms.

This chapter starts with a brief history of photovoltaics, introduces the basic principles of the PV effect in silicon and describes various ways of reducing the cost and increasing the efficiency of crystalline silicon PV cells. It then examines non-crystalline PV cell designs using thin films, and discusses various innovative and emerging PV technologies. This is followed by a brief description of the electrical characteristics of PV cells and modules. Next the chapter discusses PV systems for remote power supply, grid-connected PV systems for buildings, and large-scale PV power plants. The costs and environmental impact of energy from PV are then examined. The chapter concludes with a look at how electricity from PV can be integrated into electrical power systems, and some thoughts on how the enormous future growth potential for PV might be realized.

3.2 A brief history of PV

The term 'photovoltaic' is derived by combining the Greek word for light, *photos*, with *volt*, the name of the unit of potential difference (i.e. voltage) in an electrical circuit (see Chapter 1). The volt was named after the Italian physicist Count Alessandro Volta, the inventor of the battery. Photovoltaics thus describes the generation of electricity from light.

The discovery of the **photovoltaic effect** is generally credited to the French physicist Edmond Becquerel (Figure 3.1) who in 1839 published a paper

Figure 3.1 Edmond Becquerel, who discovered the photovoltaic effect

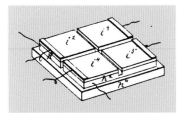

Figure 3.2 Diagram from Charles Edgar Fritts' 1884 US patent application for a solar cell

describing his experiments with a 'wet cell' battery, in the course of which he found that the battery voltage increased when its silver plates were exposed to sunlight (Becquerel, 1839).

The first report of the PV effect in a solid substance was made in 1877 when two Cambridge scientists, Adams and Day, described in a paper to the Royal Society the variations they observed in the electrical properties of selenium when exposed to light (Adams and Day, 1877). Selenium is a non-metallic element similar to sulfur.

In 1883 Charles Edgar Fritts, a New York electrician, constructed a selenium solar cell that was, superficially, similar to the silicon solar cells of today (Figure 3.2). It consisted of a thin wafer of selenium covered with a grid of very thin gold wires and a protective sheet of glass. But his cell was very inefficient. The efficiency of a solar cell is defined as the proportion of the solar energy falling on its surface that is converted into electrical energy. Less than 1% of the solar energy incident on these early cells was converted to electricity. Nevertheless, the response of selenium cells is well matched to the spectrum of visible light, and they eventually came into widespread use in photographic exposure meters.

The underlying reasons for the inefficiency of these early devices were only to become apparent many years later, in the early decades of the 20th century, when physicists like Max Planck provided new insights into the fundamental properties of materials.

It was not until the 1950s that the breakthrough occurred that set in motion the development of modern, high-efficiency solar cells. It took place at the Bell Telephone Laboratories (Bell Labs) in New Jersey, USA, where a number of scientists, including Darryl Chapin, Calvin Fuller and Gerald Pearson (Figure 3.3), were researching the effects of light on **semiconductors**. These are non-metallic materials, such as germanium and silicon, whose electrical characteristics lie between those of conductors, which offer little resistance to the flow of electric current, and insulators, which block the flow of current almost completely. Hence the term *semi*conductor.

A few years before, in 1948, three other Bell Labs researchers, Bardeen, Brattain and Shockley, had produced another revolutionary device using semiconductors – the transistor. Transistors are made from semiconductors (usually silicon) in extremely pure crystalline form, into which tiny quantities of carefully selected impurities, such as boron or phosphorus, have been deliberately diffused. This process, known as **doping**, dramatically alters the electrical behaviour of the semiconductor in a very useful manner which will be described in detail later.

In 1953 the Chapin–Fuller–Pearson team, building on earlier Bell Labs research on the PV effect in silicon (Ohl, 1941), produced 'doped' silicon slices that were much more efficient than earlier devices in producing electricity from light.

By the following year they had produced a paper on their work (Chapin et al., 1954) and had succeeded in increasing the conversion efficiency of their silicon solar cells to 6%. Bell Labs went on to demonstrate the practical use of solar cells, for example in powering rural telephone amplifiers, but at that time they were too expensive to be an economic source of power in most applications.

In 1958, however, solar cells were used to power a small radio transmitter in the second US space satellite, Vanguard I. Following this first successful

Figure 3.3 Bell Laboratories' pioneering PV researchers Pearson (left), Chapin (centre) and Fuller (right) measure the response of an early solar cell to light

demonstration, the use of PV as a power source for spacecraft has become almost universal (Figures 3.4 and 3.5).

Rapid progress in increasing the efficiency and reducing the cost of PV cells has been made over the past few decades. Their terrestrial uses are now widespread, not only in remote locations, where they provide power for telecommunications, lighting and other electrical uses, but also in a rapidly increasing number of domestic, commercial and industrial buildings which now incorporate grid-connected PV arrays supplying a substantial proportion of their electricity needs. In addition, a number of large, multi-megawatt sized PV power plants are now connected to electricity grids in such countries as Germany, Italy, Spain, Portugal, France, Canada and the USA.

Figure 3.4 The radio transmitter in the second US space satellite, Vanguard I, launched in 1958, was powered by six very small solar PV panels

The efficiency of the best crystalline silicon solar *cells* has now reached 25% in standard test conditions (see Box 3.1). The best silicon PV *modules* now available commercially have an efficiency of around 20% (Green et al., 2011).

Figure 3.5 The International Space Station is powered by large arrays of PV panels with a combined output of around 130 kW

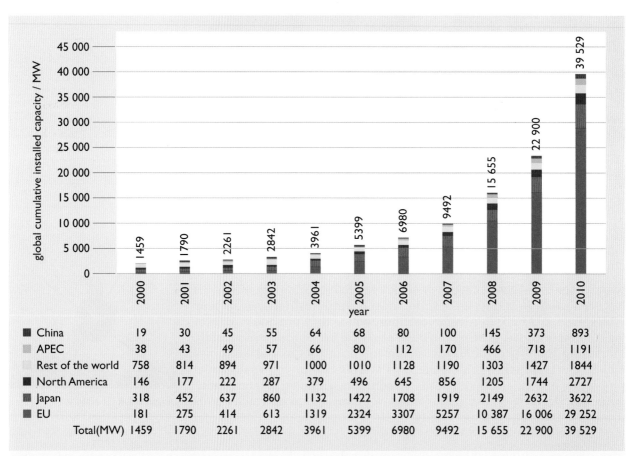

	2000	2001	2002	2003	2004	2005	2006	2007	2008	2009	2010
■ China	19	30	45	55	64	68	80	100	145	373	893
▨ APEC	38	43	49	57	66	80	112	170	466	718	1191
▤ Rest of the world	758	814	894	971	1000	1010	1128	1190	1303	1427	1844
■ North America	146	177	222	287	379	496	645	856	1205	1744	2727
▥ Japan	318	452	637	860	1132	1422	1708	1919	2149	2632	3622
■ EU	181	275	414	613	1319	2324	3307	5257	10 387	16 006	29 252
Total(MW)	1459	1790	2261	2842	3961	5399	6980	9492	15 655	22 900	39 529

Figure 3.6 Evolution of global cumulative PV installed capacity (in MW) in various regions and the world as a whole, 2000–2010 (source: EPIA, 2011) Notes: (a) APEC refers to Asia–Pacific Economic Cooperation – an inter-governmental grouping of 21 countries; (b) any discrepancies between overall and individual yearly totals are due to rounding

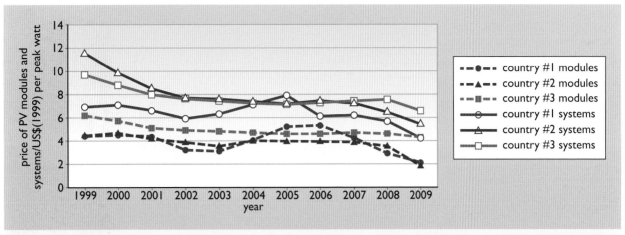

Figure 3.7 Evolution of price of PV modules and small-scale PV systems in three anonymized International Energy Agency (IEA) reporting countries, 1999–2009. Prices per watt are shown in constant US$ (1999), i.e. they are corrected for the effects of inflation. (source: IEA, 2010a). Note: data refers to the IEA reporting countries, mainly 'developed' nations

BOX 3.1   Standard test conditions for PV cells and modules

There is widespread international agreement that the performance of PV cells and modules should be measured under a set of standard test conditions.

Essentially, these specify that the temperature of the cell or module should be 25 °C and that the solar radiation incident on the cell should have a total power density of 1000 watts per square metre, with a spectral power distribution known as **Air Mass 1.5 Global (AM1.5G)**. (The term 'Global' is added here to denote the inclusion of diffuse as well as direct radiation.)

The spectral power distribution describes in graphical form the way in which the power contained in the solar radiation varies across the spectrum of wavelengths.

The concept of 'Air Mass' (sometimes called 'Air Mass Coefficient') relates to the way in which the spectral power distribution of radiation from the Sun is affected by the distance the Sun's rays have to travel though the atmosphere before reaching an observer (or a PV module).

In space, there is no atmosphere to affect the Sun's radiation. Immediately above the atmosphere, it has a power density of approximately 1365 watts per square metre. The characteristic spectral power distribution of solar radiation as measured in space is described as the **Air Mass 0 (AM0)** distribution.

At the Earth's surface, the various gases of which the atmosphere is composed (oxygen, nitrogen, ozone, water vapour, carbon dioxide, etc.) attenuate the solar radiation selectively at different wavelengths. This attenuation increases as the distance that the Sun's rays have to travel through the atmosphere increases.

When the Sun is at its zenith (i.e. directly overhead), the distance which the Sun's rays have to travel through the atmosphere to a PV module is at a minimum. The characteristic spectral power distribution of solar radiation observed under these conditions is known as the **Air Mass 1 (AM1)** distribution.

When the Sun is at a given angle θ to the zenith (as perceived by an observer at sea level), the Air Mass is defined as the ratio of the path length of the Sun's rays under these conditions to the path length when the Sun is at its zenith. By simple trigonometry (Figure 3.8), this leads to the definition:

$$\text{Air Mass} \approx \frac{1}{\cos\theta}$$

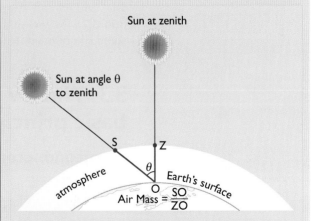

Figure 3.8 Air Mass is the ratio of the path length of the Sun's rays through the atmosphere when the Sun is at a given angle (θ) to the zenith, to the path length when the Sun is at its zenith

An Air Mass distribution of AM 1.5 therefore corresponds to the spectral power distribution observed when the Sun's radiation is coming from an angle to overhead of about 48°, since cos 48° = 0.67 and the reciprocal of this is 1.5.

The spectral power distributions for Air Masses 0 and 1.5 are shown in Figure 3.9.

In practice, the power rating of a PV cell or module, expressed in peak watts (Wp) of power output, is determined by measuring the maximum power it will supply when exposed to radiation from lamps designed to reproduce the AM 1.5 Global spectral distribution at a total power density of 1000 W m^{-2}.

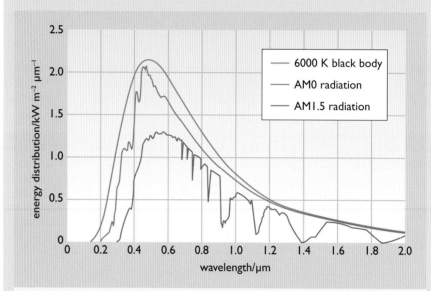

Figure 3.9 The spectral power distributions of solar radiation corresponding to Air Mass 0 and Air Mass 1.5. Also shown is the theoretical spectral power distribution which would be expected, in space, if the Sun were a perfect radiator (a 'black body') at 6000 °C

From 2000 to 2010, the total installed power capacity of PV systems increased more than 20-fold, module costs dropped to around US$2 per peak watt and overall system costs fell to around US$4 per watt (see Figures 3.6 and 3.7). As we shall see, improvements in the cost-effectiveness of PV seem likely to continue.

3.3 The PV effect in crystalline silicon: basic principles

Semiconductors and 'doping'

Conventional PV cells consist, in essence, of a junction between two thin layers of dissimilar semiconducting materials, known respectively as 'p' (positive)-type semiconductor, and 'n' (negative)-type semiconductor. These semiconductors are usually made from silicon, so for simplicity we shall initially consider only silicon-based semiconductors – although, as we shall see later, PV cells can be made from other materials.

n-type semiconductors are made from crystalline silicon that has been 'doped' with tiny quantities of an impurity (usually phosphorus) in such a way that the doped material possesses a *surplus of free electrons*. **Electrons** are sub-atomic particles with a negative electrical charge, so silicon doped in this way is known as an **n (negative)-type** semiconductor.

p-type semiconductors are also made from crystalline silicon, but are doped with very small amounts of a different impurity (usually boron) which causes the material to have a *deficit of free electrons*. These 'missing' electrons are called **holes**. Since the absence of a negatively charged electron can be considered equivalent to a positively charged particle, silicon doped in this way is known as a **p (positive)-type** semiconductor (see Box 3.2).

The p–n junction and the PV effect

We can create what is known as a **p–n junction** by joining these dissimilar semiconductors. This sets up an **electric field** in the region of the junction. (This electric field is like the electrostatic field you can generate by rubbing a plastic comb against a sweater. It will cause negatively charged particles to move in one direction, and positively charged particles to move in the opposite direction.) However, a p–n junction in practice is not a simple mechanical junction: the characteristics change from 'p' to 'n' gradually, not abruptly, across the junction.

What happens when light falls on the p–n junction at the heart of a solar cell?

Light can be considered to consist of a stream of tiny particles of energy, called **photons**. When photons from light of a suitable wavelength fall within the p–n junction, they can transfer their energy to some of the electrons in the material, so 'promoting' them to a higher energy level. Normally, these electrons help to hold the material together by forming so-called 'valence' bonds with adjoining atoms, and cannot move. In their 'excited' state, however, the electrons become free to conduct electric current by moving through the material. In addition, when electrons move they leave behind holes in the material, and these can also move (Box 3.2). The 'car parking' analogy shown in Figure 3.10 may be helpful in visualizing the processes involved.

When the p–n junction is formed, some of the electrons in the immediate vicinity of the junction are attracted from the n-side to combine with holes on the nearby p-side. Similarly, holes on the p-side near the junction are attracted to combine with electrons on the nearby n-side.

The net effect of this is to set up around the junction a layer on the n-side that is more positively charged than it would otherwise be, and, on the

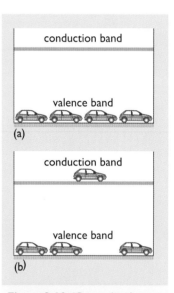

Figure 3.10 'Car parking' analogy of conduction processes in a semiconductor: (a) The ground floor of the car park is full: the cars there (representing electrons in the 'valence band') cannot move around. The first floor is empty (b) A car (electron) is 'promoted' to the first floor (representing the 'conduction band'), where it can move around freely. This leaves behind a 'hole' that also allows cars on the ground floor (valence band) to move around (source: Green, 1982)

BOX 3.2 **Crystalline silicon, doping, p-n junctions and PV cells**

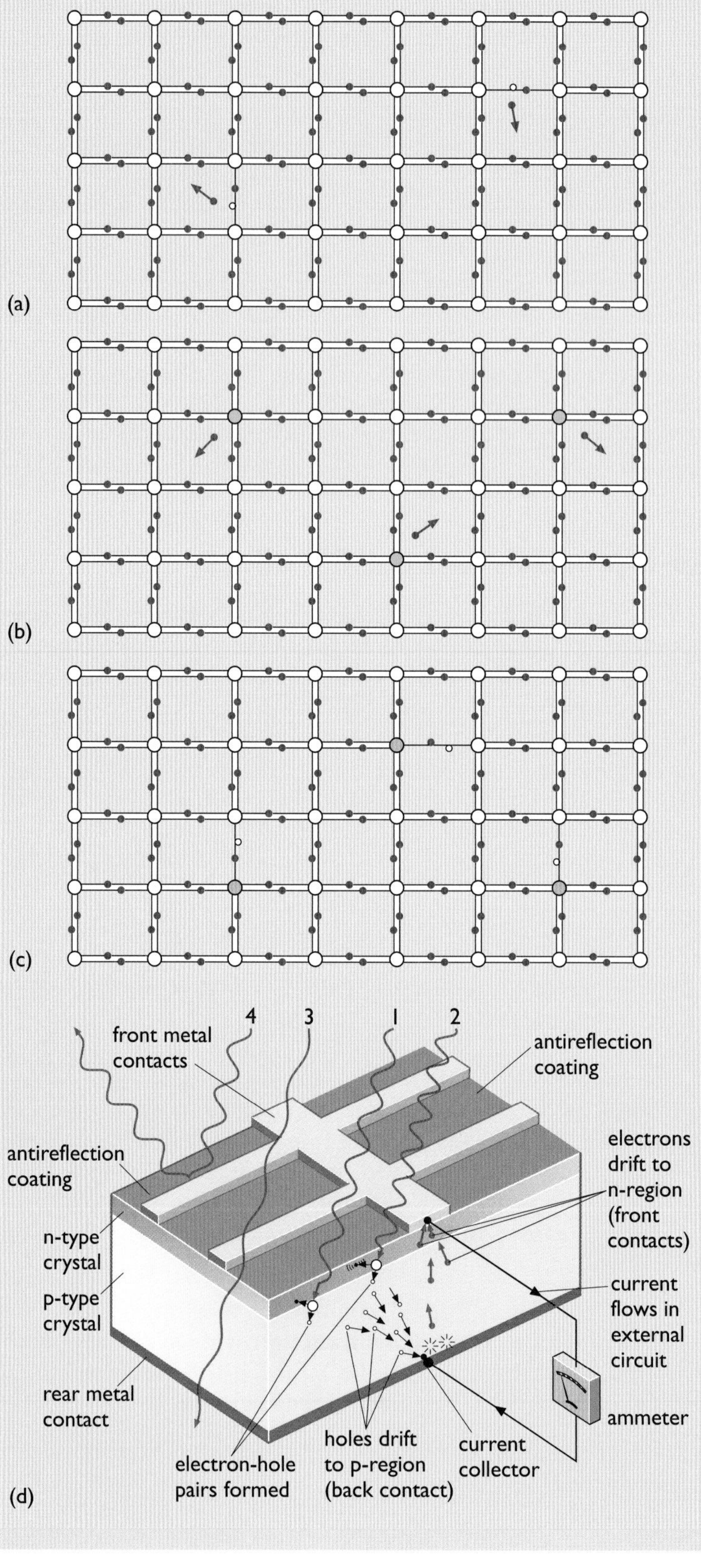

(a)

(b)

(c)

(d)

front metal contacts

antireflection coating

antireflection coating

n-type crystal

p-type crystal

rear metal contact

electron-hole pairs formed

holes drift to p-region (back contact)

current collector

electrons drift to n-region (front contacts)

current flows in external circuit

ammeter

Figure 3.11 (a) A crystal of pure silicon has a cubic structure, shown here in two dimensions for simplicity. The silicon atom has four 'valence' electrons. Each atom is firmly held in the crystal lattice by sharing two electrons (small red dots) with each of four neighbours at equal distances from it. Occasionally thermal vibrations or a photon of light will spontaneously provide enough energy to promote one of the electrons into the energy level known as the conduction band, where the electron (red dot) is free to travel through the crystal and conduct electricity. When the electron moves from its bonding site, it leaves a 'hole' (small white dot), a local region of net positive charge

(b) A crystal of n-type silicon can be created by doping the silicon with trace amounts of phosphorus. Each phosphorus atom (shown in green) has five valence electrons, so that not all of them are taken up in the crystal lattice. Hence an n-type crystal has an excess of free electrons

(c) A crystal of p-type silicon can be created by doping the silicon with trace amounts of boron. Each boron atom (shown in blue) has only three valence electrons, so that it shares two electrons with three of its silicon neighbours and one electron with the fourth. Hence the p-type crystal contains more holes than conduction electrons

(d) A silicon solar cell is a wafer of p-type silicon with a thin layer of n-type silicon on one side. When a photon of light with the appropriate amount of energy penetrates the cell near the junction of the two types of crystal and encounters a silicon atom (1), it dislodges one of the electrons, which leaves behind a hole. The energy required to promote the electron into the conduction band is known as the band gap (see Fig. 3.12). The electron thus promoted tends to migrate into the layer of n-type silicon, and the hole tends to migrate into the layer of p-type silicon. The electron then travels to a current collector on the front surface of the cell, generates an electric current in the external circuit and then re-emerges in the layer of p-type silicon, where it can recombine with waiting holes. If a photon with an amount of energy greater than the band gap strikes a silicon atom (2), it again gives rise to an electron–hole pair, and the excess energy is converted into heat. A photon with an amount of energy smaller than the band gap will pass right through the cell (3), so that it gives up virtually no energy along the way. Moreover, some photons are reflected from the front surface of the cell even when it has an antireflection coating (4). Still other photons are lost because they are blocked from reaching the crystal by the current collectors that cover part of the front surface. (Source for all above text and figures: adapted from Chalmers, 1976)

p-side, a layer that is more negatively charged than it would otherwise be. In effect, this means that a *reverse* electric field is set up around the junction: negative on the p-side and positive on the n-side. The region around the junction is also depleted of charge carriers (electrons and holes) and is therefore known as the **depletion region**.

When an electron in the junction region is stimulated by an incoming photon to 'jump' into the conduction band, it leaves behind a hole in the valence band. Two charge carriers (an **electron–hole pair**) are thus generated. Under the influence of the reverse electric field around the junction, the electrons will tend to move into the n-region and the holes into the p-region.

The process can be envisaged in terms of the energy levels in the material (Figure 3.12). The electrons that have been stimulated by incoming photons to enter the conduction band can be thought of as 'rolling downwards', under the influence of the electric field at the junction, into the n-region; similarly, the holes can be thought of as 'floating upwards', under the influence of the junction field, into the p-region.

The flow of electrons to the n-region is, by definition, an electric current. If there is an external circuit for the current to flow through, the moving electrons will flow out of the semiconductor via one of the metallic contacts on the top of the cell. The holes, meanwhile, will flow in the opposite direction through the material until they reach another metallic contact on the bottom of the cell, where they are then 'filled' by electrons entering from the external circuit.

In order to produce power, the PV cell must generate voltage as well as the current provided by the flow of electrons. This voltage is, in effect, provided by the internal electric field set up at the p–n junction.

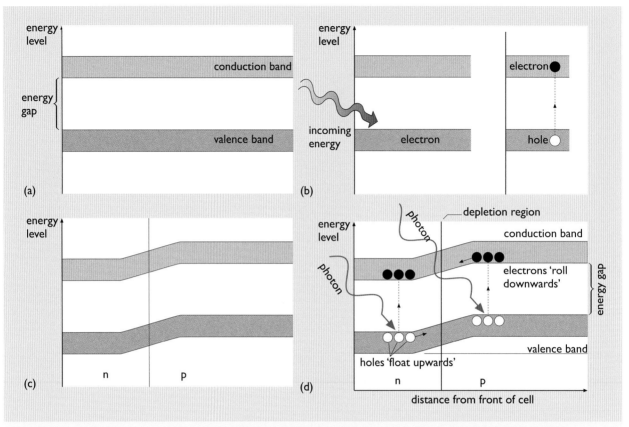

Figure 3.12 (a) Energy bands in a normal ('intrinsic') semiconductor (b) An electron can be 'promoted' to the conduction band when it absorbs energy from light (or heat), leaving behind a 'hole' in the valence band (c) When n-type and p-type semiconductors are combined into a p–n junction, their different energy bands combine to give a new distribution, as shown, and a built-in electric field is created (d) In the p–n junction, photons of light can excite electrons from the valence band to the conduction band. The electrons 'roll downwards' to the n-region, and the holes 'float upwards' to the p-region

Individual crystalline silicon PV cells are typically about 150×150 mm in size, produce a voltage of just over 0.5 volts and give a peak power of approximately 4 watts.

Monocrystalline silicon cells

Until fairly recently, the majority of solar cells were made from extremely pure **monocrystalline** silicon (Si) – that is, silicon with a single, continuous crystal lattice structure (Figure 3.11) having virtually no defects or impurities. Monocrystalline silicon is usually grown from a small seed crystal that is slowly pulled out of a molten mass, or 'melt', of polycrystalline silicon, in the sophisticated but expensive **Czochralski process** developed initially as part of the process for manufacturing 'silicon chips' for the electronics industry. This process of crystal growth is known as an **epitaxial** process and can be used for other PV semiconductors. The entire process of monocrystalline silicon solar cell and module production is summarized in Figure 3.13.

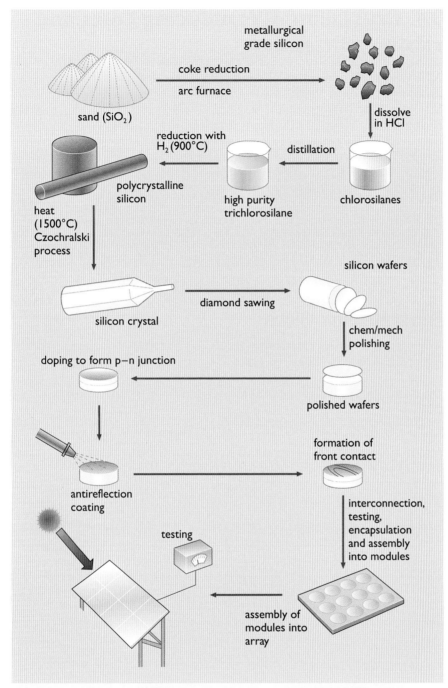

Figure 3.13 The overall process of monocrystalline silicon solar cell and module production. Note that most cells are finally trimmed to be square, or semi-square with rounded corners

3.4 Crystalline PV: reducing costs and raising efficiency

Although monocrystalline silicon PV modules are highly efficient, they are also expensive because the Czochralski process is slow, requires highly skilled operators, and is labour- and energy-intensive. Another reason for their high cost is that until recently almost all such cells were fabricated from extremely pure 'electronic-grade' silicon, including recycled leftovers from silicon chip manufacture. However, PV cells are now commonly made from slightly less pure, but less costly, 'solar-grade' silicon, with only a small reduction in conversion efficiency.

A number of approaches to reducing the cost and/or increasing the efficiency of crystalline PV cells and modules have been under development during the past 30 years or so. These include cells using polycrystalline rather than single-crystal material, growing silicon in ribbon form, and the use of other crystalline PV materials such as gallium arsenide.

Polycrystalline silicon

Polycrystalline silicon essentially consists of small grains of monocrystalline silicon. Solar cell wafers can be made directly from polycrystalline silicon in various ways. These include the controlled casting of molten polycrystalline silicon into cube-shaped ingots that are then cut, using fine wire saws, into thin square wafers and fabricated into complete cells in the same way as monocrystalline cells (Figure 3.14).

Another approach to polycrystalline silicon PV manufacture involves drawing a thin 'ribbon' of polycrystalline silicon from a silicon 'melt'. The main process used is known as 'edge-defined, film-fed growth' (EFG), and was originally developed by the US firm Mobil Solar.

Polycrystalline PV cells are easier and cheaper to manufacture than their monocrystalline counterparts, but they tend to be less efficient because light-generated charge carriers (i.e. electrons and holes) can recombine at the boundaries between the grains within polycrystalline silicon. However, it has been found that their efficiency can be substantially increased by processing the material in such a way that the grains are relatively large in size and oriented in a top-to-bottom direction to allow light to penetrate deeply into each grain. These and other improvements have enabled commercially available polycrystalline PV modules (sometimes called 'semi-crystalline' or 'multi-crystalline') to reach efficiencies of over 15%.

Polycrystalline silicon film

Conventional silicon solar cells need to be around 150–200 microns (millionths of a metre) thick in order to ensure that most of the photons incident upon them can be absorbed. But some companies have developed advanced 'light trapping' techniques to maximize the interaction of photons with the material, allowing much thinner layers or 'films' of silicon to be used. They have also developed ways of depositing such **polycrystalline films** on to ceramic or glass substrates, forming the basis of PV modules of around 8% efficiency (Green et al., 2011). The films used are somewhat thicker than

Figure 3.14 Polycrystalline silicon consists of 'grains' of monocrystalline silicon

in other 'thin film' PV cells (as discussed in Section 3.5), so cells made in this way are sometimes known as '*thick* film' polycrystalline cells.

Gallium arsenide

Silicon is not the only crystalline material suitable for PV applications. Another is gallium arsenide (GaAs), a so-called **compound semiconductor**. GaAs has a crystal structure similar to that of silicon, but consisting of alternating gallium and arsenic atoms. In principle it is highly suitable for use in PV applications because it has a high light absorption coefficient, so only a thin layer of material is required. GaAs cells also have a band gap wider than that of silicon, one close to the theoretical optimum for absorbing the energy in the terrestrial solar spectrum (see Box 3.3). Cells made from GaAs are therefore more efficient than those made from monocrystalline silicon. They can also operate at relatively high temperatures without the appreciable performance degradation from which silicon and many other semiconductors suffer. This makes them well suited to use in *concentrating* PV systems (see Section 3.6). On the other hand, cells made from GaAs are substantially more expensive than silicon cells, mainly because they require more expensive *epitaxial* crystal growth techniques. (These essentially involve growing monocrystalline GaAs by deposition of GaAs onto a supporting material, in this case a single GaAs crystal. The GaAs support defines the orientation of the new growth while the doping of the

BOX 3.3 **Band gaps and PV cell efficiency**

According to the quantum theory of matter, the quantity of energy possessed by any given electron in a material will lie within one of several levels or 'bands'. Those electrons that normally hold the atoms of a material together (by being 'shared' between adjoining atoms, as we saw in Figure 3.11) are described by physicists as occupying a lower-energy state known as the **valence band**.

In certain circumstances, some electrons may acquire enough energy to move into a higher energy state, known as the **conduction band,** in which they can move around within the material and thus conduct electricity (Figure 3.12). There is a so-called **energy gap** or **band gap** between these bands, the magnitude of which varies from material to material, and which is measured using an extremely small energy unit: the electron volt (eV) (See Appendix A.).

Metals, which conduct electricity well, have many electrons in the conduction band. Insulators, which hardly conduct electricity at all, have virtually no electrons in the conduction band. Pure (or 'intrinsic') semiconductors have some electrons in the conduction band, but not as many as in a metal. But doping pure semiconductors with very small quantities of certain impurities can greatly improve their conductivity.

If a photon incident on a doped, n-type semiconductor in a PV cell is to succeed in transferring its energy to an electron and 'exciting' it from the valence band to the conduction band, it must possess an energy at least equal to the band gap. Photons with energy less than the band gap do not excite valence electrons to enter the conduction band and their energy is 'wasted'. Photons with energies significantly greater than the band gap do succeed in 'promoting' an electron into the conduction band, but any excess energy is dissipated as heat. This wasted energy is one of the reasons why PV cells are not 100% efficient in converting solar radiation into electricity. (Another is that not all photons incident on a cell are absorbed: a small proportion are reflected.)

Because the energy of a photon is directly proportional to the frequency of the light associated with it, photons associated with shorter wavelengths (i.e. higher frequencies) of light, near the blue end of the spectrum, have a greater energy than those of longer wavelength near the red end of the visible spectrum.

The spectral distribution of sunlight varies considerably according to weather conditions and the elevation of the Sun in the sky (see Box 3.1). For maximum efficiency of conversion of light into electric power, it is clearly important that the band gap energy of the material used for a PV cell is reasonably well matched to the spectrum of the light incident upon it. In general, semiconductor materials with band gaps between 1.0 and 1.9 eV are reasonably well suited to PV use. Silicon has a band gap of 1.1 eV.

The maximum theoretical conversion efficiency attainable in a *single-junction* silicon PV cell has been calculated to be about 30%, if full advantage is taken of 'light trapping' techniques to ensure that as many of the photons as possible are usefully absorbed (Wenham et al., 2007). However, *multi-junction* cells have also been designed in which each junction is tailored to absorb a particular portion of the incident spectrum. Theoretically, such cells should have a much higher efficiency, possibly as high as 66% for an infinite number of junctions – though the efficiencies so far achieved by multi-junction cells in practice have been considerably lower than this (see Section 3.6).

In practice, the highest efficiency achieved in commercially available single-junction monocrystalline silicon PV *modules* (as distinct from individual PV *cells*) is currently around 20% (Green et al., 2011). The efficiency of PV modules is usually lower than that achieved by cells in the laboratory because:

■ it is difficult to achieve as high an efficiency consistently in mass-produced devices as in one-off laboratory cells under optimum conditions;

■ laboratory cells are not usually glazed or encapsulated;

■ in a PV module there are inactive areas, both between cells and due to the surrounding module frame, which are not available to produce power;

■ there are small resistive losses in the wiring between cells and in the diodes used to protect cells from short circuiting;

■ there are losses due to mismatching between cells of slightly differing electrical characteristics connected in series.

GaAs layer being deposited can be also be controlled.) GaAs cells have often been used when very high efficiency is required – as in many space applications and solar racing cars (see Figure 3.15).

Figure 3.15 The winner of the 2009 World Solar Challenge race across Australia was the Tokai Challenger, a solar car designed and tested by students from Tokai University together with several Japanese automotive companies. It covered 2998 km in 29 hours 49 minutes at an average speed (during daylight) of 100.54 km per hour. The car was driven by electric motors powered by an array of 2174 Sharp triple-junction III–V PV cells (see section 3.6) of a type used in space applications, with a peak power output of 1.8 kW and an efficiency of 30% (source: Sharp Solar, 2009)

3.5 Thin film PV

Crystalline wafers are not the only materials suitable for photovoltaics. PV cells can also be made from 'thin films' of various kinds, the most common of which are amorphous silicon, copper indium (gallium) diselenide and cadmium telluride.

Amorphous silicon

Solar cells can be made from very thin films of silicon in a form known as **amorphous silicon (a-Si)**, in which the silicon atoms are much less ordered than in the crystalline forms described above. In a-Si, not every silicon atom is fully bonded to its neighbours, which leaves so-called 'dangling bonds' that can absorb any additional electrons introduced by doping, so rendering any p–n junction ineffective.

However, this problem is largely overcome in the process by which a-Si cells are normally manufactured. A gas containing silicon and hydrogen (such as silane, SiH_4), and a small quantity of dopant (such as boron), is decomposed electrically in such a way that it deposits a thin film of a-Si on a suitable substrate (backing material). The hydrogen in the gas combines with the dangling silicon bonds. The dopant that is also present in the gas can then have its usual effect of contributing charge carriers to enhance the conductivity of the material.

Solar cells using a-Si have a somewhat different form of junction between the p- and the n-type material. A so-called 'p–i–n' junction is usually

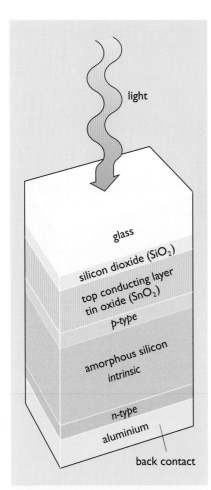

Figure 3.16 Structure of an a-Si cell. The top electrical contact is made of an electrically conducting, but transparent, layer of tin oxide deposited on the glass. Silicon dioxide forms a thin 'barrier layer' between the glass and the tin oxide. The bottom contact is made of aluminium. In between are layers of p-type, intrinsic and n-type a-Si

formed, consisting of an extremely thin layer of p-type a-Si on top, followed by a thicker 'intrinsic' (i) layer made of un-doped a-Si, and then a very thin layer of n-type a-Si. The structure is shown in Figure 3.16. The operation of the PV effect in a-Si is generally similar to that in crystalline silicon, except that in a-Si the band gap, although wider, is less clearly defined.

Amorphous silicon cells are cheaper to produce than those made from crystalline silicon. a-Si is also a better absorber of light, so thinner (and therefore cheaper) films can be used. The manufacturing process operates at a lower temperature than that for crystalline silicon, so less energy is required; it is suited to continuous production; and it allows quite large areas of cell to be deposited on to a wide variety of both rigid and flexible substrates, including steel, glass and plastics.

But a-Si cells are currently less efficient than their single-crystal or polycrystalline silicon counterparts: maximum efficiencies achieved with small, single-junction cells in the laboratory are currently around 10% (Green et al, 2011). Moreover, the efficiency of a-Si modules can degrade by about 20% over the first few months, but then it stabilizes. Manufacturers sell modules with power ratings that correspond to their estimate of the degraded, stabilized performance.

a-Si cells are widely used as power sources in a variety of applications where the requirement is not so much for high efficiency as for low cost.

Strenuous attempts have been made by manufacturers to improve the efficiency of a-Si cells and to solve the degradation problem. One of the most promising approaches involves the development of multiple-junction a-Si devices (see Section 3.6), which has resulted in reduced degradation and improved module efficiency – to around 10%.

Copper indium (gallium) diselenide

Amorphous silicon is by no means the only material suited to thin film PV. Other thin film technologies include those based on compound semiconductors, in particular copper indium diselenide ($CuInSe_2$, usually abbreviated to CIS), and copper indium gallium diselenide (CIGS).

Thin film CIGS cells have attained the highest laboratory efficiencies of all thin film devices, around 19%, and CIGS modules with stable efficiencies of 15% are available (Green et al., 2011).

Cadmium telluride

Thin film PV modules can also be made using cadmium telluride (CdTe), using a relatively simple and inexpensive electroplating-type process. The band gap of CdTe is close to the optimum, and module efficiencies of over 12% have been achieved (Green et al., 2011), without the performance degradation that occurs in a-Si cells. However the modules contain cadmium, a toxic substance, so stringent precautions need to be taken during their manufacture, use and eventual disposal or recycling (see Section 3.11).

3.6 Other PV technologies

Multi-junction PV cells and modules

One way of improving the overall conversion efficiency of PV cells and modules is the 'stacked' or **multi-junction** approach, in which two (or more) PV junctions are layered one on top of the other, each layer extracting energy from a particular portion of the spectrum of the incoming light.

The band gap of a-Si, for example, can be increased by alloying the material with carbon, so that the resulting material responds better to light at the blue end of the spectrum. Alloying with germanium, on the other hand, decreases the band gap so the material responds to light at the red end of the spectrum.

Typically, a wide band gap a-Si junction would be on top, absorbing the higher-energy photons at the blue end of the spectrum, followed by other thin film a-Si junctions, each having a band gap designed to absorb a portion of the lower light frequencies nearer the red end of the spectrum. Multi-junction modules using a-Si are available with stable efficiencies of around 8%.

Much higher efficiencies are attainable with multi-junction modules based on so-called 'III–V' materials, based on elements from groups III and V of the periodic table. Such III–V semiconductors contain a combination of one or more of aluminium, gallium or indium (elements from group III) with one or more of nitrogen, phosphorus, arsenic and antimony (elements from group V).

As shown in Figure 3.17, such cells typically have a gallium indium phosphide (GaInP$_2$) layer uppermost that absorbs light in the blue part of the spectrum; a gallium arsenide (GaAs) layer in the middle that absorbs the yellow part of the spectrum, and a germanium (Ge) layer at the bottom that absorbs the red portion of the spectrum.

Such cells are expensive as their manufacture requires an epitaxial crystal growth process, but they are very efficient. Modules in volume production achieve efficiencies of around 28%, and in laboratory tests efficiencies exceeding 40% have been achieved. Such cells are mainly used in space applications and in concentrating PV systems (see below), and are designed to resist radiation exposure and high operating temperatures.

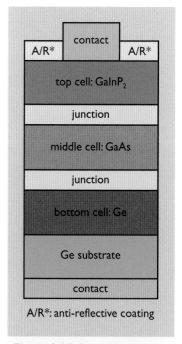

Figure 3.17 Structure of Spectrolab's 'ultra triple junction' III–V PV cell designed for space and concentrator applications

Concentrating PV systems

Another way of getting more energy out of a given number of PV cells is to use mirrors or Fresnel lenses to concentrate the incoming solar radiation onto the cells. (The approach is similar to that described in Chapter 2, Section 2.10, on solar thermal engines.) This has the obvious advantage that a substantially smaller area of cells is required – to an extent depending on the concentration ratio, which can vary from as little as two to several hundred or even thousand times. The concentrating system must have an aperture equal to that of an equivalent flat plate array to collect the same amount of incoming energy.

Systems with the highest concentration ratios use sensors, motors and controls to allow them to track the Sun in two axes (azimuth and elevation), ensuring that the cells always receive the maximum amount of solar radiation (See Fig 2.42). This also means, however, that the cells are exposed to much higher temperatures than in a flat plate array, which is one reason why multi-junction III–V cells are often used. Systems with lower concentration ratios often track the Sun only on one axis and can have simpler tracking mechanisms.

Most concentrators can only utilize direct solar radiation. This is a problem in countries like the UK where nearly half the solar radiation is diffuse, but in climates with higher proportions of direct solar radiation they can be effective.

Silicon spheres

In the 1970s, the US firm Texas Instruments (TI) developed an ingenious way of making PV cells using tiny, millimetre-sized, spheres of doped silicon embedded at regular intervals between thin sheets of aluminium foil. This approach has recently been further developed by the Japanese firm Clean Venture 21 (CV21) by enclosing each tiny silicon sphere in its own hexagonal aluminium micro-reflector cup, which concentrates light on the sphere and reduces the total amount of silicon required, in comparison with conventional crystalline silicon cells, Figure 3.18 (see Graham-Rowe, 2007; CV21, 2008). This technology is currently at a very early stage of commercialization.

Photoelectrochemical cells

A photoelectrochemical approach to producing low-cost electricity from solar energy has been developed by researchers at the Swiss Federal Institute of Technology in Lausanne. (Strictly, photoelectrochemical devices are not photovoltaic: this term normally implies a solid-state device. Photoelectrochemical devices, however, use liquids or gels.) The idea of harnessing photoelectrochemical effects to produce electricity

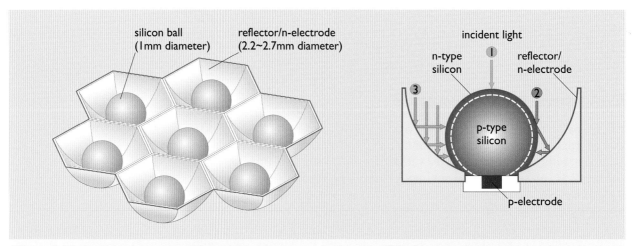

Figure 3.18 Principle of operation of Spheral PV technology as developed by CV21. Tiny spheres of doped crystalline silicon are embedded in hexagonal micro-reflectors. Ray traces 1, 2 and 3 illustrate how incident light may reach the silicon sphere

from sunlight is not new – Becquerel's pioneering experiments were with liquid-based devices.

The Swiss researchers have reported moderately high efficiencies in a device that could be extremely cheap to manufacture. It consists essentially of two thin glass plates, both covered with a thin, transparent, electrically conducting tin oxide layer (Figure 3.19). To one plate is added a thin layer of titanium dioxide (TiO_2), which is a semiconductor. The surface of the TiO_2 has been treated to give it exceptionally high roughness, enhancing its light-absorbing properties. Immediately next to the roughened surface of the titanium dioxide is a layer of 'sensitizer' dye only one molecule thick, made of a proprietary 'transition metal complex' based on ruthenium or osmium. Between this 'sensitized' TiO_2 and the other glass plate is a thicker layer of iodine-based electrolyte.

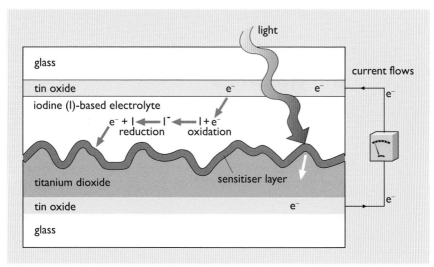

Figure 3.19 Principles of operation of the photoelectrochemical PV cell, developed at the Swiss Federal Institute of Technology, Lausanne

On absorption of a photon of suitable wavelength, the sensitizer layer injects an electron into the conduction band of the TiO_2. Electrons so generated then move to the bottom electrically conducting layer (electrode) and pass out into an external circuit where they can produce electrical power. They then re-enter through the top electrode, where they drive a reduction–oxidation process in the iodine solution. This then supplies electrons to the sensitized TiO_2 layer in order to allow the process to continue.

The Swiss researchers have reported efficiencies of 10% in full (AM1.5) sunlight and to have partially overcome the stability problems that initially were a deficiency of this approach. PV cells based on this technology are being manufactured on a small scale (see Grätzel, 1991, 2001; O'Regan and Grätzel, 1991; O'Regan et al., 1993; Jayaweera et al., 2008).

Emerging and novel PV technologies

Emerging on to the market are a number of novel PV technologies. These include the dye-sensitized cells described above and 'organic PV' cells (OPV) using low-cost, flexible polymeric materials.

Still at the research and development (R&D) stage is a range of PV technologies based on 'nanotechnology' – that is, technology which aims to manipulate molecules and atoms at extremely small scales, measured in billionths of a metre, or nanometres (nm). These tiny particles and structures are called 'nanoparticles' and 'nanostructures'; crystals of such tiny sizes are termed 'nanocrystals'. Nanoparticles consisting of extremely small collections of atoms (ranging from just a few up to around 10 000) of semiconducting material are called 'quantum dots' (QDs). The band gaps of quantum dots can be 'tuned' to different parts of the solar spectrum by changing their size. By incorporating QDs with a variety of sizes, more of the energy in the spectrum can be absorbed.

PV cells using quantum dots are theoretically capable of very high efficiencies but are still at the R&D stage: their efficiencies as measured in laboratory tests are currently low – around 5%.

3.7 Electrical characteristics of silicon PV cells and modules

One very simple way of envisaging a typical silicon PV cell is as a solar-powered battery that produces a voltage of around 0.5 V and delivers a current, proportional to cell size and sunlight intensity, of several amps.

In order to use PV cells efficiently we need to know a little more about how they behave when connected to electrical loads. Figure 3.20 shows a single silicon PV cell connected to a variable electrical resistance R, together with an ammeter to measure the current (I) flowing in the circuit and a voltmeter to measure the voltage (V) developed across the cell terminals. Let us assume the cell is being tested under standard conditions (see Box 3.1).

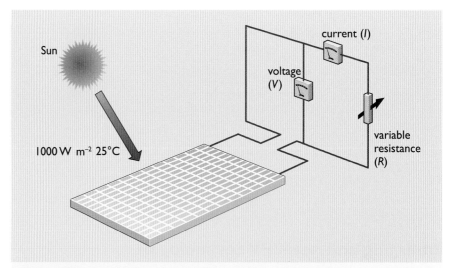

Figure 3.20 PV cell connected to variable resistance, with ammeter and voltmeter to measure variations in current and voltage as resistance varies

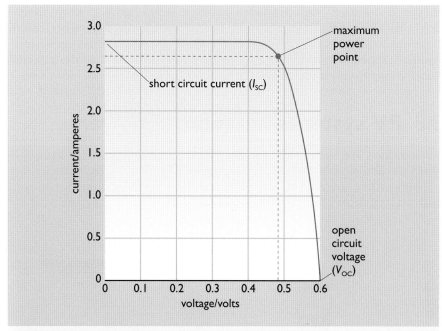

Figure 3.21 Current–voltage (I–V) characteristics of a typical silicon PV cell under standard test conditions

When the resistance is infinite (i.e. when the cell is 'open circuited') the current in the circuit is at its minimum (zero) and the voltage across the cell is at its maximum, known as the **'open circuit voltage'** (V_{OC}). At the other extreme, when the resistance is zero, the cell is in effect 'short circuited' and the current in the circuit then reaches its maximum, known as the **'short circuit current'** (I_{SC}).

If we vary the resistance between zero and infinity, the current (I) and voltage (V) will be found to vary as shown in Figure 3.21, which is known as the 'I–V characteristic' or 'I–V curve' of the cell. It can be seen from the graph that the cell will deliver maximum power (i.e. the maximum product of voltage and current) when the external resistance is adjusted so that its value corresponds to the **maximum power point** (MPP) on the I–V curve.

At lower levels of solar radiation than the maximum (1000 W m^{-2}) assumed in Figure 3.21, the general shape of the I–V characteristic stays the same, but the area under the curve decreases, and the maximum power point moves downwards and to the left.

The short circuit current is directly proportional to the intensity of solar radiation on the cell, whilst the open circuit voltage is only weakly dependent on the solar radiation intensity. The open circuit voltage also decreases linearly as cell temperature increases.

When PV cells are delivering power to electrical loads in real world conditions, the intensity of solar radiation usually varies substantially over time. Many PV systems therefore incorporate a so-called 'maximum power point tracking' device, a specialized electronic circuit that automatically varies the load 'seen' by the PV cell in such a way that it is always operating around the maximum power point.

A typical silicon PV cell produces a voltage of around 0.5 volts. Current crystalline PV *modules* are around 1.4 to 1.7 square metres in area (though some are larger), usually incorporate 60–72 PV cells connected in a series – parallel combination, and have a peak power output of some 120–300 watts, depending on the design and technology.

3.8 PV systems for remote power

PV modules are now widely used in developed countries to provide electrical power in locations where it would be inconvenient or expensive to connect to conventional grid supplies. They often charge batteries to ensure continuity of power. Some examples are shown in Figure 3.22.

(a)   (b)   (c)

Figure 3.22 (a) PV parking meter (b) PV navigation buoy (c) PV telemetry system

---

BOX 3.4 Off grid PV system sizing

To specify accurately how many PV modules would be required for a particular application, such as powering an off-grid house, and what the battery capacity should be, a PV energy system designer needs to be able to answer the following questions.

What are the daily, weekly and seasonal variations in the electrical demand of the house?

What are the daily, weekly and seasonal variations in the amount of solar radiation in the area where the house is situated?

What are the options for the orientation and tilt angle of the PV array?

For how many sunless days will the battery need to be able to provide back-up electricity?

Computer programs have been developed to help engineers calculate the size and cost of PV systems to meet clearly specified energy requirements in given locations and climatic conditions (see IEA, 2011 for a review of these design and simulation tools, and Treble, 1999 for an easy to follow step-by-step procedure for PV system sizing and design).

In many 'developing' countries electricity grids are often non-existent or rudimentary, particularly in rural areas, and all forms of energy are usually very expensive. In such countries PV can be highly competitive with other forms of energy supply – especially in countries with high solar radiation levels – and its use is growing very rapidly. Applications include: PV water pumping; PV refrigerators to help keep vaccines stored safely in health centres; PV systems for homes and community centres, providing energy for lights, radios, audio and video systems; PV-powered telecommunications systems; and PV-powered street lighting.

In thousands of villages in developing countries householders use kerosene lamps and candles at night. These are not only very inefficient light sources but can cause unhealthy smoke emissions and accidental fires. To address this problem, in 2008 the Energy and Resources Institute in Delhi, India set up the 'Lighting a Billion Lives' project (TERI, 2011). This aims to replace the traditional light sources by battery-powered solar lanterns using efficient compact fluorescent lamps (CFL) or light-emitting diode (LED) lamps. The solar lantern batteries can be recharged at solar PV charging stations in villages, run by selected local entrepreneurs (Figure 3.23).

Figure 3.23 PV-powered lanterns being charged in India, as part of the 'Lighting a Billion Lives' project

3.9 **Grid-connected PV systems**

PV systems for homes

In most parts of the developed world, grid electricity is easily accessible as a convenient backup to PV or other fluctuating renewable energy supplies. Here it makes sense for PV energy systems to use the grid as a giant 'battery'. The grid can absorb PV power which is surplus to current needs (say, on sunny summer afternoons), making it available for use by other customers and reducing the amount that has to be generated by conventional means; and at night or on cloudy days, when the output of the PV system is insufficient, it can provide backup energy from conventional sources.

Figure 3.24 In this solar house in Oxford, UK, a 4 kW grid-linked array of monocrystalline PV modules forms an integral part of the roof, alongside solar water heating panels. The PV array supplies the house's annual electricity requirements, plus a small surplus used to provide some of the power for a small electric car

In these grid-connected PV systems, a so-called 'grid commutated inverter' (or 'synchronous inverter') transforms the direct current (DC) power from the PV arrays into alternating current (AC) power at a voltage and frequency that can be accepted by the grid, while 'debit' and 'credit' meters measure the amount of power bought from or sold to the utility. In many countries, 'Feed-in Tariffs' (FiTs) have been introduced. These provide for premium payments to be made for power produced by grid-connected PV arrays and other renewable sources.

In the UK, surveys have suggested that about half of all roofs are oriented in a direction sufficiently close to south to enable them to be used for solar collection purposes. PV arrays can be added to the roofs of existing dwellings, normally by mounting them on rails on top of the roof. In the roofs of new dwellings they can in some circumstances replace all or part of the conventional roof (Figure 3.24).

Energy yield from PV systems

The annual amount of energy that will be produced by a PV system depends on various conditions, including:

- the annual total quantity of solar radiation available at the site. This can be estimated from meteorological data giving the measured number of kWh (or GJ) per square metre per year incident on a horizontal surface at the nearest meteorological station. In typical UK conditions, this is around 1000 kWh m^{-2} y^{-1} (as described in section 2.3) but in very sunny countries the figure can rise to well over 2000 kWh m^{-2} y^{-1}
- the orientation (azimuth) and tilt (elevation) of the PV arrays. For maximum energy output, the arrays should be oriented close to south and with an elevation roughly equal to the latitude of the site – although departures from these optima do not have a marked effect, and the choice of tilt angle can be varied according to the seasonal output profile desired
- the peak power rating of the arrays (or, alternatively, the area of the arrays in square metres)
- the energy conversion efficiency of the PV modules
- the variation in the efficiency of the modules with temperature
- the extent to which module efficiency is affected by the spectral distribution of the solar radiation. This varies according to the elevation of the Sun and the extent of clouds and water vapour in the atmosphere
- the power-reducing effects of array shading by trees, nearby buildings, accumulation of dirt etc. (see Figure 2.39).
- the efficiency of the inverter used to convert the DC power from the PV arrays into AC, and any losses that occur in the wiring between the PV system and the final consumer (these can amount to about 10% of the electricity generated).

The European Union's Joint Research Centre at Ispra, Italy, has produced PVGIS (PV Geographical Information System), a very useful online tool giving solar radiation data and estimated PV outputs for Europe and surrounding countries. Users can input the geographic location, type of module, its power rating, array orientation etc, and the software will calculate the expected monthly and annual energy yield. (See EC JRC, 2011.) Key aspects of this information have been summarized in a Solar Radiation Map of Europe (Figure 3.25).

Photovoltaic Solar Electricity Potential in European Countries

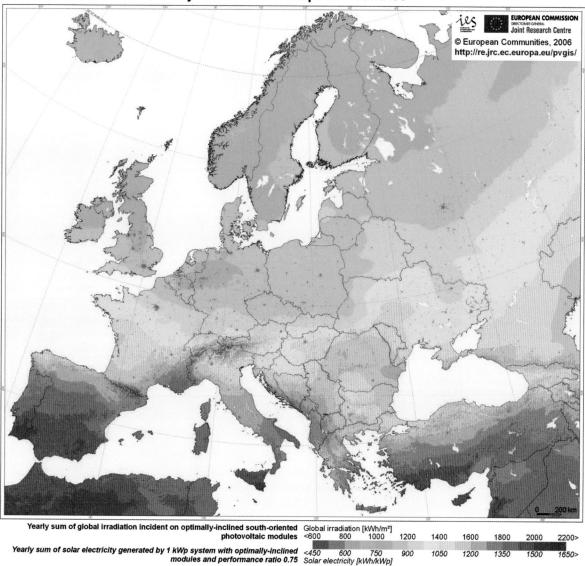

Figure 3.25 Solar radiation map of Europe, showing annual energy yields of optimally oriented PV arrays in various European locations. 'Performance Ratio' is the ratio of actual to theoretical maximum PV array output

PV systems for non-domestic buildings

PV arrays can also be integrated into the roofs and walls of commercial, institutional and industrial buildings, replacing some of the conventional wall cladding or roofing materials that would otherwise have been needed and reducing the net costs of the PV system. In the case of some prestige office buildings, the cost of conventional cladding materials can exceed the cost of cladding with PV.

Commercial and industrial buildings are normally occupied mainly during daylight hours, which correlates well with the availability of solar radiation. In countries such as Germany and the UK which operate a Feed-in Tariff scheme, users are paid a premium rate per kWh for all PV power generated.

In addition, PV power when available replaces power that would otherwise have to be purchased from the grid at the normal 'retail' price (around 15p per kWh in the UK in 2011), and during periods of low demand any surplus PV power can be sold to the grid (in the UK normally at a low price of around 3p per kWh).

There are now many examples of non-domestic buildings incorporating grid-connected PV systems, in countries like Germany, Japan, the Netherlands, Italy, the UK and the USA (see Figures 3.26, 3.27 and 3.28).

Figure 3.26 This large development at Amersfoort in the Netherland has a total of 1 MW of PV array capacity installed on the roofs of houses, schools and community buildings. The PV arrays are owned by the local electricity company, which pays home owners for the electricity they produce

Figure 3.27 This solar office building at Doxford, near Sunderland in the UK, has a 73 kW PV array integrated into its south-facing façade. The building incorporates energy-efficient and passive solar design features to minimize its need for heating and lighting

Figure 3.28 This 30 kW crystalline PV array is installed on the roof of the central catering 'Hub' at The Open University in Milton Keynes, UK

Large, grid-connected PV power plants

Large, centralized PV power systems, at the multi-megawatt scale, have also been built to supply power for local or regional electricity grids in a number of countries, including Germany, Switzerland, Italy and the USA (Figure 3.29).

Compared with building-integrated PV systems, large stand-alone PV plants can take advantage of economies of scale in purchasing and installing large numbers of PV modules and associated equipment, and can be located on sites that are optimal in terms of solar radiation. On the other hand, the electricity they produce is not used onsite and has to be distributed by the grid. This involves transmission losses, and the price paid for the power by a local electricity utility may only be the 'wholesale' price at which it can buy power from other sources (although, as noted earlier, in many countries such as Germany, PV power is purchased at premium prices from both building-mounted and land-based arrays).

Large plants also require substantial areas of land, which has to be purchased or leased, adding to costs – although low-value 'waste' land, for example alongside motorways, can be used (Figure 3.30). The land can often be compatible with other purposes as well as PV generation. For example the 1 MW 'Sun farm' shown in Figure 3.31 has been designed to allow 'wildlife-friendly' plants to live underneath the PV arrays. The need to avoid overshading of one ground-mounted panel by another, particularly in winter, means that they may only occupy 30–50% of the ground area, particularly at high latitudes, leaving plenty of space for wildlife.

Large PV power plants are more attractive in those areas of the world that have substantially greater annual total solar radiation than northern Europe. Areas such as north Africa or southern California not only have annual solar radiation totals more than twice those in Britain, but also have clearer skies. This means that the majority of the radiation is direct, making tracking and concentrating systems effective and further increasing the annual energy output. The cost of electricity from such PV installations is likely to be considerably lower than that from comparable non-tracking installations.

Figure 3.29 A 48 MW solar facility in Boulder City, Nevada, USA (source: Sempra generation)

Figure 3.30 A PV array installation on low-value land beside a motorway in Switzerland

Figure 3.31 The 1 MW 'Sun farm' located in Lincolnshire, UK, inaugurated by Ecotricity in 2011. It is adjacent to the company's wind farm, with the aim of enabling the higher electricity contribution from wind in winter to compensate for lower electricity output from PV, and vice versa in summer (source: Ecotricity)

Satellite Solar Power System

Probably the most ambitious – and some would say the most fanciful – proposal for a 'grid-connected' PV plant is the Satellite Solar Power System (SSPS) concept, first suggested some 40 years ago (see Glaser, 1972 and 1992). The basic idea was to construct huge PV arrays, each over 50 km² in area and producing several GW of electrical power, in geostationary orbit around the Earth (Figure 3.32). The power would be converted to microwave radiation and beamed from 1 km diameter transmitting antennas in space to 10 km² receiving antennas on Earth. The received power would then be converted to conventional alternating current and fed into a national electricity grid.

In theory, the advantages of the SSPS are very substantial. In space, the PV arrays would receive a full 1365 W m⁻² of solar power, instead of the 1000 W m⁻² that is the maximum available at the Earth's surface. Moreover, this high power would be available on a near-24 hour basis, and in the weightless and airless space environment it may be possible to construct extremely large, but very light, structures to support the PV arrays, without having to worry about the effects of wind and weather (although meteorites would be a problem).

On the other hand, the engineering challenges in constructing an SSPS, and the associated capital costs, would be enormous. One early US study estimated that a prototype would cost US$79 billion at 1979 prices and would take 30 years to complete.

In 2001 the US space agency NASA reviewed the SSPS concept and found that, whilst great progress had been made in both photovoltaics and space technologies since the 1970s, further dramatic cost reductions and technological progress, involving greatly increased R&D funding, would be needed to make the concept viable.

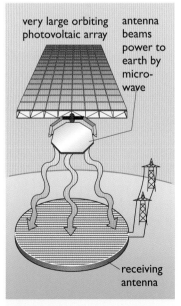

very large orbiting photovoltaic array

antenna beams power to earth by micro-wave

receiving antenna

Figure 3.32 The Satellite Solar Power System (SSPS) concept

In 2009 it was reported that the Japan Aerospace Exploration Agency (JAXA) was aiming to have launched by 2020 a satellite to demonstrate the beaming of electrical power to Earth, and planned to have an orbiting satellite solar power system operational some time in the 2030s (Edwards, 2009; Poupee, 2009).

However, there is a number of compelling arguments against developing such systems. Huge amounts of capital would need to be raised to fund them, restricting their use to a few rich nations. Building such systems also implies centralizing an energy source that many consider attractive precisely because of its decentralized nature. Concerns also persist about the health effects of the microwave beams that would be employed – especially if there were a malfunction in the 'fail safe' control system that is meant to ensure the beam always points at the receiving antenna. Interference with communications and radio astronomy could also be a serious problem.

3.10 Costs of energy from PV

As with any energy source, the cost per kilowatt-hour of power from PV cells consists essentially of a combination of the *capital cost repayment* including interest, and the *operation and maintenance (O&M)* cost as outlined in Appendix B.

The *capital* cost of a PV energy system is usually considered to be proportional to its rated power output. It is thus quoted in £ per peak kW ($£\,kW_p^{-1}$. It includes not only the cost of the PV modules themselves, but also the 'balance of system' (BOS) costs, i.e. the costs of the interconnection of modules to form arrays, the array support structure, land and foundations (if the array is not building-mounted), the costs of cabling, charge regulators, switching, metering and inverters, plus the cost of either storage batteries or connection to the grid.

Although the initial capital costs of PV systems are still relatively high, their operation and maintenance costs are extremely low in comparison with those of other renewable or non-renewable energy systems. Not only does a PV system not require any fuel, but also, unlike most other renewable energy systems, it has no moving parts (except in the case of tracking systems) and should require far less maintenance than, say, a wind turbine. (PV systems including batteries, however, have additional maintenance requirements.)

Over the past decade in many countries there has been a steady decline in the price of electricity from PV and a steady increase in the price of electricity from conventional sources. As Figure 3.33 suggests, in some countries with high solar radiation levels the price of PV power is already competitive with expensive day-time 'peak' power from electric utilities. If current trends continue, PV power could be competitive even with cheaper off-peak 'base load' power from utilities soon after 2020.

For 'grid parity' to be reached, i.e. for the price of PV power to be comparable with that for a conventional utility supplier, the installed cost per peak kilowatt of PV systems needs to decrease still further. The key to this lies in mass production. Historically, PV production costs have dropped by more than 20% for every doubling of production quantity and in the past decade PV production has doubled every 2–3 years (see below).

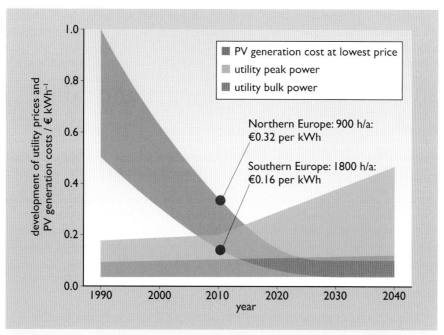

Figure 3.33 Progress towards 'grid parity': convergence of utility electricity prices and PV costs, 1990 to 2010 and projection to 2040 (source: EPIA/Greenpeace, 2011). Note: h/a is hours of Sun per annum: 900 h/a corresponds to the northern European countries, 1800 h/a corresponds to the southern European countries

In addition, the BOS costs need to be reduced. This should be feasible, given volume production of BOS components such as inverters, and the likelihood of substantially reduced installation costs when the industry has gained additional experience.

Figure 3.34 shows a breakdown of the capital costs of a large UK PV system in 2011, together with long-term projections of likely costs in 2020, 2030 and 2040. It suggests that capital costs could be halved by 2030 and fall

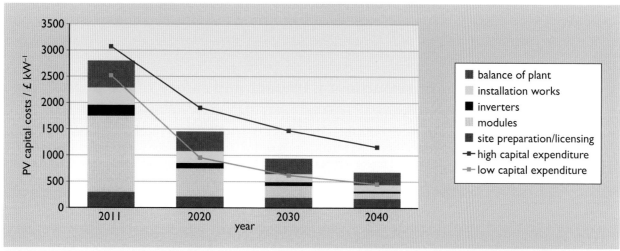

Figure 3.34 Capital costs of solar PV systems in the UK in 2010, and projected capital costs in 2020, 2030 and 2040 (source: CCC, 2011). Note: costs relate to a 10 MW ground-mounted system; lines indicate range of variation of cost estimates.

to perhaps a quarter of current levels by 2040. The capital costs of smaller PV systems are expected to fall in similar fashion (Committee on Climate Change, 2011).

3.11 Environmental impact and safety

Environmental impact and safety of PV systems

The environmental impact of PV is probably among the lowest of all renewable or non-renewable electricity generating system.

In normal operation, PV energy systems emit no gaseous or liquid pollutants, and no radioactive substances. Crystalline silicon PV cells contain only minuscule quantities of dopants such as boron and phosphorus. (In the case of CIS or CdTe modules, which include very small quantities of indium, cadmium and tellurium, there is a slight risk that a fire in an array might cause small amounts of toxic material containing these elements to be released into the environment.)

PV modules have no moving parts, so they are also safe in the mechanical sense, and they emit no noise. However, as with other electrical equipment, there are some risks of electric shock – especially in larger systems operating at voltages substantially higher than the 12–48 volts employed in most small PV installations. But the electrical hazards of a well-engineered PV system are, at worst, no greater than those of other comparable electrical installations.

PV arrays do, of course, have some visual impact. Rooftop arrays will normally be visible to neighbours, and may or may not be regarded as attractive, according to aesthetic tastes. Several companies have produced special PV modules in the form of roof tiles which blend into roof structures more unobtrusively than conventional module designs.

As already mentioned, PV arrays on buildings require no additional land, but large, multi-megawatt PV arrays will usually be installed on land specially designated for the purpose, and this will entail some visual impact. In some countries such as Switzerland, the authorities are installing large PV arrays as noise barriers alongside motorways and railways. Arguably PV is here *reducing* the overall environmental impact.

Environmental impact and safety of PV production

The environmental impact of manufacturing silicon PV cells is unlikely to be significant – except in the unlikely event of a major accident at a manufacturing plant. The basic material from which the vast majority of PV cells are made, silicon, is not intrinsically harmful. However, small amounts of toxic chemicals are used in the manufacture of some PV modules. For example, cadmium is obviously used in the manufacture of CdTe modules; and small amounts of cadmium are used in manufacturing CIS and CIGS modules – although new processes are now available which allow this to be eliminated.

As in any chemical process, careful attention must be paid to plant design and operation, to ensure the containment of any harmful chemicals in the event of an accident or plant malfunction.

Even though PV arrays are potentially very long-lived devices, eventually they will come to the end of their useful life and will have to be disposed of – or, preferably, recycled. Many European manufacturers are already beginning to recycle PV modules at the end of their working lives under the auspices of the voluntary 'PV Cycle' initiative, and draft EU regulations on PV module recycling under the Waste Electrical and Electronic Equipment (WEEE) directive are in preparation.

The energy balance of PV systems and potential materials constraints

A common misconception about PV cells is that almost as much energy is used in their manufacture as they generate during their lifetime. This might have been true in the early days of PV, when the refining of monocrystalline silicon and the Czochralski process were very energy-intensive, and the efficiency of the cells produced was relatively low, leading to low lifetime energy output.

However, with the more modern PV production processes introduced in recent years, and the use of thinner cells with improved efficiency, the energy balance of PV is now much more favourable. Recent research (see Fthenakis et al., 2009) suggests that the time taken for a PV array to produce as much energy as was used in its production, its energy payback time, is relatively short. Energy payback times for modules range from approximately 1.8 years for monocrystalline and polycrystalline PV to 1.2 years for silicon ribbon and 1.1 to 0.8 years for cadmium telluride (in southern European insolation conditions of approximately 1700 kWh per year). The use of materials with low embodied energy in PV array support structures can also improve the overall energy payback time of PV systems.

Could the availability of certain exotic materials be a constraint on the future growth of PV? As noted above, the material from which conventional PV cells are made, silicon, is extremely abundant, but some have expressed concern that supplies of indium and tellurium, used in thin film PV cells, could be limited. These materials are 'secondary metals', extracted during the production of primary metals such as zinc and copper. A report by the UK Energy Research Centre addresses this issue. It concludes that, while data on reserves and resources is inadequate:

> there is no immediate cause for concern about availability of either indium or tellurium. PV occupies a small fraction of current markets and there is evidence of considerable potential to increase the extraction of both metals, because a sizeable proportion of the material potentially available from primary metal extraction is not currently utilised.
>
> Speirs et al., 2011

3.12 PV integration, resources and future prospects

Integration

If PV energy systems continue to improve in cost-effectiveness compared with more conventional sources, as seems likely, in what way would national energy systems need to be modified to cope with the long-, medium- and short-term fluctuations in the output of PV arrays?

In the UK, most PV power would be produced in summer, when electricity demand is relatively low: much less would be produced in winter when demand is high. Furthermore, although PV power is quite reliable (during daylight hours) in climates with mainly clear skies, it can be more intermittent in countries like the UK, where passing clouds can reduce the output of individual arrays substantially. As long as the capacity of variable output power sources such as PV is fairly small in relation to the overall capacity of the grid (most studies suggest a proportion of 10–20%), there should not be a major problem in coping with their fluctuating output. The grid is, after all, designed to cope with massive fluctuations in *demand*, and similarly fluctuating sources of *supply* like PV can be considered equivalent to 'negative loads'. Fluctuations would also, of course, be substantially smoothed out if PV power plants were situated in many different locations subject to widely varying solar radiation and weather patterns.

But if PV power stations, and other fluctuating renewable energy sources, such as wind power, were in future to contribute a more significant proportion of electricity supplies, then the 'generating mix' supplying the grid would have to be changed to include a greater proportion of 'fast response' power plant, such as hydro or gas turbines, and increased amounts of short-term storage and 'spinning reserve'. (The subject of integration of PV and other renewables into existing energy systems will be covered in more detail in Chapter 10 Integrating renewable energy.)

The growing world photovoltaics market

Between 2000 and 2010, as shown in Figure 3.6, world installed PV capacity grew extremely rapidly, from some 1.5 GW in 2000 to just under 40 GW in 2010 (EPIA, 2011). If this growth rate continues, it implies a doubling of world PV production every two to three years. Approximately three quarters of this capacity is installed in the EU, with other countries and regions currently having considerably smaller shares, as Figure 3.35 shows.

Crystalline silicon still dominates PV technology in the marketplace, with monocrystalline and polycrystalline modules accounting in 2010 for some 27 GW (two thirds) of world production capacity in 2010. As the EPIA reports:

> Almost 50% of this capacity is located in China. The rest is produced in Taiwan (over 15%), the EU (over 10%), Japan (slightly less than 10%) and the USA (less than 5%).

> The global production capacity of Thin Film (TF) modules reached around 3.5 GW in 2010. This is likely to increase to more than 5 GW in 2011 and might reach 6 to 8.5 GW in 2012. Today, Copper (Gallium) Indium (di)Selenide (CI(G)S) modules represent about 15% of the TF total capacity, with the remainder equally shared between Cadmium Telluride (CdTe) and Silicon TF. However, by 2012, EPIA expects that each of the TF technologies will represent an equal share in terms of production capacity. While a large part of c-Si (crystalline silicon) modules are assembled in

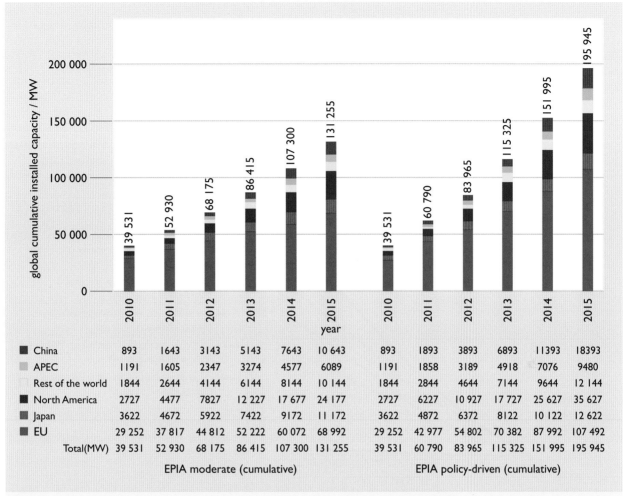

	EPIA moderate (cumulative)						EPIA policy-driven (cumulative)					
■ China	893	1643	3143	5143	7643	10 643	893	1893	3893	6893	11393	18393
■ APEC	1191	1605	2347	3274	4577	6089	1191	1858	3189	4918	7076	9480
Rest of the world	1844	2644	4144	6144	8144	10 144	1844	2844	4644	7144	9644	12 144
■ North America	2727	4477	7827	12 227	17 677	24 177	2727	6227	10 927	17 727	25 627	35 627
■ Japan	3622	4672	5922	7422	9172	11 172	3622	4872	6372	8122	10 122	12 622
■ EU	29 252	37 817	44 812	52 222	60 072	68 992	29 252	42 977	54 802	70 382	87 992	107 492
Total(MW)	39 531	52 930	68 175	86 415	107 300	131 255	39 531	60 790	83 965	115 325	151 995	195 945

Figure 3.35 Projected evolution of global cumulative installed PV capacity in MW, 2010–2015 (source: EPIA, 2011). Notes: (a) APEC refers to Asia–Pacific Economic Cooperation – an inter-governmental grouping of 21 countries (b) any discrepancies between overall and individual yearly totals are due to rounding

China, most of the TF manufacturing plants are located in other parts of the world; the leaders being the USA, the EU, Japan and Malaysia.

(EPIA, 2011)

Future long-term PV prospects: realizing the global potential

Medium-term projections of the world PV market by the European Photovoltaics Industry Association (EPIA, 2011) suggest that strong growth is likely to continue to 2015, resulting in between 131 GW and 196 GW of installed capacity by 2015, depending on whether 'moderate' or stronger 'policy-driven' growth is achieved (Figure 3.35). By 2015, the EU's share is likely to have reduced to around 40%, with strong growth occurring in China, the Asia-Pacific region, the USA and Japan.

The European Photovoltaics Industry Association, in association with the environmental group Greenpeace, has since 2000 been producing a regularly updated series of *Solar Generation* reports, projecting the long-term future prospects for solar photovoltaics.

The actual performance of the world PV industry since 2000 has in many cases surpassed the various previous *Solar Generation* projections. The most recent *Solar Generation* report (EPIA, 2011) suggests that the contribution of PV to world energy supplies could be as high as 21% by 2050, in a 'paradigm shift' scenario involving strong promotion of PV coupled with major improvements in the efficiency of electricity use (Figure 3.36).

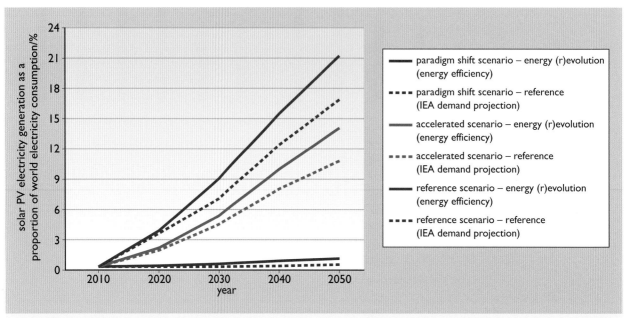

Figure 3.36 Projections of PV share of global electricity production, 2010–2050, according to various assumptions. The share of PV generation could reach 21% by 2050 in the Energy (r)evolution 'paradigm shift' scenario, if coupled with major improvements in energy efficiency (source: EPIA/Greenpeace, 2011)

The International Energy Agency in its *Energy Technology Perspectives* report (IEA, 2010b), is somewhat less optimistic, envisaging the PV share of world electricity production reaching some 11% of an assumed demand of 40,000 TWh by 2050 (see Figure 3.37).

Long-term expert projections such as those of the UK Climate Change Committee, the IEA and the EPIA are obviously subject to a very wide range of uncertainty. Nevertheless, the future prospects of photovoltaics as a clean, renewable source of energy for the world by mid-century do seem bright.

An inspiring – and literally uplifting – symbol of the rising hopes of the global PV industry is the 'Solar Impulse' project. This involves the design and construction of a solar PV-powered single-seater aircraft that has successfully completed flights of increasing duration in Europe and has the ultimate aim of a PV-powered round-the-world flight (Figure 3.38).

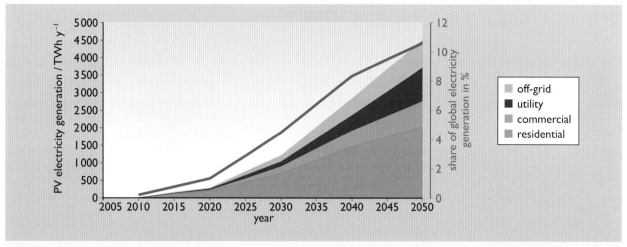

Figure 3.37 International Energy Agency projections for global growth of PV 2010–2050. The share of PV generation (blue line) could reach 11% by 2050 (source: IEA, 2010b)

The leaders of the project, Bertrand Piccard and André Borschberg, do not see PV powered aircraft replacing conventional aircraft in the foreseeable future. On landing after Solar Impulse's first international flight Piccard indicated that the project's goal was not to cause a revolution in aviation: its aim was to promote a revolution in the mindset of people when they think about renewable energy, energy saving and new technologies (Solar Impulse, 2011a).

Figure 3.38 The Solar Impulse PV-powered single-seater aircraft, which flew from Switzerland to Belgium on 11 May 2011. The 480 km flight achieved an average speed of 50 km per hr and an average altitude of 1828 metres. Solar Impulse is powered by some 11 625 monocrystalline PV cells of 22% efficiency, mounted on the wings and tail. Four 7.5 kW electric motors drive the propellers and electrical storage is provided by lithium polymer batteries (Solar Impulse, 2011b)

3.13 **Summary**

This chapter started with a brief look at the history and basic principles of photovoltaic energy conversion, concentrating initially on devices using monocrystalline silicon.

It then reviewed various ways of reducing the currently high costs of energy from photovoltaics. The electrical characteristics of PV cells and modules were then described, followed by a discussion of the various roles of PV energy systems in supplying power to homes and other buildings and in feeding power into local or national electricity grids. The concluding sections reviewed the economics and the environmental impact of photovoltaic electricity, examined how PV might be increasingly integrated into electricity supplies, and looked at the world market and future prospects for photovoltaics.

References

Adams, W. G. and Day, R. E. (1877) 'The action of light on selenium', *Proceedings of the Royal Society*, London, Series A, vol. 25, p. 113.

Becquerel, A. E. (1839) 'Recherches sur les effets de la radiation chimique de la lumière solaire au moyen des courants électriques' and 'Mémoire sur les effets électriques produits sous l'influence des rayons solaires', *Comptes Rendus de l'Académie des Sciences*, vol. 9, pp. 145–9 and pp. 561–7.

Chapin, D. M., Fuller, C. S. and Pearson, G. L. (1954) 'A new silicon p–n junction photocell for converting solar radiation into electrical power', *Journal of Applied Physics*, vol. 25, pp. 676–7.

Chalmers, R. (1976) 'The photovoltaic generation of electricity', *Scientific American*, October, pp. 34–43.

CV21 (2008) *Spherical Silicon Solar Cells with Micro Reflector Cups* [online], Clean Venture 21 http://www.cv21.co.jp/en/images/pdf/a401en.pdf (accessed 15 June 2011).

CCC (2011) *The Renewable Energy Review*, London, The Committee on Climate Change, available at http://hmccc.s3.amazonaws.com/Renewables%20Review/The%20renewable%20energy%20review_Printout.pdf (accessed 3 October 2011).

EC JRC (2011) *PV Potential Estimation Utility* [online], http://re.jrc.ec.europa.eu/pvgis/apps4/pvest.php (accessed 11 November 2011).

Edwards, L. (2009) '$21 billion orbiting solar array will beam electricity to Earth', *Physorg.com* [online], 15 September http://www.physorg.com/news172224356.html (accessed 13 September 2011).

EPIA (2011) *Global Market Outlook for Photovoltaics to 2015*, Brussels, European Photovoltaics Industry Association; available at http://www.epia.org/publications/photovoltaic-publications-global-market-outlook/global-market-outlook-for-photovoltaics-until-2015.html (accessed 4 October 2011).

EPIA/Greenpeace (2011) *Solar Generation 6: Solar Photovoltaic Electricity Empowering the World*, Brussels, European Photovoltaic Industry Association and Amsterdam, Greenpeace International; available at http://www.greenpeace.org/international/Global/international/publications/climate/2011/Final%20SolarGeneration%20VI%20full%20report%20lr.pdf (accessed 4 October 2011).

Fthenakis V. M., Kim H. C., Raugei M. and Kroner J. (2009) 'Update of PV energy payback times and life-cycle greenhouse gas emissions', *24th European Photovoltaic Solar Energy Conference and Exhibition*, Hamburg, September 21–5, 2009.

Glaser, P. (1972) 'The case for solar energy', *Proceedings of the Annual Meeting of the Society for Social Responsibility in Science, Conference on Energy and Humanity*, Queen Mary College, London, September 3, 1972.

Glaser, P. (1992) 'An overview of the solar power satellite option', *IEEE Transactions on Microwave Theory and Techniques*, vol. 40, no. 6, pp. 1230–8.

Graham-Rowe, D. (2007) 'Focusing light on silicon beads'. *Technology Review*, 13 November [online], http://www.technologyreview.com/Energy/19696/ (accessed 15 June 2011).

Grätzel, M. (1991) "The Artificial Leaf: Molecular Photovoltaics Achieve Efficient Generation of Electricity from Sunlight', Research Report, Coordination Chemistry Reviews, Vol 111, 6 December, pp. 167–174.

Grätzel, M. (2001) 'Photoelectrochemical Cells' *Nature*, vol. 414, pp. 338–344.

Green, M. (1982) *Solar Cells*, New York, Prentice-Hall.

Green, M. A., Emery, K., Hishikawa, Y., Warta, W. and Dunlop, E. (2011) 'Solar cell efficiency tables (Version 38)', *Progress in Photovoltaics: Research and Applications*, vol. 19, no. 5, pp. 565–72.

IEA (2010a) *Trends in Photovoltaic Applications*, International Energy Agency, Report IEA PVPS T1-19:2010; available at http://www.iea-pvps.org/index.php?id=92&eID=dam_frontend_push&docID=823 (accessed 5 October 2011).

IEA (2010b) *Energy Technology Perspectives, 2010*, Paris, International Energy Agency.

IEA (2011) *World-wide Overview of Design and Simulation Tools for Hybrid PV Systems*, International Energy Agency, Report IEA-PVPS T11-01:2011.

Jayaweera, P. V. V., Perera, A. G. U. and Tennakone, K. (2008) 'Why Grätzel's cell works so well' *Inorganica Chimica Acta*, vol. 361, no. 3, pp. 707–11.

Ohl, R. S. (1941) *Light Sensitive Device*, US Patent No. 2402622; and *Light Sensitive Device Including Silicon*, US Patent No. 2443542: both filed 27 May.

O'Regan, B. and Grätzel, M. (1991) 'A low cost, high efficiency solar cell based on dye-sensitised colloidal TiO_2 films', *Nature*, vol. 235, pp. 737–40.

O'Regan, B., Nazeeruddin, M. K. and Grätzel, M. (1993) 'A very low cost, 10% efficient solar cell based on the sensitisation of colloidal titanium dioxide films', *Proceedings of 11th European Photovoltaic Solar Energy Conference*, Montreux, 1992, Gordon and Breach.

Poupee, K. (2009) 'Japan eyes solar station in space as new energy source', *Physorg.com*, 8 September [online] http://www.physorg.com/news176879161.html (accessed 13 September 2011).

Sharp Solar (2009) http://sharp-world.com/corporate/news/091029.html (accessed 5 November 2011).

Speirs, J., Gross, R., Candelise, C. and Gross, B. (2011) *Material Availability: Potential constraints to the future low-carbon economy*, UK Energy Research Centre, UKERC/WP/TPA/2011/002 [online], http://www.ukerc.ac.uk/support/tiki-download_file.php?fileId=1687 (accessed 13 September 2011).

Solar Impulse (2011a) *From Payerne to Brussels* (video) [online], http://www.solarimpulse.com/sitv/index.php?lang=en (accessed 13 September 2011).

Solar Impulse (2011b) *Solar Impulse: around the world in a solar airplane* [online], http://www.solarimpulse.com (accessed 13 September 2011).

TERI (2011) *Lighting a Billion Lives* [online], http://labl.teriin.org/ (accessed 13 September 2011).

Treble, F. C. (1999) *Solar Electricity: a layman's guide to the generation of electricity by the direct conversion of solar energy* (2nd edn), Oxford, The Solar Energy Society.

Wenham, S., Green, M., Watt, M. and Corkish, R. (2007) *Applied Photovoltaics* (2nd edn), London, Earthscan.

Chapter 4

Bioenergy

By Dick Morris and Jonathan Scurlock (based on earlier edition by Steven Larkin, Janet Ramage and Jonathan Scurlock)

4.1 Introduction

Bioenergy is the general term for energy derived from materials such as wood, straw, oilseeds or animal wastes which are, or were recently, living matter, referred to collectively as **biomass**. Wood pellets, charcoal, bioethanol and biodiesel are all examples of energy-rich materials derived from biomass.

All the Earth's living matter, its total biomass, exists in the thin surface layer called the **biosphere**. It forms only a tiny fraction of the total mass of the Earth, but represents an enormous store of chemical energy (see Chapter 1). Although the majority of this is unavailable for human use, it is a store which is continually replenished by the flow of energy from the Sun, through the process of **photosynthesis** (from *photo*, to do with light, and *synthesis*; putting together). This in effect takes in carbon dioxide from the air, and uses it to make living material, releasing oxygen in the process (see Figure 1.9 in Chapter 1). Although only a small fraction of the solar energy reaching the Earth each year is fixed in this way, the amount fixed annually as chemical energy in biomass is nevertheless equivalent to between two and five times the world's total primary energy consumption, as defined in Chapter 1.

It is important to appreciate that biomass is also vital in maintaining the Earth's atmosphere. If something were to sweep away all the plant life on Earth, the resulting loss of mass would be no more than one part in a billion – like blowing the dust off a school globe. Yet the physical consequences of this infinitesimal change would be enormous. There would no longer be a supply of oxygen to the atmosphere, and it is the particular mixture of nitrogen, oxygen and trace gases such as CO_2 in the atmosphere which maintains the surface conditions on the Earth.

In nature, the energy that has been stored in the biomass of plants is dissipated through a series of conversions. These involve **metabolic** processes such as **respiration** (effectively the reverse of photosynthesis) in living matter, and physical processes such as re-radiation of heat energy and the **evaporation** of water (Figure 4.1). These transfer energy to the surrounding atmosphere and eventually the energy is radiated away from the Earth as low-temperature heat. Some biomass will be metabolized within a year, but it can also accumulate over decades in the wood of trees and as the **humus** component of the soil organic matter. A small fraction may accumulate over centuries as **peat**, traditionally burned for heating, and over millions of years a tiny proportion has become the major fossil fuels: coal, oil and gas.

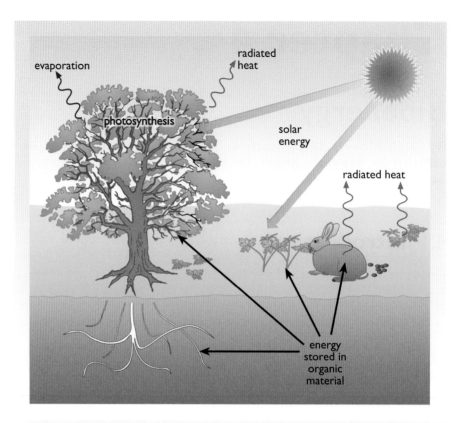

Figure 4.1 The bioenergy cycle on a local scale. In addition, a small fraction of the carbon from decomposing animal and plant residues (including roots) is transformed into soil organic matter, a significant carbon store.

The significance of these processes is that if we can intervene and 'capture' some of the biomass at the stage where it is acting as a store of chemical energy, we can use it as a **fuel**, that is, a material that can release useful energy through a change in its chemical composition, usually through burning. Material such as firewood, rice husks and other plant or animal residues can simply be burned in air to produce heat, and in many developing countries this **traditional biomass** continues to account for a large part of their energy consumption for cooking, heating and light. Biomass can also be burned to raise steam for electricity generation, or can be converted into intermediate **biofuels** such as charcoal, biogas, bioethanol and biodiesel (note that some authorities restrict the use of the term 'biofuel' to liquid fuels – in this chapter, we use it to describe solid, liquid and gaseous forms). These biofuels can be used to replace conventional heating fuels or to power some form of engine for motive power or electricity generation. In recent decades, terms such as '**new**' or **modern bioenergy** have come into use to characterize biomass materials that are processed on a large, commercial scale to produce fuel, usually in the more industrialized countries.

Provided our consumption does not exceed its natural level of production, the burning of biomass *should generate no more heat and create no more carbon dioxide* than would have been formed in any case by natural processes. So it seems that here we have a truly sustainable energy source, with no deleterious global environmental effects. However, this is not always the case in practice, particularly where fossil fuels are used in the production process, and there may be other effects to consider.

Box 4.1 summarizes some very approximate basic data about the quantities of biomass worldwide and related flows of energy. Freshly harvested biomass has a moisture content (m.c.) of around 50–80% of the total mass. If left to air-dry (as when turning fresh grass into hay) the moisture level drops naturally to around 20–25%. To avoid the confusion caused by materials having varying moisture contents, biomass yields are often quoted as 'dry tonnes' which assumes 0% m.c. as if the biomass has been dried in an oven.

BOX 4.1 Biomass basic data

Note that almost all the data here is subject to considerable uncertainty.

World totals

Total mass of living matter[1]	Approx. 1800 billion dry tonnes
Total mass in land plants[1]	1800 billion dry tonnes
Total mass in oceans[1]	4 billion dry tonnes
World population (2009)	6.8 billion
Per capita terrestrial plant biomass	270 dry tonnes
Energy stored in terrestrial biomass[2]	25 000 EJ
Net annual production of terrestrial biomass[2]	130 billion dry tonnes y^{-1}

World energy comparisons

Rate of energy storage by land biomass[2]	2400 EJ y^{-1} (76 TW)
Rate of global primary energy consumption (2009)[3]	502 EJ y^{-1} (15.9 TW)
Biomass energy consumption[4]	50 EJ y^{-1} (1.6 TW)
Energy consumed as food[5]	29 EJ y^{-1} (0.9 TW)

Sources: 1 Slesser and Lewis, 1979; 2 Haberl et al., 2007; 3 BP, 2010; 4 IEA Bioenergy, 2009; 5 Derived from WHO, 2003.

4.2 Bioenergy past and present

From wood to coal

Until recent times, the history of human energy use was essentially the history of bioenergy. Although there is evidence of coal-burning as early as 3000 years ago, its contribution remained relatively small until about 1800. Indeed, bioenergy was still dominant in many areas of life well into the Industrial Revolution, with wood for heat, whale oil or tallow candles from animal fats for light, and grass and other agricultural crops as 'fuel' for the main means of land transport and haulage, horses and oxen.

The move from bioenergy to fossil fuel was a key feature of the Industrial Revolution. For many centuries, the high temperatures needed for iron smelting could be achieved only in furnaces using charcoal made from wood. The impurities and variable nature of coal made it unsuitable for smelting, and attempts to convert it to a type of charcoal had little initial success. But in the early 1700s an effective 'coal charcoal', was produced and within a few decades, this new material, now called **coke**, was replacing charcoal throughout the growing industrial sector.

The increased demand for coal led to deeper mines, and the need to pump flood water from great depths led to the first steam engines – powered of course by coal. By the end of the nineteenth century, coal was dominant in the world's industrialized countries. The twentieth century saw the rise of oil and natural gas, but it is worth noting that coal consumption also increased seven-fold between 1900 and 2008 (EIA, 2010).

Will the twenty-first century see the reversal of this process with a transition from coal to biomass? The data in Box 4.1 shows that this is theoretically possible since the annual energy storage in biomass is approximately five times annual world total primary energy consumption. However, there are many practical and economic factors that may limit the uptake of modern bioenergy processes.

Present biomass contributions

Obtaining a reliable estimate of the total worldwide energy contribution from the many sources of bioenergy is a task fraught with difficulties. Unlike the fossil fuels, bioenergy has few global companies producing detailed reports on production and consumption. Indeed, the use of traditional biomass often involves no financial transaction at all, or the trading is local, informal and largely unrecorded. In recent years, the International Energy Agency has endeavoured to collect national data based on agreed categories of renewable energy, but they warn that their figures are still subject to a great deal of uncertainty. Recent estimates of the annual contribution to world primary energy from traditional biomass fall in the range 40–60 EJ (IEA Bioenergy, 2009). Figure 1.2 in Chapter 1 is based on a contribution of 30 EJ, with industrially processed biomass adding a further 9 EJ or so (BP, 2010).

Although the details may be uncertain, biomass is the fourth largest source of energy over much of the world, accounting for at least 10% of total human energy use (El Bassam, 2010). It accounts for about a third of total primary energy consumption in the developing countries (up to 90% in some of the poorest countries such as Uganda), and its total annual contribution continues to rise. Its *percentage* contribution to world primary energy is currently falling slightly as developing countries industrialize and move towards fossil fuel use, but it remains important even in the more advanced of these, accounting for about 10% of primary energy in China and 35% in India (El Bassam, 2010).

In the industrialized world, the energy contribution from biomass is significant and rising, particularly in countries with large forestry industries, or well-developed technologies for processing residues and wastes. In Sweden and Finland, where biomass contributes at least 20% of primary energy, the use of residues in the pulp and paper industries is important. Advanced systems for domestic heating, district heating and CHP (combined heat and power) have helped Sweden towards a fivefold increase in the use of bioenergy since the 1980s, reaching nearly 50 GJ per head of population in 2006 (Björheden, 2006). In the UK, the bioenergy contribution rose at about 7% per year during the 1990s, and had reached a total of 230 PJ, or 3.8 GJ per capita in 2009 (DECC, 2010a, see also Figure 1.6 in Chapter 1).

4.3 **Biomass as a solar energy store**

The key mechanism in the use of bioenergy is photosynthesis, in which plants take in carbon dioxide and water from their surroundings and use energy from sunlight to convert these substances into plant biomass.

The essential features of the process can be represented by the chemical equation:

$$6CO_2 + 6H_2O + \text{light energy} \rightarrow C_6H_{12}O_6 + 6O_2$$

The first product on the right of the equation is glucose ($C_6H_{12}O_6$), a **carbohydrate** (see Box 4.2). The second product in the equation is oxygen which is released to the atmosphere. Although glucose is not necessarily the final 'vegetable matter' it is the crucial building block in subsequent biomass producing reactions. Thus a plant *grows* by using solar energy to convert carbon dioxide and water into carbohydrate or similar material, with a release of oxygen into the atmosphere.

BOX 4.2 Carbon-containing compounds in biomass

Carbon is a key element in the biosphere and is found combined with a range of other elements to form a vast array of simple and complex molecules that make up the bulk of all living tissues.

The term **carbohydrate** refers to a compound consisting only of carbon, hydrogen and oxygen having a general form $C_x(H_2O)_y$ (*x* and *y* can be different but crucially there is a hydrogen : oxygen ratio of 2:1). Of importance to the discussions in this chapter are carbohydrates such as the following.

Glucose – a simple sugar, which has the chemical formula $C_6H_{12}O_6$; these small carbohydrate molecules can be linked together into polymeric chains.

Starch – a polymer consisting of many glucose units, it can be represented by the formula $(C_6H_{10}O_5)_n$, where the subscript *n* indicates that there are many identical $C_6H_{10}O_5$ units joined together.

Cellulose – another polymer with the formula $(C_6H_{10}O_5)_n$ but having a different overall structure that allows the formation of fibrous structures and is thus an important component in plant cell walls.

Hemicellulose – a complex polymer containing a variety of different sugar units (i.e. a polymer consisting of more than just glucose base units), found with cellulose in plant cell walls.

Other organic (carbon containing) compounds discussed include the following.

Lignin – a polymeric organic compound containing carbon, oxygen and hydrogen having a complex structure and found in wood.

Lipids – a class of carbon, hydrogen and oxygen containing compound: the generic term for the oils and fats contained in living tissues.

Proteins – large, highly complex organic molecules which also contain nitrogen and may consist of a number of polymeric chains. Proteins have complex folded structures.

In living organisms, these different types of compound are synthesized and broken down by the operation of specific **enzymes** (specialist proteins that act as catalysts) that enable the reactions to occur at ambient temperatures and pressures.

In an energy context, **hydrocarbons** which comprise only carbon and hydrogen are also important. They are mostly formed from previously living material by reactions occurring at high temperatures and pressures below the Earth's surface.

The glucose molecule contains more chemical energy than the sum total in the molecules of carbon dioxide and water from which it was formed. Within the plant glucose can be converted into more complex carbohydrates including starches and cellulose, or it can be combined with nitrogen and other elements to form proteins and other components.

Some of the energy-rich carbohydrate formed is broken down elsewhere in the plant through the process of respiration, releasing energy to power the synthesis of proteins and other components. Respiration is in effect the reverse of photosynthesis, with oxygen taken in, carbon dioxide given off and energy released. Plant material that has been consumed by an animal or allowed to decay also undergoes respiration with the energy released being used by the different organisms involved. All the energy involved in processes within living organisms ultimately appears as heat. So we have a continuous process, with energy from the Sun being stored in the form of carbon-based biomass which may persist for a time, or be consumed by another organism. Over time, nearly all this biomass is respired, releasing energy which is ultimately re-radiated back to outer space, and carbon dioxide which returns to the atmosphere (Figure 4.2). So carbon cycles around the biosphere while energy flows through it.

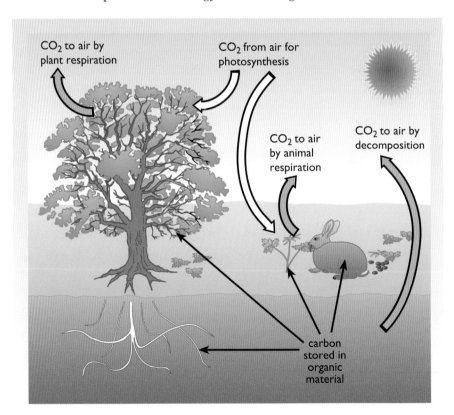

Figure 4.2 The carbon cycle on a local scale

Conversion efficiencies

The **yield** of a crop is the mass of biomass produced per hectare per year. For an **energy crop**, that is, one grown specifically for biomass energy, yield is obviously the primary determinant of the energy we can obtain from that crop. Yields depend on many factors: the location, climate and weather, the nature of the soil, supplies of water, nutrients, etc., and the choice of plant. For energy crops, the air-dry mass of plant matter produced annually on an area of one hectare can be as little as one dry tonne or, in favourable circumstances, as much as thirty dry tonnes. In energy terms, this represents a range from perhaps 15 GJ to 400 GJ per hectare per year.

These yields actually represent an extremely low conversion efficiency from solar energy to energy in biomass. Consider, for instance, northern Europe, where the average solar energy delivered in a year is about 1000 kWh per square metre (see Chapter 2). An energy crop in this region might yield perhaps 200 GJ per hectare per year: a solar-to-bioenergy conversion efficiency of about two-thirds of one percent (see Box 4.3).

BOX 4.3 Conversion of solar energy

Consider one hectare (ha) of land, in an area such as southern England where the annual energy delivered by solar radiation is 1000 kWh m^{-2} y^{-1}.

1000 kWh is 3.6 GJ and 1 ha is 10 000 m^2,
so the total annual energy is   36 000 GJ

After losses about an eighth of this reaches the crop at the right time. Say

12% of the annual energy reaches growing leaves	4320 GJ
50% of this is photosynthetically active radiation	2160 GJ
85% of which is captured by the growing leaves	1836 GJ
21% of which is converted into stored chemical energy	386 GJ
40% of which is consumed in respiration to sustain the plant	
or lost in photorespiration leaving	231 GJ

This is about 5.3% of the solar radiation reaching the growing plant, and only 0.64% of the original total annual energy.

The first significant point is that much of the annual solar radiation may be ineffective for photosynthesis. Some misses the plant altogether or arrives during the wrong season. There may not be enough water or nutrients for the plant to use the solar energy that does reach it, then there are the effects of diseases, pests (reducing the leaf area of the crop able to intercept light) and weeds (competing for light or nutrients). The remaining losses shown in Box 4.3 are specific to the interaction between the plant and sunlight. A leaf responds only to a part of the solar spectrum – the pattern of absorption is shown in Figure 4.3. Note the dip in the green region: a leaf looks green because it absorbs less green light than red or blue. Only about 85% of the 'useful' radiation that falls on a leaf surface is captured in the leaf for use in photosynthesis. Plants of the temperate regions are referred to as **C3 plants**, because the first products of photosynthesis actually contain three

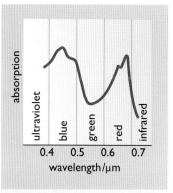

Figure 4.3 Relative absorption of different parts of the solar spectrum by a leaf

carbon atoms rather than the six shown for glucose in the general reaction described above. The fundamental biochemistry of C3 photosynthesis means that 21% of the energy of absorbed light is converted into the energy content of fixed carbohydrate. Finally, 40% of this fixed energy is used in respiration within the plant to drive the synthesis of proteins and other components of its biomass.

In the C3 plants, too much light relative to the CO_2 available in the leaf can actually lead to damage through the main photosynthetic enzyme producing a toxic molecule called **phosphoglycolate** rather than useful three-carbon compounds. The process of photorespiration protects the plant from damage by using previously stored energy to break down the toxic phosphoglycolate, releasing carbon dioxide but effectively wasting as much as 30% of the stored energy in doing so.

Plants from tropical areas are at greater risk from light-induced damage because of the higher light levels. Some tropical plants, especially those in the grass family (e.g. maize, sugar cane and miscanthus) effectively concentrate CO_2 in the cells where photosynthesis is carried out, so there is never a shortage and no risk of damage. They produce a four-carbon molecule as the first product of photosynthesis. While the fundamental efficiency of photosynthesis in these **C4 plants** is lower than the C3 process, such plants can potentially produce higher yields of biomass than C3 plants by avoiding the need for wasteful photorespiration. (For a more detailed treatment of photosynthesis, see Hall and Rao (1999).)

For C4 plants grown in tropical regions, the 5.3% of Box 4.3 might increase to perhaps 6.7%, giving an overall efficiency closer to 1%. Furthermore, good crop management can increase the fraction of solar radiation that is captured and used. Nevertheless, plants remain much less efficient solar energy converters than, for example photovoltaic (PV) cells which can achieve solar to electrical energy conversions of 10 to 20%. However, plants are currently much cheaper per square metre of light intercepting surface!

4.4 Biomass as a fuel

What are fuels?

Fuels are those materials from which useful energy can be extracted. In using biomass as fuel, this release of energy usually involves burning (combustion). Some essential features of combustion are:

- it needs air – or to be more precise, oxygen
- the fuel undergoes a major change of chemical composition
- heat is produced, i.e. energy is released.

Consider, as an example, methane – the principal component of the fossil fuel natural gas and also one form of gaseous biofuel. Each methane molecule consists of one carbon and four hydrogen atoms: CH_4. Atmospheric oxygen has molecules consisting of two atoms (O_2), so in full combustion each methane molecule reacts with two oxygen molecules:

$$CH_4 + 2O_2 \rightarrow CO_2 + 2H_2O + energy.$$

The heat energy released in this process is the *difference* between the chemical energy of the original fuel plus oxygen, and the chemical energy of the resulting carbon dioxide plus water. In practice, it is usual to refer to it as the **energy content** (or *heat content*, or *heat value*) of the fuel – the methane in this case.

The reaction shown above shows the essential features of the combustion of many common fuels which contain carbon and hydrogen. The fuel interacts with oxygen from the air to produce carbon dioxide and water, the latter usually as water vapour in the form of steam. The process of respiration in living organisms is similar, but takes place at near-ambient temperature. If we know the composition of the fuel and the relative masses of the chemical elements making up its molecules, we can predict how much carbon dioxide will be produced in burning a given amount of fuel (see Box 4.4 and Table 4.1).

BOX 4.4 CO_2 from fuel combustion

As an example, consider the combustion of methane (CH_4). The masses of the atoms of carbon and oxygen are respectively 12 times and 16 times the mass of a hydrogen atom, so we can associate masses with the items in the combustion equation:

$$CH_4 + 2O_2 \rightarrow CO_2 + 2H_2O$$
$$12 + (4 \times 1) + 2 \times (2 \times 16) \rightarrow 12 + (2 \times 16) + 2 \times (2 \times 1 + 16)$$

We can see, therefore, that burning 16 tonnes of CH_4 releases 44 tonnes of CO_2.

The energy content of methane is 55 GJ t^{-1} if the water vapour produced is condensed but only 50 GJ t^{-1} if it isn't (see Box 4.5). This is the amount of heat produced by burning one tonne of methane and thus releasing 2.75 tonnes (2750 kg) of CO_2.

Other fossil fuels, apart from coal, are chemically more complex than methane (although many are hydrocarbons like methane in that they consist almost entirely of hydrogen and carbon), but their combustion follows a similar process. The heat produced per tonne is rather less, however, and they also produce more CO_2 per tonne because they have a higher ratio of carbon to hydrogen atoms, so the CO_2 per unit of heat output is greater (see Table 4.1). Coal, which is largely carbon, produces one of the highest outputs of CO_2 per unit of heat output. Coal and oil may also contain sulfur which burns to sulfur dioxide. This sulfur can be refined out or collected from flue gases and used in fertilizers.

Table 4.1 Heat content (net calorific value, see Box 4.5) and CO_2 emissions

Fuel	Heat content/GJ t^{-1}	CO_2 released/kg GJ^{-1}
Coal	24	94
Fuel oil	41	79
Natural gas	48	57
Air-dry wood	~15	~80*

Note that the composition of coal, oil and wood and the moisture content can vary significantly, so the figures are typical values.

* If the wood is grown sustainably, so that trees are planted to replace those harvested and combustion is complete, its *life cycle* CO_2 emission should be close to zero.

Sources: DECC, 2011, AEA, 2011.

The energy produced in burning one tonne or one cubic metre of various biological materials is shown in Table 4.2, with the main fossil fuels for comparison. Note, again that the composition of coal and most biofuels is variable (Box 4.5) and the energy content per tonne can differ significantly from the averages shown here. The energy *per cubic metre* depends on the density of the material and can vary even more widely.

Table 4.2 Typical heat energy content (net calorific value, see Box 4.5) per unit of dry mass and volume of various forms of biomass and fossil fuel

Fuel	Energy content	
	GJ t^{-1}	GJ m^{-3}
Wood (green, 60% moisture)	6	7
Wood (air-dried, 20% moisture)	15	9
Wood (oven-dried, 0% moisture)	18	9
Charcoal	30	*
Paper (stacked newspapers)	17	9
Dung (dried)	16	4
Grass (fresh-cut)	4	3
Maize grain (air-dried)	19	14
Straw (as harvested, baled)	15	1.5
Sugar cane residues (bagasse)	17	10
Domestic refuse (as collected)	9	1.5
Commercial wastes (UK average)	16	*
Domestic heating oil	43	36
Coal (domestic heating, average)	28	50
Natural gas (at supply pressure)	48	0.04

* Indicates dependence on specific types of material.
Source: includes data from Biomass Energy Centre, 2011a.

BOX 4.5 Energy, moisture content and calorific values

We have seen that water is an inevitable combustion product of most common fuels. It can appear as a result of the combustion of hydrogen in a fuel or the vaporization of moisture contained in the fuel. It appears as water vapour at a high temperature (steam), and some extra useful heat can be obtained by cooling and condensing this.

Fuels are often quoted with two different calorific values giving their potential heat output when burned. Values for biomass, industrial wastes, coal and oil are normally quoted as the lower heat value (LHV) or net calorific value (NCV) of the fuel. This assumes that any water vapour produced is not condensed. However, if it is condensed, then a higher heat value (HHV), also called the gross calorific value (GCV) is appropriate. For fuels such as coal and oil the difference between NCV and GCV is small enough to be ignored in rough calculations, but this is not the case for natural gas. This is largely methane whose combustion in described in Box 4.4. In this case the GCV is about 10% higher than the LCV. For hydrogen (to be discussed in Chapter 10) the difference is 17%.

Unlike fossil fuels, biomass often has a high water content prior to combustion, which adds to the mass but contributes no energy. This water also absorbs energy for heating and evaporation. In general, each 10% increase in moisture content reduces the NCV by roughly 11%. Condensing all the water can increase the heat output by 50% or more when burning fresh green plants, but is difficult in practice.

Moisture content also affects the rate at which biomass materials decay naturally. Drier materials such as straw or air-dried wood can be stored for relatively long periods, whereas wet materials will decay rapidly unless treated.

Making use of biomass

An important issue in the use of biomass as an energy source is the need to process it into a suitable form. The input to these processes may be purpose-grown plants in the form of **energy crops**, but can also be organic **wastes**, **residues** or **byproducts/coproducts** including straw, animal manures and human sewage. The output may be useful heat, or one of a range of solid, liquid or gaseous biofuels. Figure 4.4 shows a very general outline of the primary sources of biomass, the conversion processes and products involved.

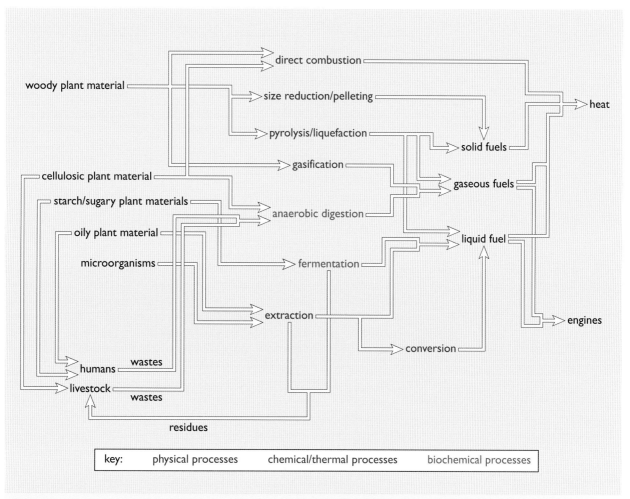

Figure 4.4 An outline of some of the major categories of biomass and the processes involved in using biomass as an energy source (figure does not show all possible combinations of input and process)

In Figure 4.4 the sources of biomass are divided into primary materials, derived directly from plants, and secondary 'wastes', the otherwise unwanted or unutilized products of human activities. To qualify as biomass energy, the wastes need to have been themselves derived ultimately from renewable, plant-based materials. The primary sources are divided into four categories of plant-derived material, **woody**, **cellulosic**, **starchy** and **oily** plus a separate group of **microorganisms**. These categories are not exact, but are useful as a general guide. Processing the biomass materials can involve **physical** (grinding, pelleting etc), **thermochemical** (involving heat) and **biochemical** (using other living organisms) methods.

The next two sections will look at what types of materials are considered primary and secondary biomass energy sources, respectively. Then Sections 4.7–4.9 will discuss the physical, thermochemical and biochemical processing of biomass.

4.5 Primary biomass energy sources: plant materials

Energy crops, in the form of plants grown specifically to provide bioenergy, have attracted increasing attention in recent years. This has been mainly in response to the need for alternatives to fossil fuels, to reduce net CO_2 emissions, but also as an import saving device for countries without fossil fuel resources, reducing their dependence on other countries. The relative importance of these factors varies between different countries, depending on their particular circumstances and the constraints imposed by the local climate, soil, etc. Figure 4.4 categorized energy crops broadly according to the physical and chemical nature of the biomass involved, and examples of each of these are given below.

- Woody biomass – trees are the main source of woody biomass. Wood itself is a complex mixture of the carbohydrates, lignin, cellulose and hemicellulose. It provides the structural strength of the tree and is relatively resistant to decay.

- Cellulosic biomass – cellulose together with related hemicelluloses are the main components of plant cell walls. Cellulose is the most abundant organic large molecule on Earth. The cell walls of seaweeds are primarily cellulose and hemicellulose while terrestrial herbaceous plants like grasses, the cereal crops such as wheat and maize (corn) also contain small amounts (relative to woody plants) of lignin and lignocellulose (a combination of lignin, cellulose and hemicellulose).

- Starch/sugary biomass – starch is a relatively simple carbohydrate which forms the energy store in many seeds including the cereals and in the tubers of crops like potatoes. High concentrations of simpler sugars occur in the stems of sugar cane and the roots of sugar beet.

- Oily biomass – oils form the basis of the storage material in some seeds such as oilseed rape, soya and oil palms.

- Microorganisms – a possible novel 'energy crop' is represented by microalgae, which can have high concentrations of oils in their cells.

This general categorization is summarized and expanded in Table 4.3, which also gives structural and energy yield information. It should be remembered that the categories overlap to a considerable extent so some crops may be grown primarily for their sugar or starch content, but their supporting lignocellulosic structural materials may also be used. The mechanical strength and susceptibility of the different forms of biomass to natural breakdown affects the storage and processing of the crops.

Table 4.3 A broad generalized classification of primary bioenergy sources

Category	Major energy-rich components	Structural strength /resistance to natural decay	Examples	Typical yields of dry matter /t ha^{-1} y^{-1}
Woody biomass	Lignin/lignocellulose (complex carbohydrates)	High	Trees (deciduous or hardwoods)	10 (temperate) to 20 (tropics)
Cellulosic biomass	Cellulose/lignocellulose (complex carbohydrates)	Medium	Grasses (e.g miscanthus), water hyacinth, seaweeds	10 (temperate) to 60 (tropical aquatics)
Starch/sugar crops	Simpler carbohydrates	Low	Cereals (maize, sugar cane, wheat)	10 (temperate cereals) to 35 (sugar-cane)
Oily crops	Lipids (i.e. oils/fats)	Low	Oilseeds (rape, sunflower, oil palm, jatropha)	8 to 15
Microorganisms	Oils	Low	Microalgae	Unknown – still speculative

The values in Table 4.3 for the gross biomass yields per hectare are measured as tonnes of oven-dry mass of all above-ground matter. The yield of particular components, such as starch or oil, will be less than the figures presented, and will depend on the processing route from crop to product. Gross yields depend on growing conditions, and vary widely within the limits shown.

Many non-woody plants are in current agricultural use to provide foods, but could also be used for bioenergy, providing starch/sugar, oils or non-specialized, mainly cellulosic, tissue. As an indication of the spectrum of possibilities, El Bassam (2010), lists over 100 species that could be used for ethanol production. He also lists about 70 species that are particularly suited to arid and semi-arid areas, areas which may currently be problematic for growing food crops. The various species have differing characteristics (water requirement, ability to fix nitrogen, etc.) which could make them useful candidates for specific situations.

Woody biomass

In the absence of major human activity, trees would form the bulk of the vegetation over much of the higher latitudes and the equatorial regions. Well-managed forests can provide a sustainable fuel source, reducing atmospheric CO_2 as the trees grow, storing it for up to a few centuries and providing a substitute for fossil fuel when felled. In conventional forestry, energy production has usually been incidental, although some countries use the forest and process residues (branches, bark, sawdust, etc.) to provide heat and power (Section 4.6). Domestic use of wood fuel, logs and pellets is well developed in central Europe, Scandinavia and other regions and wood is still a vital fuel for cooking and heating in many developing countries. In **modified conventional forestry**, whilst biomass is not the main product, it can be an integral part of the production process. For example, coniferous trees can be planted at higher than usual density and vigorously thinned after a few years, using integrated harvesting techniques to produce chipped wood. The remaining trees grow to maturity in the normal way. An integrated harvesting system that simultaneously extracts logs for timber products and wood chips for bioenergy purposes can also contribute.

Using trees as an energy source occupies an ambiguous position in environmental debates. In the worst case, 'forestry' can involve clear felling of the world's major natural forests with the immediate release of large volumes of carbon dioxide from the soil and from the uncollected residues. The uncontrolled removal of wood can bring economic as well as environmental problems, reducing productive capacity in countries where it contributes appreciably to industrial energy consumption – in Brazil, for instance, where over 12 million tonnes of charcoal is produced per year (Kato et al., 2005). Fourteenth-century England had a similar problem, but it was 'solved' temporarily when the Black Death killed a third of the population. The past decade or so has seen a move away from the earlier view that forests in some developing countries were being destroyed mainly by increasing local populations cutting wood for domestic use. The evidence is that most of these small-scale users collect wood from local scattered areas, and that the large-scale disappearance of forest trees is the result of commercial forestry, charcoal production or just clearance to establish other crops (UNDP, 2000). The wood in this last instance is not cut primarily for fuel, so it does not really fall within the remit of this book – although one might ask whether the *residues* of unsustainable forestry should be included as potential 'renewables'.

In recent years there has been increased interest in woody crops planted and harvested entirely for energy production. The commonest management for these is **short rotation forestry (SRF)**, the 'rotation' being the periodic cutting of the wood every 8–20 years, with the trees then being re-planted or, for suitable species, left to regrow from the stump. This latter approach is also known as **short rotation coppicing (SRC)**, referring to the centuries-old practice of coppicing hazel, willow or other fast-growing trees. (Some of the resulting 'copses' remain as attractive features of the countryside after the coppicing management ceases.) In the modern version, cuttings are planted at 3–15 thousand stems per hectare, depending on the species. Cut back close to the ground after one to four years, they re-grow, either with multiple stems or as single-stem trees. The crop is allowed to grow for two to seven more years, depending on species and location and then harvested by cutting the stems back to a low stump. The stumps re-grow and the cycle is repeated, for perhaps 30 years, although traditional coppice woodlands of ash, hazel or willow in the UK were often harvested for much longer periods. Annual yields of 10 dry tonnes per hectare should be achievable in northern Europe and up to 30 dry tonnes in more favourable regions. Adequate water, appropriate fertilizers and weed control are required for maximum continued yield. Modern harvesting machines can reduce the stems *in situ* to short lengths suitable for transporting, storage and future use.

Figure 4.5 Short rotation coppicing

An obvious use of SRC products is for the provision of heat, and one of its most successful applications has been in Sweden where around 18 000 ha of willow coppice supplies energy for district heating. The generation of electricity using SRC, either as the sole fuel or co-fired with coal, has received considerable attention in Europe, where some 30 MW of capacity is fuelled by energy crops.

Outside Europe, there has been some development of SRC in the Americas, India, Australia and New Zealand where *Eucalyptus* species are grown. There is interest in its potential for improving degraded land as well as its value as an energy source. In Australia, millions of hectares of cereal lands have been going out of production due to rising salt levels as a result of injudicious irrigation during arable cultivation. Growing *Eucalyptus* in strips around 100 m apart for short rotation coppice can help lower the water table, since the trees can tolerate the salt and have a high water demand. Lowering the water table carries the salts back down to lower levels. However, in other regions such as Portugal, there are concerns about the crops' high water usage in areas where water supplies are marginal.

Cellulosic materials

The structural integrity of non-woody (i.e. herbaceous) plants is provided mainly by cellulose. Humans are unable to digest cellulose, but it forms a major food source for sheep, cattle and other ruminant livestock. It is a major component of the straw and other supporting tissues of all the major food crops. The most commonly considered cellulosic bioenergy crop suitable for temperate climates is **miscanthus**, a C4 grassy plant (Figure 4.6). It originated in Asia and Africa, but some forms will grow in northern Europe and research in the 1990s suggested that it could yield up to 18 dry tonnes per hectare per year under UK conditions. It reproduces vegetatively by means of a dense underground network of specialized organs called

rhizomes, short sections of which can be used for planting new crops. The thick stems of miscanthus are suitable for direct combustion since they have a low water content (20–30%) when harvested in late winter. As it does not set seed in northern Europe, it is not likely to become an invasive weed.

Figure 4.6 Some possible plants for production of cellulosic biomass: (a) bracken, (b) miscanthus, (c) *Gunnera*, (d) kelp, (e) water hyacinth

Herbaceous crops have an advantage over woody biomass in that they can be grown using normal farming techniques, and it is relatively easy to plough in the crop when no longer required, which allows for flexibility in land use. Agricultural crops that were primarily grown for their sugar or starch content, such as maize and new varieties of sugar beet are increasingly used whole for bioenergy, along with fodder beet and grasses. When used as animal feed these crops were often conserved as **silage** by storing the fresh material in nearly airtight conditions. Such silages can be used as high-energy feedstocks for anaerobic digestion (Section 4.9) together with animal manures.

Other possible non-agricultural materials in this general category include bracken (*Pteridium aquilinum*), regarded in the UK as a weed although the dry fronds can be used as a horticultural mulch and as livestock bedding. In the 1980s, various other herbaceous species including the seriously invasive Japanese knotweed (*Reynoutria japonica*) and *Gunnera maricata*, a more benign giant rhubarb-like plant, were investigated as possible bioenergy crops.

Miscanthus and another grass, reed canary grass (*Phalaris arundinacea*), have generally been grown for bioenergy using agricultural land, equally suitable for food crops. In contrast, weedy species such as bracken usually grow on land that is difficult to cultivate and thus produces little food. The spread of weedy species needs controlling, but if done successfully their use for biomass energy should not affect food production through diversion of land from food crops.

Water hyacinth (*Eichornia crassipes*), a very fast growing aquatic weed that causes problems in reservoirs and irrigation channels in tropical areas, could possibly be controlled by harvesting for bioenergy use, providing a dual benefit. The giant seaweeds, such as kelp, which grow in shallow waters and again do not compete for use of land are more remote contenders to be a cellulose source.

Figure 4.7 Harvesting miscanthus using conventional agricultural machinery

Starchy/sugar crops

The major biomass energy component of these is their sugar or starch, which, as shown in Figure 4.4, can be fermented to produce ethanol as a liquid fuel. Their supporting cellulosic tissues may also be used in the same way as the mainly cellulose crops. Globally, at the time of writing, the most important crops for bioenergy purposes are sugar cane and maize. Both are high yield C4 crops with favourable energy balance (that is, a good ratio of energy input to final energy output – see Section 4.10). Sugar cane, as its name implies, is widely grown for the sugary sap (12–16% sugar content) which is stored in its 5cm diameter stems.

Sugar cane is restricted to areas of the world between 37° north and 31° south, with Brazil producing 500 million tonnes out of a world total of around 1500 million tonnes in 2008 (El Bassam, 2010). India also produces a substantial quantity of sugar cane (about 300 million tonnes per year). This is primarily used for edible sugar with only the molasses (the liquid remaining after crystallization of the sugar) used for ethanol production. The cane is propagated in a similar way to miscanthus, but using above-ground stem sections, not rhizomes, and several successive harvests of stems can be taken before yield declines and the crop has to be replanted. In contrast, almost all the other crops in this category are annuals that provide a single harvest from each sowing. Maize can be grown over a wide range of latitudes, with the USA growing over 300 million tonnes of grain out of a world production of 785 million tonnes in 2007 (El Bassam, 2010). Other cereals such as wheat (total world production 725 million tonnes in 2007), rice, barley and sorghum could potentially be used to produce starch, as could the starchy tubers of potatoes. World production of potatoes in 2006 was 315 million tonnes, with China the largest producer. Sweet sorghum, which is also a sugar-rich crop, requires less water than sugar cane and can grow in semi-arid regions. Two crops per year are possible, and there is genetic potential for improving yields or developing hybrids suitable for the European climate. Another sugar-rich crop of interest in Europe is sugar beet, where the swollen, fleshy roots contain 15–20% sugar.

Sugar-cane bagasse (the sugar cane's fibrous residue) and the straw from maize (stover) and other cereals can also be used for energy in the same way as cellulosic crops. Straw and bagasse are considered in the next section as 'waste' products.

(a)   (b)   (c)

Figure 4.8 Starch/sugar crops: (a) sugar cane, (b) maize harvesting, (c) sugar beet

Oilseed crops

Sunflowers, oilseed rape and soya beans are grown widely for the oil in their seeds, while in tropical areas, palm (*Elaeis guineensis*) is a major crop. The worldwide total production of vegetable oils in 2005–6 was around 110 million tonnes, of which 34 million tonnes came from palm and a similar amount from soya (El Bassam, 2010).

The oils from these various crops are extracted by pressing, with or without additional chemical treatments and can be converted to a diesel substitute known as biodiesel (Section 4.8). As with the starchy/sugar crops considered above, there is a residue left from these processes that is currently often used as animal feed, but which could be a bioenergy feedstock. The oil from all these crops has a vast number of uses, in foods, cosmetics etc. There are already widespread misgivings about the conversion of mature forests to palm oil plantations in Indonesia and the Philippines, and use of palm oil for bioenergy could lead to increases in such problems.

The vivid yellow of oilseed rape (*Brassica napus*, also known as canola in Canada) has become familiar in many countries over recent decades. There are two general types – winter varieties that are sown in the autumn to overwinter and spring varieties, sown and harvested in the same year to avoid frost damage but with a shorter total growing period and slightly lower yield. World production of rapeseed was 47 million tonnes in 2006 (El Bassam, 2010) making it one of the smaller volume crops, but one that has well-developed biofuel use in Europe. Sunflower (*Helianthus annuus*) is another crop which originated in the Americas, but is widely grown in southern Europe for its oil and seed. Some non-edible oil yielding plants such as *Jatropha curcus* and *Pongamia pinnata* have been recently used in government programmes in India. These can be grown with relatively fewer inputs and also used as fencing material, but results of ongoing programmes were not available at the time of writing.

Soya beans are extensively grown in the middle latitudes, with the USA, Brazil and Argentina together producing some 180 million tonnes out of a global total of 220 million tonnes in 2006 (El Bassam, 2010). Soya oil, like palm oil, is in widespread use in a whole range of products and is also a major source of protein for human and livestock feed. It has an advantage over other herbaceous crops in that it is a *leguminous* crop. It is therefore less dependent on external supplies of nitrogen fertilizer since it can fix nitrogen from the air through the agency of bacteria present in its roots. The extent of soya cultivation worldwide is limited by the requirement for a particular day length for it to flower at the correct time (although this could potentially be overcome in a breeding programme). Soya was one of the first crops to undergo direct genetic modification (GM), incorporating a bacterial gene that gives the crop resistance to the widely used herbicide glyphosate. This simplifies cultivation, but has been a continuing source of controversy mainly in the UK and Europe with regard to use of GM crops in foodstuffs.

Most of the cellulosic, starchy and oilseed crops considered are familiar in current agriculture, requiring no new agricultural technology for their exploitation. They have in the past been bred primarily for food production and to respond to high levels of manufactured fertilizer. The production of fertilizers is a relatively energy-expensive process, requiring 30–40 MJ per kilogram of the major plant nutrients applied (NNFCC, 2010). This, together with the recurrent energy costs for soil preparation, sowing and harvesting, affect the *net* energy yield of these crops (Section 4.10). Their existing widespread cultivation also means that pests and diseases have had time to co-evolve with them, and could spread relatively easily between food and biomass crops. Pesticide production and use will therefore also have to be taken into account. There may be advantages in considering more exotic plants that are not currently in widespread use, such as algae.

Microalgae and other microorganisms

Seaweeds are one form of large algae, but there are also single-celled aquatic microalgae and cyanobacteria that photosynthesize. These could potentially make a very attractive bioenergy source as:

- they grow in water, and are tolerant of wide ranges of salinity and temperature
- they do not occupy land that could be used for other products
- under appropriate conditions, the cells of the algae can contain high percentages of oils, and the cyanobacteria can actually excrete oil into their surroundings: the microbial residues after oil extraction can also be used as an energy source
- there are suggestions they can be used simultaneously to clean up waters polluted with plant nutrients and the resulting biomass, including the oil, can be used for energy (Clarens et al., 2010)
- some forms are also seen as candidate material for the capture of carbon dioxide from power plants.

Although there has been considerable recent publicity concerning the potential of such material and research in the area has taken place for several decades, at the time of writing no commercial oil production facilities exist (despite the numerous design proposals for **bioreactors** for algal production and some pilot plants). There are algae 'farms' for producing the microalga *Spirulina* as a nutritional supplement, with the largest in California – experience gained here could be used in the development of large scale outdoor systems. However, it is possible that there is an inherent problem with microalgae in that the cells only produce large amounts of oil when under stress, through lack of nutrients or similar restrictions on their growth (Griffiths and Harrison, 2009). There would therefore be a trade-off between total algal growth and oil content that would restrict the oil yield from a given area. The use of cyanobacteria is even more speculative, but is the subject of intensive investment and research at the time of writing.

Figure 4.9 A possible bioreactor design for use of microalgae as an energy source

4.6 Secondary biomass sources: wastes, residues and co-products

Materials such as straw or rice husks resulting from 'non-energy' uses of biomass are sometimes discarded as 'wastes' (at which point they may become subject to environmental regulations) but they are also potential sources of biomass for energy. Since they are co-products and often collected and transported as part of the primary product supply chain, they tend to be much cheaper than purpose-grown forests or energy crops. Another potential biomass source includes the many kinds of mixed urban and industrial waste, much of which is carbon-containing material and will release energy if burned. It is debatable whether this industrial waste material should be regarded as a *renewable* resource. In recent years it has become customary to regard only wastes of biological origin as true renewables, excluding, for instance, polymeric materials ('plastics') that have been synthesized from fossil fuels.

This section looks first at the co-products and residues arising from existing uses of biomass in forestry, arable agriculture and animal husbandry. It then moves on to household, or more generally, municipal wastes, and finally to the specific wastes associated with industrial processes.

Wood residues

Around 15% of the standing tree crop is left behind as **forestry residues** during operations such as thinning plantations and delimbing felled trees. At present these residues are often left to rot on site, with some being used as a form of matting to prevent soil damage by the harvesting machinery. This has the environmental merit of retaining nutrients and a relatively slow release of carbon dioxide as the residues decay. Their bulk and inconvenient form makes transporting the residues for wider use generally uneconomic. However, with the development of integrated harvesting techniques, some fraction of the residues is increasingly being used for heat and/or power generation in many countries. Sweden has been using fuel chips from forest residues for some 30 years and the rate of utilization has been claimed to be increasing at 10% per annum, towards an estimated potential of 20 million m³ per year. Most of the residues (more than 71%) are derived during the final felling of the trees (Kuiper and Oldenburger, 2006).

The quantity of residues used as firewood is uncertain, but was thought to be about a million dry tonnes per year in the UK in 1999 (ETSU, 1999). For comparison, total UK sales of **roundwood** (tree trunks and limbs) for fuel increased from around 0.4 million tonnes fresh weight in 2005 to over 1 million tonnes in 2009 (Forestry Commission, 2010).

Possibly the largest use of wood wastes as fuel is in the pulp and paper industries, where the production plant has ready access to such wastes. Some 3% of the US total energy demand is supplied in this way (Haq, 2002).

Temperate crop co-products

Worldwide, residues from wheat and maize (corn), the two main temperate cereal crops, amount to more than a billion tonnes per year, with an estimated energy content of 15–20 EJ. The residues have many uses including as bedding and feed, and hence may be described as **co-products** rather than residues, but in major cereal-growing regions substantial quantities are ploughed back into the soil. In the 1970s much of the straw in Britain was burned in the field, but air pollution concerns led to a ban on field burning from the end of 1992. There has been similar legislation across Europe and elsewhere. China experienced a similar pollution problem in the 1990s, when residues that had been used for heating and cooking were replaced by 'modern' fuels, leaving a surplus of residues that was burned in the fields. In this case, the solution was the introduction of biomass-fuelled village-scale gasifiers (see Section 4.8) distributing gas to households. Emissions were reduced and the conversion efficiency was better than direct combustion in individual straw-fuelled stoves.

For use as a bioenergy source, straw must be baled, removed from the fields, stored in a dry atmosphere and transported to its point of use. Although straw has a reasonable energy density of up to 15 GJ t^{-1}, it has a relatively low mass density. One tonne of bales can occupy a volume of up to 6 cubic metres, which makes transport and storage expensive.

Several European countries already have wide experience of straw burning, and Denmark (Box 4.6) has a programme to use 1.2 Mt per year in combined heat and power (CHP) plants with district heating. It has been estimated that straw could provide up to 3.2% of UK energy demand by 2020 (DTI, 2004). The first straw-fired power station in the UK, commissioned in 2000, was the world's largest single plant at the time, using 0.2 Mt per year. Co-firing of straw with other solid fuels occurs in the USA, with for example a 35 MW plant in Tacoma that derives 40% of its energy from biomass (Haq, 2002).

BOX 4.6 The Avedøre 2 straw-fired combined heat and power station

In 1996 the Danish authorities banned the burning of coal on its own in power stations. The 570 MW Avedøre-2 power plant in Denmark (Figure 4.10) is designed to burn biomass, and is located adjacent to an existing coal-fired 250 MW CHP plant. As a condition of project approval, SK Power had to decommission three older coal-fired power plants to reduce net emissions. The Avedøre-2 plant uses a combination of gas turbines, a fossil fuel boiler and a biomass boiler. The unit can supply district heat to about 180 000 homes and provide electricity consumption for 800 000 households.

When the plant was originally built, natural gas was expected to contribute 85% of total fuel consumption. However, the price of natural gas then rocketed, so in early 2001 biofuel was decided upon for the main fuel source. The system requires 300 000 tonnes of straw pellets, at least half of which come from local sources. After transport to the plant, the pellets are re-ground to a small particle size (the improvement in logistics and transport costs for pelleted feedstocks more than compensates for the energy and economic cost of this process).

The bioenergy plant comprises the straw storage facility, a boiler, an ash separator and ash and fly-ash handling equipment. A flue-gas filter restricts particle discharge into the atmosphere. Bottom ash is recycled as a fertilizer.

Figure 4.10 The Avedøre-2 straw-fired power station

Tropical crop residues

The total energy content of the annual residues of two major tropical food crops, **sugar cane** and **rice**, is estimated as about 18 EJ (similar to the total for temperate crops), and significant quantities of tropical crop residues are already being used as fuels.

Bagasse, the fibrous residue of sugar cane, is used in sugar factories as a fuel for raising steam and to produce electricity for use in the plant, in much the same way as wood residues are used in pulp and paper mills. During the 6 to 7 month cane-crushing period there is often a surplus, and whilst transporting the bagasse from the crushing site may not be currently economic, selling surplus electricity can be an opportunity for business diversification. In the past, the problem of selling power that is available for only half the year has limited investment, but liberalization of energy markets in many countries has improved prospects and efficient boilers have now been installed in many sugar production facilities. The potential for year-round generation using wood or other crop residues in the non-crushing season has also created interest, as has the pelleting of the bagasse to make it more easily transported and utilized. However, the moisture content of the material is still a problem for both options.

Increased recovery of wastes, combined with improved efficiency of conversion to electricity, could result in up to 50 GW of generating capacity from the sugar industry worldwide. In Brazil, it has been predicted that 15% of electricity generation could be derived from bagasse by 2015, compared with about 2–3% today. In India, a bagasse co-generation programme for producing surplus power from sugar mills by upgrading the boilers has been run successfully. The estimated potential for bagasse cogeneration in India is about 5000 MW, out of which about 1500 MW is already installed (Purohit and Michaelowa, 2007).

Rice husks are among the most common agricultural residues in the world, making up about one-fifth of the dry weight of unmilled rice. Around 40 million tonnes of rice husks were produced in China in 1999 (Jin et al., 2002), out of a world total of some 140 million tonnes. Although they have a high silica (ash) content that can lead to combustion problems compared with other biomass fuels, their uniform texture makes them suitable for technologies such as gasification. Rice-husk power plants have been deployed in China, India, Thailand and other rice-growing countries.

Other important agricultural residues are ground-nut shells, mustard stalks, jute sticks etc. which are currently used inefficiently in developing countries like India for cooking and can be used better by deploying technologies such as gasification (Kishore et.al., 2004).

Animal wastes

Animal manure can be a major source of greenhouse gases. Manure from grazing livestock that is deposited in the field decomposes **aerobically**, through respiration by a whole range of organisms that feed on it, taking in oxygen and releasing mainly carbon dioxide. This is effectively carbon neutral and allows the return of plant nutrients, especially nitrogen, to the soil. However, wet slurry (from housed livestock) that is stored in bulk decomposes **anaerobically** (in the absence of air) releasing methane rather than carbon dioxide. Manure management has been estimated to account for 8% of methane emissions in the USA (EPA, 2010). When not correctly managed, farm slurries can seriously pollute local watercourses and anaerobic decomposition also results in emissions of nitrous oxide, another greenhouse gas. The combination of increased housing of livestock, stricter environmental controls on odour and water pollution, and incentives for renewable energy production, are encouraging farmers to invest in controlled **anaerobic digestion** (Section 4.9) where the biogas generated can be collected and used on-farm or for export of energy services.

Poultry litter, a mixture of chicken droppings and material such as straw, wood shavings etc., has a relatively low moisture content compared to other manures, and an energy content in the range 9–15 GJ t^{-1}, enabling its direct combustion for electricity generation (Box 4.7).

BOX 4.7 Power from poultry litter

At the time of writing £1 = €1.15.

The world's largest biomass power plant to run exclusively on poultry manure was opened in 2008 in Moerdijk, in the Netherlands. The €150 million plant converts roughly 440 000 tonnes of chicken manure into energy annually. Running at a capacity of 36.5 MW it can generate more than 270 TWh of electricity per year. The plant is intended to convert one-third of the Netherlands' chicken waste into energy while also reducing the environmental problems associated with chicken manure in the country.

The first UK poultry manure power plant, with a 12.7 MW output capacity, started operating in 1992 at Eye in Suffolk, using some 140 000 tonnes of litter per year from surrounding poultry farms. A second 13.5 MW plant was commissioned a year later, followed by a 38.5 MW plant, the largest in the UK, in 1998, which consumes 420 000 tonnes of litter each year. Residues from combustion are sold as fertilizer and one plant has been converted to burn material from animal carcass disposal. All three projects were supported by the NFFO (the Non-Fossil Fuel Obligation, a mechanism of Government incentives for producing low-carbon electricity which ran from 1990 to 1998).

Sewage sludge, the semi-solid residue from the initial settlement of raw sewage, can be treated anaerobically, as has been done in the UK since the first large 'sewage farms' were built in the last century. Originally, much of the resulting methane-containing biogas was simply flared (burnt off), but an increasing proportion is now used for heat and electricity production on-site. The sludge residue can then be dewatered with mechanical presses and incinerated, producing further heat and electricity. In 2005–6, the water industry in the UK generated approximately 500 GWh of electricity (POST, 2007), although this only represents about 6% of the total energy use in that industry.

Municipal solid waste

The average household in an industrialized country generates rather more than a tonne of solid waste per year. Up to 20 million tonnes of food waste is generated in the UK each year, with each household throwing away 25% of all food purchased (WRAP, 2009). Households in the USA throw away some 40 million tonnes annually (Kantor et al., 1997). Most household waste is collected as Municipal Solid Waste (MSW) with an energy content of about 9 GJ per tonne (Figure 4.11). It appears therefore that the average UK household throwing away a tonne of waste could, in principle, supply one tenth of its total annual energy consumption of about 90 GJ from its own wastes.

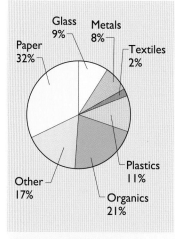

Figure 4.11 The average composition of municipal solid wastes in the UK. About 55% of the total is organic material

In continental Europe and elsewhere, refuse incineration with heat recovery, or energy-from-waste (EfW), is an important part of waste management. The heat may be used directly for district heating, or for power production (often in CHP plant). Countries with successful recycling and composting programmes have often seen parallel growth in EfW, which accounts for 30–60% of MSW disposal in most western European countries. In 2009 around 390 MW of electricity was generated in the UK by incinerating 2.8 million tonnes municipal waste (DECC, 2010a). Some MSW combustion plants in the UK now also make use of their heat output (see, for instance, Figure 4.12). World installed capacity is over 3 GW, about half of it in Europe.

Landfilling of waste, using suitable holes in the ground such as old quarries, has been the main disposal method for MSW in a number of countries (including the USA and the UK where 75% of MSW is currently landfilled). Other countries landfill a smaller fraction (or none at all in some instances) and in continental Europe up to 60% or so of MSW is incinerated (Figure 4.13). Legislation and taxation in the UK are gradually decreasing the fraction of MSW going to landfill, as authorities diversify into other methods of waste treatment including materials recovery and energy recovery. The extent to which MSW can be regarded as a renewable energy source due to the varying share of non-organic and fossil-fuel derived components is debatable.

Figure 4.12 The SELCHP (South-East London Combined Heat and Power) plant, commissioned in 1994, was designed to incinerate 420 000 tonnes of MSW per year, and produce steam for a 31 MW turbo-generator and heat for a local district heating scheme (although the latter is not currently implemented)

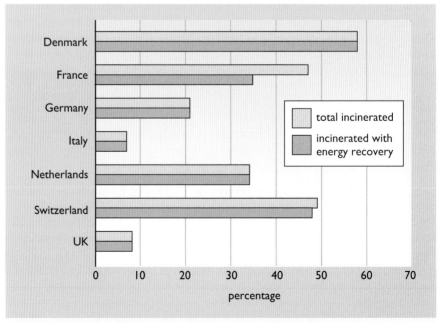

Figure 4.13 Share of incineration of MSW, with and without energy recovery, in selected European countries. Most of the balance of the waste is landfilled

The composition and characteristics of MSW in some developing countries, such as India, may be quite different. Wastes like paper and plastic are almost totally removed by informal, unorganized rubbish-pickers and are recycled, leaving mainly wet wastes plus construction debris, for which incineration is not a suitable option.

Commercial and industrial wastes

Commercial and industrial wastes of organic origin can be used as fuel. The UK generates about 36 Mt of specialized wastes each year, about two-thirds of which are combustible (for example the UK furniture industry has been claimed to burn 200 000 tonnes of off-cuts and sawdust per year (BFM, 2005), although DEFRA (2008) suggests a much lower figure of 70 000 tonnes).

The wastes from commercial food processing facilities must be treated before discharge to prevent water pollution. Much food waste was traditionally used as animal feed, particularly for pigs, but problems of disease transmission have led to restrictions on this use in Europe and elsewhere. However, the nature of general food waste makes it a useful substrate for anaerobic digestion.

Fats, either in the form of used cooking oils or unwanted fatty tissue removed during meat processing, have a high energy density and are potentially suitable either for direct combustion or conversion to biodiesel.

Hospital wastes, all of which must be incinerated, are increasingly subject to energy recovery as health authorities upgrade their waste-handling equipment. Approximately 200 000 tonnes of healthcare waste is produced annually in the UK, of which 26 450 tonnes, with an energy content of some 15 GJ tonne^{-1}, requires high-temperature treatment (DEFRA, 2007).

The majority of the 50 million tyres discarded in Britain every year are unsuitable for reuse, but with an energy content of 32 GJ per tonne, they constitute a major fuel resource. Under EU legislation, whole tyres have been banned from landfill sites since 2003 and chipped tyres from 2006. Lafarge Cement consumed over 75 000 tonnes of waste tyres in 2008, using them as a partial substitute for coal or coke (Sapphire Energy Recovery, 2009).

4.7 Physical processing of biomass

We now turn from the biomass sources to the next stages shown in Figure 4.4: the ways in which biomass is treated and used or might be used. The premium fuels – oil and natural gas – are valued because of their high energy density and the fact that they can be easily stored, made available where and when needed, and used in a wide range of existing appliances or machinery. Biomass resources come in a variety of physical forms, with widely varying energy content. They are likely to require processing to make them more acceptable to users or easier to transport, and then may require special equipment to release their energy in a useful form. This can involve physical, thermochemical or biochemical processes either singly or in combination.

Separation, size reduction and pelleting

As anyone who uses an open fire or wood stove will recognize, wood is not the most convenient fuel. The felled timber has to be converted into logs small enough to feed the fire and their often irregular shape means they need a large space for storage. Chipping the material produces a more homogeneous material that is easier to handle in bulk, but the energy required for chipping is not insignificant.

To form the material into consistently sized pellets (Figure 4.14) that can be bagged for retail sale, or transported in bulk, further stages of processing are required firstly to reduce the size even further, then the fine material is passed through a small (about 6–10mm diameter) **die** under pressure. Similar machinery can be used to pelletize miscanthus or bagasse feedstocks, and there is now a significant world trade of about 7 million tonnes of pellets per annum for applications ranging from domestic stoves to industrial plant. Canada has dominated this market until recently, exporting around 1.5 million tonnes annually, but EU trade is growing from about 1.3 million tonnes in 2003. Production in the UK in 2010 was estimated to be in excess of 200 000 tonnes y^{-1}, mainly in Scotland (European Pellets Centre, 2008).

(a)   (b)

Figure 4.14 (a) A wood pelleting plant and (b) a sample of the pellets produced

Systems have been developed, as with the Avedøre plant, for producing high-density pelleted straw, with a typical density in excess of 1 t m^{-3} (denser than wood). These allow boilers to be fed automatically and transport and storage costs to be reduced. There has been some commercial success with the production and sale of straw briquettes for domestic heating.

Size reduction and pelleting requires energy inputs of between 0.1 and 0.4 GJ per tonne for miscanthus (Jannasch et al., 2002) or 0.7 GJ per tonne for wood (Risović; et al., 2008). Where the feedstock has to be dried, the total energy cost can be as high as 4 GJ per tonne (Mani, 2006). However, this energy cost is partly offset by an increased efficiency of combustion of 83% as opposed to 77% for air-dried wood (USDA, 2007).

Raw household and commercial refuse can also be pelleted. However, its contents are variable, its moisture content tends to be high (20% or more) and its energy density is about a thirtieth of that of coal. Thus it is expensive to transport, and requires combustion plant designed specifically for this type of fuel.

The term **refuse-derived fuel** (RDF) refers to a range of products derived from municipal and industrial wastes. In a mechanical biological pre-treatment plant (MBT) metals and inert materials are separated out from the waste and organic fractions are screened out for further stabilization using biological processes. This produces a residual fraction which has a high energy content but low mass density as it is composed mainly of dry residues of paper, plastics and textiles. Packaging derived fuel (PDF) or process engineered fuel (PEF) is usually of higher quality than RDF as it is a source-separated dry combustible fraction which cannot be used for recycling, for example cardboard drink containers or PE/PET bottles contaminated by PVC. The most fully processed product, known as densified refuse-derived fuel (or d-RDF), is the result of separating out the combustible part which is then pulverized, compressed and dried to produce solid fuel pellets which have perhaps twenty times the energy density of the original material. In 2000 there were some 50 plants in the EU that produced about 3 Mt of fuel per annum (WRc, 2003).

Extraction of oils

The oil in crops such as rapeseed, soya and oil palms is contained within the seed, which also contains the plant embryo and may have a structural shell or coat for protection. The oil has to be separated from these surrounding tissues, either by pressing or by using a **solvent** which dissolves the oil making it less viscous and easier to separate from the remaining material. Commercial oilseed extraction involves a number of steps including:

- seed cleaning – removal of foreign matter
- tempering – pre-heating of the seed to improve ease of oil extraction
- dehulling – removal of seed coat
- flaking – to increase surface area
- conditioning – re-heating the flaked seed
- mechanical extraction – by pressing and extrusion and/or expansion
- solvent extraction – for maximum extraction of oil.

A typical press for the mechanical extraction stage uses a metal screw to force the material over a sieve, which allows the oil to pass out but retains the fibrous material for separate ejection (Figure 4.15).

Transport and storage

The transport and storage of biomass energy materials may offer some of the largest challenges to its further development.

Primary biomass materials, by definition, require an area (of land or water) on which to intercept light in order to grow, and so require harvesting, collecting and transporting from a wide area to the point of use, for further processing or for storage. Since many biomass materials have a relatively low mass density the amount of material carried, or stored, in a given volume is less than it would be for other energy rich materials so adding to the energy cost of using biomass. Furthermore, as transport is currently predicated on the use of fossil fuels, this also affects the sustainability of biomass energy systems.

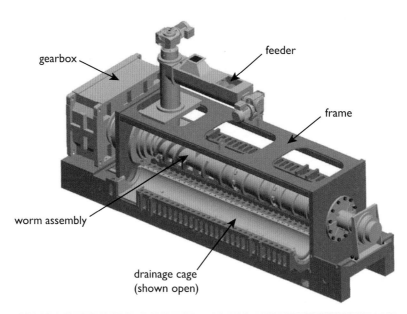

Figure 4.15 Schematic of a screw press for oilseed extraction

As noted earlier, biomass often has a high water content, and depending on its precise nature, may be more or less susceptible to natural decay. Some materials such as cereals, straw and wood can be rendered stable by air drying, but wetter materials may require forced air drying or the use of heat to drive off moisture, which again requires energy. An alternative for cellulosic materials like fresh grass is storage in a sealed container as silage. Sealing the container excludes air, and halts the normal decay process after a short time. This approach requires less additional energy input than forced drying.

4.8 Thermochemical processing

Themochemical processing involves the use of heat (thermo-) and possibly the use of chemical reagents, to convert biomass into energetically more useful forms. The output from such processes may be heat, or intermediate gaseous or liquid fuels.

Combustion of solid biomass

Most biomass is initially solid, and can be burned in this form to produce heat for use *in situ* or in close proximity, although it may first require relatively simple physical processing, sorting, chipping, compressing and/or air-drying, as discussed above.

Direct combustion is a simple process to release useful energy. Unfortunately, it can be very inefficient (as exemplified in Box 4.8) and can potentially produce a wide range of pollutants.

BOX 4.8 **Boiling a litre of water over a wood fire**

How much wood is needed to bring one litre of water to the boil?

The starting point for answering this question is to apply the following equation:

Heat energy required = mass × temperature rise × specific heat capacity
of water

Note: the **specific heat capacity** (sometimes called the specific heat) of a substance is the amount of energy in joules that has to be transferred to 1 kg of material to raise its temperature by 1 kelvin.

Data

Specific heat of water	$= 4200 \text{ J kg}^{-1}\text{ K}^{-1}$
Mass of 1 litre of water	$= 1 \text{ kg}$
Heat value of air-dry wood (Table 4.1)	$= 15 \text{ MJ kg}^{-1}$
Density of air-dry wood	$= 600 \text{ kg m}^{-3}$
1 cubic centimetre (1 cm³)	$= 10^{-6} \text{ m}^3$

Calculation

Heat energy needed to heat 1 litre of water from 20 °C to 100 °C	$= 80 \times 4200 \text{ J}$
	$= 336 \text{ kJ}$
Heat energy released in burning 1 cm³ of wood	$= 15 \times 600 \times 10^{-6} \text{ MJ}$
	$= 9.0 \text{ kJ}$
Volume of wood required	$= 336 \div 9.0 \text{ cm}^3$
	$= 37 \text{ cm}^3$. (say, two 200 mm sticks)

Experience suggests that on an open fire many more than two thin 200 mm sticks would be needed. However, a well-designed enclosed stove using small pieces of wood could boil the water with as little as four times this 'input' – implying an efficiency of 25%.

Designing a stove or boiler that will make good use of valuable fuel requires an understanding of the series of processes involved in combustion. The first process, which consumes rather than produces energy, is the evaporation of any water in the fuel. Then more energy must be supplied to raise the temperature of the material to its ignition point, when it starts to burn. Once it starts burning, there are two stages, because any solid fuel contains two combustible constituents. The **volatile matter** is released as a mixture of gases and vapours as the temperature of the fuel rises. The combustion of these produces the little spurts of flame seen around burning wood or coal. The solid matter consists of the **char** together with any inert matter. The char, mainly carbon, burns to produce CO_2, whilst the inert matter becomes ashes, slag or clinker.

It is essential that the design of any stove, furnace or boiler ensures that the vapours that are released burn and don't just disappear up the chimney. Air must also reach all the solid char, which is best achieved by burning the fuel in small particles. This can raise a different problem, because finely divided fuel may produce finely divided fly ash or **particulates** that must be removed from the flue gases. The air flow should also be controlled: too little oxygen means incomplete combustion and leads to the production of poisonous carbon monoxide, while too much air is wasteful because it carries away heat in the flue gases. This is the reason that burning wood in an open fire is only about 10% efficient.

Modern systems for burning solid biofuels are as varied as the fuels themselves, ranging in size from small stoves through domestic space and water heating systems to large boilers producing megawatts of heat. Combustion efficiencies (heat output/heat content of biofuel) of over 85% can be achieved in well designed modern systems.

Pollutants from biomass combustion

Biomass combustion potentially produces a wide range of pollutants. The past few decades have seen many programmes in developing countries for the design and dissemination of improved stoves, with the joint aims of reducing both fuel consumption and smoke emissions inside houses. Emissions are still a major concern in the UK, particularly where Municipal Solid Wastes are burned.

The flue gases from solid biomass combustion may require specific treatment to remove unburnt particles, although the emissions to air from clean biomass sources such as wood are generally lower than for coal. The use of biomass through **co-firing** with coal in boilers for electricity generation, in effect 'dilutes' greenhouse gas and particulate emissions and so avoids the need to install additional pollution control equipment that would be needed for the continued combustion of coal alone (Faaij, 2006). Typically, around 5–10% of the input to these co-firing systems is biomass, but rates of up to 40% have been used, e.g. in Belgium and the Netherlands. Although fuel delivery systems need to be adapted or retro-fitted, only modest changes to existing coal boilers are required for co-firing, so the practice is highly cost effective.

In the case of biomass derived from municipal or industrial wastes, there are strict standards for the amounts of a range of pollutants that are allowed to escape from the plant. Pollutants of concern include sulfur dioxide, hydrogen chloride and fluoride, heavy metals such as cadmium and mercury, and complex organic compounds such as dioxins. In Europe, emissions from such heat and electricity plants are regulated under the Waste Incineration Directive.

Pyrolysis and gasification

These related processes involve heating biomass materials in such a way that they decompose to produce useful products, without complete combustion. **Pyrolysis** is the simplest and almost certainly the oldest method of processing one fuel in order to produce a better fuel, in the form of charcoal. Other, more controlled forms of pyrolysis, and similar techniques of gasification, have since been developed to produce solid, gaseous and liquid biofuels.

Charcoal and torrefaction

Charcoal is traditionally produced in the forest where the wood is cut. The 'kiln', consisting of stacked wood covered with an earth layer, is allowed to smoulder for a few days in the near absence of air, typically at 300–500 °C, a process now called **slow pyrolysis**. The volatile matter is driven off, leaving the charcoal (the 'char' component mentioned above). Charcoal is almost pure carbon with about twice the energy density of the original wood, making it easier and more efficient to transport and store. It burns at a much higher temperature, so it is much easier to design a simple and efficient stove for its use. From 4 to 10 tonnes of wood are needed for each tonne of charcoal, and if no attempt is made to collect the volatile matter, up to three-quarters of the original energy content can be lost. The process also releases vaporized tars and oils and the products of incomplete combustion into the atmosphere, making this charcoal fuel cycle potentially highly polluting (numerous projects and studies have attempted to improve the charcoal fuel cycle's energy efficiency and reduce emissions).

Although traditional charcoal production is highly inefficient, the product does at least have the great benefit of burning cleanly and efficiently. This is particularly important in the developing world where food may be cooked over an open fire or in an inefficient stove inside a home with poor ventilation. Various projects to introduce locally produced improved designs of charcoal stoves, such as the Kenyan 'jiko', are considered to have been successful. (For more detailed accounts, see UNDP, 2000 and Anderson et al., 1999.)

Torrefaction is the industrial-scale thermochemical treatment of biomass in the 200–340 °C range. In this mild pyrolysis process, the biomass (particularly the hemicellulose) only partly decomposes, giving off some of the combustible volatiles. The resulting solid torrefied biomass (sometimes referred to as biocoal) has an approximately 30% higher heat energy content, increased density and reduced content of undesirable elements such as chlorine in comparison to the original biomass. Torrefied biomass is also **hydrophobic,** that is, it does not absorb moisture from the air. It is particularly suitable for co-firing with coal, because it can be stored in the open for long periods in the same way as coal. It can also be transported more easily and cheaply than wood chips or pellets. Mitchell et al. (2007) speculated that torrefied pellets from wood residues could cost 4–6 € per GJ (£4–6 per GJ), compared to 12 € per GJ (£10.7 per GJ) for heating oil (assuming £1 = 1.15 €). Although the economics of the process have yet to be fully proved, torrefaction pilot plants and demonstration units are being commissioned in many parts of the world. There are also significant prospects for large-scale production and trade in torrefied biomass for co-firing in the world's existing fleet of coal power stations.

Pyrolysis

The traditional slow pyrolysis process that reduces wood to charcoal wastes a great deal of energy but the term pyrolysis is now normally applied to processes where the aim is to collect the *volatile components* and condense them to produce a liquid fuel or **bio-oil**. Since it is a characteristic of biomass that the volatile matter carries more of the energy than the char, this process should be more efficient. Pilot studies using pyrolysis of MSW and plastics wastes have suggested relatively high energy efficiencies.

The method involves heating the bio-material with a carefully controlled air supply. It must not burn, of course, and as the aim is a liquid product, gasification (see below) must be minimized. The resulting reactions are complex and hard to predict, giving a range of oils, acids, water, solid char and uncondensed gases, depending on the feedstock and operating conditions. The bio-oil product usually has about half the energy content of crude oil, and contains acid contaminants that must be removed. However, it can be used as an oil substitute for heating or power generation, or could be refined to produce a range of chemicals and fuels.

Variations on the basic process include **solvolysis**, the use of organic solvents at 200–300 °C to dissolve the solids into an oil-like product, and **fast pyrolysis**, requiring temperatures of 500–1300 °C and high pressures (between 50 and 150 atmospheres). Fast pyrolysis, with capture of the volatiles, is used in the production of commercial charcoal, but pyrolysis processes for liquid fuel production are mostly at the pilot or demonstration stage (Sims, 2002).

Gasification

As the name implies, **gasification** is a thermochemical process where a gaseous fuel is produced from a solid fuel. It is not a new process, as 'coal gas' or 'town gas', the product of coal gasification, was widely used in the UK and elsewhere for many decades. 'Wood gas' was used for heating, lighting, and even as vehicle fuel during the coal shortages of World War II, but both were superseded by natural gas by the 1970s. Rice husk gasifiers have been successfully operated in Indonesia, China and Mali (Manurung and Beenackers, 1990) and are increasingly promoted in India.

There are many different designs of modern gasifier but essentially two basic processes. First, the combustion of biomass in a restricted supply of air carries out the pyrolysis described above, releasing the volatile matter from the heated solid together with a mixture of methane, carbon monoxide and hydrogen, plus nitrogen from the combustion air. The tars in the gas stream can be condensed out, leaving **producer gas**. The energy content of the gas may only be 3–5 MJ m^{-3}, about a tenth of that of natural gas. However this is sufficiently good to run internal combustion engines.

This basic low temperature gasification process may leave a residue of char, mostly pure carbon. This can also then be gasified by burning it at a very high temperature (about 1000 °C) and injecting steam. This temperature is sufficiently high to break down the steam into hydrogen and oxygen (which combines with the carbon), producing a mixture of carbon monoxide, hydrogen and nitrogen. This is known as **water gas**. In practice, depending on the design of the gasifier, its operating temperature and its precise fuel, both producer gas and water gas can be made simultaneously, the necessary steam coming from the fuel itself.

Even higher quality gas can be made by using oxygen rather than air for combustion, but this has to be produced by the low temperature liquefaction of air. A complete industrial gasification process using oxygen can produce a stream of carbon monoxide and hydrogen, with any other impurities such as tars, ammonia and sulfur compounds being chemically removed. This is known as **synthesis gas**, or **syngas**.

From syngas, almost any hydrocarbon compound may be synthesized, including premium liquid fuels such as methanol. The first stage in the synthesis is to adjust the proportions of H_2 and CO to the ratio required in the desired product. For example, methanol is CH_3OH, and therefore needs two H_2 molecules for each CO converted.

In the **Fischer-Tropsch process**, named after the chemists who developed it in the 1920s, the two components are passed over a suitable catalyst at high temperature and pressure, and the product, initially formed as a gas, is condensed. (A catalyst is a substance that influences a chemical reaction without itself being changed.) The result, depending on the syngas composition and plant conditions, is a mixture of liquid and gaseous hydrocarbons.

The overall conversion efficiency from the energy of the solid fuel to that of the resulting materials varies widely, from as little as 40% or so in relatively simple systems, to 75% or more in well designed plant. Small gasification plants (<300 kW) are available commercially, often combined with gas engines driving small generators. Fuelled with locally sourced agricultural or forest wastes, demonstration plants in the range 10–30 MW have been in operation since the mid-1990s. However, such plant requires carefully standardized fuel, with a low moisture content in order to avoid production of tars and this has limited their popularity (Faaij, 2006).

If the output from a biomass gasifier can be cleaned and used to run the gas turbine, the possibility arises of a self-contained biomass integrated gasification combined cycle (BIGCC) system for local power generation (Box 4.9). Pilot demonstration plants fuelled by wood, bagasse, rice husks or maize stalks (corn stover), with electrical outputs in the 5–25 MW range, have been tested since the mid-1990s in the USA, Europe and China. For more detailed accounts of steam and gas turbines and gasification processes in general, see Everett et al., 2012. For descriptions of some biomass gasification plants, see Sims, 2002.

BOX 4.9 Power station turbine systems

Most of the world's power stations use the heat from burning fuel to produce hot, high-pressure steam for the **steam turbines** that drive the generators (Figure 4.16). Steam temperatures are limited to about 600 °C for technical reasons, so the maximum Carnot efficiency is about 65%, and the actual efficiency perhaps 45% (see Box 2.4 in Chapter 2).

Gas turbines, driven directly by the combustion products of a burning gas at 1000 °C or more, should have higher efficiencies, but the most significant improvement is achieved in the **combined cycle gas turbine (CCGT)** system. The gases leaving the gas turbine are still hot enough to raise steam for a secondary, steam turbine. In order not to corrode or foul the gas turbine blades, the fuel burned must be very clean. Nearly all present gas turbine and CCGT plants burn natural gas or vaporized light heating oil.

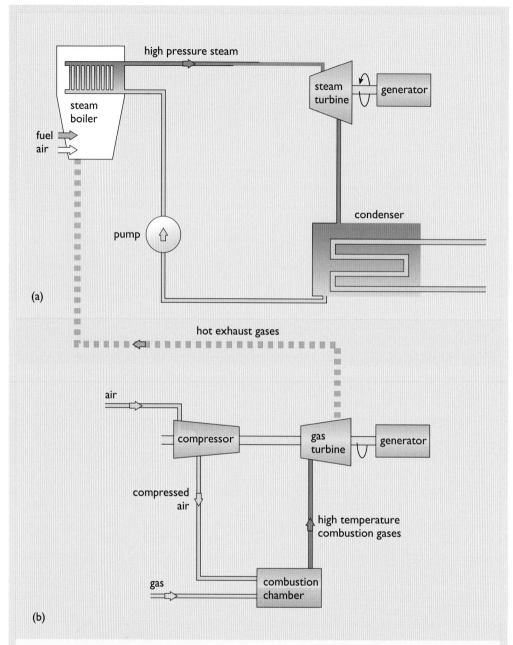

Figure 4.16 Types of generating system (a) conventional steam turbine, (b) simple gas turbine. In a CCGT plant the gas turbine exhaust gases (dotted line) replace the fuel/air input as the heat source for the boiler of the steam turbine

Hydrothermal processing

As Table 4.3 suggested, the complex carbohydrates, cellulose and lignocellulose, are relatively resistant to breakdown by biological processes, so are not readily accessible for further biological processing to more valuable biofuels. Mechanical treatment such as grinding increases the available surface area for biochemical processes and hydrothermal processing, using hot water under pressure, further breaks apart the lignified structures. **Steam explosion** is a pretreatment method aimed at removing lignin from the biomass to expose the cellulose and other polymeric carbohydrates (such as hemicellulose) for subsequent breakdown. Steam explosion usually involves heating the biomass with water under pressure to a temperature of around 200 °C. Then the pressure is suddenly released, causing the biomass to break and explode, exposing much of the lignin. The material after steam explosion still has a large amount of cellulose combined with lignin, which has to be further treated by processes such as **acid hydrolysis** and **peroxide** treatment to separate the remaining cellulose and lignin.

Obviously, such treatment requires energy, but can be valuable as part of an integrated process to increase the useful energy value of straw and other lignocellulosic materials. The lignin, which is relatively unaffected by hydrothermal treatment, can be dried and used as a solid fuel as part of the process (Inbicon, 2010).

Transesterification

Following his invention of the compression ignition engine that bears his name, Dr Rudolf Diesel demonstrated the use of a variety of vegetable oils. More have been tried since, but as in other cases, cheap crude oil came to dominate the market.

After extraction from the parent plant as outlined above, certain vegetable oils can be burned directly in some modern diesel engines, either pure or blended with diesel fuel, but most applications require minor modifications to the engine and fuel system. Upgrading of vegetable oils to **biodiesel** results in a fuel that can blend with or replace petroleum diesel in unmodified engines.

Vegetable oils and most fats are compounds called **triglycerides**, a form of **ester** whose large molecules are effectively various organic acids combined with **glycerol** (an alcohol). In biodiesel the main components are esters where the organic acids are combined with other, lighter alcohols rather than glycerol. The conversion process from vegetable oil to biodiesel, called **transesterification**, involves adding the alcohols, usually methanol or ethanol (both potentially derived from biomass) to the vegetable oil in the presence of a catalyst. This converts the triglycerides into esters of methanol or ethanol, together with free glycerol. Every tonne of biodiesel produces about 100 kg of glycerol, which is removed, and any excess alcohol extracted for recycling, leaving the biodiesel. Although no engine modification is required to use biodiesel its slightly lower energy content means that fuel consumption may be a few per cent higher. To avoid any potential problems of *waxing* where some components solidify in cold conditions, it may be used in blends (for example B10 indicates a blend of 10% biodiesel blended with 90% diesel).

Global production of biodiesel, from a wide variety of plant and animal oils and fats, is around 9 EJ per year and growing rapidly (IEA, 2007). In Europe, the potential of rape methyl ester (RME), which is made by combining rapeseed oil with methanol, has generated particular interest, with Europe accounting for around half of world production. In France all diesel contains 5% RME, and oil from rapeseed grown in the UK has been exported to both France and Germany. In the USA, production is mainly based on oil from soya beans and recycled cooking oil. In the UK, government policy has also encouraged biodiesel production from commercially-collected used cooking oils and animal rendering fats.

Figure 4.17 A car powered by vegetable oil against a background of oilseed rape fields

The estimated cost of producing biodiesel from rape seed at £300–350 per tonne was roughly 60p per litre, in 2010 (Scurlock, 2011) including credits for the glycerol and cattle-feed by-products. However, EU member states have been allowed since 1993 to adjust fuel tax levels to make biofuels more attractive. The UK government's Renewable Transport Fuels Obligation target for 2009/10, set at 3.3% of all UK transport fuels, was comfortably exceeded, with diesel containing around 4.5% biodiesel. Germany previously exempted biodiesel from fuel duty altogether, but has since scaled back its level of incentives. The European Commission Biofuels Directive proposed a target of 5.75% for the share of biofuels in the transport sector by 2010, although only a handful of member states achieved this. At the time of writing, the EU Renewable Energy Directive obliges all 27 member states to attain a 10% contribution of renewable energy in transport by 2020.

In countries with warmer climates, blends of up to 30% vegetable oil with diesel have been used without the need for transesterification. Coconut oil is used in tractors and lorries in the Philippines, palm and castor oil in Brazil and sunflower oil in South Africa. However, since the food and cosmetics industries can usually pay a better price, these applications are usually limited to places where diesel fuel is expensive and in short supply, or where fuel standards are less stringent.

The glycerol formed during biodiesel production is a potentially valuable by-product used in the food and cosmetic industries, although the price varies depending on whether there is a world surplus or not. An alternative use for the glycerol is as a source of biomethanol, using a thermochemical process similar to that involving syngas, described above. Biomethanol can be used to produce gasoline blends to replace petrol and, with four hydrogen atoms per molecule, methanol is also in effect a means of storing hydrogen. It can be converted to hydrogen for use in hydrogen-powered cars, or used in the car itself, where direct methanol **fuel cells** can convert methanol to hydrogen and generate electricity (see Chapter 10 for comments on fuel cells).

4.9 **Biochemical processing**

Biochemical processes rely on the use of microorganisms to convert biomass into more useful forms for bioenergy. The processes may also involve some conventional chemical and physical stages, but the essential stage is biological.

Anaerobic digestion

The process of anaerobic digestion (AD) is complex, but in outline, bacteria break down organic material into sugars and then into various organic acids which are further decomposed to produce **biogas**, a mixture of methane, carbon dioxide and trace gases, including hydrogen sulfide. The feedstock used may include dung or sewage, food processing wastes or discarded food, agricultural residues or specially grown silage crops that are harvested green. Digestion can take place in either wet or dry systems. In **wet** systems, the raw feedstock is usually converted to a slurry with up to 80–95% water, and fed into a purpose-built **digester** whose temperature can be controlled. The high throughput of water in wet anaerobic digestion systems may be a disadvantage, and in '**dry**' digestion systems the moisture content in the digester is much lower. This requires a higher input of energy to mix the material, but avoids the need to dispose of large volumes of water.

Digesters can range in size from perhaps one cubic metre (roughly 200 gallons) for a small 'household' unit in rural India or China, to around 100–1000 m³ for a typical farm plant (Figure 4.18) and more than 10 000 m³ for a large installation. The input may be continuous or in batches, and digestion is allowed to continue for a period of between ten days and a few weeks.

The bacterial action itself generates a small amount of heat, but in cooler climates additional heat is normally required to maintain the ideal process temperature of around 35 °C for **mesophilic digestion**, or 55 °C for more rapid **thermophilic** processing (the two processes involving different groups of microorganisms). In extreme cases all the gas may be burned for this purpose, but while the net energy output is then zero, the plant may still pay for itself through the saving in fossil fuel and other costs which would have been incurred to process the wastes by other means. The composition of the solid residue remaining after digestion (the **digestate**) depends on the system and the original feedstock, but it contains much of the nitrogen

Figure 4.18 On-farm digestion plant in Scotland

and other plant nutrients from the original feedstock, incorporated into the mass of the bacteria present and can be used as a fertilizer and soil conditioner.

In a well-run digester, each dry tonne of input will produce 200–400 m³ of biogas with 50% to 75% methane content, an average energy output of perhaps 8 GJ per tonne of input. This is only about half the fuel energy content of the dry dung or sewage, but the process nevertheless offers a valuable combination of a relatively high yield of clean fuel, reduction in environmental nuisance (such as odours, risks to water resources) and efficient recycling of valuable plant nutrients.

The biogas produced by a digester can be burned to produce heat or to fuel an internal combustion engine to drive a generator for electric power. Heat from the engine cooling water and exhaust gases can be used to heat the digester if necessary, or for other uses. If the biogas is upgraded by scrubbing and filtration to remove the carbon dioxide and hydrogen sulfide, the resulting **biomethane** is similar to natural gas and can be used as transport fuel or injected into natural gas pipelines for wider distribution. Cars and trucks using spark ignition engines can be converted to use both petrol and biomethane, as some water companies have done for their own vehicles.

Likely future developments in the AD industry include pre-processing of feedstocks to increase their biogas yield, and the post-processing of digestate to increase its value, e.g. by separation into solid and liquid fractions containing different proportions of nutrients. Research interest includes the use of microbes adapted to higher temperatures: extreme thermophiles offering faster throughput. Other research involves the use of **psychrophiles**, that is organisms adapted to lower temperatures that would reduce the need for digester heating in cold climates.

Anaerobic digestion in 2011

Large-scale centralized AD plants set up by European farmers' co-operatives digest manures and slurries together with other biodegradeable wastes, receiving income from both energy sales and gate fees for waste disposal. Such plants already supply some 3 PJ (0.8 TWh) per year in Denmark, while installed capacity has reached about 130 MW in the Netherlands (IEA, 2010b). The first large-scale AD plant of this kind in Britain was commissioned in Devon in 2002. Based on a design that had proved successful in Denmark and Germany, the Holsworthy plant initially used an annual 146 000 tonnes of slurry from 28 nearby farms, together with wastes from local food processors, to supply the heat input for a generating capacity of 2.1 MW. Digestate was transported back to the contributing farms and others for landspreading, but the relatively high cost of transporting manures meant that the Holsworthy plant subsequently changed to handling mostly food processing wastes.

The UK Department for Environment and Rural Affairs Implementation Plan for Anaerobic Digestion (DEFRA, 2009) gave a target of 1000 medium-sized, 500 kW, on-farm plants by 2020. These could utilize some 20–25% of UK recoverable manures, and would require about 100–125 000 ha of land in crop rotations for growing additional crops to provide silage feedstock. The area required represents a relatively moderate proportion of the total 18 Mha agricultural area. An additional 200 waste-based AD plants, typically of around 1.5 MW electrical capacity, could process about 10 million tonnes of organic feedstocks by 2020, or at least half of total UK food waste arisings.

Germany is the largest European producer of biogas, with 4500 AD plants in operation at the end of 2009, amounting to a total electrical generating capacity of around 1.7 GW. The majority of these AD plants have been developed in the agricultural sector, using manures and silage crops such as maize. Larger waste-based AD plants feeding biomethane into the natural gas network using small-scale equipment for upgrading and pressurizing the gas are of growing interest. Motor vehicle use includes both dedicated and dual-fuel vehicles adapted for compressed biogas (CBG) or cryogenic liquefied biogas (LBG). Ahrer (2008) has suggested that biomethane produced from the output of one hectare of agricultural land would power a suitable vehicle for twice as far as biodiesel derived from the same area.

Fiscal support for European agriculture has been a major contributing factor here. Employing around 11 000 people, the German biogas sector provides a useful model for the development of AD in other countries.

The developing world has seen many schemes for biogas plants during the past few decades. A major Chinese programme in the 1970s initially resulted in more than 7 million small-scale digesters, but suffered from many failures. A later drive, with better technology and supporting infrastructure, resulted in some 5 million domestic plants operating successfully by the mid-1990s. In India some 4 million biogas plants were installed by the end of 1998 (Figure 4.19) and a potential for 12 million has been identified (Ministry of New and Renewable Energies, 2010). However, in many developing countries the capital cost of a digester is out of reach of the typical small farmer, and attempts to introduce community biogas plants have met with mixed fortunes, largely due to difficulties with balancing 'ownership' of animal dung against credit in the form of biogas consumption. The Indian government provides cash subsidies for family-size biogas plants, with higher subsidies for poorer farmers. A typical rural digester costs about £200

Figure 4.19 Small scale biogas plant in India

(without subsidy) at 2011 prices and requires dung from four cattle. There are also similar programmes for biogas development in Nepal, Sri Lanka and other developing countries. With the introduction of the Clean Development Mechanism (CDM) under the Kyoto protocol, several large plants for power generation are also being developed in India and other countries.

Landfill gas

At present, the majority of the UK's AD capacity is actually in the landfill sector. A large proportion of municipal solid waste is material of biological origin (Figure 4.11), and its disposal in deep landfills furnishes suitable conditions for anaerobic digestion to occur naturally. It was known for decades that landfill sites produced methane, and systems were fitted to burn it off safely, but the idea of collecting and using this **landfill gas (LFG)** developed only in the 1970s.

The natural digestion process in a landfill (Figure 4.20(a)) takes place over years, rather than the days or weeks of in-vessel systems. In developing a landfill gas site, each area is covered with a layer of impervious material after it is filled, and the gas is collected by an array of interconnected perforated pipes placed at depths of up to 20 metres in the refuse (Figure 4.20(b)). In a large well-established landfill there can be several kilometres of pipes, with as much as 1000 m³ per hour of gas being pumped out.

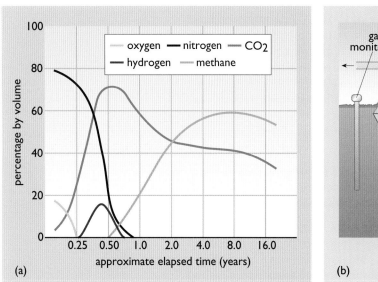

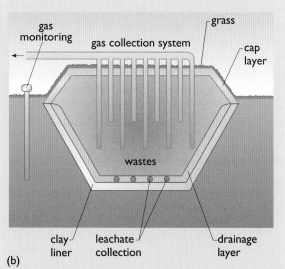

Figure 4.20 (a) The changing gas composition in a landfill site, (b) Extraction of landfill gas

In theory, the lifetime yield per tonne of waste in a good site should lie in the range 150–300 m³ of gas, with between 50% and 60% by volume of methane. This suggests a total energy output of 5–6 GJ per tonne of refuse, but in practice, at the average UK gas extraction rate, the heat energy output per tonne of wastes (as collected) is rather less than 2 GJ, with extremely variable and often unpredictable yields depending on the exact composition of the wastes and conditions in the site.

The gas may be used directly, to fire kilns, furnaces or boilers, but there are rarely enough large users close to a landfill site, so the gas is increasingly used for electricity generation. The generators are driven either by internal combustion engines or by gas turbines. Using several smaller engine-generators on one site allows better matching of gas supply to generating capacity, with individual units being removed as gas output falls (Figure 4.20(a)). Assuming a gas-to-electricity energy efficiency of 25%, this brings the overall energy efficiency of the system below 10%. A site containing a million tonnes of MSW might support an electrical capacity of perhaps 2 MW over a 15–20 year generating lifetime. Despite the low energy conversion efficiency, LFG plants have proven to be the most financially attractive of the systems in the UK and in 2008 were the largest single biomass energy source, supplying 66 PJ (Slade et al., 2010). However, the European Landfill Directive requires a reduction in the amount of waste going to landfill, so that in the longer term, the prospects for landfill gas look less favourable.

Rather than being landfilled, Municipal Solid Waste (MSW) can also be subject to more controlled anaerobic digestion. Under these conditions gas yields are much higher and digestion is complete within a matter of weeks rather than years. Feedstock for the digester is the organic fraction of MSW diluted into a slurry, possibly mixed with sewage. Figure 4.21 shows the full complexity of a system which first recovers useful materials from the MSW, then produces methane by digestion and finally generates electric power using the combustion heat of the residual solids.

Fermentation to produce ethanol

Fermentation is another anaerobic biological process but here the simple sugars from the biomass feedstock are converted to alcohol and carbon dioxide by the action of a different set of microorganisms, usually yeasts. The required product, **ethanol** (C_2H_5OH), is then separated from other components using heat to distil the mixture, so that the ethanol boils off and can then be cooled and condensed back to liquid. Bio-based ethanol (bioethanol) is most commonly used as an extender in gasohol, that is petrol (gasoline) containing a percentage of ethanol. Higher blends (typically 85% ethanol, 15% gasoline – designated E85) are used in 'flexible-fuel vehicles' (FFVs) in countries such as Sweden, the USA and Brazil – in Brazil many cars also run on 100% bioethanol. Ethanol can be burned directly in suitably modified and retuned petrol engines – most FFVs involve minor changes to the fuel-injection system, generally using a fuel mixture sensor communicating with the engine control unit. Fuels containing methanol use a similar nomenclature – M100 is a pure methanol fuel used in some racing cars.

Since fermentation requires sugars, an obvious starting point is sugar-cane, and this is the basis of Brazil's PRO-ALCOOL gasohol programme. Cereals, where the main carbohydrate is starch, require initial processing (**malting**) to convert the starch to sugar. This conversion occurs naturally

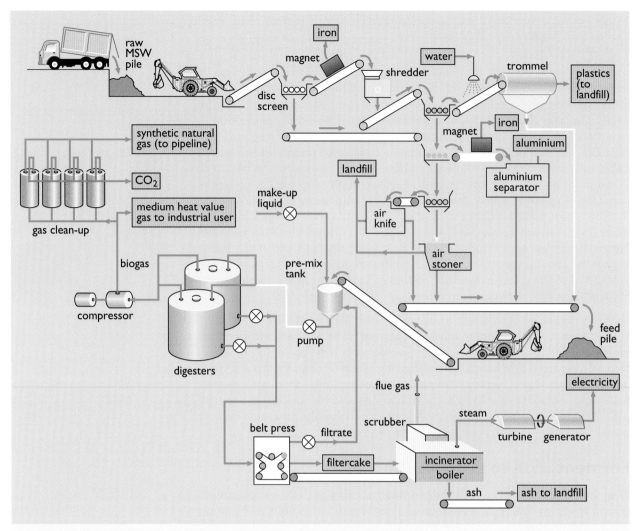

Figure 4.21 Integrated waste materials plant with facilities for recovery of metals and removal of plastics, followed by anaerobic digestion of the remainder. The solid residue from the digester serves as fuel for power production

when seeds germinate, so the seed is dampened to start germination and sugar formation, then dried rapidly to prevent decomposition.

The liquid resulting from fermentation contains about 10% ethanol. Distillation to increase the concentration requires a considerable heat input, usually supplied by crop residues. The energy content of ethanol is about 30 GJ t^{-1}, or 0.024 GJ per litre. The 360 litres of ethanol produced from a tonne of maize (Table 4.4) therefore has an energy content of 8.6 GJ. Comparing this with the 19 GJ t^{-1} gross energy present in the grain (Table 4.2) shows that the conversion efficiency of fermentation is relatively poor, but the technology is well developed and can be adapted to convert a range of inputs into a useful product. The only gaseous by-product is pure carbon dioxide which can be captured and stored for use in products such as carbonated drinks. It can also be used in greenhouses to stimulate the production of horticultural crops or even new sources of biomass.

Table 4.4 Ethanol yields from a range of crops

Raw material	Litres per tonne[1]	Litres per hectare per year[2]
Sugar cane (harvested stalks)	70	400–12 000
Maize (grain)	360	250–2000
Cassava (roots)	180	500–4000
Sweet potatoes (roots)	120	1000–4500
Wood	160	160–4000[3]

[1] This depends mainly on the proportion of the raw material that can be fermented
[2] The ranges reflect worldwide differences in yield
[3] The upper figure is a theoretical maximum

Sugar beet was the main feedstock for the first bioethanol plant in the UK , with an annual capacity of 70 million litres (NNFCC, 2007). About 650 000 tonnes of sugar beet from the 2006 UK harvest of 7.1 Mt was needed from existing growers (NNFCC, 2007). Two grain-based UK bioethanol plants each have the capacity to produce around 400 million litres annually from about 1 million tonnes of locally grown wheat grain. The 2006 UK wheat harvest was 14.7 million tonnes (NNFCC 2007; Royal Society, 2008).

Brazil's PRO-ALCOOL programme, producing ethanol from sugar residues, is the world's largest commercial biomass system. It was established in 1975, when oil prices were high and sugar prices low, and during its first 25 years, the fossil fuel imports that it replaced saved $40 billion directly in hard currency, with further savings from reduced interest on foreign debt. Production stagnated around 12–15 billion litres per year in the late 1980s and 1990s, but increased again following the introduction of flexible-fuel vehicles from 2003 onwards. In 2009, Brazil's ethanol production of around

(a)

(b)

Figure 4.22 (a) Sugar cane, (b) Brazilian ethanol production plant

25 billion litres was surpassed only by the USA (40 billion litres) and greatly exceeded EU production (just 4 billion litres). In 2008, Brazilians consumed more ethanol fuel than petrol for the first time, and Brazil now exports up to 4 billion litres annually to the USA, Europe and other markets.

The viability of ethanol production for bioenergy depends critically on world prices for sugar and crude oil, and on government policies. All of these factors have varied widely – and often rapidly – over the past 40 years. The USA has a long history of bioethanol production from **maize** (encouraged by Henry Ford, amongst others). But as with many other bioenergy enterprises, cheap petroleum destroyed the market. More recently, low prices for maize and the high price of oil triggered a major revival, centred on the main grain-growing states of the Midwest. Annual world production of ethanol for fuel has risen dramatically since 2000, from about 20 to nearly 90 billion litres (equivalent to about one million barrels of oil per day). Blends of ethanol with petrol are now either permitted, or even mandated, in many countries. An E10 blend containing 10% ethanol is possibly the most widespread and is mandatory in some US states.

Bioethanol can be used to produce **ethanol gel**, a clean-burning fuel that consists of bioethanol bound in a hydrated cellulose thickening agent. Cooking stoves specially designed for use with ethanol gel have been developed for sale both in developing countries and in European leisure/camping applications (there are also ethanol gel burners that can be retrofitted into several kinds of traditional African cooking stoves). In such appliances, ethanol gel is a highly controllable, easily lit cooking fuel with a heating efficiency of roughly 40%. Initial market penetration has taken place in several African countries, including Nigeria, Zimbabwe, Malawi and South Africa. Ethanol gel can act as a substitute for wood fuels and kerosene, reducing CO_2 emissions and indoor air pollution, thereby addressing the serious problem of clean, controllable, renewable heat for cooking in the developing world (UN Foundation, 2008). It is ironic, but perhaps industrially and economically necessary, that ethanol stoves are simultaneously growing in popularity as aspirational focal points for Western homes.

Another possible fermentation product that is attracting attention is **biobutanol**. This is produced from similar feedstocks and uses similar processes to ethanol fermentation. It also has some advantages over ethanol to the end user: it can be blended at higher concentrations with gasoline, is less susceptible to separation from the mixture in the presence of water and is better suited to current engines (BP, 2006).

Enzymatic conversion

Fermentation to produce ethanol requires some form of soluble sugar, but as we have seen, much biomass consists mainly of cellulose, hemicellulose and lignin. Following a pre-treatment of such material by, for example, hydrothermal processing, the watery suspension of cellulose and hemicellulose that results can be treated with enzymes, which are biological catalysts. These are usually derived from microorganisms such as bacteria or fungi, and are capable of breaking down the cellulose and hemicellulose to simpler carbohydrates that can be used in traditional fermenters. A combination of thermochemical and enzymatic conversion processes which allow cellulosic biomass to be used for the production of ethanol are now generally referred to as advanced or *second/next generation* bioethanol technologies. These new technologies could greatly increase the potential supply of feedstocks, while avoiding competition with food or fodder supply and more easily meeting emerging bioenergy sustainability standards.

An example of a pilot-scale second generation plant is the system being set up by Inbicon in Kalundborg, Denmark (Inbicon, 2010). This uses 4 tonnes per hour of straw as its feedstock, with a combination of mechanical and hydrothermal treatment to separate the cellulose and lignin. The cellulose is then subjected to enzymatic hydrolysis and fermentation to ethanol. After distillation to separate ethanol from the lignin and unfermented molasses, the lignin is dried for use as fuel in a co-generation boiler. Figure 4.23 is a schematic diagram of the process, which is designed to produce 4300 tonnes of ethanol, 13 000 tonnes of lignin and 11 000 tonnes of molasses per year.

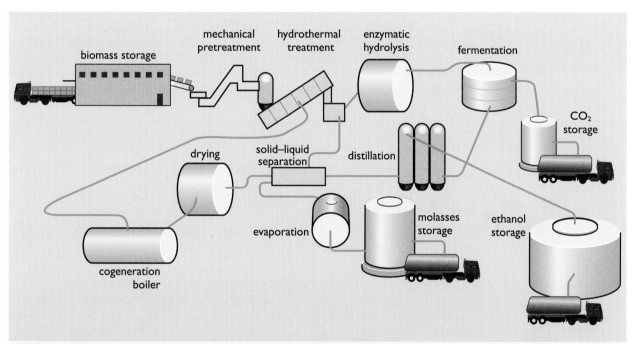

Figure 4.23 Schematic of the Inbicon integrated process for ethanol production from straw. The cogeneration boiler supplies steam to the hydrothermal treatment.

4.10 Environmental benefits and impacts

As we saw at the start of this chapter, the world's biomass plays a very basic role in maintaining the environment and providing our food, animal feed and fibre, so it is important to consider not only the *benefits* of using it for bioenergy but also the possible *deleterious* effects, global or local, of the harnessing of these natural processes. The following sections concentrate on some of the more significant benefits and impacts, considering first atmospheric emissions and then land use and energy balance.

Atmospheric emissions

Emissions to atmosphere are a major concern for energy systems. Emissions of carbon dioxide and methane are associated with climate change and other gases such as the oxides of nitrogen and sulfur also act as pollutants. Bioenergy systems have both positive and negative aspects with respect to these atmospheric emissions.

Carbon dioxide

The concept of 'fixing' atmospheric CO_2 by planting trees on a very large scale has attracted much attention. Halting deforestation would bring many environmental benefits, as would replanting large areas with trees. However, absorption of carbon dioxide by a new forest plantation is a once-and-for-all measure, 'buying time' by fixing atmospheric CO_2 while the trees mature, say for 30–60 years, after which the CO_2 would probably be released. A wider bioenergy strategy, concentrating on the substitution of biofuels for fossil fuels, may be a more effective lasting solution.

To analyse the benefits of substitution, it is essential to assess all the effects in a **life-cycle analysis (LCA)**. This involves considering all the inputs and outputs of carbon dioxide from the whole biofuel system, including growing the biomass, its conversion to useful product(s), transport (if necessary) to the site of end use, and effects associated with this end use. Figure 4.24 shows one possible interpretation of this, using various possible different

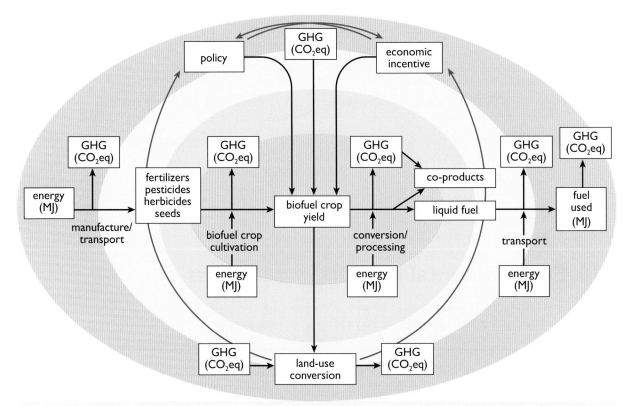

Figure 4.24 Different system boundaries for the life-cycle analysis (LCA) of biomass energy systems (Davies et al., 2009) (GHG = Greenhouse Gases, measured in terms of the amount of CO_2 that would have an equivalent greenhouse effect, CO_2eq)

system boundaries. So, we could analyse the inputs and outputs of just the biofuel crop system (the innermost coloured area in Figure 4.24), or we can widen the boundary to include the effects involved in producing the inputs for the crop, and in using the biofuel. The widest boundary includes the changes in carbon dioxide emissions that may result from land use changes caused by the use of biomass for energy, but there is also 'credit' for the CO_2 *removed* from the atmosphere by the growing crop. A life-cycle analysis tool of this kind, the Biomass Environmental Assessment Tool, is under development by government agencies for use in bioenergy sustainability reporting in the UK (Biomass Energy Centre, 2011b).

As an example of LCA, we can consider electricity generating plants that are either operating or near to commercial implementation at the time of writing. Table 4.5 shows the emissions of carbon dioxide and also of the two main contributors to **acid rain**, sulfur dioxide and nitrogen oxides. The data is *life-cycle emissions per unit of electrical output*, taking into account all the processes involved.

Even the best systems are not entirely carbon-neutral, but all the bioenergy systems, even MSW combustion, have lower CO_2 emissions than any of the fossil fuel plants.

Table 4.5 Net life cycle gaseous emissions from electricity generation systems in the UK

	Emissions[1]/t GW h^{-1}		
	CO_2	SO_2	NO_x
Combustion, steam turbine			
Poultry litter	10	2.42	3.90
Straw	13	0.88	1.55
Forestry residues	29	0.11	1.95
MSW (EfW)	364	2.54	3.30
Anaerobic digestion, gas engine			
Sewage gas	4	1.13	2.01
Animal slurry	31	1.12	2.38
Landfill gas	49	0.34	2.60
Gasification, BIGCC[2]			
Energy crops	14	0.06	0.43
Forestry residues	24	0.06	0.57
Fossil fuels			
Natural gas: CCGT[2]	446	0.0	0.5
Coal: with minimal pollution abatement	955	11.8	4.3
Coal: Flue Gas Desulfurization and low NO_x[3] burner	987	1.5	2.9

[1] Note that 1 g kWh^{-1} is the same as 1 t GWh^{-1}
[2] Biomass Integrated Gasification Combined Cycle. Wood is gasified and the gas used to feed a gas turbine to generate electricity, with the hot exhaust gas used to raise steam to power a further, steam turbine.
[3] Flue gas desulfurization is a process for removal of sulfur compounds after combustion, and special forms of burners can be used that minimize emissions of oxides of nitrogen.
Source: Adapted from ETSU, 1999

Other emissions

Nitrogen oxides (NO_x) are an inevitable product of the combustion, in air, of any fuel, because four-fifths of the air is nitrogen. High temperatures – in furnaces or internal combustion engines – increase NO_x production, and bioenergy systems will need to meet the same 'clean-up' requirements as those using fossil fuels. This also applies to the removal of particulates. The emissions of sulfur oxides depend on the sulfur content of the biofuel, which will vary with the particular characteristics of the feedstock concerned. Many biomass feedstocks contain very little sulfur, so emissions of SO_2 may be greatly reduced compared with potentially high-sulfur fossil fuels such as coal and oil. There may also be liquid effluents, from flue gas cleaning, for instance, that must be treated before release. Other sources of pollution include the fly ash residue from MSW combustion, which has a relatively high concentration of heavy metals and needs special disposal (e.g. in controlled or hazardous waste landfill sites).

Dioxins (complex, carcinogenic compounds formed during combustion) are a continuing source of public concern. However, it has been estimated that EfW accounts for only 0.1% of UK dioxin emissions, and a Swiss study (EDIE, 1999) found that domestic bonfires were a greater source than controlled MSW incineration in Switzerland. Both the UK and the EU are enforcing increasingly stringent emission standards and the installation of pollution control technology. However, there are concerns that the standards are not always maintained – and history shows that, as data improves, the accepted 'safe' levels of such pollutants tend to become lower and lower.

Methane

Methane is a powerful greenhouse gas and is produced from the anaerobic breakdown of biomass (either naturally or by human intent). A molecule of CH_4 is about 22 times as effective as a molecule of CO_2 in trapping the Earth's radiated heat. Collection of the methane released from anaerobic manure and slurry stores, and from landfills, and subsequent combustion effectively replaces each CH_4 molecule by a CO_2 molecule. The combustion of landfill gas was estimated to have reduced UK greenhouse gas emissions by the equivalent of some 70 Mt of carbon dioxide in 2002. Without this, total UK greenhouse gas emissions in that year would have been more than 10% higher. Likewise, the deployment of on-farm AD plants, processing about one-fifth of UK agricultural manures and slurries, could potentially reduce methane emissions by 0.6 Mt.

Table 4.5 showed that life-cycle CO_2 emissions are higher for the EfW route than for landfill gas, but with careful storage and efficient combustion, the methane emissions from MSW should be low. With landfill, it is never possible to collect all the gas, and there are inevitably methane emissions to the atmosphere. Depending on the collection efficiency, these could add the equivalent of another 100–200 g kWh^{-1} to the actual CO_2 emission shown in Table 4.5.

Land use

At some stage, all forms of biomass require a surface area of land or water for plant growth and using this land for non-biological forms of renewable energy rather than for biomass may do more to mitigate the impacts of CO_2 (Smith et al., 2000). Consider, for instance, the area of land needed for an annual electrical output of 10 GWh, the equivalent of a small 1.5 MW thermal power station running with a 75% capacity factor. Under typical UK conditions, an array of photovoltaic (PV) modules as a 'solar farm' might need an area of some 30–35 ha to provide this, and a small wind farm might need approximately 100 ha, (including necessary separation distances). Apart from the built structures, most of the land could still be used for grazing, for free-range poultry or other purposes. With reasonable yields and conversion efficiencies, the land area of energy crops required to fuel this power plant would be in the range 600–900 ha (6–9 km^2).

Land area is not the only consideration, and in any case, the above three systems are unlikely to be competing for the same land. PV arrays could be deployed to considerable advantage in semi-arid and desert areas with high solar input, or on rooftops in urban areas.

In the case of energy crops, particularly maize and some of the oilseeds, there have been widespread concerns about their use for biofuel rather than for food. With the explosion of interest in the US use of maize for ethanol production in the first decade of this century, it was claimed that this diversion reduced the supply of maize flour as a staple food in Mexico, leading to food riots. It has subsequently appeared that the shortage of maize flour may have been more the result of hoarding and speculation by large investors, but the suspicion remains that there may be conflict between food and biofuel. A reduction in biological diversity through conversion of existing vegetation to fuel crops is another concern. Any such direct losses could be amplified by the use of pesticides in bioenergy crops, although such crops generally do not require such rigorous quality control as do food crops. Some bioenergy systems such as short rotation forestry or coppice can actually increase biodiversity compared to conventional agriculture. Field et al. (2008) suggest that, to preserve biodiversity, only land that has been abandoned from agricultural use should be used for bioenergy production.

Effects on soil also need to be considered. Soils can contain large amounts of organic matter or humus, formed from the *recalcitrant* remains of crops etc. that are hard to break down, including material such as lignin. The humus acts as a store of large amounts of carbon over long periods (tens to hundreds of years) if the soil is undisturbed. Increasing the proportion of stable carbon-containing material in soils has been suggested as one means of reducing atmospheric CO_2 levels. Soil organic matter is also essential to the maintenance of **soil structure**. The humus fraction of soil includes gums and other compounds that bind together smaller soil particles and improve the drainage characteristics of the soil as well as acting as a store of plant nutrients. However, cultivating the soil exposes the organic material to the air, and it then begins to break down more rapidly, releasing carbon dioxide. This loss of carbon is more of a problem with annual crops than with perennials such as trees, because of the higher frequency of soil disturbance. This may be partly addressed by newer reduced-tillage forms of cultivation. The emerging national and international sustainability standards for liquid transport fuels and solid bioenergy incorporate measures to exclude the use of land with previously high carbon stocks, whether in the form of soil organic matter (e.g. tropical peat soils) or in vegetation (typically forest).

A concept that has received considerable recent attention is the use of **biochar** that can be produced as a co-product of pyrolysis or gasification of biomass. Rather than using this char as a fuel, it can be added directly to soils, and there are claims that this both acts as a long-term store of carbon and improves soil fertility. The value and practicability of this approach are subject to continuing research (Shackley et al., 2011). Indeed, in 2011 a specific biochar research centre was opened at the University of Edinburgh. An essential problem is that, at the time of writing, there is no financial incentive to store carbon from biochar rather than use it as an energy source.

Energy balance

The terms **energy balance, energy payback ratio**, or **fuel energy ratio** are used to describe the relationship between the energy output of a system and the energy inputs needed to operate it (usually from fossil fuels). The concept came to the fore when doubts arose concerning some of the early fuel-from-biomass projects introduced following the oil price increases of the 1970s. There were claims that, when a full life-cycle analysis was undertaken, the fossil-fuel energy input for some schemes was actually greater than their bioenergy output.

The ratio of output to input will of course depend on the type of system, and the extent of the processing involved. In particular, ratios will normally be lower if the final 'output' is electricity, because of the inherent limit on the Carnot efficiency (see Box 2.4) of heat engines used in generation.

Davies et al. (2009) examined the fuel energy ratio (FER) for a selection of biofuel crops that had been reported in the literature up to 2008 (Table 4.6). The FER is the ratio of the useful energy in the fuel to the total amount of fossil fuel used in producing that fuel including that used to construct equipment. Where the ratio is less than one, more energy is used to produce the biofuel than is available from that fuel. Ratios vary from over 5 to less than one but the authors also make the point that there is little consistency among LCA estimates of biofuel energy efficiency.

Over the full range of renewable sources, the energy ratio of output to input can vary from less than 1:1 to as much as 300:1 (this for some hydroelectric plants). Woody energy crops perform well, with ratios between 10:1 and 20:1 on a heat output basis, but biodiesel may achieve only 3:1. Ethanol from grain may barely break even at just over 1:1 in the worst case. When wastes are the input, the question arises of how much of the energy input to the whole process should be attributed to the energy extraction system as opposed to waste disposal. Disposal of the waste necessarily involves expenditure of energy and the additional expenditure for energy recovery may be small. This results in a high energy ratio in terms of output to the additional energy used. A value of 30:1 for electricity from woody sawmill wastes is an example.

Table 4.6 The range of fuel energy ratios for selected bioenergy systems reported in the literature to 2008

| | Fuel energy ratio | |
	Lowest	Highest
Lignocellulosic crops (generalized)	1.8	5.6
Switchgrass	0.44	4.43
Corn	0.69	1.95
Miscanthus (combustion)	1.16	1.16
Miscanthus (gasification)	0.99	0.99

Only one set of data for miscanthus systems was available.
Source: derived from Davies et al., 2009

Energy ratios can be improved by good design of the production and processing systems to ensure that no energy is wasted. The co-products, such as straw or bagasse, associated with biofuels can be used to replace fossil fuels in supplying heat for the processes involved. The use of bagasse instead of coal or natural gas to provide process heat for ethanol production from sugar cane is a classic example.

The energy balance of a biomass energy system also affects its environmental impact. The greater the final energy outputs, the greater the quantity of fossil fuel displaced. The lower the fossil fuel inputs, the lower the extra demands put upon the environment by the biomass system.

4.11 Economics

It is not possible to treat the economics of all of the various biomass materials and conversion processes in any detail, so this discussion is limited to some key examples which illustrate specific issues relating to biofuels.

Energy prices

Any energy source must have a cost to the final user that is at least comparable with that of competing sources (which currently means fossil-fuel based systems), although where the biomass system offers significant advantages such as a reduction in pollutants, it may be able support a higher price. Table 4.7 offers a summary of UK energy prices in 2010, priced per unit of energy delivered.

Table 4.7 gives a baseline against which any bioenergy system needs to be compared. At the time of writing, many Governments across the world are introducing policies that will tend to increase the price of fossil fuels and favour the adoption of renewables, including biomass. For example, in 2010, the UK introduced favourable feed-in tariffs for electricity generated from renewables, and a similar scheme for renewable sources of heat began in November 2011. The succeeding sections look at the costs of some different bioenergy schemes, but all such costings are liable to very rapid change.

Table 4.7 Average UK retail fuel prices, 2010

	Large industrial users (excluding tax)[1]/p kWh^{-1}	Domestic consumers (including taxes)/p kWh^{-1}
Coal	1	3.5
Oil	3.25	3.2
Natural gas	1.5	4.2
Petrol, diesel	No data	~8
Electricity	8.5	13

Source: DECC, 2010b

[1] The fossil fuel prices paid by large consumers are directly related to bulk world prices. Competition between UK electricity generators means that industrial consumers are able to negotiate electricity contracts at low prices (see Chapter 10). The subject of comparative fuel prices is discussed in more detail in Everett et al. (2012)

Costing bioenergy

The four main factors that determine the cost of energy from any system are described in Appendix B. For most renewable energy systems, the initial *capital cost* (including the cost of borrowing the money) is a major component. Unlike many other renewable energy technologies, bioenergy systems can also have significant *fuel* costs. Energy crops, for example, must be planted, fertilized, protected against weeds and pests, harvested and transported. On the other hand, EfW may have *negative fuel costs* in the form of savings in payments for disposal of wastes.

The remaining two factors are common to most energy systems. *Operation and Maintenance* (O&M) costs are usually proportional to the output of the plant, and will depend on the type of fuel – in particular, the nature of its emissions and the residues it leaves. It is usually assumed that the *decommissioning costs* of bioenergy plant will be covered by the scrap value of equipment.

It is also worth noting that for any electrical plant, the cost per kWh of output depends on the annual output, so it is important to maximize the capacity factor (Box 5.1 in Chapter 5).

Electricity from wastes

The economics of electric power from **landfill gas** are relatively simple. The gas itself is a waste product that must in any case be collected and flared off to protect the environment and prevent explosions. So the marginal additional cost of piping it to a gas engine is likely to be small. The main costs are thus the capital cost of the engine and generator and connection to the grid, together with a modest allowance for O&M.

Owners of sites can charge increasingly high **gate fees** to accept waste, particularly near large cities, which operators can offset against the additional costs of generating electricity from landfill gas.

With landfill sites in ever-shorter supply, the cost of transport to distant sites is also rising, with some UK councils paying more than £20 per tonne to transport material to landfill. Many countries also impose a **landfill tax**, levied on every tonne taken to landfill. The EU waste directive is intended to reduce the use of landfill, and in the 2007 UK Budget the Chancellor announced that the landfill tax would increase more quickly and to a higher level than previously planned. Failure to meet targets for the reduction of waste going to landfill after 2013 is expected to attract fines of up to £150 per tonne. These changes together should, in the longer term, reduce the amount of waste going to landfill. This suggests that the potential for landfill gas in Europe will probably reduce in the long term.

The obvious, but generally unpopular, alternative to landfill is incineration of wastes. In 2010, there were 21 waste incinerators in operation in the UK, with another 11 approved for construction. Modern EfW generating plant must meet strict requirements on emissions, residues and other environmentally sensitive issues. This requires complex and expensive flue gas cleaning equipment, making up 30–60% of the capital cost of the plant, and UK estimates in recent years have suggested capital costs of over £4000 per kW for 65 MW electricity generating capacity (Rawlinson and Hicks, 2010). Such a plant could process 600 000 tonnes of refuse per annum. Incinerators dispose of refuse producing saleable electricity. However, the electricity sales do not cover the full overall cost. The overall cost of incineration using EfW ranges from £65 to £136 per tonne of refuse (equivalent to an extra cost of about 10p per kWh). This could be offset in part by the gate fees charged for accepting waste for disposal. The use of waste heat from the EfW plant for district heating may be a useful addition.

Public acceptability and the enthusiasm of local authorities for alternative waste strategies may be at least as important as the simple economics. In almost all cases where EfW plant has been proposed, there has been local opposition based around health fears, particularly associated with dioxins. The most recent statement on this from the UK Health Protection Agency is as follows.

> While it is not possible to rule out adverse health effects from modern, well-regulated municipal waste incinerators with complete certainty, any potential damage to the health of those living close-by is likely to be very small, if detectable.
>
> (Health Protection Agency, 2009)

Despite these reassurances, and the fact that emissions from large incinerators are now a tiny fraction of the total dioxin exposure, poor public acceptability and the delays in planning approval that this causes were cited as the major barrier to increased use of EfW technology.

The capital costs of power plants based on conventional boiler technology for clean forestry wastes or straw range from £2500 to £3500 per kW at current prices. For plants burning specialized waste-derived fuels, such as poultry litter, the costs are higher. There are also fuel costs, for transportation, storage, etc., which are obviously lower for smaller-scale plants close to the waste sources. Economies of scale, including higher generation efficiencies, can counterbalance this for larger plants.

Ethanol from biomass

Although this is probably the most established biofuel technology, especially in Brazil, the economics (and even the energy efficiency) of the process have been the subject of much controversy, especially in the USA.

US Government policy in the early twenty-first century sought strongly to encourage the production of ethanol from corn (maize), and there was a major response to this for a period. However, by 2008, it appeared that many of the plants that were producing ethanol were barely breaking even and opponents of corn ethanol were vocal in condemning the whole programme. It is difficult to obtain authoritative data for the economics of the process, and the actual data changes rapidly with changes in grain and fossil fuel prices, but Table 4.8 gives one set of figures for the Midwest in 2006/7. The major components of cost are the maize feedstock, followed by the cost for natural gas used in the process.

Table 4.8 Cost of production at seven Midwest corn ethanol plants, 2006/07 ($ per US gallon)

Item	Cost/$
Electricity	0.025
Natural gas	0.190
Denaturant	0.070
Enzymes etc.	0.063
Labour and management	0.051
Maintenance and repair	0.019
Miscellaneous	0.037
Total processing	0.450
Feedstock	1.063
Distillers grains (for animal feed)	−0.229
Operating costs	1.288
Estimated capital cost	0.350
Estimated total cost	1.640

Source: Perrin, 2009

By 2008, a posting on The Oil Drum website suggested that the average cost per US gallon of bioethanol had increased as a result of increases in fossil fuel price and particularly an increase in the maize price, to $2.68 per US gallon of ethanol produced (The Oil Drum, 2008). The problem of volatility of prices is particularly acute for ethanol, because of changes in US agricultural and energy policy and the direct competition between food and fuel uses of maize. Other biomass sources are likely to be less affected, but the tie-in between fossil-fuel prices, food prices and the economics of biomass energy may still be problematic.

Energy crops

The issue of land for bioenergy crops has already been mentioned, and assuming that efficient technologies for conversion or direct use of biomass can be developed, the future development of bioenergy crops is then likely to depend on the economics of alternative land uses. The obvious competitor on land that is easily cultivated is food crops. Given an expanding world population, and changes in diet such as increased meat consumption, that may reduce the effective output of food per hectare, it seems unlikely that this competition will decrease. At the time of writing, there is widespread concern about world food supplies, although it has to be said that such concerns surface regularly, and so far the situation has usually improved in the interim. However, it may be sensible for increased use of biomass for energy to rely on land that is marginal for conventional agriculture, or is currently not utilized for production. For this to be brought into use, the costs and potential income need to be realistic.

An example is provided by an analysis of the potential of various tree species for short rotation forestry in the UK (LTS International, 2006). Expenditure is needed for preparation of the ground for the crop, for fencing and for herbicides etc. that need to be used during the growth of the trees. With trees planted on a 2 metre by 2 metre grid, the total cost for establishing the crop was estimated at £2841 per hectare. For short rotation coppice, planted more closely and so needing less weed control in the early stages, the cost of establishment was about £1200 per hectare. These costs represent a major investment for the operator and the profitability of the resulting crop depends critically on the yield per hectare, the time taken before the crop can be harvested, any government financial support that is available and the final price for the crop.

At a wood price to the producer of £10 per fresh tonne, the LTS study suggested that none of the potential tree species would provide any profit without a substantial government subsidy to cover establishment costs. The situation at £25 per fresh tonne was better, but without government funding, the return on the growers' initial investment was on average only around 3% per annum. Given the risks involved in such an investment, it is unlikely that many would take this risk without an initial subsidy. For an acceptable return on the initial investment, subsidy levels of around £2000 per hectare to cover initial costs would probably be needed. The lower establishment costs of short rotation coppice suggest that the economics of this should be more positive.

To set these figures in context, the average annual expenditure per hectare on larger cereal and livestock farms in 2003–4 in the UK was around £300 (Centre for Rural Policy Research, 2005). Total sales were approximately £900 per hectare, leaving an annual margin of £600 per hectare.

Some of the risks inherent in this area are illustrated by the ARBRE (Arable Biomass Renewable Energy) power station in North Yorkshire, UK. This would have been one of the world's most advanced energy crop systems, a biomass integrated gasification combined cycle (BIGCC) plant running on the gas from a wood gasifier (see Box 4.9). Fuel for the initial 8 MW pilot plant was planned to come from some 2000 ha of SRC. Unfortunately, despite an EU grant and support under the UK Renewables Obligation scheme, the project folded due to financial and technical problems. This left investors with no immediate market for the coppice material. The development of energy crops is thus critically dependent on the price realized for the product, which will depend on the price of competitive fossil fuels, and on government policy and financial support. The politics of this are difficult to predict!

The general picture

In 2008, the UK Royal Society produced estimates of current and possible future costs of a range of biofuels, as shown in Table 4.9. Ethanol from sugar cane is much the cheapest biofuel, with biodiesel from surplus animal fats a reasonably close second. The use of animal fats, only available as a co-product from meat processing and catering, raises some ethical and energetic issues, but does make the point that use of residues is generally economically as well as energetically favourable. The price of fossil fuels in 2030 is unlikely to be less than the quoted 2006 price, which suggests that many of the biofuels could reach a comparable basic price to the consumer, depending on the taxation regime that is imposed.

Table 4.9 Estimated costs of biofuels and the costs of fossil fuels. Biofuel prices in US cents per litre exclude any possible taxes

Fuel type	2006 (price/¢ l^{-1})		2030 (price/¢ l^{-1})	
	Low	**High**	**Low**	**High**
Petrol excluding tax	35	60		
Petrol including tax (Europe)	150	200		
Petrol including tax (USA)	80	80		
Ethanol from sugar cane	25	50	25	35
Ethanol from maize	60	80	35	55
Ethanol from sugar beet	60	80	40	60
Ethanol from wheat	70	95	45	65
Ethanol from lignocellulose	80	110	25	65
Biodiesel from animal fats	40	55	49	50
Biodiesel from oilseeds	70	100	40	75
Fischer-Tropsch synfuels	90	110	70	85

Derived from Royal Society, 2008.
Only one figure for US petrol price was used.

The Royal Society data suggests that ethanol from lignocellulose is ultimately likely to be comparable in cost to ethanol from sugar cane. Lignocellulose is available in most areas, unlike sugar cane, which suggests that it may be the biofuel with the greatest potential. The science and technology of lignocellulose conversion is relatively well developed and full-scale demonstration and commercial plants are in operation.

4.12 Future prospects for bioenergy

In the 27 countries of the European Union, which are probably typical of the **industrialized countries** of the world, about 12% of final energy in 2009 came from renewable sources (REN21, 2011). In many of these EU countries, large-scale hydroelectricity accounts for the majority of renewable supply, but the next largest contribution is usually from bioenergy, including the combustion and anaerobic digestion of wastes, and accounts for most of the remainder (with the exception of those countries with a significant geothermal resource).

Hydroelectricity has limited scope for expansion in much of the industrialized world (see Chapter 5). Although wind and solar photovoltaic power is expanding rapidly, the present total contribution from solar, wind, tidal or wave energy is still low. Growth in renewable energy supply in the *near* future may come from increased direct combustion of a wide variety of biomass feedstocks for heat and electricity generation, together with a substantial increase in the supply of 'first generation' liquid transport biofuels. In the longer term, newer biomass technologies, such as electricity from high-efficiency combined-cycle plants or fuel cells, and next generation liquid biofuels, could play an increasing role – but biomass may by then be overtaken in the developed world by fast-developing renewables such as wind and solar power. New sources of large-scale demand for biomass may continue to be significant, in particular, demand for biocoal formed by torrefaction of a range of biomass feedstocks. Coupling biomass or biocoal power plants to carbon capture and storage could offer a form of carbon-negative electricity production, since the growth of the biomass would remove carbon dioxide from the atmosphere.

In the UK, a study by the Imperial College Centre for Energy Policy and Technology (Slade et al., 2010) reviewed available estimates for the potential contribution of biomass resources to UK energy supplies. Figure 4.25 shows the range of estimates they encountered for different time periods.

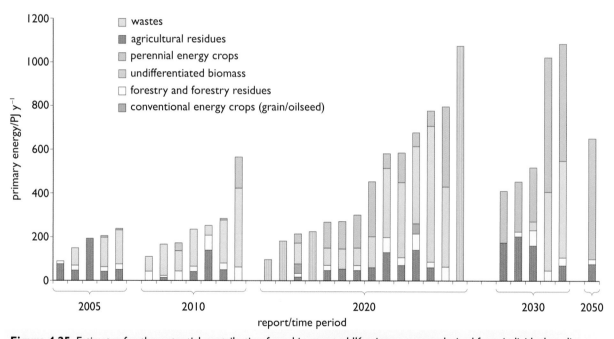

Figure 4.25 Estimates for the potential contribution from biomass to UK primary energy derived from individual studies (each bar represents a different data set). Estimates have been grouped by time period. Studies include different resource categories and encompass many definitions of the potential (source: Slade et al., 2010)

There was a wide spread in estimates of the biomass contribution, even for 2005 and 2010 where real data was available. Many of the studies suggest that it may be difficult to increase the potential supply of biomass energy to the UK by 2020, although more optimistic studies suggest a possible doubling. A rise from some 250 PJ per annum to perhaps 600 PJ by 2030 seems to be regarded as possible. The two sources suggested as having most potential for increases in production are wastes and perennial energy crops such as SRF/SRC. (More optimistic estimates of the potential bioenergy contribution to UK energy needs were given in the UK Government's *Bioenergy Strategy* (DECC, 2012).)

In the developing countries, future demand for biomass energy is likely to be determined mainly by two opposing factors: the rise in overall energy demand due to increasing population, and the reduction in the demand for traditional forms of biomass due to a shift to more 'modern' forms such as ethanol gel. Most projections suggest that biomass consumption will continue to rise during the next few decades, but at a lower rate than renewables in general or total primary energy. Nevertheless, it is estimated that traditional biomass will remain the sole domestic fuel for 2.8 billion people, in 2030 (IEA, 2010c). In purely quantitative terms, therefore, it is arguable that the improvement of traditional wood and charcoal stoves should be the most important technical aim in the field of bioenergy today. Better and more efficient utilization of non-wood biomass, e.g., rice hulls, stalks etc is another important technical aim, because of the large quantities of these available.

In the medium term, population growth and changing diets are likely to increase competition for land between needs for energy and for food, although much of this tension could be mitigated by investment in better infrastructure, technology, inputs and other resources for both agriculture and bioenergy. It has been suggested that, after allowing for increased food production for a growing world population, as much as 400–700 million additional hectares could be available for energy crops worldwide in the year 2050, without unacceptable loss of biodiversity (UNDP, 2000). Other authors (e.g. Field et al., 2008) suggest a much lower figure. Table 4.10 summarizes the potential bioenergy contributions in future years according to a range of authors. Most foresee a diminishing contribution from traditional biomass, i.e. the direct combustion of wood, dung, etc., with the growth coming from increasing biomass-fuelled generating capacity, and in the longer term, from liquid biofuels, or even a biomass-based hydrogen economy.

Table 4.10 Potential future world roles for bioenergy

Source, date	Time frame (year)	Bioenergy contribution	
		EJ per year	Percentage of total global primary energy
UNDP[1], 2000			
Shell, 1994	2060	220–222	15–22%
WEC, 1994	2050	94–157	~15%
	2100	132–215	11–15%
IPCC, 1996	2050	280	50%
	2100	325	46%
Hoogwijk et al., 2003	2050	33–1130	5–190%
Moreira, 2006	2100	164	15%
Ladanai and Vinterbäck, 2009	2050	1135–1548	~200%
Beringer et al., 2011	2050	126–216	13 to 22%

[1] UNDP, 2000 contained estimates from a number of sources.

The most optimistic recent forecast suggests over 1500 EJ per year, way beyond the total energy use at the time of writing. A more widely accepted figure might be of the order of 100–200 EJ of bioenergy per year. As we have seen, traditional biomass currently contributes an estimated 30 EJ to world primary energy, with a further 9 EJ or so from 'new' bioenergy (mainly in the form of electricity). 200 EJ would therefore represent a five-fold increase, and a new contribution equal to over a fifth of present world total primary energy consumption.

The future for renewable energy on the world scale is discussed in more detail in Section 10.10 of this book, and it is worth noting that each of the scenarios described there sees an increasing contribution from biomass.

4.13 **Summary**

Bioenergy is the general term for the use of living, or recently living, material as a source of energy for human activities. All bioenergy use depends ultimately on the process of photosynthesis, whereby solar energy is used to take carbon dioxide from the air and convert it into energy-rich carbon compounds. Subsequently, these initial products of photosynthesis can be converted into other energy-rich forms.

Materials used for bioenergy can be subdivided broadly into woody, cellulosic and oil-rich plant materials, microbial products and carbon-containing wastes derived from living materials.

Physical processes of size-reduction, pelleting or extraction are used to convert these raw materials into more useful forms. They can also be subject to thermochemical processes, either directly to produce heat energy, or through pyrolysis or gasification to produce more desirable solid, liquid or gaseous biofuels. Biochemical processing, involving microorganisms or enzymes, can also be used to convert some biological materials into liquid or gaseous biofuels. Biofuels can be used to produce heat, or to power engines.

Bioenergy processes are inherently carbon neutral, since all the carbon that is released into the atmosphere when they are used was itself taken from the atmosphere only a relatively short time before. The energetic and economic effectiveness of different processes varies, as does the usefulness and acceptability of different products to consumers.

The various processes and products are illustrated with a range of practical examples, examining the factors likely to affect the uptake of bioenergy systems – although the technologies likely to be adopted will, in practice, depend crucially on governments' fiscal policies, on the physical/biological/economic situation in a given country and on the world fossil fuel price.

References

AEA (2011) *2011 Guidelines to Defra/DECC's Greenhouse Gas Conversion Factors for Company Reporting*, Excel spreadsheet, available at http://www.defra.gov.uk (accessed 14 November 2011).

Ahrer, W. (2008) *Bio-methane and bio-hydrogen from organic by-products and their applicability as transport fuels* [online], Profactor, http://www.a3ps.at/site/images/stories/a3ps_allgemein/A3PS_TPF_2008/Biofuels/02_ahrer_tpf2008_05_28.pdf (accessed 20 January 2011).

Anderson, T., Doig, A., Rees, D. and Khennas, S. (1999) *Rural Energy Services: A Handbook for Sustainable Energy Development*, London, Practical Action.

Beringer T. D., Lucht, W. and Schaphoff, S. (2011) 'Bioenergy production potential of global biomass plantations under environmental and agricultural constraints', *GCB Bioenergy*, vol. 3 [online] doi: 10.1111/j.1757-1707.2010.01088.x (accessed 4 April 2011).

BFM (2005) *Benchmarking wood waste combustion in the UK furniture manufacturing sector* [online], London, Association for British Furniture Manufacturers, BFM Ltd   http://www.bfmenvironment.co.uk/images/BFM%20Wood%20combustion%20benchmarking%20-%20full1.pdf (accessed 31 March 2011).

Biomass Energy Centre (2011a) *Typical calorific values of fuels* [online], http://www.biomassenergycentre.org.uk/portal/page?_pageid=75,20041&_dad=portal&_schema=PORTAL (accessed 18 May 2011).

Biomass Energy Centre (2011b) *BEAT2* [online], http://www.biomassenergycentre.org.uk/portal/page?_pageid=74,153193&_dad=portal&_schema=PORTAL (accessed 12 January 2011).

Björheden, R. (2006) 'Drivers behind the development of forest energy in Sweden', *Biomass and Bioenergy*, vol. 30, no. 4, pp. 289–95.

BP (2006) *Biobutanol fact sheet* [online] http://www.bp.com/liveassets/bp_internet/globalbp/STAGING/global_assets/downloads/B/Bio_biobutanol_fact_sheet_jun06.pdf (accessed 16 June 2011).

BP (2010) *BP Statistical Review of World Energy 2010*, London, The British Petroleum Company; available at http://www.bp.com (accessed 3 October 2011).

Centre for Rural Policy Research (2005) *Farm Management Handbook 2005*, Exeter, University of Exeter.

Clarens, A. F., Resurreccion, E. P., White, M. A., Colosi, L. M. (2010) 'Environmental life cycle comparison of algae to other bioenergy feedstocks', *Environ. Sci. Technol.*, vol. 44, no. 5, pp. 1813–19.

Davies, S. C., Anderson-Teixeira, K. J. and DeLucia, E. H. (2009) 'Life-cycle analysis and the ecology of biofuels', *Trends in Plant Science*, vol. 14, no. 3, pp. 140–6.

DECC (2010a) *Energy statistics: renewables* [online], http://www.decc.gov.uk/en/content/cms/statistics/source/renewables/renewables.aspx (accessed 29 March 2011).

DECC (2010b) *Quarterly Energy Prices, December 2010* [online], http://www.decc.gov.uk/en/content/cms/statistics/publications/prices/prices.aspx (accessed 4 April 2011).

DECC (2011) *Energy statistics: calorific values* [online], http://www.decc.gov.uk/en/content/cms/statistics/source/cv/cv.aspx (accessed 13 May 2011).

DECC (2012) *UK Bioenergy Strategy* (online), http://www.decc.gov.uk/en/content/cms/meeting_energy/bioenergy/strategy/strategy.aspx (accessed 26 April 2012).

DEFRA (2007) *Waste strategy for England 2007 Annexes* [online], http://archive.defra.gov.uk/environment/waste/strategy/strategy07/documents/waste07-annexes-all.pdf (accessed 16 June 2011).

DEFRA (2008) *Waste Wood as a Biomass Fuel: Market information report*, London, DEFRA; also available at http://archive.defra.gov.uk/environment/waste/topics/documents/wastewood-biomass.pdf (accessed 16 June 2011).

DEFRA (2009) *Developing an Implementation Plan for Anaerobic Digestion: Report of the Anaerobic Digestion Task Group July 2009* [online], http://archive.defra.gov.uk/environment/waste/ad/documents/implementation-plan.pdf (accessed 16 June 2011).

DTI (2004) *Renewables Innovation Review: Biomass* [online], Department of Trade and Industry http://webarchive.nationalarchives.gov.uk/+http://www.berr.gov.uk/files/file22017.pdf (accessed 31 March 2011).

EDIE (1999) *Switzerland: illegal waste burning is largest source of dioxin says environment agency* [online], http://www.edie.net/news/news_story.asp?id=1693 (accessed 18 May 2011).

EIA (2010) *World Coal Consumption 1980–2008* [online], US Energy Information Administration, http://www.eia.doe.gov/aer/txt/ptb1115.html (accessed 29 March 2011).

El Bassam, N. (2010) *Handbook of Bioenergy Crops: A Complete Reference to Species, Development and Applications*, London, Earthscan.

EPA (2010) *Methane: Sources and emissions* [online], http://www.epa.gov/methane/sources.html (accessed 31 March 2011).

ETSU (1999) *New and Renewable Energy: Prospects in the UK for the 21st Century: Supporting Analysis*, Energy Technology Support Unit, ETSU R-122; also available at http://webarchive.nationalarchives.gov.uk/+http://www.berr.gov.uk/files/file21102.pdf   (accessed 31 March 2011).

European Pellets Centre (2008) *2nd Newsletter of the Pellets@las project* [online], http://www.pelletsatlas.info/pelletsatlas_docs/showdoc.asp?id=090313113002&type=doc&pdf=true (accessed 11 Nov 2010).

Everett, B., Boyle, G. A., Peake S. and Ramage, J. (eds) (2012) *Energy Systems and Sustainability: Power for a Sustainable Future* (2nd edn), Oxford, Oxford University Press/Milton Keynes, The Open University.

Faaij, A. P. C. (2006) 'Bio-energy in Europe: changing technology choices', *Energy Policy*, vol. 34, no. 3, pp. 322–42.

Field, C. B., Campbell, J. E. and Lobell, D. B. (2008) 'Biomass energy: the scale of the potential resource', *Trends in Ecology and Evolution*, vol. 23, no. 2, pp. 65–72.

Forestry Commission (2010) *UK Wood Production and Trade (provisional figures)* [online], Edinburgh, Economics & Statistics, Forestry Commission http://www.forestry.gov.uk/pdf/trprod10.pdf/ $FILE/trprod10.pdf (accessed 31 March 2011).

Griffiths, M. J. and Harrison, S. T L. (2009) 'Lipid productivity as a key characteristic for choosing algal species for biodiesel production', *Journal of Applied Phycology*, vol. 21, no. 5, pp. 493–507.

Haberl, H., Erb, K. H., Krausmann, F., Gaube, V., Bondeau, A., Plutzar, C., Gingrich, S., Lucht, W., Fischer-Kowalski, M. (2007) 'Quantifying and mapping the human appropriation of net primary production in Earth's terrestrial ecosystems', *Proceedings of the National Academy of Sciences of the USA*, vol. 104, no. 31, pp. 12942–7; also available from http://www.pnas.org/content/104/31/12942.full.pdf+html (accessed 29 March 2011).

Hall, D. O. and Rao, K. K. (1999) *Photosynthesis (Studies in Biology)* (6th edn), Cambridge, Cambridge University Press.

Haq, Z. (2002) *Biomass for energy generation* [online], Washington, Energy Information Administration http://www.eia.gov/oiaf/analysispaper/biomass/pdf/biomass.pdf (accessed 31 March 2011).

Health Protection Agency (2009) *The impact on health of emissions to air from municipal waste incinerators* [online], http://www.hpa.org.uk/web/HPAwebFile/HPAweb_C/1251473372218 (accessed 4 April 2010).

Hoogwijk, M., Faaij, A., van den Broek, R., Berndes. G., Gielen, D. and Turkenburg, W. (2003) 'Exploration of the ranges of the global potential of biomass for energy', *Biomass and Bioenergy*, vol. 25, no. 2, pp. 119–33.

IEA (2007) 'Biofuel production', *IEA Energy Technology Essentials*, ETE02, January 2007 [online], http://www.iea.org/techno/essentials2.pdf (accessed 9 February 2011).

IEA (2010a) *Key World Energy Statistics 2010*, International Energy Agency; also available at: http://www.iea.org/textbase/nppdf/free/2010/key_stats_2010.pdf (accessed 29 March 2011).

IEA (2010b) *Member Country Reports* [online], http://www.iea-biogas.net/_content/publications/member-country-reports.html (accessed 3 January 2010).

IEA (2010c) *Energy Poverty: How to make modern energy access universal?*, International Energy Agency; also available at: http://www.worldenergyoutlook.org/docs/weo2010/weo2010_poverty.pdf (accessed 4 April 2011).

IEA Bioenergy (2009) *Bioenergy – a Sustainable and Reliable Energy Source (Executive Summary)* [online], IEA Bioenergy, www.ieabioenergy.com/MediaItem.aspx?id=6360 (accessed 29 March 2011).

Inbicon (2010) *Background: Bringing cellulosic ethanol production to industrial scale at Kalundborg* [online], http://www.inbicon.com/projects/kacelle/pages/background.aspx (accessed 7 January 2011).

Jannasch, R., Quan, Y. and Samson, R. (2002) *A process and energy analysis of pelletizing switchgrass* [online], REAP-Canada http://www.reap-canada.

com/online_library/feedstock_biomass/11%20A%20Process.pdf (accessed 31 March 2011).

Jin, J-Y., Wu, R. and Liu, R. (2002) 'Rice production and fertilization in China', *Better Crops International*, vol. 16, Special Supplement May 2002, pp. 26–9; also available online at: http://www.ipni.net/ppiweb/bcropint. nsf/$webindex/BD1D60B1CA86D2D185256BDC007762DC/$file/BCI-RICEp26.pdf (accessed 31 March 2011).

Kantor, L. S., Lipton, K., Manchester, A. and Oliveira, V. (1997) 'Estimating and addressing America's food losses', *Food Review*, vol. 20, no. 1, pp. 1–12; also available at http://www.ers.usda.gov/Publications/FoodReview/Jan1997/Jan97a.pdf (accessed 9 February 2011).

Kato, M., DeMarini, D. M., Carvalho, A. B., Rego, M. A. V., Andrade, A. V., Bonfim, A. S. V. and Loomis, D. (2005) 'World at work: Charcoal producing industries in northeastern Brazil', *Occupational and Environmental Medicine*, vol. 62, pp. 128–32.

Kishore, V. V. N., Bhandari, P. M. and Gupta, P. (2004) 'Biomass energy technologies for rural infrastructure and village power – opportunities and challenges in the context of global climate change concerns', *Energy Policy*, vol. 32, no. 6, pp. 801–10.

Kuiper, L. and Oldenburger, J. (2006) *The harvest of forest residues in Europe* [online], Biomassa-upstream Stuurgroep, Report on bus ticket no. D15a, http://www.probos.net/biomassa-upstream/pdf/reportBUSD15a.pdf (accessed 31 March 2011).

Ladanai, S. and Vinterbäck, J. (2009) *Global potential of sustainable biomass for energy*, SLU, Swedish University of Agricultural Sciences, Department of Energy and Technology Report 013; also available at: http://www.bioenergypromotion.net/bsr/publications/global-potential-of-sustainable-biomass-for-energy (accessed 4 April 2011).

LTS International (2006) *A review of the potential impacts of short rotation forestry* [online], LTS International http://www.forestry.gov.uk/pdf/SRFFinalReport27Feb.pdf/$FILE/SRFFinalReport27Feb.pdf (accessed 4 April 2011).

Mani, S. (2006) 'Simulation of biomass pelleting operation' [online], *Bioenergy Conference & Exhibition 2006*, Prince George, May 31, http://www.bioenergyconference.org/docs/speakers/2006/Mani_BioEn06.pdf (accessed 31 March 2011).

Manurung, R. and Beenackers, A. A. C. M. (1990) 'Field test performance of open core downdraft rice husk gasifiers', *Biomass for Energy and Industry, 5th EC Conference: Proceedings of the International Conference on Biomass for Energy and Industry*, Lisbon, October 9–13, London, Elsevier, pp. 2.512–2.523.

Ministry of New and Renewable Energies (2010) *Implementation of National Biogas and Manure Management Programme (NBMMP) during 11th Five Year Plan* [online], Government of India, Ministry of New and Renewable Energy (Bio-energy technology development group), No. 5-5/2009/BE; available at http://www.mnre.gov.in/adm-approvals/biogasscheme.pdf (accessed 9 February 2011).

Mitchell, P., Kiel J., Livingston, B. and Dupont-Roc, G. (2007) *A foresighting study into the business case for pellets from torrefied biomass as a new solid fuel* [online], http://www.all-energy.co.uk/UserFiles/File/2007PaulMitchell. pdf (accessed 31 March 2011).

Moreira, J. R. (2006) 'Global biomass energy potential', *Mitigation and Adaptation Strategies for Global Change*, vol. 11, pp. 313–42.

NNFCC (2007) *NNFCC Position Paper. Biorefineries: Definitions, examples of current activities and suggestions for UK development* [online], http:// www.nnfcc.co.uk/publications/NNFCC-position-paper-biorefineries-definitions-examples-of-current-activities-and-suggestions-for-UK-development (accessed 16 June 2011).

NNFCC (2010) *Environmental Assessment Tool for Biomaterials* [online], http://www.nnfcc.co.uk/chemicals-materials/our-services/lca-of-chemicals-and-materials/environmental-assessment-tool-for-biomaterials (accessed 30 March 2011).

Perrin, R. K. (2009) *Is Corn Ethanol Economically Viable in the Long Run?* [online], Lincoln, NE, Department of Agricultural Economics, University of Nebraska-Lincoln http://digitalcommons.unl.edu/ageconfacpub/79/ (accessed 4 April 2011).

POST (2007) 'Energy and Sewage' *Parliamentary Office of Science and Technology postnote* April 2007, Number 282 [online], http://www. parliament.uk/documents/post/postpn282.pdf (accessed 31 March 2011).

Purohit, P. and Michaelowa, A. (2007) 'CDM potential of bagasse cogeneration in India', *Energy Policy*, vol. 35, no. 10, pp. 4779–98.

Rawlinson S. and Hicks M. (2010) *Cost model: Energy from waste* [online], Building.co.uk, http://www.building.co.uk/data/cost-model-energy-from-waste/3162156.article (accessed 4 April 2011).

REN21 (2011) *Renewables 2011 Global Status Report,* REN21 Secretariat, Paris, 216pp. http://www.ren21.net/Portals/97/documents/GSR/GSR2011_Master18.pdf (accessed 18 November 2011).

Risović, S., Djukić, I. and Vučković, K, (2008) 'Energy analysis of pellets made of wood residue', *Croatian Journal for Engineering*, vol. 29, no. 1, pp. 95–108; also available at http://crojfe.sumfak.hr/v29no1/11risovic95. pdf (accessed 31 March 2011).

Royal Society (2008) *Sustainable Biofuels: Prospects and Challenges*, London, Royal Society; also available at: http://royalsociety.org/Sustainable-biofuels-prospects-and-challenges/(accessed 16 June 2011).

Sapphire Energy Recovery (2009) *Press Release: Lafarge becomes sole owner of Sapphire Energy Recovery* [online], http://www.sapphirerecovery. co.uk/media/pr_2009/Lafarge%20Cement%20UK%20becomes%20 sole%20owner%20of%20Sapphire%20Energy%20Recovery.pdf (accessed 31 March 2011).

Scurlock, J. (2011) *Personal communication.*

Shackley, S., Carter, S., Sims, K. and Sohi, S. (2011) 'Expert perceptions of the role of biochar as a carbon abatement option with ancillary agronomic and soil-related benefits', *Energy & Environment*, vol. 22, no. 3, pp. 167–88.

Sims, R. E. H. (2002) *The Brilliance of Bioenergy: In Business and in Practice*, London, James and James.

Slade, R., Bauen, A., and Gross, R. (2010) *The UK Bio-energy Resource Base to 2050: Estimates, Assumptions, and Uncertainties*, UK Energy Research Centre, UKERC/WP/TPA/2010/002; also available at http://www.ukerc.ac.uk/support/tiki-download_file.php?fileId=727 (accessed 1 April 2011).

Slesser, M. and Lewis, C. (1979) *Biological Energy Resources*, London, E.& F.N. Spon.

Smith, P., Powlson, D. S., Smith, J. U., Falloon, P. and Coleman, K. (2000) 'Meeting Europe's climate change commitments: quantitative estimates of the potential for carbon mitigation by agriculture', *Global Change Biology*, vol. 6, pp. 525–39.

The Oil Drum (2008) *Updated Corn Ethanol Economics* [online], http://www.theoildrum.com/node/4208 (accessed 4 April 2011).

UNDP (2000) *World Energy Assessment*, New York, United Nations Development Programme; also available online at http://www.undp.org/energy/activities/wea/drafts-frame.html (accessed 29 March 2011).

United Nations Foundation (2008) *Sustainable Bioenergy Development in UEMOA Member Countries* [online], http://www.globalproblems-globalsolutions-files.org/gpgs_files/pdf/UNF_Bioenergy/UNF_Bioenergy_exec_summary.pdf (accessed 16 June 2011).

USDA (2007) *Fuel Value Calculator* [online], Madison, WI, Forest Products Laboratory http://www.fpl.fs.fed.us/documnts/techline/fuel-value-calculator.pdf (accessed 31 March 2011).

WHO (2003) *Diet, Nutrition and the Prevention of Chronic Diseases: Report of a Joint WHO/FAO Expert Consultation*, WHO Technical Report Series 916.

WRAP (2009) *Household Food and Drink Waste in the UK*, Banbury, WRAP; also available online at http://www.wrap.org.uk/downloads/Household_food_and_drink_waste_in_the_UK_-_report.70fec881.8048.pdf (accessed 11 January 2010).

WRc (2003) *Refuse Derived Fuel, Current Practice and Perspectives (B4-3040/2000/306517/MAR/E3)*, WRc plc, CO 5087-4; also available online at http://ec.europa.eu/environment/waste/studies/pdf/rdf.pdf (accessed 31 March 2011).

Chapter 5

Hydroelectricity

By Janet Ramage

5.1 Introduction

> How to promote socio-economic development and eradicate
> poverty, whilst simultaneously halting environmental degradation,
> is one of the greatest challenges at the start of the 21st century. This
> challenge is most conspicuous in the policy for water and energy,
> as both are essential elements for human life.
>
> (Sustainable Hydropower, 2011a)

Water power, like most renewable energy sources, is indirect solar power,
and like others such as the wind, it has been contributing to local energy
supplies for many centuries. It is, however, unique in that it became a major
'modern' energy source over a hundred years ago, supplying the input for
some of the earliest power stations. Hydroelectricity has become a well-
established technology, delivering (at the time of writing) about a sixth of
the world's annual electricity supply (Figure 5.1).

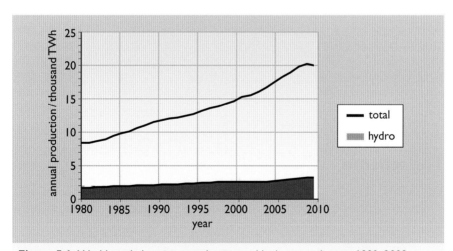

Figure 5.1 World total electricity production and hydro contribution, 1980–2009
(source: BP, 2010)

To introduce the terminology and the main features of hydroelectric
systems, this chapter starts with an account of one relatively modest
hydro scheme. Commissioned over three-quarters of a century ago and
still operating with much of its original plant, this group of power stations
in Scotland exemplifies both the technical and the economic aspects of
hydroelectricity.

The chapter continues with a discussion of the nature of the hydro resource and its present contribution to world energy. This is followed by a summary of the basic science and a brief history of the development of water power, leading to the modern turbine systems that are the subject of Sections 5.7 and 5.8.

The remaining sections are concerned with the problems and the potentialities of hydroelectricity. We find the familiar issues of cost, variability of output and integration into a grid system which arise for all renewable sources, but for large-scale hydroelectricity the main questions are rather different: whether there are limits to growth, what determines these limits, and whether we are already reaching them.

5.2 The Galloway Hydros

The Galloway Hydroelectric Scheme on the River Dee in south-west Scotland makes an interesting study for several reasons. Initially commissioned in 1935, it was the first major UK scheme designed specifically to provide extra power at times of peak demand. Its six power stations are controlled as one integrated system. Several of its dams are on major salmon-fishing rivers, raising environmental issues common to many hydro schemes. It is also technically interesting because significant differences between the site conditions at the power station locations mean that across the system several types of turbo-generator are used.

Origins

The Galloway Hydros owe their origin to local pride and individual enthusiasm, and an Act of Parliament (ScottishPower, 2011). The first proposals to use the rivers and lochs of south-west Scotland for hydropower appeared in the 1890s, but the scheme became feasible only with the establishment of a National Grid in the 1920s. This meant that the great industrial conurbation of Glasgow became a potential customer, and also that the hydro system could meet the need for plant that could quickly respond to daily and seasonal peaks in the otherwise coal-dominated grid system.

The scheme

The system (Figure 5.2) has three main elements. The first is Loch Doon, which provides the main long-term seasonal storage. Its natural outflow is to the north, but a dam now restricts this and the main flow is diverted eastwards through a 2 km tunnel into the upper valley of the Water of Ken. An interesting feature is the Drumjohn Valve: when demand for power is low, this directs the flow from two eastern tributaries of the Water of Ken through the tunnel in the 'reverse' direction, *into* Loch Doon, adding to the stored volume. The level in the loch can vary by 12 metres, releasing 80 million cubic metres of water. Falling through the 200 metre height difference down to the final outflow at Tongland, this represents a gross release of some 150 million MJ of energy – over 40 million kWh.

The second element of the system, for fast response to short-term demand variations over the course of a day, is the series of dams and power stations

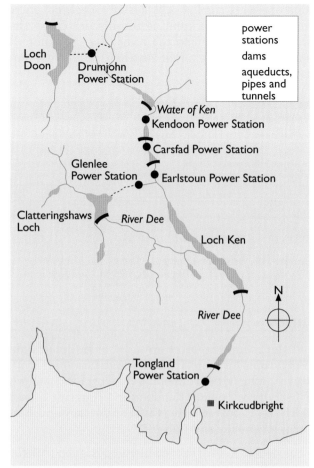

Figure 5.2 The Galloway Hydros (source: adapted from Hill, 1984)

along the course of the Water of Ken: Kendoon, Carsfad and Earlstoun, and the Tongland plant near the mouth of the River Dee.

Clatteringshaws Loch, the third element, is fed by the tributaries of the Dee, rising to the west of the main valley. This is the only completely artificial reservoir in the system, and its main outflow, through a tunnel nearly 6 km long and pipes with a fall of over 100 metres, supplies the 24 MW Glenlee power station before joining the Ken below Earlston. The remaining natural flow of the Dee and its tributaries is south-east into Loch Ken (into which the Ken also flows).

Figure 5.3 Carsfad power station and dam

Power

The essential characteristics of a hydro site are the **effective head** (*H*), the height in metres through which the water falls, and the **flow rate** (*Q*), the number of cubic metres of water passing through the plant per second. As we shall see in Section 5.5, there is a simple approximate relationship between these two quantities and the power delivered by the water (*P*), measured in kilowatts (kW):

$$P \text{ (kW)} = 10 \times Q \times H$$

The conversion of energy carried by water into electrical energy is carried out by the **turbo-generator**: a rotating turbine driven by the water and connected by a common shaft to the rotor of a generator. (The electric power *output* will of course be rather less than the above input, for reasons discussed in Section 5.5.)

BOX 5.1 Capacity factor   a reminder

As we saw in Chapter 1, the annual capacity factor of any plant is equal to its actual annual output divided by its maximum possible output, usually expressed as a percentage.

So, if Carsfad, with its capacity of 12 000 kW, generates 30 million kWh of electricity in a year (remembering that there are 8760 hours in a year), its capacity factor will be:

$$30 \times 10^6/(12\ 000 \times 8760) = 0.285, \text{ or } 28.5\%$$

Of course, in practice, the annual capacity factor of a plant depends on both the demand for its output and the power that it can produce at any given time.

The turbines

The head and the required power are critical in determining the most suitable type of turbine for a site. Glenlee's high head puts it at one extreme in the Galloway system, with Drumjohn's very low head and power rating at the other. Of the four river plants, Kendoon and Tongland have intermediate heads and fairly high power ratings whilst Carsfad and Earlstoun have almost identical low heads and powers.

Any turbine consists of a set of curved blades designed to deflect the water in such a way that it gives up as much as possible of its energy. The blades and their support structure make up the turbine **runner**, and the water is directed on to this either by channels and guide vanes or through a jet, depending on the type of turbine. The Galloway plants include two types of runner: 'propellers' and Francis turbines (see Figure 5.17). As discussed in more detail later, 'propeller' types are most suitable for large flows at low heads, and Francis turbines for medium to high heads. Comparison of Tables 5.1 and 5.2 shows that this is true for the Galloway scheme.

Table 5.1 The Galloway power stations

Power station	Average head/m	Maximum flow/m³ s⁻¹	Output capacity /MW	Number of turbines
Drumjohn	11	16	2	1
Kendoon	46	55	24	2
Carsfad	20	73	12	2
Earlstoun	20	71	14*	2
Glenlee	116	26	24	2
Tongland	32	127	33	3

* Although the *average* head and flow are similar to Carsfad, a slightly higher dam at Earlstoun allows it to take advantage of seasonal variations in flow.
Source: ScottishPower, 2011

Table 5.2 The turbines

Power station	Turbine rating/MW	Turbine type	Rate of rotation*/rpm
Drumjohn	2	Propeller	300
Kendoon	12	Francis	250
Carsfad	6	Propeller	214.3
Earlstoun	7	Propeller	214.3
Glenlee	12	Francis	428.6
Tongland	11	Francis	214.3

*The significance of these rates is explained in Section 5.7.
Source: ScottishPower, 2011

The salmon

The principal environmental issue raised during the approval process was the possible effect on salmon. Several dams blocked rivers below their salmon spawning pools, and concern was expressed about the fate of adult salmon making their way upstream and young smolt on the reverse journey. The response was the incorporation of fish ladders at four dams. These are a series of stepped pools with a constant downward flow of water to attract the fish, which leap up from pool to pool (Figure 5.4). The Doon dam had insufficient space for a long series of pools, so the fish ladder there is partly inside a round tower – it is claimed that the fish do not find this spiral staircase a problem. The issue does not arise at Glenlee since salmon do not use the man-made Clatteringshaws Loch. (It is worth noting that a much more detailed environmental impact statement would be required today.)

Figure 5.4 The fish ladder at Tongland

Economics

The Galloway scheme was built to supply extra power to the grid at times of peak demand. In other words, it was assumed that it would generate for only a few hours a day, resulting in an annual capacity factor of no more than perhaps 25%. This would suggest a poor return on the initial investment, which was in any case higher than the cost of a coal-fired plant of similar capacity. However, a number of circumstances made the scheme financially attractive.

- The company was able to assume a firm demand for the planned output.

- During its first three years the scheme received an annual treasury grant of £60 000 (over a million pounds at present-day values), reducing the cost per unit by about 20%.

- The company also argued successfully that the local taxes levied on the hydro power plant should be lower than those for an equivalent coal-fired plant.

From the start, demand, and the consequent economic performance, exceeded expectations, and only in a few years of serious drought did output fall below the planned level. This remains the case today (ScottishPower, 2010). After over 70 years, the original five plants are still generating power, joined in 1984 by the 2 MW plant at the Drumjohn Valve. The five original plants are all operated by the engineer at Glenlee, the only permanently manned plant – and the original construction costs were repaid many years ago.

In 2009 ScottishPower announced a £20 million investment programme to upgrade equipment at three hydroelectric sites, including the Galloways and also Cruachan (see Figure 5.8), which is intended to safeguard their operation until at least 2035.

5.3 The resource

For hydroelectricity, as for other renewables, the resource is basically an amount of *power* – a rate at which energy is delivered. However, for comparison with other major energy sources, where the resource is effectively a quantity of *stored energy* (tonnes of coal, barrels of oil, etc.), it is usual to express a hydro resource in terms of the total energy it delivers in the course of a year. The customary unit on the national or world scale is **terawatt hours per year**, **TWh y^{-1}** (1 TWh = 3.6 PJ, or 1 EJ = 278 TWh).

The world resource

We saw in Figure 1.9 that almost a quarter of the 5.4 million EJ (1.5 billion TWh) of solar energy reaching the Earth's atmosphere each year is consumed in the evaporation of water. The water vapour in the atmosphere therefore represents an enormous, constantly replaced, store of energy. Unfortunately most of it is not available to us. When the water vapour condenses into water, most of its stored energy is released into the atmosphere as heat, and ultimately re-radiated into space. But a tiny fraction, about 200 000 TWh y^{-1}, reaches the Earth as rain or snow. Roughly one fifth of this precipitation falls on hills and mountains, descending ultimately to sea level as the world's streams and rivers. The 40 000 TWh y^{-1} of energy carried by this flowing water can be regarded as the world's **total hydro resource**.

It is obviously not possible – or desirable – to build hydro plants on every river or stream, so the usable fraction of this flow will be significantly lower, and at the time of writing, the world's **technical hydro potential** is estimated to be about 16 000 TWh y^{-1}, or roughly two-fifths of the above total resource. It is characteristic of such calculations that sixty years ago the estimated world technical potential was only 6000 TWh y^{-1}, and the figure may well rise again in the future, in part due to the different ways countries assess their resource. Some, for instance, include in their 'technical potential' only those sites that have been fully surveyed.

There remains one important question in the assessment of the realistic resource: 'How much hydroelectricity is available at a cost that is competitive with power from other sources?' In other words, what is the **economic potential** for hydroelectricity? The issues involved in financing hydro plants are discussed in Section 5.11, and Chapter 10 considers the relative costs of electricity from different sources; but the economic potential for any country or region may depend on many other factors. One definition suggests that it is the fraction of the total resource that...

can be exploited within the limits of current technology under present and expected local economic conditions. The figures may or may not exclude economic potential that would be unacceptable for social or environmental reasons.

(WEC, 2010a)

This obviously leaves open many questions, and we should not be surprised to find that estimates of the economic potential for hydropower in different countries or regions are generally regarded as much less reliable than the estimates of the total or technical potential. The following discussion of resources therefore considers only the *total resource* and the *technical potential*, leaving the financial aspects to Section 5.11.

Regional resources

The first four columns of Table 5.3 show the estimated total hydro resource and technical potential for different regions of the world, together with their percentage share of the technical potential. As we have seen above, the world's total technical potential is about two-fifths of its total resource, and the table shows a not dissimilar relationship for most of the individual regions.

Table 5.3 Regional and world hydro potential and output, 2009

Region	Resources			Output		
	Total resource /TWh y^{-1}	Technical potential /TWh y^{-1}	% of world technical potential	Installed capacity/GW	Annual output /TWh y^{-1}	Percentage of world total output
North America	5500	2420	15%	168	670	21%
South America	7500	2840	18%	132	730	22%
Europe	4900	2760	17%	221	750	23%
Middle East	690	280	1.7%	22	11	<1%
Africa	3900	1890	12%	22	97	3%
Asia	16600	5590	35%	307	980	30%
Oceania	700	230	1.5%	13	36	1%
World	39 800	16 000		874*	3272	

* The total here is mainly based on national data, which may or may not include small-scale hydro plants. Some sources suggest that world total capacity, including small-scale hydro (see Section 5.4), is close to 1000 GW.
Sources: WEC, 2010a; BP, 2010

However, when we look at the extent to which the regions have developed their hydro potential, the situation is very different. The table shows that world hydro output in the year 2009 was roughly one fifth of the technical potential, suggesting that a fourfold increase in output might be possible in future. But this is by no means the case for individual regions.

The regions with significant hydro resource can be divided into three categories. In the Americas and Europe current output is already between a quarter and a third of the estimated technical potential, perhaps implying lower limits to future growth. For Asia, the factor is still less than one fifth, but this region includes China and India, whose current rates of development are rapidly changing the situation, as we shall see.

At the other extreme is Africa, contributing only 3% of world output from an estimated 12% of world technical potential. There has been some development of the continent's river resources: the Volta in Ghana, the Nile in Egypt and the Zambezi in Zimbabwe and Mozambique, but it is not surprising that the continent features increasingly in discussions of the future for hydropower. A major element in these discussions has been the Grand Inga project, the controversial proposal for a pan-African electricity system based on hydro plants on the river Congo. (See Section 5.12 for more detail.)

National resources

Table 5.4 shows data for the eleven countries whose hydro output in 2009 was greater than 50 TWh, together with a few other European countries of interest. It is worth noting that several countries appear to have developed more than half their technical potential already; but as pointed out above, some may limit their estimates of 'technical potential' to sites that have been studied in detail. This is known to be the case for Canada, with its enormous unexplored land area, and for Switzerland, which limits its 'technical potential' to sites that have already passed a full environmental assessment.

The output from any hydroelectric plant will of course depend on the available flow of water, but the overall output for an individual country will

Table 5.4 National hydro potential and contributions, 2009

Country	Technical potential /TWh y⁻¹	Installed capacity/GW	Annual output /TWh y⁻¹	Average capacity factor	Percentage of nation's electricity
China	2500	171	616	41%	17%
Canada	830	73	399	62%	63%
Brazil	1250	78	391	58%	84%
USA	1340	77	275	41%	7%
Russia	1670	50	176	40%	18%
Norway	240	30	127	49%	96%
India	660	38	106	32%	12%
Venezuela	260	15	86	67%	69%
Japan	136	28	74	30%	7%
Sweden	130	16	76	54%	49%
France	100	21	58	32%	11%
Austria	75	8	37	50%	53%
Italy	65	18	46	28%	16%
Switzerland	43	14	36	31%	52%

Sources: WEC, 2010a; BP, 2010

also depend on the role played by hydroelectricity in the national power system. As Table 5.4 shows, the countries with the highest capacity factors tend to be those where hydropower makes a significant contribution but is not the only major source of electricity – a situation that allows the relatively cheap hydropower to be used to its full potential, with an alternative source of electricity available when hydropower cannot meet demand. (Brazil's fairly high capacity factor despite its almost total dependence on hydropower is in part the result of its unique power-sharing arrangements with Paraguay, described in Box 5.12.)

In general, if almost all a nation's electricity comes from hydro plants, annual capacity factors are usually lower, because the installed capacity must be large enough to meet the maximum demand experienced during any day (or year). Conversely, it is interesting to note that countries with significant nuclear power, such as Japan, France and Switzerland, all have low hydro capacity factors, a consequence of the use of hydroelectricity mainly for providing power at times of high demand.

Compared with the countries in Table 5.4, the hydro resource of the UK is small, and the installed capacity in 2009 was about 1.5 GW (DECC, 2010a), mainly in Scotland. The annual output has varied in recent years between about 3.3 and 5.3 TWh – reflecting year-on-year weather variations throughout this relatively small area.

World output

Figure 5.5 shows the changing world annual hydro output over a longer timescale and in more detail than Figure 5.1. In 1900, about two decades after the first commercial hydroelectric plants, world annual output had reached an estimated 3.7 TWh from an installed capacity of about 1.3 GW. (Roughly the 2010 hydro capacity and output of the UK.)

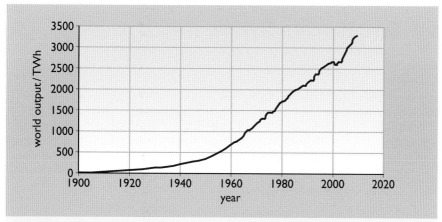

Figure 5.5 World annual hydroelectricity output, 1900–2009 (sources: Romer, 1976; BP, 2010)

However, despite two world wars and the depression of the 1930s, world capacity then rose very rapidly. As the upward curve shows, output was increasing at an annual rate of nearly 10% per year throughout the first half of the twentieth century, leading to an output in 1950 that was nearly a hundred times that of 1900.

This exponential rate of rise did not continue, but world capacity still grew fast enough to maintain a fairly constant year-on-year increase of 50 TWh in annual output throughout the second half of the century (Figure 5.5). Remarkably, throughout the wars, economic crises and significant variations in annual rainfall, not once during the twentieth century was there a year-on-year *fall* in world hydro output.

This changed in 2001 when output fell for the first time, by about 5%, despite a further rise in installed capacity. The drop was attributed mainly to exceptionally dry conditions in the Americas, the source of about a third of world output (Table 5.3). The next year, 2002, was mixed. World output rose by about 50 TWh despite a fall of 70 TWh in the European contribution, attributed mainly to low rainfall (although the UK experienced a 'wet year', resulting in a 20% rise in output – an increase of about one tenth of a terawatt-hour).

As Figure 5.5 shows, the years from 2003 to 2008 were strikingly different again, with world output increasing by a total of 600 TWh over the five-year period. Assuming that the average capacity factor remained at about 43%, this implies an increase of about 30 GW in hydro capacity.

It is tempting to note that this period saw growing contributions from the world's two largest hydro plants, Three Gorges and Itaipú (see Boxes 5.10 and 5.12), whose combined rated output just happens to be 30 GW! However, although the increasing contributions from these large new plants were indeed relevant, the changing situation in the first decade of the present century was rather more complicated, and may have implications for the future growth of hydropower. These recent developments are therefore the starting point for the discussion of the future for hydroelectricity in Section 5.12.

5.4 Small-scale hydro

In the early days of electric power, generators with output ratings between a few kilowatts and a few megawatts were installed on streams or rivers, often using the dams and sluices of old watermills. As late as the mid-twentieth century, many towns in Europe and elsewhere were still served by hydro or other plants in the upper part of this range. However, with continually rising demand for electricity, and the growth of national transmission networks, capacities of several hundred megawatts have become the norm for modern power stations, and plant outputs below about 10 MW are now referred to as **small-scale** (Box 5.2).

BOX 5.2 What is small-scale?

The general view of most international energy organizations is that a plant should be recognized as small-scale hydro (SSH) or 'small hydro' if its capacity, or rated output, does not exceed 10 MW.

Unfortunately, this definition is not universally accepted. Whilst a number of countries (Switzerland, for instance) use the above figure, many prefer their own criteria. Notable examples are India (25 MW), the USA (30 MW) and China, whose limit of 50 MW could include a plant powering a small city (REN21, 2010; U.S. Dept of Treasury, 2010).

In publishing national energy data, the UK includes as 'renewables' all hydro output, regardless of plant size. However, as in many other countries, a much more detailed breakdown is used for different levels of government support (DECC, 2010a).

The term **micro-hydro** is commonly used for plants with capacities below 100 kW and **pico-hydro** appears occasionally for very small plants; but again there is a lack of agreement between countries on these definitions.

The past few decades have seen growing interest in smaller power plants, for several reasons. In the industrialized countries, environmental issues have increasingly limited the potential for further major hydro development, whilst small-scale schemes, considered to produce fewer deleterious effects, have received growing encouragement (see for instance Box 5.3). More recently, a market has begun to emerge for micro-hydro plants, generating a few tens of kilowatts for an isolated house or farm.

At the time of writing, small-scale hydro is more costly than electricity from large hydro or other conventional sources, although technical improvements are expected to bring small-scale plants to a level where, in suitable locations, they will be competitive with other options. Nevertheless, in many European countries, investment in electricity from renewables during first decade of the twenty-first century has concentrated on wind and solar PV rather than small-scale hydropower.

In an entirely different context, small-scale plants are a practicable independent option for electricity in developing countries without extensive grid systems. Large areas of Nepal, for instance, have no electricity supply and only mules or human porters for transport, but do have many mountain streams suitable for 'high-head' plants. This has led to the development of local industries producing extremely small-scale systems, transportable by a single person on foot. The Peltric turbo-generator set, for example, consists of a tiny Pelton wheel (see Section 5.8) driving a simple generator. Operating under heads of 50–70 metres, it produces a kilowatt or so of output. With cheap and relatively simple 'civil works', these tiny systems have proved sufficiently popular to be copied in other countries.

BOX 5.3 Elan Valley

The Elan Valley scheme in the Cambrian Mountains in mid-Wales (Figure 5.6) is unusual in that one government contract in 1997 under the UK's Non-Fossil Fuel Obligation (NFFO) low-carbon electricity incentive scheme supported the development of five power stations. Like the Galloway Hydros they are in sequence along one river; but these plants use the dams and reservoirs of an existing water-supply system – and their total output capacity is only 4.2 MW.

Table 5.5 gives some details, and the contribution from Foel Tower on the Garreg Ddu reservoir is worth noting: its Kaplan turbine (see Section 5.8) sits inside one of the 14 metre diameter pipes supplying water to the city of Birmingham.

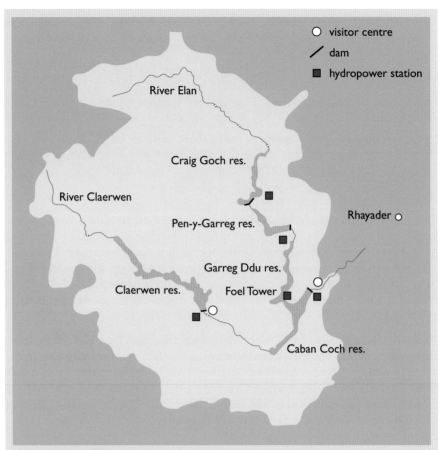

Figure 5.6 The Elan Valley hydro scheme

Table 5.5 The Elan Valley plants

Site	Craig Goch	Pen-y-Garreg	Caban Coch	Foel Tower	Claerwen
Head/m	36.5	37.5	37	13.5	56
Capacity/kW	480	810	950	300	1680
Turbine	Francis	Francis	2 × Francis	Kaplan	Francis

All the reservoirs and dams date from the early twentieth century – the latter are listed historic monuments. Caban Coch was already a hydro plant. One of its old turbines was replaced some time ago and the other as part of this scheme. In a region of great natural beauty and with 90% of its area designated as Sites of Special Scientific Interest, the four new plants have required careful siting. To meet environmental constraints, their turbines make use of existing discharge pipes, and their associated buildings are mainly below ground (Elan Valley Trust, 2011).

Assessing small-scale contributions

It is generally agreed that world output from small-scale hydro plants is rising; but it is also generally agreed that it is impossible to estimate either the total output or the installed capacity with any reasonable precision. A World Energy Council survey in 2010 (WEC, 2010a) suggests that the global total small-scale capacity (<10 MW) at the end of 2009 was about 60 GW (including an estimated 33 MW from those plants in China that met the <10 MW criterion). This is about 6% of the world's total hydro capacity: a proportion that seems to have remained much the same for several decades.

Why differentiate?

All hydroelectric plants, regardless of size, operate on the same principles: moving water causes a turbine to rotate, driving a generator. They differ significantly only in scale and therefore output. The little Peltric plant mentioned above uses the same type of turbine as a plant with ten thousand times its output. There seems, therefore, to be no *technical* reason for differentiating between the two types of plant.

The REN21 report already cited explains its reasons for presenting separate figures for large and small hydropower:

> Small hydro is counted, reported, and tracked separately from large hydro in a variety of policy and market contexts around the world, for example, as an eligible technology for Renewable Portfolio Standards, feed-in tariffs, tax credits, and in portfolio tracking by financiers and development assistance agencies.
>
> Some policy targets ... count only small hydro in calculating share of electricity from renewables and exclude large hydro from policy targets. In addition, many countries separate small and large hydro when tracking renewables development. Further, because it represents such a large portion of total renewable energy capacity on a global basis and in many individual countries, large (or total) hydropower masks the dynamic growth and features of ongoing markets for wind, biomass, solar, and other "new renewables" if it is not separated out.
>
> (REN21, 2010)

In other words, it treats large and small hydropower separately because most of the countries sending detailed data do so, mainly for the reasons briefly summarized above. Not everyone agrees with this dual system, however. The two extreme positions may be summarized as follows:

- Large-scale hydro may be renewable but is not a sustainable source, since its environmental and social consequences are such that it fails to fulfil the essential requirements for sustainability outlined at the beginning of Chapter 1, but small-scale plants are inherently free of the main deleterious consequences that accompany large-scale plants.

- Hydro should simply be treated as one renewable resource. The technology is fundamentally the same regardless of scale, and there is no basis for inclusion or exclusion at an arbitrary size. Moreover, the energy storage potential of large hydro plants can play an essential role in the increased use of renewable resources with variable output.

This debate seems likely to continue.

BOX 5.4 **Direct uses of water power**

Although the generation of electricity is by far the major use of water power today, other uses still remain. The kinetic energy of moving water (or the potential energy of water at a height) can be used directly to drive machines – indeed this was the sole use of water power until the mid-nineteenth century.

Old watermills do still operate in a few places in today's industrialized countries, grinding corn or sawing wood; and some mountain railways use a counter-balancing tank filled with water at the top and emptied at the foot of the hill. But these are now rarities.

In the developing countries, direct water power plays a slightly greater role. In Nepal, for instance, simple turbines which can be produced locally are used to drive machinery, and in the Middle East and Asia the use of a flowing stream to raise water for irrigation has by no means disappeared. Nevertheless, the worldwide energy contribution from direct use is negligibly small compared with the hydroelectric power output.

5.5 Stored energy and available power

Potential energy

Water held at a height represents stored energy – the gravitational potential energy discussed in Chapter 1. As we saw, if m kilograms are raised through H metres, the stored potential energy in joules is given by the following simple equation:

potential energy $= m \times g \times H$

where g here is the acceleration due to gravity (about 9.81 m s^{-2}, but for rough calculations, 10 m s^{-2} is often used). Thus about 10 joules of energy input are needed to lift one kilogram of anything vertically through one metre against the gravitational pull of the Earth.

However, we are concerned here with large reservoirs, whose capacities are almost always given in *cubic metres*, rather than kilograms. The necessary conversion is simple, because one cubic metre of fresh water has a mass of 998 kg: so close to 1000 kg that the tiny difference is not significant in most calculations. We can therefore say that, within this degree of precision, the energy stored by a volume of V cubic metres of water raised through a height H is:

stored energy $= 1000 \times V \times g \times H$ $\qquad\qquad\qquad$ (1)

and this is therefore the energy that will be released when this volume of water *falls* through a vertical distance H.

Pumped storage

The growing interest in the *storage* of electrical energy in recent decades, particularly in the industrialized countries, is a consequence of increasing contributions from two very different sources: nuclear power and renewables. The output of a nuclear reactor cannot easily follow large hour-by-hour variations in demand, so if the nuclear input to a national supply rises above the base-load level, it becomes important to find a means of storing the excess output at times of low demand. (Indeed, many other types of generating plant also run most efficiently at their full rated power, and the cost of constantly varying the output can be significant.) The second reason is the increasing use of renewables for electricity, and here the problem is that the output of many renewables *does* vary – sometimes suddenly and often for reasons outside the operators' control. Back-up power is therefore needed which can be brought on stream very quickly.

At present the only practicable and economically viable way to store electrical energy in very large quantities is to use it to pump water up a mountain. Pumped storage has thus become increasingly important, with installed capacity worldwide having grown from 78 GW in 2005 to reach 127 GW in 2009 – nearly a seventh of world total hydro capacity (REN21, 2010).

The principle is simple. Electrical energy is converted into gravitational potential energy when the water is pumped from a lower reservoir to an upper one, and the process is reversed when it is released to run back down, driving a turbo-generator on the way (Figure 5.7). The economic viability of the method depends on two nice technological facts.

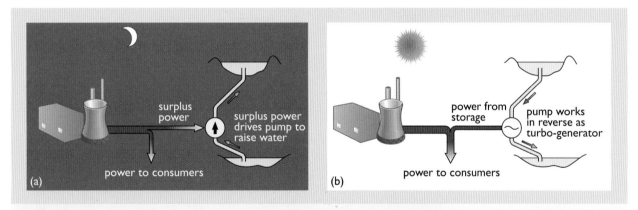

Figure 5.7 Pumped storage system (a) at time of low demand, (b) at time of high demand

- A suitably designed generator can be run 'backwards' as an electric motor: the machine which converts mechanical energy into electrical energy can perform the reverse process.
- A suitably designed turbine (see Section 5.8) can also run in either direction, either extracting energy *from* the water as a turbine or delivering energy to the water as a pump.

The complete reversal is thus **turbo-generator** to **electric pump**. The machines must of course be designed for this dual role, but the cost saving is obviously significant.

There will, as always, be losses associated with the conversion processes, but turbines and generators are very efficient, and nearly 80% of the input electrical energy can be retrieved as electrical output when needed. The value of the system is enhanced by its speed of response: any of the six 300 MW Francis turbines of the Dinorwig storage plant in Wales can be brought to full power in just 12 seconds if initially spinning in air, and even from complete standstill the process takes only a few minutes (Dinorwig, 2011). Pumped storage is thus particularly useful as back-up in case of sudden changes in demand, or failure elsewhere in a grid system.

The location must of course be suitable. A low-level reservoir of at least the capacity of the upper one must be available, or must be constructed. Sites such as Cruachan in Scotland (Figure 5.8), where the mountains rise from a large loch or lake, are obviously ideal. The high-level reservoir, behind a large dam, provides an operating head of 365 metres. Running the four 100 MW reversible machines for one hour at full capacity, as electric pumps or turbo-generators, raises or lowers the reservoir level by about a metre, storing or releasing about 400 000 kilowatt-hours of energy (Cruachan, 2011).

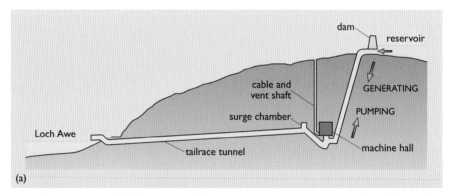

Figure 5.8 Cruachan pumped storage plant (a) the installation, (b) the dam

Pumped storage can be combined with 'normal' hydroelectric generation in locations where the potential exists. The upper reservoir will in any case have a local catchment area, so there may be a positive net output from the plant.

There are however trade-offs. Very high heads have the advantage of needing smaller reservoirs for a given amount of stored energy, but the types of turbine most suited to high heads (see Section 5.8) cannot be run 'backwards' as pumps. At the other extreme, however, very low heads need much greater volumes of stored water, but the switch from pumping to generating may be achieved simply by reversing the pitch of the propeller-type turbines used in such circumstances.

Power, head and flow rate

In estimating the value of any proposed hydroelectric plant, the *power* available at any time is probably the most important factor. The power supplied by a plant, the number of *watts*, is the rate at which it delivers energy: the *number of joules per second*. This will obviously depend on the **volume flow rate** of the moving water. Note that this is not just the speed of the water; it is the number of *cubic metres per second* passing through the plant, usually represented by the symbol Q (think quantity).

It then follows from Equation (1) above that the power P (in watts), which is the energy per second, will be

$$P = 1000 \times Q \times g \times H \qquad (2)$$

However, resource estimates must take into account energy losses. In any real system the water falling through a pipe will lose some energy due to frictional drag and turbulence, and the **effective head** will thus be less than the actual or **gross head**. These flow losses vary greatly from system to system: in some cases the effective head is no more than 75% of the actual height difference, in others as much as 95%. Then there are energy losses in the plant itself. Under optimum conditions, a hydroelectric turbo-generator is extremely efficient, converting all but a few percent of the input power into electrical output. Nevertheless, the **efficiency** – the ratio of the output power to the input power, usually expressed as a percentage – is always less than 100%. With these factors incorporated, the output power becomes:

$$P = 1000 \times \eta \times Q \times g \times H \qquad (3)$$

where H is now the effective head and η (Greek *eta*) is the turbo-generator efficiency. (Note that although efficiency is often quoted as a percentage, in an equation like this it will be a number between 0 and 1. For example, 85% becomes 0.85.)

If we now express P in kilowatts, and use the approximation $g = 10$ m s^{-2}, we obtain a very useful simple expression:

$$P \text{ (kW)} = 10 \times \eta \times Q \times H \qquad (4)$$

Box 5.5 shows how this can be used for rough calculations of power output.

BOX 5.5 Available power

As examples of power calculations, we can consider two systems, each with a plant efficiency of 83%, but of very different sizes.

The first site is a mountain stream with an effective head of 25 metres and a modest flow rate of 600 litres a minute, which is 0.010 cubic metres per second. Using Equation (4), we find that the power output will be approximately 2 kW.

In contrast, suppose that the effective head is 100 m and the flow rate is 6000 cubic metres per second – roughly the total flow over Niagara Falls. The power is now about 5 million kW, or 5 GW.

5.6 **A brief history of water power**

The prime mover

Moving water was one of the earliest energy sources to be harnessed to reduce the workload of people and animals. No one knows exactly when the waterwheel was invented, but irrigation systems existed at least 5000 years ago and it seems probable that the earliest water power device was the *noria*, used to raise water for this purpose (Figure 5.9(a)). It appears to have evolved over a long period from about 600 BC, perhaps independently in different regions of the Middle and Far East (Strandh, 1989).

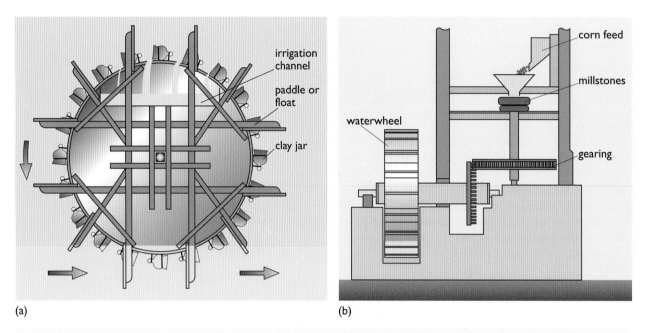

(a)   (b)

Figure 5.9 (a) A noria – in this earliest waterwheel the paddles dip into the flowing stream and the rotating wheel lifts a series of jars, raising water for irrigation, (b) A Roman mill – this corn mill with its horizontal-axis wheel was described by Vitruvius in the first century BC (note the use of gears)

The earliest **watermills** were probably corn mills, which seem to have appeared during the first or second century BC in the Middle East, and a few centuries later in Scandinavia. In the following centuries, increasingly sophisticated watermills were built throughout the Roman Empire and beyond its boundaries in the Middle East and Europe (Figures 5.9(b) and 5.10). In England, the Saxons are thought to have used both horizontal- and vertical-axis wheels. The first documented mill was in the eighth century, but three centuries later the Domesday Book of 1086 AD recorded about five thousand, suggesting that every settlement of any size had its mill.

Raising water and grinding corn were by no means the only uses of the watermill, and during the following centuries the applications of this power source kept pace with the developing technologies of mining, iron working, paper-making, and the wool and cotton industries. Water was the main source of mechanical power, and by the end of the seventeenth century England alone is thought to have had some 20 000 working mills.

There was much debate on the relative efficiencies of different types of waterwheel (Figure 5.11), and the period from about 1650 until 1800 saw some excellent scientific and technical investigations of various designs. These revealed output powers ranging from about one horsepower to

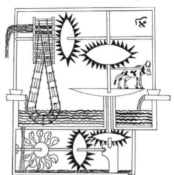

Figure 5.10 Medieval saqiya. The diagram comes from the Book of Knowledge of Ingenious Mechanical Devices of al-Jahazi, written in Mesopotamia 700 years ago. (The ox is a wooden cut-out designed to fool the public, who can't see the hidden waterwheel and gears below.)

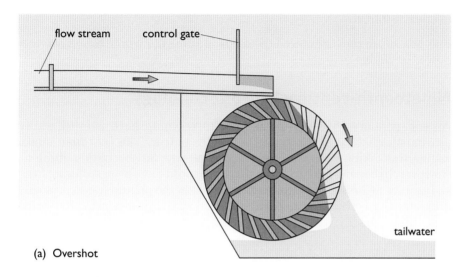

(a) Overshot

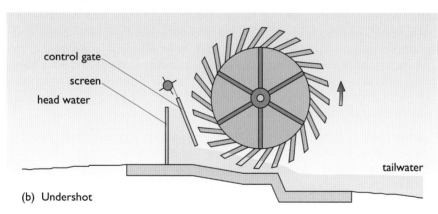

(b) Undershot

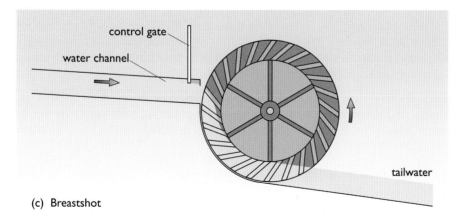

(c) Breastshot

Figure 5.11 Types of waterwheel (a) overshot – water falls onto blades with closed sides, (b) undershot – driven by water pressure against lower blades, (c) breastshot – water strikes paddles at about the level of the wheel axle

perhaps 60 for the largest wheels (in modern terms roughly 1 – 50 kW) with overshot wheels being in principle the most efficient. They also confirmed that for maximum efficiency the water should pass across the blades as smoothly as possible and fall away with minimal speed, having given up almost all its kinetic energy: two features that were to be important in modern turbines.

But then steam power entered the scene, putting the whole future of water power in doubt.

Nineteenth-century hydro technology

An energy analyst writing in the year 1800 would have painted a very pessimistic picture of the future for water power. The coal-fired steam engine was taking over and the waterwheel was fast becoming obsolete. However, like many later experts, this one would have been suffering from an inability to foresee the future. A century later the picture was completely different: the world now had an electrical industry and a quarter of its generating capacity was water powered.

The growth of the power industry was the result of a remarkable series of scientific discoveries and developments in electro-technology during the nineteenth century, but significant changes in what we might now call *hydro technology* also played their part. In 1832, the year of Faraday's discovery of the principle of the electric generator, a young French engineer, Benoît Fourneyron, patented a new and more efficient waterwheel: the first successful water **turbine**. (The name, from the Latin *turbo*: something that spins, was coined by Claude Burdin, one of Fourneyron's teachers.) The traditional waterwheel, essentially unaltered for nearly two thousand years, had finally been superseded.

Fourneyron's turbine (Figure 5.12) incorporates many new features. It is a vertical-axis machine, itself something of a novelty. But the important innovations are that the turbine runs *completely submerged* and that the water enters the turbine vertically, *along the axis*, and is directed outwards by fixed **guide vanes** on to the *blades* mounted on the *runner*, which is of course free to rotate. These are the features that ensure the smooth flow of water essential for high efficiency. The water entering at the centre is travelling horizontally outwards almost parallel to the faces of the runner blades as it reaches them. Deflected as it crosses the faces, it exerts a sideways pressure which transmits energy to the runner. Having given up its energy, it then falls away into the outflow.

Tests showed that Fourneyron's turbine converted as much as 80% of the energy of the water into useful mechanical output – an efficiency previously equalled only by the best overshot wheels. The rotor could also spin much faster, an advantage in driving 'modern' machines. The first pair of these turbines to come into use were installed in 1837 in the small town of St Blasien in the Grand Duchy of Baden (now part of southern Germany). Development did not stop there, and within a few years the American engineer James Francis started his experiments on *inward-flow radial* turbines which ultimately led to the modern machines known by his name (see Section 5.8).

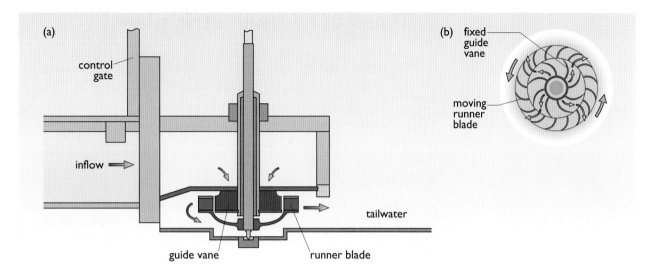

Figure 5.12 Fourneyron's turbine – the runner consists of a circular plate with curved blades around its rim and a central shaft (a) vertical section, (b) the flow across guide vanes and runner blades

These turbines were of course used to provide *mechanical* power. Half a century of development was needed before Faraday's discoveries were translated into a major electrical industry. In 1878 Lord Armstrong, a wealthy Victorian engineer and inventor, installed one of the first 'hydro-electric' plants to provide electric lighting in 'Cragside', his house in the north of England (National Trust, 2011). About three years later and 300 miles south, the small town of Godalming opened the world's first public electricity supply, powered by the river Wey. Unfortunately, the source proved unreliable – a problem common to many forms of renewable energy – and the waterwheel was soon replaced by a steam engine. Credit for the first *successful* public hydroelectricity scheme therefore usually goes to a little 12 kW plant on the Fox River in Wisconsin, USA, providing lighting for local paper mills from 1882.

From these primitive beginnings, the electrical industry grew during the final decades of the nineteenth century at a rate far exceeding that of any earlier technology. The capacities of individual power stations, including many hydro plants, were also rising, increasing tenfold, from about 100 kW to over 1 MW, during the 1890s. This growth was to continue. The data in Section 5.3 reveals that world total hydro *output* rose from 3.7 TWh in 1900 to 3272 TWh in 2009 – an almost nine-hundredfold increase. Over the same period, the outputs of the largest turbo-generators rose to 700 MW, and with twenty-six of these, the Three Gorges plant in China (see Box 5.10) has a rated capacity of 18.2 GW, the world's highest at the time of writing.

It is interesting to note, however, that not only the modes of operation but the main components of today's enormous turbo-generators are very similar to those of their much smaller precursors of over a hundred years ago. These components, mainly the turbines, are the subject of the next section.

5.7 Types of hydroelectric plant

Present-day hydroelectric installations range in capacity from a few hundred watts to more than 10 GW – a factor of some hundred million between the smallest and the largest output. We can classify installations in different ways:

- by the effective head of water
- by the capacity – the rated power output
- by the type of turbine used
- by the location and type of dam, reservoir, etc.

These categories are not of course independent of one another. The available head is an important determinant of the other factors, and the head and capacity together largely determine the type of plant and installation. We start therefore with the customary classification in terms of head, but shall soon see that it is really the fourth criterion that matters.

Low, medium and high heads

Two hydroelectric plants with the same power output could be very different: one using the huge volume flow of a slowly moving river and the other a relatively low volume of high-speed water from a mountain reservoir. Sites, and the corresponding hydroelectric installations, can be classified as *low*, *medium* or *high* head. The boundaries are fuzzy, and tend to depend on whether the subject of discussion is the civil engineering work or the choice of turbine; but **high head** usually implies an effective head of appreciably more than 100 metres and **low head** less than perhaps 10 metres. Figure 5.13 shows the main features of the three types.

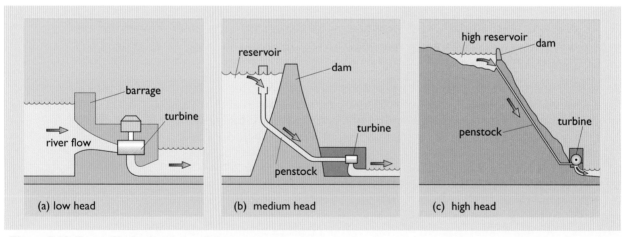

Figure 5.13 Types of hydroelectric installation

The low dam or barrier of the installation in Figure 5.13(a) serves to maintain a head of water and also houses the plant. It may incorporate locks for ships (or as we have seen above, a fish-ladder for salmon). '*Run-of-river*' power stations of this type, having relatively little storage capacity, are dependent on the prevailing flow rate and can present problems of reliability if the flow varies greatly with the time of year or the weather. The large volume flow through a low-head plant means that the plant and the associated civil engineering works are likely to be massive, which means high capital cost, although this may be ameliorated where there is a second function such as flood control or irrigation.

The plant in Figures 5.13(b) and 5.14 is typical of the very large hydroelectric installations with a dam at a narrow point in a river valley. The large reservoir behind the dam provides sufficient storage to meet demand in all but exceptionally dry conditions. (It will also have flooded an extensive area and may not have been entirely welcomed by the population (see Section 5.10)). The USA has some of the world's largest dams of this type, including the Grand Coulee, which when completed had the distinction of being the first artificial bulk structure with a volume greater than the Great Pyramid! On this scale, the civil engineering costs are obviously considerable, but the large reservoir normally ensures a reliable supply. Systems of this type don't of course have to be on a gigantic scale, and quite small reservoirs can provide power for hydroelectric plant located below their dams.

To call the 220 metre head of the Hoover Dam 'medium' may seem rather surprising, but it illustrates the fact that the distinction between this and high-head systems lies more in the type of installation. Figure 5.13(c) shows the difference. In the high-head plant the entire reservoir is well above the outflow, and the water flows through a long **penstock** – possibly passing through a mountain – to reach the turbine. (The penstock was originally the wooden gate or 'stock' which controlled the flow of 'penned-up' water. It later came to mean the channel, or the pipe, carrying the flow.)

With a high head, the flow needed for a given power is much smaller than for a low-head plant, so the turbines, generators and housing are more compact. But the long penstock adds to the cost, and the structure must be able to withstand the extremely high pressures below the great depth of water – as much as 100 atmospheres for a 1000 m head (Box 5.6).

Figure 5.14 The Hoover Dam, 1936. This dam on the Colorado River (originally called the Boulder Dam) is 220 metres high and its reservoir, Lake Mead, holds 35 billion cubic metres of water. The 2.1 GW power plant is at the foot of the dam.

BOX 5.6 Height and pressure

The pressure in a liquid (or gas) is the force with which it presses on each square metre of surface of anything submerged in it.

Atmospheric pressure, due to the weight of the air above us, is equivalent at sea level to the weight of a 10 tonne mass acting on each square metre of any surface. As you move up through the atmosphere, the pressure decreases, initially by about 1% per 100 m vertically.

As you move down through any body of water, the pressure increases due to the increasing weight of water above. Since water is several hundred times denser than air, the change is much more noticeable – at a depth of about 10 m the pressure is twice that at the surface: a pressure of two atmospheres. This increase of about one atmosphere per 10 m continues as the depth becomes greater (Figure 5.15).

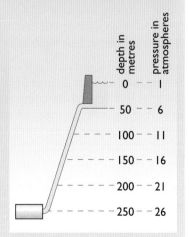

Figure 5.15 Depth and water pressure

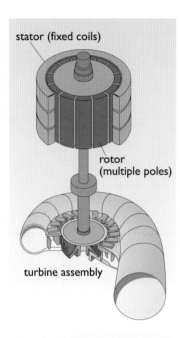

stator (fixed coils)

rotor
(multiple poles)

turbine assembly

Figure 5.16 Multi-pole
generator in a hydro power plant

Rates of rotation

The **rate of rotation (*n*)** of a turbine is the number of completed revolutions per minute (rpm), and, as we saw for the Galloway plants in Table 5.2, this can vary appreciably depending on the site and the turbine type. However, if the turbine drives the generator directly, only certain rates of rotation are permitted, for the following reason.

The alternating voltages from all the power stations contributing to any grid system must have the same frequency. In European countries and many others, the agreed frequency is 50 Hz (hertz, or cycles per second). So in a very simple generator, consisting of a magnet and a pair of coils, the magnet would need to spin at 50 cycles per second, which is $50 \times 60 =$ 3000 rpm. In the usual terminology, this simple system would be called a **two-pole generator**.

However, a large power-station generator will have several pairs of electromagnets on its spinning rotor, closely surrounded by a stator containing many coils of fine copper wire in which the voltage is generated (Figures 5.16 and 5.22). A 20-pole machine, for instance, would need to rotate at only 300 rpm, a tenth of the above rate, to produce the required 50 Hz. A little arithmetic reveals that all the rates of rotation in Table 5.2 are sub-multiples of 3000 rpm.

Note however that in the USA and other countries where the supply frequency is 60 Hz, the rates of rotation are sub-multiples of 3600 rpm. (Generators are discussed in more detail in Chapter 9, *Electricity*, of Everett et al., 2012.)

Estimating the power

Reliable data on water flow rates and, equally important, their variations, is essential for the assessment of the potential capacity of a site. Stopping the flow and catching the water for a measured time is hardly practicable for large flows or as a routine method. The preferred techniques depend on establishing empirical relationships between flow rate and either water depth or water speed at chosen points. Simple depth or speed monitoring then provides a record of flow rates. For many major rivers, particularly in developed countries, such data has been accumulated for years.

Where such records are not available, an entirely different approach is to determine the annual precipitation over the catchment area. This gives the total flow into the system and is particularly suitable for large systems. However, allowance must be made for losses due to processes such as re-evaporation, take-up by vegetation or leakage into the ground, and as these could account for as much as three-quarters of the original total they are hardly negligible corrections.

Dealing with time variations adds further problems. In most areas there will be seasonal changes, but these at least come at known times. The more serious problems are with changes over very long or very short periods. Year-to-year variations can be large: the average annual precipitation on the catchment area of the River Severn in the UK, for instance, is 900 mm but it can range from as little as 600 mm to as much as 1200 mm. For countries which depend heavily on hydroelectric power a succession of dry years can mean a serious supply shortage.

At the other extreme, the installation must be designed to survive the '100-year flood', the sudden rush of water following unusually heavy rain. As in any power system, the need to guard against rare but potentially catastrophic events adds to the cost.

5.8 Types of turbine

Present-day turbines come in a variety of shapes (Figure 5.17). They also vary considerably in size, with runner diameters ranging from as little as a third of a metre to some 20 times this. In the next four subsections we look at how they work, the factors that determine their efficiency, and the site parameters that determine the most suitable turbine.

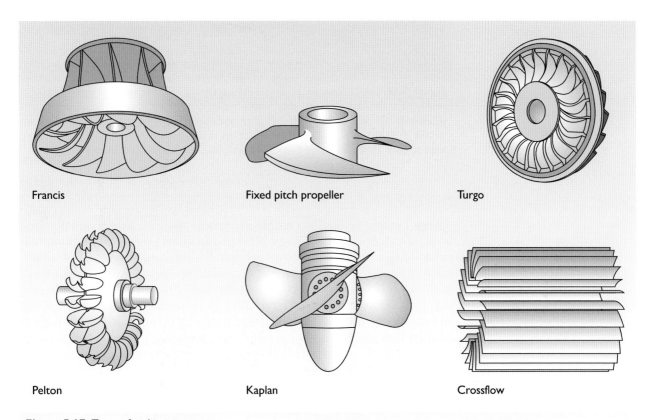

Figure 5.17 Types of turbine runner

Francis turbines

Francis turbines (Figures 5.17 to 5.20) are by far the most common type in present-day medium- or large-scale plants, being used in locations where the head may be as low as 2 m or as high as 300 m. They are radial-flow turbines, and although the water flow is inwards towards the centre instead of the outward flow of Fourneyron's turbine, the principle remains the same.

Figure 5.18 The photograph shows six Francis runners and behind them three Turgo runners. The different sizes and shapes reflect their outputs and also the 'head' under which each will operate. The largest of the Francis turbines, for instance, is designed for an output of 10 MW at a head of 280 m, whilst the smallest one generates only 600 kW at a head of 80 m. (The subsection Ranges of Application, below, discusses these factors.)

Action of the turbine

As the Francis turbine is completely submerged, it can run equally well with its axis horizontal (Figure 5.19) or vertical (Figure 5.20). In medium- or high-head turbines the flow is channelled in through a scroll case (also called the volute) a curved tube of diminishing size rather like a snail shell, with the guide vanes set in its inner surface. Directed by the guide vanes, the water flows in towards the runner. The shapes of the guide vanes and runner blades and the speed of the water are critical in producing the smooth flow that leads to high efficiency (see below). Francis turbines run most efficiently when the blade speed is only slightly less than the speed of the water meeting them.

As it crosses the curved runner blades the water is deflected sideways, losing its whirl motion. It is also deflected into the axis direction, so that it finally flows out along the central draft tube to the tail race. That the water exerts a force on the blades is obvious because it has changed direction in passing through the turbine. In being deflected by the blades, it pushes on them in the opposite direction – the way they are travelling – and this reaction force transfers energy to the runner and maintains the rotation. For this reason, these are called **reaction turbines**. An important feature of this type is that the water arrives at the runner under pressure, and the pressure drop through the turbine accounts for a large part of the delivered energy.

Maximizing the efficiency

Although, as we saw above in Section 5.5, there are always energy losses, modern turbines can achieve efficiencies as high as 95% – but only under

Figure 5.19 The 450 MW horizontal-axis Francis turbine of a small-scale plant in Scotland, commissioned in 1993. The inflow (at lower right) is 2.1 m³ s⁻¹ at a head of 25 m. Part of the generator casing can be seen on the left

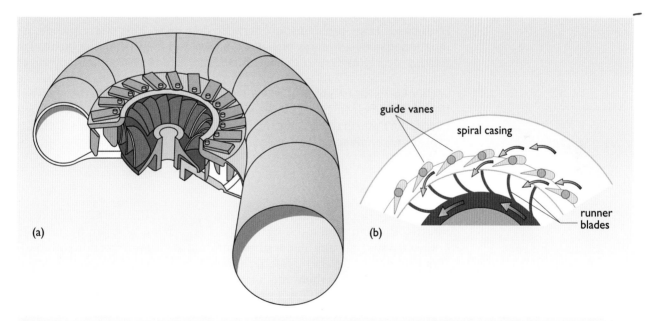

Figure 5.20 Structure of a Francis turbine, showing the central runner blades, the pivoting guide vanes and the surrounding volute

optimum conditions. Maintaining exactly the right speed and direction of the incoming water relative to the runner blades is important, and this leads to a problem. Suppose demand falls: the output power can be reduced by reducing the water flow, and in a Francis turbine this is done by turning the guide vanes; but this changes the angle at which the water hits the moving blades and the efficiency falls. This is a characteristic which must be accepted with this type of turbine. (As we'll see below, some 'propeller' types allow adjustment of the pitch, changing the angle of the blades to match the new conditions.)

A rather different cause of less than 100% efficiency is that the water flowing out carries away kinetic energy. A partial remedy is to flare the draft tube. If the tube becomes larger but the volume flow stays the same, the actual speed of the water must decrease, reducing the energy loss. It may seem strange that this change *after* the water has left the turbine makes any difference to its efficiency, but the effect of the deceleration is to reduce the pressure back at the exit from the turbine, increasing the pressure drop across it and therefore the energy it extracts.

Limits to the Francis turbine

The available head is an important factor in selecting the best turbine for a particular site. If the head is low a large volume flow is needed for a given power. But a low head also means a low water speed, and these two factors together mean that a much larger input area is required. Attempts to increase this area whilst adapting the blades to the reduced water speed and at the same time deflecting the large volume into the draft tube led to turbines with wide entry and blades which were increasingly twisted. Ultimately the whole thing began to look remarkably like a propeller in a tube, and this is indeed the type of turbine now commonly used in low-head situations (see below).

High heads bring problems too, because they mean high water speeds. As mentioned above, Francis turbines are most efficient when the blades are moving nearly as fast as the water, so high heads imply high speeds of rotation. A look at Tables 5.1 and 5.2 in Section 5.2 reveals that the turbine at Glenlee, with its much higher head, rotates at up to twice the speed of the other Francis turbines in the Galloway system. For sites with very high heads, the Francis turbine becomes unsuitable, and yet another type takes over, as we shall see.

'Propellers'

In the 'propeller' or **axial-flow** turbines shown in Figures 5.17 and 5.21, the area through which the water enters is as large as it can be: it is the entire area swept by the blades (these turbines are, again, reaction turbines). Axial-flow turbines are therefore suitable for very large volume flows and have become usual where the head is only a few metres. They have the advantage over radial-flow turbines in that it is technically simpler to improve the efficiency by varying the angle of the blades when the power demand changes. Axial-flow turbines with this feature are called **Kaplan turbines**.

An important feature of 'propeller' turbines is that their optimum blade speed for maximum efficiency is appreciably greater than the water speed – as much as twice as fast. This allows the high rate of rotation needed by the generator even with relatively low water speeds. (Note that because the outer parts of the blade move faster than the more central parts, the blade angle needs to increase with distance from the axis. This is why a propeller has its familiar twisted shape.)

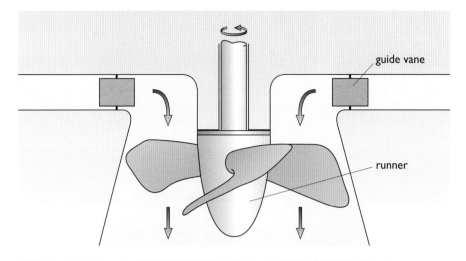

Figure 5.21 A 'propeller' or axial-flow turbine

With axial flow there is no need to feed the water in from the side, and it is obviously simpler to let it flow in along the axis instead of being deflected through a right angle. However, this raises the problem of where to position the generator: if located directly along the axis of the turbine it will either get in the way and/or get wet! Several different solutions to this problem – rim generators and tubular turbines – are shown in Chapter 6.

Pelton wheels

For sites of the type shown in Figure 5.13(c), with heads above 300 metres or so (or lower for small-scale systems) the **Pelton wheel** is the preferred turbine. It evolved during the gold rush days of late nineteenth century California, was patented by Lester Pelton in 1880, and is entirely different from the types described above. It is, in contrast to the reaction turbines discussed previously, an **impulse turbine**. One important difference between the turbine types is that whereas a reaction turbine runs fully submerged and with a pressure difference across the runner, impulse turbines essentially operate in air at normal atmospheric pressure.

A Pelton wheel is basically a wheel with a set of double cups or 'buckets' mounted around the rim (Figures 5.22 and 5.23(a)). A high-speed jet of water, formed under the pressure of the high head, hits the splitting edge between each pair of cups in turn as the wheel spins (Figure 5.23(b)). The water passes round the curved bowls, and under optimum conditions gives up almost all its kinetic energy. The power can be varied by adjusting the jet size to change the volume flow rate, or by deflecting the entire jet away from the wheel.

The efficiency of a Pelton wheel is greatest when the speed of the cups is half the speed of the water jet (Box 5.7). As the cup speed depends on the rate of rotation and the wheel diameter, and the water speed depends on the head, there is an optimum relationship between these three factors.

(a)   (b)

Figure 5.22 The 16.5 MW Finlarig power station, on the shores of Loch Tay, draws its water from Lochan na Lairige at a head of 415 metres. Its average annual output is 70 GWh (SSE, 2011) (a) the power station, (b) the original twin-jet Pelton wheels and rotor of the horizontal-axis 30 MW multi-pole generator

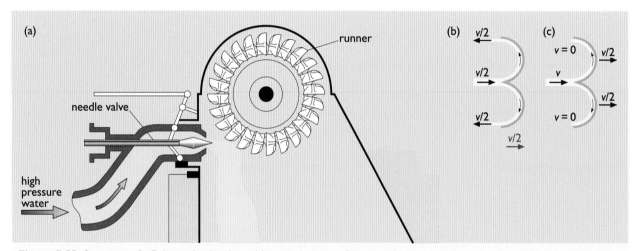

Figure 5.23 Structure of a Pelton wheel turbine: (a) vertical section, (b) water flow as seen from moving cup, (c) actual motion of water and cup

BOX 5.7 Optimum speed for a Pelton wheel

The following informal argument shows that a Pelton wheel extracts the maximum energy from the water if the cups move at half the speed of the water jet.

Consider the situation when water at speed v approaches a cup which is already moving in the same direction at half this speed ($v/2$). As seen from the cup, the water will be approaching at the difference between these two speeds, which in this case is also $v/2$.

Suppose now that the water passes smoothly round inside the curved cup until it leaves travelling in the opposite direction – as seen from the cup. You now have water moving backwards at speed $v/2$ relative to a cup that is moving forwards at just this speed. A person on the ground would see the cup moving on while the water simply falls vertically out of it. The water has given up all its kinetic energy to the wheel. 100% efficiency, in principle!

In practice this is only approximately true, and the best cup speed is a little less than $v/2$.

Input power

The power input to a Pelton wheel is determined, as usual, by the effective head and the flow rate of the water. Box 5.8 shows that, ideally, the volume rate of flow (Q) corresponding to an effective head H is:

$$Q = A \times \sqrt{(2gH)}$$

where A is the area of the jet.

Equation (2) in Section 5.5 shows that the input power to a turbine is:

$$P = 1000 \times Q \times g \times H$$

so substituting for Q we find that:

$$P = 1000 \times A \times \sqrt{(2gH)} \times g \times H$$

Using the approximate value of g (10 m s^{-2}), the power in kilowatts becomes

$$P\,(\text{kW}) = 45A\sqrt{(H^3)}$$

If adjacent cups are not to interfere with the flow, the wheel diameter needs to be about ten times the diameter of the jet. But two or even four jets can be spaced around the wheel to give greater output without increasing the size. If the number of jets is j, the power equation becomes

$$P\,(\text{kW}) = 45jA\,\sqrt{(H^3)}$$

BOX 5.8 Effective head, water speed and flow rate

Although there are in practice always energy losses in forming a jet, we'll assume here that the water leaves the jet at the speed that it would have gained in 'free fall' through the effective head.

We know from Equation (1) in Section 5.5 that the potential energy lost by M kg of water in falling through H metres is given by:

potential energy $= MgH$

In Chapter 1 we saw that the kinetic energy of a moving object is proportional to its mass and the square of its speed:

kinetic energy $= \dfrac{1}{2}Mv^2$

So if all the lost potential energy is converted into kinetic energy, we have:

$$\frac{1}{2}Mv^2 = MgH$$

so $v^2 = 2gH$ and

$$v = \sqrt{(2gH)} \tag{5}$$

If this water flows as a jet with a circular area of A square metres, the *volume* flowing out in each second (Q) will be equal to A times v. So the volume flow rate for an effective head H is given by

$$Q = A \times \sqrt{(2gH)}$$

Turgo and cross-flow turbines

A variant on the Pelton wheel is the **Turgo turbine** (Figures 5.17 and 5.24), developed in the 1920s. The double cups are replaced by single, shallower

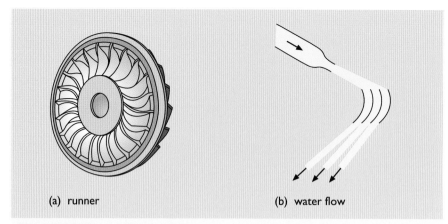

(a) runner   (b) water flow

Figure 5.24 Flow in a Turgo turbine

ones, with the water entering on one side and leaving on the other. The water enters as a jet at a low angle to the plane of the turbine, striking the cups in turn, so this is still an impulse turbine (and the relationships in Box 5.8 still apply). However, its ability to handle a larger volume of water than a Pelton wheel of the same diameter gives it an advantage for power generation at medium heads.

The cross-flow turbine shown in Figure 5.17 (also known as the **Mitchell-Banki**, or **Ossberger turbine**) is yet another impulse type. The water enters as a flat sheet rather than a round jet. It is guided on to the blades, then travels across the turbine and meets the blades a second time as it leaves. As we see in the next section, cross-flow turbines are often used instead of Francis turbines in small-scale plants with outputs below 100 kW or so, and some ingenious technological ideas have gone into the development of simple types of generator which can be constructed (and maintained) without sophisticated engineering facilities and are therefore suitable for remote communities.

Ranges of application

We have seen that, in general, Pelton wheels are most suitable for high heads, propellers for low heads and Francis turbines for the intermediate ranges. But the effective head is not the only factor determining the most appropriate type for a given situation. The available power also matters.

Figure 5.25 represents one way to display the ranges of application of the different turbines. It shows the ranges of head, flow rate and corresponding power which best suit each type. It should be noted however that the boundaries are by no means clear-cut, and other technical issues, or criteria such as cost, simplicity in manufacture or ease of maintenance can lead to choices outside these ranges. Locating the site data for the Galloway plants (and others in this chapter) on Figure 5.25 and comparing the turbine types actually used with those suggested by the diagram shows that this is indeed often the case.

Small-scale installations can also be classified by the available head and flow rate, but the ranges may be very different from those of large plants. Many of these plants are run-of-river, with heads of only a few metres, and

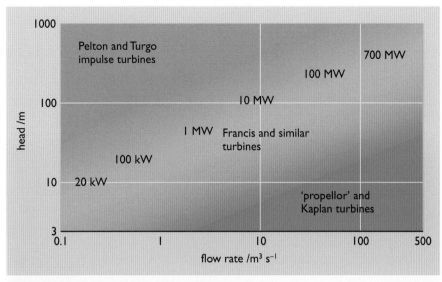

Figure 5.25 Ranges of application of turbines

as little as 10 m could be regarded as 'high head' for a very small plant. The corresponding choices of turbine type may vary appreciably from those indicated in Figure 5.25.

5.9 **Hydro as an element in a system**

Even if a potential source of electric power is acceptable environmentally and financially, other factors remain that affect its viability. Few large power stations operate in isolation, and the extent to which the proposed plant can form a useful part of a supply *system* is important.

From the point of view of the operator of the system, the characteristics of the ideal power station would be:

(1) constant availability

(2) a reserve energy store to buffer variations in input

(3) no correlation in input variations between power stations

(4) rapid response to changing demand

(5) an input which matches annual variation in demand

(6) no sudden and/or unpredictable changes in input

(7) a location which does not require long transmission lines.

Few, if any, sources meet all these criteria, but each compromise with the ideal adds to the effective cost.

Almost all hydroelectric plants score well on item 4, and in regions with cold, dark, wet winters, on item 5 as well – unless the water is locked up as ice. Furthermore, sudden unplanned fluctuations in input (item 6) are rare, at least in large plants.

How well hydro performs on items 1, 2 and 3 depends in part on the type of plant. A high-head installation with a large reservoir will normally have

Figure 5.26 The Grand Coulee Dam on the Columbia River, constructed in 1942, is 170 metres high and about 1 km long. The total generating capacity of its 30 turbo-generator sets is 6.8 GW, of which 300 MW is pumped storage plant (USBR, 2011)

little difficulty in maintaining its output over a dry period, whereas the water held behind the low dam of a run-of-river plant may not be sufficient to compensate for periods of reduced flow. A serious drought can of course affect all hydro plant over a wide region, so it cannot be said fully to satisfy the third requirement.

The final criterion is the real hurdle. Hydro locations are determined by geography, and whilst run-of-river plant may sometimes be near major centres of population, this is rare for high-head systems.

Is the case different for small-scale hydro? Smaller plants are predominantly run-of-river, or perhaps served by relatively small reservoirs, in either case a less reliable supply. On the other hand, scattered sites with different rainfall patterns could result in increased reliability. Local small-scale plant can reduce the need for long-distance transmission, reducing energy losses and costs; although the generating cost per unit of output may be greater.

Overall, hydroelectricity ranks reasonably well in terms of the above criteria. And it may also offer a bonus. A hydro plant with a large reservoir not only maintains its own reserve of energy – it might, as we have seen earlier, provide a store for the surplus output of other power stations.

5.10 Environmental considerations

The environmental issues associated with hydroelectricity are no less controversial than those for other energy sources. We might usefully start by briefly summarizing the environmental *benefits* of hydroelectricity compared with other types of power plant:

- in operation it releases no CO_2, and negligible quantities of the oxides of sulfur and nitrogen that lead to acid rain
- it produces no particulates or chemical compounds such as dioxins that are directly harmful to human health
- it emits no radioactivity
- dams may collapse, but they will not cause major fires.

Moreover, hydroelectric plant is often associated with positive environmental effects such as flood control or irrigation, and in some cases, its development leads to a valued amenity or even a visual improvement to the landscape.

However, during the twentieth century, the construction of large dams has led to the displacement of many millions of people from their homes, and dam failures have killed many thousands. We'll consider these and other deleterious effects under three headings:

- hydrological effects – water flows, groundwater, water supply, irrigation, etc.
- other physical effects of large hydro plants
- social effects.

Hydrological effects

The three categories of effect listed above are not of course independent. Any hydrological change will certainly affect the ecology and thus the local community. A hydroelectric scheme is not basically a *consumer* of water, but the installation does 'rearrange' the resource. Diverting a river into a canal, or a mountain stream into a pipe, may not greatly change the total flow, but it can have a marked effect on the environment (see for instance Box 5.9). Furthermore, evaporation from the exposed surface of a large reservoir may appreciably reduce the available water supply.

BOX 5.9 Gabcikovo-Nagymaros

The River Danube has provided hydroelectric power in several countries along its length for many years, but a proposed development on the Slovak-Hungarian border has been a controversial issue for over three decades. The original 880 MW scheme, agreed in 1977, involved the construction of a large dam downstream from Bratislava (Figure 5.27). The reservoir formed behind this would define the upstream level of a new canal carrying diverted water to a hydro plant at Gabcikovo before rejoining the Danube further downstream. There was also to be a second barrage and power plant at Nagymaros.

Work proceeded on the upstream project for about a decade, but the political changes of the late 1980s brought increasingly vocal opposition on environmental grounds, particularly in Hungary. Construction on the Hungarian side effectively stopped in 1989, and in May 1992 the government announced its cancellation of the 1977 agreement.

The newly established Slovak state, committed to the scheme, declared that the cancellation was illegal. In October 1992 Slovak engineers used the period when the river was at its lowest to complete the diversion into the new canal, 18 km long and with concrete walls rising 15 metres above the surrounding countryside.

The resulting fall in the natural water table, with wells drying up, vegetation dying and unique forms of wildlife in danger, reinforced calls for legal limits to the diversion of water. An artificial irrigation scheme only slightly alleviated the situation, and in 1993 the Hungarian and Slovak governments took the dispute to the International Court of Justice at the Hague. In 1995, before the Court ruled, Slovakia agreed to reduce the water diverted.

In September 1997, the Court ruled that both countries had acted illegally (ICJ, 1997), and demanded that they should jointly negotiate a new solution (without the Nagymaros plant). Pointing out that the requirements for continuous assessment of environmental risks had become stronger since the original treaty, the Court noted that '*both Parties agree on the need to take environmental concerns seriously and to take the required precautionary measures, but they fundamentally disagree on the consequences this has for the joint Project*'.

The two governments have met for negotiations, but at the time of writing have not agreed a solution. The present half-scheme, with only one power station to generate income, is reported to be a financial disaster.

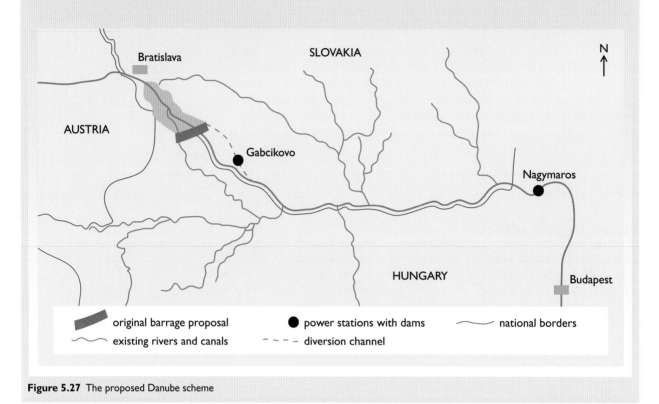

Figure 5.27 The proposed Danube scheme

Other physical effects

Any structure on the scale of a major hydroelectric dam will affect its environment in many ways in addition to the hydrological changes discussed above. The construction process itself can cause widespread disturbance, and although the building period may be only a few years, the effect on a fragile eco-system can be long-lasting. In the longer term, a large reservoir is bound to bring other significant environmental changes. Whether these are seen as catastrophic, beneficial or neutral will depend on the situation – in the geographical and biological sense – and certainly on the points of view and interests of those concerned.

Dam failures

The available records suggest that worldwide during the half-century since 1960 there have been about 35 major dam failures – defined as those resulting in serious material damage and/or deaths (Wikipedia, 2011a; 2011b). Many dams are of course for purposes other than hydroelectricity: flood control, water supply, irrigation or recreation. A US study found, for instance, that only 2400 of the country's 80 000 dams had hydroelectricity plants (U.S. Dept of Energy, 2005). It is not therefore surprising to find only five or six hydro plants in the world list of major dam failures. There is however one caveat: the relatively meagre data on China. It is known that severe flooding in 1975 led to many dam failures, and estimates of the total fatalities vary between about 70 000 and over a quarter of a million (New Scientist, 2011). If the figure is indeed in this range, and hydro plants made up a significant fraction of the destroyed dams, then hydropower must rank high in any list of energy sources in terms of deaths per kWh of useful energy.

Plant failures

Hydroelectricity, with no high temperatures or combustible or radioactive fuels, has historically compared very well with other types of power plant in terms of serious damage due to plant failures. It does however involve high *pressures*, and even a head as low as 200 metres implies a pressure of 20 atmospheres. This was the head driving the ten 640 MW turbo-generators of the Sayano-Shushenskaya power station on the Yenisei River in Western Siberia, Russia, a plant whose annual output of about 25 TWh supported its claim to be the world's sixth-largest hydroelectric plant (Wikipedia, 2011c).

At 8 a.m. on Monday 17 August 2009, Turbine No.2 at Sayano-Shushenskaya disintegrated, destroying or seriously damaging the remaining nine turbines and the building, killing 75 workers, releasing 40 tonnes of transformer oil into the river and depriving the major industrial plants in the surrounding region of power (which resulted in the loss of an estimated half a million tonnes of aluminium output). Subsequent investigation found that a year or so earlier an inspection had declared all ten turbines unfit for operation due to age and poor maintenance.

Silt

Silt accumulation behind dams has been a known problem for many years – the build-up reduces the volume of stored water and consequently the hydro potential of a site. The Hoover Dam (Figure 5.14), for instance, lost about one sixth of its useful storage volume in its first thirty years, although the loss rate was reduced when the Glen Canyon dam was built 370 miles upstream.

The Aswan High Dam, built in the 1960s, also supplies rather less electricity than originally planned, but a concern attracting greater attention than silt accumulation is that the land downstream no longer receives the soil and nutrients previously carried by the annual Nile floods. An agricultural system in existence for millennia has largely been destroyed, to be replaced by irrigation and the use of fertilizers (El-Sayad and van Dijken, 1995).

Fish

The account of the Galloway Hydros in Section 5.2 mentions the effect on the salmon passing up the rivers to spawn as a major environmental issue – solved in that case by fish ladders. France has many dams constructed during the early twentieth century on rivers previously used by Atlantic fish, and as licences became due for renewal in the 1990s, stringent requirements were introduced for the construction of fish ladders or similar passages. One consequence was the *decommissioning* of dams deemed unsuitable for renewal, on environmental or economic grounds (ERN, 2000).

The decommissioning schemes implemented for a number of relatively small French dams were designed in part as pilot schemes: to identify some of the issues that might arise across Europe as more of the early and mid-twentieth century dams built for hydroelectricity become due for renewal or renovation. Recent decades have also seen calls in the USA for the decommissioning of dams on environmental grounds, leading to interesting debates with others who, also citing environmental concerns, favour hydro development (ORNL, 1993).

Methane

For many years, hydroelectricity was regarded as one of the renewables that produce virtually no greenhouse gases, but this view came into question in the early years of the present century. It had long been known that vegetable matter that would normally decay in the air to produce carbon dioxide (CO_2) could decay anaerobically under water to produce methane (CH_4). When methane was identified as a much more potent greenhouse gas than carbon dioxide, the question arose as to whether hydro schemes that flooded land previously covered in vegetation should join the fossil fuels as significant contributors to global warming.

Following this concern, some detailed studies of individual reservoirs were carried out. One, by a team from the Swiss aquatic research institute EAWAG, was prompted by the local observation that the reservoir lake of a hydroelectric plant on the river Aare in northern Switzerland was bubbling during the summer. The detailed study (EAWAG, 2010) estimated that each square metre of the lake surface was releasing about 0.15 grams of methane per day – much more than would be expected for its temperate location, and equivalent to a total annual methane release of about 150 tonnes. The suggested cause was the anaerobic digestion of the particularly large annual quantity of vegetable matter (leaves, etc.) brought down by the river. The report also observed that a coal-fired power plant producing the same electrical output would release approximately 40 times more greenhouse gases, expressed in CO_2 equivalents – a calculation ignored in the frequently expressed view that the greenhouse gas emissions from the reservoirs serving hydro plants are comparable to those from fossil fuel plants of similar capacity,

Although this issue features in current overall assessments of the environmental consequences of hydro development, it seems at the time of writing to play a relatively small role. The *2010 Survey of Renewable Energy Resources* of the World Energy Council (WEC, 2010a) includes 'greenhouse-gas footprint', together with hydrological vulnerability, climate-change mitigation and adaptation, as aspects of environmental impact that are being studied, but anticipates that 'the hydropower projects sector will continue to be one of the main contributors to the carbon credits market'. The methane issue also features in the new protocol towards which the hydroelectricity industry is currently working (see below).

Social effects

From a child's book on energy:

> They built a dam and made a lake in the place where Ahmed lived. Ahmed and his family had to leave the farm. His grandfather had lived there. Ahmed was born in that place. He was sad to go.

Just so. Cost–benefit analysis, it has been said, usually means that I pay the cost and you get the benefit.

The Aswan and Kariba dams involved the relocation of some 80 000 and 60 000 people respectively, whilst the rising water behind the Three Gorges Dam (Box 5.10) submerged about 100 towns and displaced over a million people. It is estimated that during the second half of the twentieth century, some 10 million people were displaced by reservoirs in China alone.

BOX 5.10 Three Gorges

History

China's Three Gorges plant on the Yangzi River was the largest of a series of hydro developments worldwide that attracted major opposition on environmental and social grounds. Originally proposed in 1919 by Sun Yat Sen, the project had a varied political history, culminating in its approval in 1992 against the unprecedented opposition of a third of the Congress delegates. Table 5.6 summarizes the arguments in China at the time. (The table originally appeared in ChinaOnline, in English, and the wording here is verbatim.)

Table 5.6 Summary of arguments for and against the dam

Issue	Criticism	Defense
Cost	The dam will far exceed the official cost estimate, and the investment will be unrecoverable as cheaper power sources become available and lure away ratepayers.	The dam is within budget, and updating the transmission grid will increase demand for its electricity and allow the dam to pay for itself.
Resettlement	Relocated people are worse off than before and their human rights are being violated.	15 million people downstream will be better off due to electricity and flood control.
Environment	Water pollution and deforestation will increase, the coastline will be eroded and the altered ecosystem will further endanger many species.	Hydroelectric power is cleaner than coal burning and safer than nuclear plants, and steps will be taken to protect the environment.
Local culture and natural beauty	The reservoir will flood many historical sites and ruin the legendary scenery of the gorges and the local tourism industry.	Many historical relics are being moved, and the scenery will not change that much.

(Table continues over page.)

Table 5.6 Summary of arguments for and against the dam (Continued)

Issue	Criticism	Defense
Navigation	Heavy siltation will clog ports within a few years and negate improvements to navigation.	Shipping will become faster, cheaper and safer as the rapid waters are tamed and ship locks are installed.
Power generation	Technological advances have made hydrodams obsolete, and a decentralized energy market will allow ratepayers to switch to cheaper, cleaner power supplies.	The alternatives are not viable yet and there is a huge potential demand for the relatively cheap hydroelectricity.
Flood control	Siltation will decrease flood storage capacity, the dam will not prevent floods on tributaries, and more effective flood control measures are available.	The huge flood storage capacity will lessen the frequency of major floods. The risk that the dam will increase flooding is remote.

Figure 5.28 The Three Gorges Dam

Opposition, locally and internationally, centred on the displacement of the 1.13 million people who were to lose their homes, farms and workplaces. Other serious concerns included the problem of silt that was predicted to block harbours upstream, increase flooding in some areas and ultimately reduce the plant output. Furthermore there was also the loss of one of China's most valued landscapes.

The international outcry led the World Bank and some other major financial institutions to dissociate themselves from the project, but support was found elsewhere and work started in 1993. The dam, a mile long and 181 metres high, was completed in early 2003, and the sluices were finally closed at midnight on 1 June. The river was released from its five-year diversion and by early morning the water behind the dam was over 100 m deep. Two weeks later it was almost a metre above its intended final level.

Operation of the plant

The first fourteen 700 MW turbo-generators on the left bank went into operation in September 2005 and the twelve plants on the right bank in October 2008. The entire 26 plants operated simultaneously for the first time in June 2009, but, due to the relatively low head of water during the flood season, the total output was only 16.1 GW – well under their potential maximum of 18.2 GW.

This highlights the well-known problem that the electrical output of any hydro plant depends on the head and the flow rate of the water. At this location on the Yangtze the available flow rate varies in the course of a year between 5000 and 30 000 cubic metres per second. The head, in this case the height difference between the water behind the dam and the water below it, also varies depending on the season and the flood conditions.

These factors explain why Chinese claims for records by Three Gorges concentrate on 'output in a single month', whilst the Brazilians can still claim the record *annual* output, despite Itaipú's lower rated capacity (see Box 5.12).

Sources: International Water Power & Dam Construction, 2007; International Rivers, 2011a; Hvistendahl, 2008; Ni, 2009

But even for the people immediately affected, the building of dams can have very different consequences. For those living in a valley which will become a reservoir it means the loss of your family home, possibly your livelihood, and often your entire community. In contrast, for people living on a river which periodically overflows its banks, the barrage and embankments of a hydroelectric scheme can bring freedom at last from devastating floods. On a smaller scale, the changes that mean the loss of a beloved riverside walk to some may be welcomed by others as an opportunity for new leisure activities.

Responses from the industry

Recent years have seen growing concern by the (large-scale) hydroelectricity industry over the above criticisms. The following are two responses.

Sustainable Hydropower

The website Sustainable Hydropower, supported by the International Hydropower Association (IHA), an international organization representing the industry, includes the following presentation of the issues.

> After more than a century of experience, hydropower's strengths and weaknesses are equally well understood. Hydropower's negative impacts are well understood and, although not all can be eliminated, much can be done to mitigate them. These are summarised for economic, social and environmental aspects of hydropower in the following tables:

Environmental aspects	
Advantages	**Disadvantages**
Produces no atmospheric pollutants	Inundation of terrestrial habitat
Neither consumes nor pollutes the water it uses for electricity generation purposes	Modification of hydrological regimes
Produces no waste	Modification of aquatic habitats
Avoids depleting non-renewable fuel resources (i.e. coal, gas, oil)	Water quality needs to be monitored/managed
Very few greenhouse gas emissions relative to other large-scale energy options	Greenhouse gas emissions can arise under certain conditions in tropical reservoirs
Can create new freshwater ecosystems with increased productivity	Temporary introduction of methylmercury into the food chain needs to be monitored/managed
Enhances knowledge and improves management of valued species due to study results	Species activities and populations need to be monitored/managed
Can result in increased attention to existing environmental issues in the affected area	Barriers for fish migrations, fish entrainment
	Sediment composition and transport may need to be monitored/managed
	Introduction of pest species needs to be monitored/managed

Social aspects	
Advantages	**Disadvantages**
Leaves water available for other uses	May involve resettlement
Often provides flood protection	May restrict navigation
May enhance navigation conditions	Local land use patterns will be modified
Often enhances recreational facilities	Waterborne disease vectors may occur
Enhances accessibility of the territory and its resources (access roads and ramps, bridges)	Requires management of competing water uses
	Effects on impacted peoples' livelihoods need to be addressed, with particular attention to vulnerable social groups
Provides opportunities for construction and operation with a high percentage of local manpower	Effects on cultural heritage may need to be addressed
Improves living conditions	
Sustains livelihoods (fresh water, food supply)	

Economic aspects	
Advantages	**Disadvantages**
Provides low operating and maintenance costs	High upfront investment
Provides long life span (50 to 100 years and more)	Precipitation dependent
Meets load flexibly (i.e. hydro with reservoir)	In some cases, the storage capacity of reservoirs may decrease due to sedimentation
Provides reliable service	Requires long-term planning
Includes proven technology	Requires long-term agreements
Can instigate and foster regional development	Requires multidisciplinary involvement
Provides highest efficiency rate (payback ratio and conversion process)	Often requires foreign contractors and funding
Can generate revenues to sustain other water uses	
Creates employment opportunities	
Saves fuel	
Can provide energy independence by exploiting national resources	
Optimizes power supply of other generating options (thermal and intermittent renewables)	

Source: (Sustainable Hydropower, 2011b)

The Hydropower Sustainability Assessment Protocol

The Hydropower Sustainability Assessment Protocol (HSAP) is the result of a collaboration by representatives of different sectors of the hydro industry, led by the International Hydropower Association (IHA). Essentially a list of criteria that should be satisfied by any new hydroelectricity project, it is no doubt the response of the industry to many of the problems discussed above. It was accepted by the membership of the IHA in November 2010.

The Protocol (HSAP, 2010) consists of an explanatory background document and four assessment sections, covering the four stages of any new project: *Early Stage, Preparation, Implementation*, and *Operation*. Collectively, these impose conditions designed to meet many of the criticisms of large-scale hydro, and if implemented could indeed make a contribution to the advancement of 'sustainable hydropower'.

However, a governance model for the Protocol has yet to be established, and it remains to be seen how far it will be adopted in practice.

Environmental effects of small-scale systems

There is general consensus that small-scale hydro plants have fewer deleterious effects than large systems. In some respects this is evidently true – few people have been displaced from their homes by the installation of small 5 MW plants, whilst deaths from the collapse of dams across small streams seem rare.

However, not everyone agrees with the consensus. The claim, made mainly by proponents of large-scale hydro, is that a general world view – 'small is beautiful' – has been allowed to override detailed analysis. It is true that the efficiencies and the capacity factors of small-scale plants tend to be lower, and in some cases the 'reservoir area' per unit of output is greater. But as we have seen, all these factors vary significantly from site to site, and generalization is difficult.

Comparisons

It should not be forgotten that the choice may not be hydroelectricity or nothing, but hydroelectricity or some other form of power station. Despite the 'penalties' discussed above, hydroelectricity scores relatively well in terms of many other criteria. Its overall greenhouse gas emissions, including the construction of dams, are likely to be lower than those of rival fossil-fuelled generating systems (see Chapter 10). Current issues for hydropower include the question of methane emissions and the costing of long-term compensation for the people displaced by major new hydroelectric installations. Nevertheless, on the criteria used in these studies, hydro appears amongst the least harmful sources of electricity.

5.11 Economics

> Generating plant can be broadly categorised either as being expensive machines for converting free or low cost energy into electrical energy or else lower cost machines for converting expensive fuels into electrical energy.
>
> (Mott MacDonald, 2010)

No matter how elegant the technology, few will invest in it unless it is going to make a profit. Potential investors need to know how much each kilowatt-hour of output will cost, taking all relevant factors into consideration. These factors include plant data, such as its initial capital cost, operation and maintenance costs, predicted lifetime and capacity factor, and also external factors such as the discount rate, or cost of borrowing money over a period of time.

Capital costs

Hydroelectricity is well-established, and much of the information listed above is easily available. The water-control systems, turbo-generators and output controls are standard items, covering a power range from a few hundred watts to hundreds of megawatts. The expected lifetime of the machinery is 25–50 years, and of the external structures, 50–100 years. Nevertheless, as mentioned in Section 5.3, it is difficult to generalize meaningfully about 'the cost of hydroelectric power', or to assess the economic potential for hydroelectricity in a country or a region.

The difficulty lies in the combination of the extremely site-specific construction costs and the heavy 'front-end loading' of these costs. In other words, the dominant factor in determining the cost per unit of hydro output is the initial capital cost, and a major part of this can be the civil engineering costs, which vary greatly from site to site.

Unit costs

An interesting study of hydro potential in the USA (Hall et al., 2003) assessed the costs for over two thousand sites with potential hydro capacities in the range 1–1300 MW. About half of these were green-field (blue-water?) sites, with no existing dams or hydro plants, and the estimated development costs for these, based on data for similar existing plants, fell mainly in the range $2000–$4000 per kW. (At the time of writing inflation would have increased these initial costs to about $2400–$4800 or approximately £1500–£3000 per kW.)

These studies revealed the striking influence of the *initial costs* in determining the cost per kilowatt-hour of output averaged over the life of a hydro plant.

The civil engineering works typically accounted for 65–75% of this unit cost, whilst meeting the environmental and other criteria necessary for a licence added another 15–20%. In all, 85–95% of the capital cost was 'site' cost, with the turbo-generator and control systems accounting for only 10% or so.

With no fuel costs, and relatively low operation and maintenance costs, it is the annual repayments of these initial costs that dominate the cost of the electricity. Box 5.11 reveals the unfortunate consequences for hydropower of a period of high interest rates.

BOX 5.11 Cost comparison of hydro and CCGT plants

A combined cycle gas turbine (CCGT) plant may be regarded as the opposite extreme to a hydro plant. Built quickly, using standard components, its capital costs are relatively low; but the fuel costs are high (and unpredictable for the future) and its lifetime is likely to be much less than that of a hydro plant. Nevertheless, investors may choose to put their money into CCGT plants rather than hydro – as has indeed been the case in the UK in recent years. The following reasoning shows why.

Table 5.7 has relevant data for two 100 MW plants. The main differences are obvious: the hydro plant costs much more, but the CCGT has fuel costs and a much shorter expected life. In the case considered, the high CCGT capacity factor suggests that it will be in almost constant use, whilst the hydro plant may be used mainly as backup (or is perhaps subject to flow variations). The final column of the table is calculated from the 100 MW rated output of both plants and their respective capacity factors.

The cost per kWh of electricity from each of these 100 MW plants can be calculated for different discount rates by the method described in Appendix B, which contains examples of using discounted cash flow to obtain levelized electricity costs. In Table 5.8, data from Table B1 is used to find the annual repayment for each of the above plants at four selected discount rates. Using the annual outputs and plant lifetimes in Table 5.7, these are then expressed as annual repayment costs per kilowatt-hour of output. Finally, adding the other unit costs from Table 5.7 leads to a *total cost per kWh* of electricity from each plant.

Table 5.7 Financial data and performance factors

	Capital cost /£million	Plant lifetime /years	Average capacity factor	O&M cost per kWh of output	Fuel cost per kWh of output	Average annual plant output /million kWh
CCGT	80	25	80%	0.6 p	4.4 p[1]	701
Hydro plant	200	60	50%	2.0 p[2]	-	438

[1] Assuming a gas cost of 2.2 p per kWh and a fuel-to-output efficiency of 50%.

[2] The relatively high operation and maintenance (O&M) cost for the hydro plant allows for possible machinery updating during its 60-year life.

Sources: (adapted from Mott MacDonald, 2010; Hall et al., 2003)

Table 5.8 Financial factors

Discount rate	0%	5%	10%	15%
CCGT repayment factor/£ per £1000 capital	40	71	110	155
CCGT annual repayment/£ million	3.2	5.7	8.8	12.4
Annual repayment/pence per kWh of output	0.46 p	0.81 p	1.26 p	1.77 p
CCGT total cost per kWh of output	**5.46 p**	**5.81 p**	**6.26 p**	**6.77 p**
Hydro repayment factor/£ per £1000 capital	17	53	100	150
Hydro annual repayment/£ million	3.4	10.5	20.0	30.0
Annual repayment/pence per kWh of output	0.78 p	2.42 p	4.57 p	6.85 p
Hydro total cost per kWh of output	**2.78 p**	**4.42 p**	**6.57 p**	**8.85 p**

The bottom line of Table 5.8 in Box 5.11 reveals the importance of discount rates in determining the effective unit cost of power plant output. In the two cases considered, power from the hydro plant costs more than from the CCGT plant at any discount rate greater than about 10%. It is also worth noting the effect of the very high capacity factor of the CCGT plant, suggesting that it is in almost constant use, which of course affects the unit costs. In regions where hydro plants are the main sources of electricity, providing reliable output, the unit cost could, of course, be significantly lower.

Refurbishing and upgrading

Two other options have been considered – and implemented – in recent years by countries with little remaining accessible hydro capacity:

- the installation of hydro plants at existing dams constructed for other purposes
- the refurbishing and/or upgrading of existing hydro plants.

Both approaches offer the possibility of increased hydro capacity at a lower initial cost and with fewer environmental consequences than green-field development. The US study described above noted the lower cost of these options (at least for relatively modest plants), with estimates of perhaps half the cost of green-field development when the dam already exists, and as little as a third, per 'new' kW, for refurbishment and upgrading of older plant.

The offer of financial support undoubtedly influences producers in their decisions to invest in upgrading. In the UK, the Elan Valley scheme (Box 5.3) was one small-scale example that used this option, and Scottish and Southern Energy (SSE), one of two main owners of hydro plants in Scotland, developed a programme for the refurbishment of five plants which qualified for government support as they all rated at under 20 MW. Finlarig (Figure 5.22) was one of the five, gaining new runners and nozzles for its Pelton turbines (Wilson, 2003).

Other countries are adopting similar policies. The USA and Canada have extensive refurbishment and upgrading programmes for large-scale hydro, and in Europe hydro capacity is expected to rise by over 20 GW between 2010 and 2020, largely through upgrading and modernization of existing schemes. This is to be achieved mainly by a 10% increase in output from the major producers: Austria, Italy, Norway, Spain, Sweden and Switzerland.

5.12 Future prospects

Given its long history and large scale, hydropower is the most mature of the renewables industries. In developed markets such as the European Union, United States, Canada, and Japan, where many hydropower plants were built 30–40 years ago, the industry is focused on relicensing and repowering as well as adding hydro generation to existing dams. In developing nations such as China, Brazil, Ethiopia, India, Malaysia, Turkey, and Vietnam, utilities and developers are focused on new hydro construction.

(REN21, 2010)

The least-cost way to increase hydro generating capacity is almost always to modernise and expand existing plants, where this is an option. Most of the hydro plant presently in operation will require modernisation by 2030. While capacity expansions are generally made at existing hydro stations, there are sometimes opportunities for installing generators at non-hydro dams. There are 45 000 large dams in the world and the majority do not possess a hydro component.

(WEC, 2010b)

World total electricity production in 2009 was 20 000 TWh, from an installed generating capacity of about 4500 GW. The corresponding figures for hydroelectricity are 3270 TWh from a capacity in the region of 900 GW – roughly 16% of world output from about 20% of world capacity. The percentage contribution from hydro had been falling for many decades, as the building of new plants failed to keep up with the rapid growth in total electricity consumption (Figure 5.1). However, the years since about 2005 have seen a different picture, with fluctuations in total electricity consumption but no similar behaviour in hydro output – perhaps a result of its being a renewable resource, with no fuel costs.

In terms of the general future for hydroelectricity, the two quotations above offer broadly similar views but with slightly different emphases. There are of course many who would reject the implied acceptance (REN21) of the construction of new large hydro plants in the less developed regions of the world (see for example Box 5.12), whilst not everyone agrees that countries such as the USA, Switzerland and other parts of Europe (including the UK) have no room at all for further hydro development. There has been growth in large-scale hydro capacity in some of these countries in recent years, but it has been relatively slow, and it may indeed be the case that for the wealthier parts of the world upgrading and refurbishment are the principal remaining options.

BOX 5.12 Large hydroelectric projects in Brazil

Itaipú

Construction of the hydroelectric plant at Itaipú, on the Paraná River between Brazil and Paraguay, started in 1975, and the last two of its twenty 700 MW generators came on line in 2006–2007.

The effective head is about 200 metres and the reservoir area 1350 km², with an average water flow of some 10 000 tonnes a second, peaking at times to over 30 000 t s⁻¹.

The plant supplies 95% of Paraguay's power and about a quarter of Brazil's, requiring a feature that must be unique in power stations: all 20 generators produce alternating current (AC) power, but half of them at a frequency of 50 Hz, and the other half at 60 Hz. Moreover, as Paraguay (50 Hz) uses only a fraction of its share, the rest is sold to Brazil, where it is rectified, transmitted as DC, and then re-converted to provide the required 60 Hz supply.

The total annual output obviously depends on the available flow in the Paraná and plant availability, and could of course be affected by varying demand. Over the five years 2006–2010 the average annual output was

Figure 5.29 The Itaipú dam

91.1 TWh, implying an average capacity factor of slightly under 75% – significantly below the originally contracted 85%.

Nevertheless, it is claimed that the sceptics who called Itaipú a white elephant, producing unwanted power in the wrong place, have themselves been proved wrong, and one target still seems within reach: to be the first power station to generate more than 100 TWh in a single year (Itaipú, 2011).

Xingu river project at Belo Monte

The first suggestions for the development of hydro power on Brazil's Xingu river, a tributary of the Amazon, appeared in 1975 (the year in which Itaipú came into operation). Initial studies led to a proposal for six large hydro plants, but the release of this plan in 1987 led to a vigorous campaign by indigenous people living in the affected areas. Their campaign attracted widespread international support, and the next two decades saw continuing opposition to a range of schemes involving between one and six dams along the Xingu. There were also objections on technical grounds – the variable flow of the river meant that the annual capacity factor of the plant could be as low as 30% (the problem experienced at Three Gorges).

In 2008, a new environmental assessment resulted in a plan involving one main power plant, the Belo Monte. However, in order to avoid inundating 'indigenous territory' in the bend of the river, this required the construction of the second, larger Pimental dam. This creates two reservoirs: one upstream along the route of the river, and a second filled by the diversion of water through two parallel 500 m wide canals on to previously dry land behind the Belo Monte dam. In February 2010, the Brazilian government environmental agency granted a provisional environmental licence to this plan.

The following year saw a series of reversals and reinstatements of this approval, culminating in the granting on 1 June 2011 of a full licence to the Norte Energia consortium, which undertook to pay $1.9 billion to address social and environmental problems. (Barrionuevo, 2011; Wikipedia, 2011d)

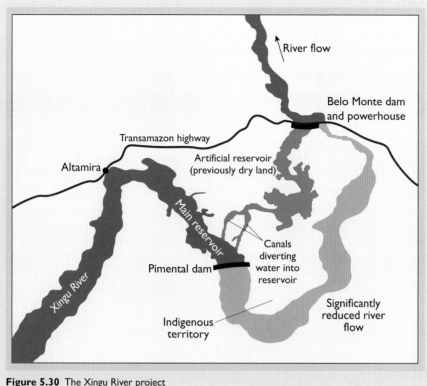

Figure 5.30 The Xingu River project

The past decade has revealed that any account of the world future for hydropower must consider China.

> China's hydroelectric resources are enormous. The country has 378 GW of technically exploitable hydropower reserves capable of producing 1,920 TWh per year and 676 GW of theoretical hydroelectric capacity, which would yield 5,900 TWh per year. Both are the largest such estimated resources in the world.
>
> (Ni, 2009)

The discrepancy between the hydro technical potential of 1920 TWh per year here and the 2500 TWh per year of Table 5.4 might be regarded as relatively small, given the doubtful nature of such data. The more significant fact is that both projections suggest a possible future Chinese output that is well over half the world's total current output. (However, the implied 60% capacity factor does seem optimistic, as China's current average for hydroelectricity is only 41%.)

In the decade to 2009, China's reported hydro output increased threefold, to reach the 600 TWh per year shown in Table 5.4. However, in the more recent years the country's desperate need for electricity led to a concentration on coal-fired plants with their shorter construction times, and the expectation is that this will continue for a considerable time. Eventually, investment in coal-fired power plants is expected to decline, mainly due to environmental concerns, and investment in nuclear power, wind power and hydropower is expected to accelerate (Ni, 2009).

Africa, as we saw in Section 5.3, has a lower proportion of its hydro technical potential developed than any other major region of the world – a situation that has led to its place at the centre of bitter controversies, typified by the proposals for and objections to the Grand Inga project on the Congo (Box 5.13).

BOX 5.13 Grand Inga

The first Inga power plant, with a rated output of 351 MW from six turbines, was commissioned in 1972 on the Congo river, in the region of the Inga rapids in the then country of Zaïre – now the province of Bas-Congo in the Democratic Republic of the Congo. In 1982, this plant (now Inga 1) was joined by Inga 2, with its rated output of 1424 MW from eight turbines.

Unfortunately, with many financial, political and structural problems, the record of these plants has not been good. (In 2008, two top directors of the company were interrogated after the disappearance of $6.5 million intended for renovation of Inga 2.) In consequence, by early 2011 only half the generators of each plant were operational, with a total output of 875 MW. In 2008, the World Bank announced a plan to renovate ten turbines at Inga 1 and 2, but the world financial crisis intervened and at the time of writing only the renovation of four turbines of Inga 2 is planned.

However, in 2008, the World Energy Council called for financing for a feasibility study of its **Grand Inga** proposal, whose main features were:

■ a third plant, Inga 3, with an output of 40 GW, requiring a 205 metre high dam forming a 15km-long reservoir;

■ a pan-African power distribution system reaching to Egypt in the north, Nigeria in the west and south to South Africa.

Many regarded this as unrealistic and/or undesirable, and considered the plan dead, but in March 2011, it was announced that a joint venture of Aecom (a US technology services firm) and Electricité de France (EDF) had secured a US$13.4 million contract to conduct a feasibility study on the hydroelectricity development of the site and the associated interconnection transmission.

Sources: BBC, 2008; BP, 2010; The Economist, 2010; HydroWorld, 2011; International Rivers 2008, 2010, 2011(a), 2011(b); WEC, 2007.

One form of hydro development that is generally expected to attract support in the coming years is *pumped storage*. World pumped storage capacity reached 127 GW in 2009 (HydroWorld, 2009), but at least 15 projects were under construction, increasing the total by about 9 GW. With rising demand for peak load balancing and the need to accommodate the growing output from intermittent sources such as wind and solar power, the demand for pumped storage is expected to increase by 60% over the next four years (WEC 2010a).

Finally, there is the large-scale hydro potential of the world's 45 000 dams currently without a power plant, mentioned in the WEC report quoted at the start of this section.

Small-scale hydro

We saw in Section 5.4 that is not easy to assess the world *output* from small-scale hydro. It is not therefore surprising that, with the lack of agreement on the range of outputs treated as 'small-scale', the world *resource* is even more uncertain, and there have been relatively few detailed estimates of the potential for small-scale hydro development in specific countries or regions. Box 5.14 summarizes the results of two studies in the UK.

BOX 5.14 Small-scale hydro in the UK

A study in 2008 of the potential for small-scale hydro in Scotland found some 36 thousand potential sites, with an estimated total power of 26 000 MW, giving an annual output of about 10 TWh per year. Only about 1000 sites were considered to be financially viable, but even this reduced number, with a total capacity of 660 MW, might contribute some 2.7 TWh per year – a potential increase equal to about half the present total UK hydro output (FHSG, 2008).

Two years later, a study of the potential for small-scale hydro in England and Wales was published (DECC, 2010b). This differed from the Scottish study in that it was based on sites identified in an earlier analysis (re-assessed in the light of technological and other changes) and also in that it excluded sites that would require the construction of new dams or weirs. Nevertheless, its estimated total potential installed capacity for England and Wales fell into the range 150–240 MW (from 'pessimistic' to 'optimistic').

5.13 Summary

After an introduction to hydro terminology in a short study of a small group of plants in Scotland, this chapter may be seen as falling into three parts:

- a survey of the present situation – the hydro resource and its use on the large and small scale
- an account of the technologies associated with the use of hydropower
- a discussion of the benefits and penalties associated with this use.

In the course of the chapter we have seen that, whilst 'large' and 'small' hydro may differ relatively little in their technologies, they may be viewed very differently in their benefits and penalties. Whether this difference broadens or narrows in the coming decades may well determine the future for this form of 'renewable' energy.

References

Barrionuevo, A. (2011) 'Brazil, After a Long Battle, Approves an Amazon Dam', *The New York Times*, 1 June [online], http://www.nytimes.com/2011/06/02/world/americas/02brazil.html?_r=2&scp=1&sq=belo%20monte&st=cse (accessed 20 October 2011).

BBC (2008) *Africa plans biggest dam project* [online], http://news.bbc.co.uk/1/hi/business/7358542.stm (accessed 8 June 2011).

BP (2010) *Statistical Review of World Energy 2010* [online], http://www.bp.com/productlanding.do?categoryId=6929&contentId=7044622 (accessed 2 January 2011).

Cruachan (2011) *Cruachan Power Station* [online], www.scottishpower.com/uploads/CruachanPowerStation.pdf (accessed 8 June 2011).

DECC (2010a) *Digest of United Kingdom Energy Statistics* [online], www.decc.gov.uk/en/content/cms/statistics/publications/dukes/dukes.aspx (accessed 12 June 2011).

DECC (2010b) *England and Wales Hydropower Resource Assessment*, Department of Energy and Climate Change and Welsh Assembly Government (WAG), URN 10D/556 [online], www.decc.gov.uk/assets/decc/what we do/uk energy supply/energy mix/renewable energy/explained/microgen/753-england-wales-hydropower-resource-assess.pdf (accessed 10 May 2011).

EAWAG (2010) *Reservoirs: a neglected source of methane emissions* [online], http://www.eawag.ch/medien/bulletin/20101011/delsontro_etal_2010_est.pdf (accessed 6 June 2011).

Elan Valley Trust (2011) *Elan* [online], http://www.elanvalley.org.uk (accessed 23 January 2011).

El-Sayad, S. and van Dijken, G. L. (1995) 'The southeastern Mediterranean ecosystem revisited: Thirty years after the construction of the Aswan High Dam', *Quarterdeck*, vol. 3, no. 1 [online], http://ocean.tamu.edu/Quarterdeck/QD3.1/Elsayed/elsayed.html (accessed 18 March 2011).

ERN (2000) *Dam decommissioning*, European Rivers Network [online], http://www.rivernet.org/general/dams/decommissioning/decom3_e.htm (accessed 22 March 2011).

Everett, B., Boyle, G. A., Peake S. and Ramage, J. (eds) (2012) *Energy Systems and Sustainability: Power for a Sustainable Future* (2nd edn), Oxford, Oxford University Press/Milton Keynes, The Open University.

FHSG (2008) *Scottish Hydropower Resource Study Final Report* [online], Hydro Sub Group of the Forum for Renewable Energy Development in Scotland (FHSG) http://www.scotland.gov.uk/Resource/Doc/917/0064958.pdf (accessed 24 March 2011).

Hall, D. G., Hunt R. T., Kelly, S. R. and Greg, R. C. (2003) *Estimation of Economic Parameters of U.S. Hydropower Resources*, Idaho National Engineering and Environmental Laboratory, USA. The main report and the related database may both be accessed at http://hydropower.inel.gov/resourceassessment/index.shtml (accessed 4 October 2011).

Hill, G. (1984) *Tunnel and Dam: the Story of the Galloway Hydros*, Glasgow, South of Scotland Electricity Board.

HSAP (2010) *Hydropower Sustainability Assessment Protocol* [online], http://hydrosustainability.org (accessed 14 May 2011).

Hvistendahl, M. (2008) 'China's Three Gorges Dam: An Environmental Catastrophe?', *Scientific American* [online], http://www.scientificamerican.com/article.cfm?id=chinas-three-gorges-dam-disaster (accessed 23 March 2011).

HydroWorld (2009) *Hydro Review*, vol. 17, issue 6 [online], http://www.hydroworld.com/index/current-issue/hydro-review-worldwide/volume-17/issue-6.html (accessed 23 March 2011).

HydroWorld (2011) *Aecom, EDF partner for Grand Inga hydropower project feasibility study in Congo* [online], http://www.hydroworld.com/index/display/article-display/3703861373/articles/hrhrw/hydroindustrynews/newdevelopment/2011/01/aecom_-edf_partner.html (accessed 14 June 2011).

ICJ (1997) *Gabcíkovo-Nagymaros Project (Hungary/Slovakia)* [online], http://www.icj-cij.org/docket/index.php?sum=483&code=hs&p1=3&p2=3&case=92&k=8d&p3=5 (accessed 22 June 2011).

International Rivers (2008) *Grand Inga Dam, DR Congo* [online], http://www.internationalrivers.org/africa/grand-inga-dam (accessed 12 June 2011).

International Rivers (2010) *World Rivers Review: Focus on the New Dam Builders – December 2010* [online], http://www.internationalrivers.org/node/6029 (accessed 24 March 2011).

International Rivers (2011a) *Three Gorges Dam Key documents* [online], www.internationalrivers.org/featured/112/461 (accessed 23 March 2011).

International Rivers (2011b) *World Economic Forum, Davos* [online], http://www.internationalrivers.org/en/blog/lori-pottinger/2011-1-31/grand-illusions-african-energy-davos (accessed 12 June 2011).

International Water Power & Dam Construction (2007) 'Beyond Three Gorges in China', *International Water Power & Dam Construction* [online], www.waterpowermagazine.com/story.asp?storyCode=2041318 (accessed 23 March 2011).

Itaipú (2011) *ITAIPÚ – largest power plant on Earth* [online], www.solar.coppe.ufrj.br/itaipu_conv.html (accessed 4 June 2011).

McKenna, P. (2011) 'Fossil fuels are far deadlier than nuclear power', *New Scientist*, 23 March [online], http://www.newscientist.com/article/mg20928053.600-fossil-fuels-are-far-deadlier-than-nuclear-power.html (accessed 25 October 2011).

Mott MacDonald (2010) *UK Electricity generation costs update* [online], http://www.decc.gov.uk/assets/decc/statistics/projections/71-uk-electricity-generation-costs-update-.pdf (accessed 22 June 2011).

National Trust (2011) *Cragside* [online], http://www.nationaltrust.org. uk/main/w-vh/w-visits/w-findaplace/w-cragsidehousegardenandestate (accessed 15 March 2011).

Ni, C. (2009) *China Energy Primer* [online], Earnest Orlando Lawrence Berkeley National Laboratory, LBNL-2860E http://china.lbl.gov/sites/ china.lbl.gov/files/China%20Energy%20Primer%20w%20LBNL%20cover. Sept2010.pdf (accessed 23 March 2011).

ORNL (1993) *Exploring the issues involved in developing hydropower resources in the United States,* Oak Ridge Environmental Sciences Division. Available at http://www.ornl.gov/info/ornlreview/rev26-34/text/hydmain. html (accessed 6 June 2011).

REN21 (2010) *Renewables 2010 – Global Status Report*, Renewable Energy Network for the 21st Century [online], http://www.ren21.net/ REN21Activities/Publications/GlobalStatusReport/tabid/5434/Default.aspx (accessed 14 March 2011).

Romer, R. H. (1976), *Energy: An Introduction to Physics*, San Francisco, W. H. Freeman and Company.

ScottishPower (2010) *Tongland Power Station Turns 75 - But Retirement Not In Sight* [online], http://www.scottishpower.com/PressReleases_1995. htm (accessed 11 March 2011).

ScottishPower (2011), *Galloway Hydroelectric Scheme*, ScottishPower; also available online at www.scottishpower.com/uploads/ GallowayHydroElectricScheme.pdf (accessed 11 March 2011).

SSE (2011) *Power from the Glens* Available as pdf at http://www.sse. com/uploadedFiles/Content/11_About_Us/How_we_run_are_business/ Community/PowerFromTheGlens.pdf (accessed 4 November 2011).

Strandh, Sigvard (1989) *The History of the Machine*, London, Bracken Books.

Sustainable Hydropower (2011a) *About Sustainability in the Hydropower Industry: Sustainability Challenges* [online], http://www. sustainablehydropower.org/site/info/aboutsustainability.html (accessed 21 June 2011).

Sustainable Hydropower (2011b) *About Sustainability in the Hydropower Industry: Hydropower Strengths and Weaknesses* [online], http:// www.sustainablehydropower.org/site/info/aboutsustainability/ strengthsweekness.html (accessed 22 June 2011).

The Economist (2010) *Dams in Africa 'Tap that water'* [online], http://www. economist.com/node/16068950 (accessed 8 June 2011).

USBR (2011) *Grand Coolee Dam Statistics and Facts*, U.S. Department of Interior, Bureau of Reclamation, Pacific Northwest Region. Available as pdf at www.usbr.gov/pn/grandcoulee/pubs/factsheet.pdf (accessed 14 May 2011).

U.S. Department of Energy (2005) *Energy Efficiency and Renewable Energy* Available at http://www1.eere.energy.gov/windandhydro/hydro_plant_ types.html (accessed 20 March 2011).

U.S. Department of Treasury (2010) Available at http://www.dsireusa.org/ incentives/index.cfm (accessed 11 June 2011).

WEC (2007) *How to make the Grand Inga Hydropower Project happen for Africa,* London, World Energy Council [online], www.worldenergy.org/documents/grandingapressfile2.pdf (accessed 24 March 2011).

WEC (2010a) *2010 Survey Of Energy Resources*, London, World Energy Council; [online], http://www.worldenergy.org/documents/ser2010_report.pdf (accessed 11 March 2011).

WEC (2010b) *Survey Of Energy Resources, Executive Summary*, London, World Energy Council; [online], http://www.worldenergy.org/documents/ser2010exsumsept8.pdf (accessed 24 March 2011).

Wikipedia (2011a) *Major dam failures* [online], http://en.wikipedia.org/wiki/Dam_failure#List_of_major_dam_failures (accessed 18 March 2011).

Wikipedia (2011b) *Hydroelectric power station failures* [online], http://en.wikipedia.org/wiki/List_of_hydroelectric_power_station_failures (accessed 14 May 2011).

Wikipedia (2011c) *2009 Sayano-Shushenskaya hydro accident* [online], http://en.wikipedia.org/wiki/2009_Sayano-Shushenskaya_hydro_accident (accessed 24 June 2011).

Wikipedia (2011d) *Belo Monte Dam* [online], http://en.wikipedia.org/wiki/Belo_Monte_Dam#cite_note-elehis-2 (accessed 20 October 2011).

Wilson, D. (2003) 'Up-grading plants in Scotland', *The Engineer* [online], http://www.theengineer.co.uk/news/upgrading-plants-in-scotland/279545.article (accessed 24 June 2011).

Chapter 6

Tidal power

By David Elliott

6.1 Introduction

The rise and fall of the seas represents a vast, and as King Canute demonstrated, relentless natural phenomenon. The use of the tides to provide energy has a long history: small tidal mills on rivers were used for grinding corn in Britain and France in the Middle Ages. Subsequently, the idea of using tidal energy on a much larger scale to generate electricity emerged, using turbines mounted in large **barrages** – essentially low dams – built across suitable estuaries. More recently interest has grown in putting free-standing turbines into tidal currents.

Figure 6.1 A view of La Rance tidal power station

This chapter looks at some examples of these tidal technologies and at their potential and limitations. It is, as yet, still a relatively undeveloped field, although a number of projects do exist. A medium-scale 240 MW tidal barrage scheme has been built at the Rance Estuary in France (Figure 6.1) and, at the time of writing, a similar scale scheme is being implemented in South Korea. Around the world, a number of smaller tidal barrages have been built, and at the other end of the scale there have been proposals for large-scale developments at several sites. For example, Figure 6.2 shows

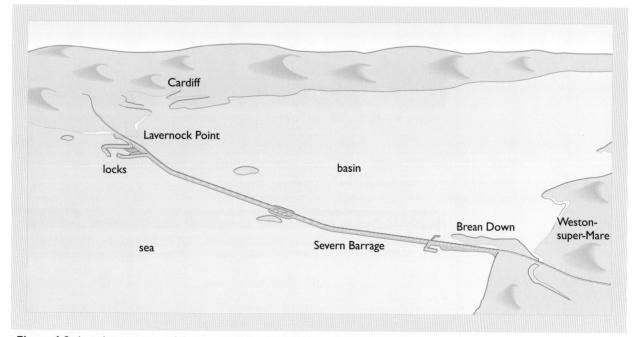

Figure 6.2 Artist's impression of the proposed Cardiff to Weston Severn Barrage

an artist's impression of the proposed 8.6 GW Severn Barrage, stretching 16 kilometres across the Severn Estuary in the UK. If built, this would generate 17 TWh per year, the equivalent of about 4.5% of the electricity generated in the UK in 2010. However, in recent years much of the emphasis has been on free-standing tidal current turbines – a range of devices are being developed, and some have already been deployed at full scale (Figure 6.3).

Figure 6.3 Marine Current Turbines' 1.2 MW 'SeaGen' concept

Box 6.1 outlines the early history of tidal power and describes some existing and proposed tidal projects around the world.

BOX 6.1 A brief history of tidal power

Small tidal mills, not unlike traditional watermills, were used quite widely on tidal sections of rivers and estuaries in the Middle Ages for grinding corn, but the idea of exploiting the full power of the tides in estuaries to generate electricity is relatively recent.

There have been a number of proposals for various types of barrage across the Severn, the UK's largest estuary, with the world's largest low- to high-tide range, but so far none have been taken forward. For example, a barrage concept (albeit with no provision for electricity generation) attributed to Thomas Telford was put forward in 1849. The first serious proposal involving electricity production came in 1920 from the Ministry of Transport. This was followed by a major study by the Brabazon Commission, which was set up in 1925. Its 1933 report focused on a barrage crossing the estuary along the 'English Stones line', not far from the modern Severn Bridges. It was to have 72 turbines with a total installed capacity of 804 MW and to incorporate road and rail crossings. The scheme was not taken up. It was reassessed in 1944, but, again, not implemented.

During the 1960s and 1970s a number of schemes were proposed, each crossing along different lines. These proposals culminated in another government-supported study by the Severn Barrage Committee, which was set up in 1978 under Professor Sir Hermann Bondi. The Committee reported in 1981 (Department of Energy, 1981), concluding that it was 'technically feasible to enclose the estuary by a barrage located in any position east of a line drawn from Porlock due North to the Welsh Coast'. Of all the possible crossing lines, three were favoured, the most ambitious being from Minehead to Aberthaw, which, it was estimated, could generate 20 TWh per year from 12 GW of installed capacity.

Subsequently, the less ambitious, but still very large, so-called 'inner barrage', on a line (first proposed by E. M. Wilson in 1966) from Brean Down, Weston-super-Mare to Lavernock Point near Cardiff, became the favourite, and was pursued by the Severn Tidal Power Group (STPG) industrial consortium in the 1980s. It was initially conceived as generating approximately 13 TWh per year from 7 GW installed, although this was later upgraded to 17 TWh per year from 8.64 GW installed.

Enthusiasm for tidal schemes was fuelled in part by the success of the French scheme on the Rance Estuary in Brittany, near St Malo (Figure 6.1). This was constructed between 1961 and 1967 and the first output from its 240 MW turbine capacity was achieved in 1966. The structure includes a road crossing. Apart from a problem with the generator mountings in 1975, it has operated very successfully. Subsequently, a much larger 15 GW scheme was proposed, to enclose a vast area of sea from St Malo in the south to Cap de Carterel in the north, the so-called 'Île de Chausey Project'. This has not been implemented.

Although large-scale schemes have also been proposed for the Bay of Fundy in Canada and at various sites in Northern Russia, the only significant tidal plants to be built to date, other than La Rance, are an 18 MW single unit, using a 'rim generator' (see Figure 6.15), at Annapolis Royal in Nova Scotia, Canada, completed in 1984; a 400 kW unit in the Bay of Kislaya, 100 km from Murmansk in Russia, completed in 1968; and a 500 kW unit at Jangxia Creek in the East China Sea.

However, at the time of writing, some medium-scale projects are being built in South Korea (including a 254 MW barrage), and a number of other barrage schemes have been considered both there and elsewhere around the world. Interest in the much larger Severn 'inner barrage' concept has continued, but there are economic and environmental challenges.

Instead, the emphasis in the UK and many other countries has moved on to smaller-scale modular tidal current turbines, operating on (horizontal) tidal flows rather than on (vertical) tidal height ranges. The world's first commercial-scale grid-linked tidal current turbine, a 1.2 MW two-bladed unit (see Figures 6.3 and 6.32) has been operating successfully in Strangford Narrows, Northern Ireland, since 2008. Dozens of other tidal current devices of various designs are under development in the UK and elsewhere and this area of activity seems likely to expand (see Sections 6.9 and 6.10).

The nature of the resource

It is important at the outset to distinguish *tidal* power from *hydro* power. As we saw in Chapter 5, *hydro* power is derived from the hydrological cycle (which is driven by solar energy) and is usually harnessed via hydroelectric dams. In contrast, *tidal* power is the result of the interaction of the gravitational pull of the Moon and, to a lesser extent, the Sun, on the seas. Schemes that use tidal energy rely on the twice-daily tides, and the consequent upstream flows and downstream ebbs in estuaries and the lower reaches of some rivers, as well as, in some cases, tidal movements out at sea.

Equally, we must distinguish between tidal energy and the energy in waves. Ordinary waves are caused by the action of wind over water, the wind in turn being the result of the differential solar heating of air over land and sea (see Chapter 8). If we consider wave energy, like hydroelectric energy, to be a form of solar energy, tidal energy could be called 'lunar energy'. Such distinctions are, unfortunately, not helped by the terminology which is often used – for example, the term 'tidal wave' is sometimes used to describe what are these days more usually called tsunami, the occasionally dramatic surges of water (which are neither wind-driven waves nor lunar-driven tides!) that can be produced by undersea earthquakes. There are also large climate-driven water flows in the oceans, such as the Gulf Stream, which are ultimately the result of solar heating. For the sake of completeness, we should also note that the gravitational pull of the Sun and the Moon also cause tidal phenomena in the atmosphere and in the Earth.

The energy in these various movements of water can, in principle at least, be tapped. The rise and fall of the tides can be exploited without the use of dams across estuaries, as was done in the traditional **tidal mills** on the tidal sections of rivers, as mentioned earlier. A small pond or pool is simply topped up and closed off at high tide and then, at low tide, the trapped water is used to drive a waterwheel, as with traditional watermills.

There is also the possibility of using turbines mounted independently in **tidal streams** (also called **tidal currents** – both terms are used, often interchangeably), that is *horizontal* tidal flows. Indeed, in the Middle Ages there were some floating 'undershot' waterwheels running on tidal currents. These flows of *kinetic energy* can be enhanced in some locations due to the effects of concentration in narrow channels, for example between islands or other constrictions, or around headlands.

In addition, it may be possible to harness some of the energy in large-scale ocean streams such as the Gulf Stream. Some recent developments in the tidal current and ocean stream areas are discussed below, from Section 6.8 onwards. As noted in Box 6.1, tidal current projects are now emerging as a leading tidal option, not least since, unlike large estuary-wide barrages, they are modular and can be developed incrementally.

Before looking at tidal current systems, the initial sections of this chapter focus on tidal barrages across estuaries. The vertical difference between the water level at low tide and high tide is usually called the **tidal range**. In most *tidal range energy generation systems*, the water carried upstream by the **tidal flow** – usually called the **flood tide** – is trapped behind a barrage across the estuary. As the tide ebbs, the water level on the downstream side of the barrage reduces and a head of water develops across the barrage. The head is used to drive the water through turbine generators and the barrage scheme operates like a low-head hydro plant (see Chapter 5). Thus the barrage can be used to harness the *potential energy* provided by the *vertical* difference between tides. The main difference from hydro, apart from the salt-water environment, is that the power-generating turbines in tidal barrages have to deal with regularly varying heads of water.

A variant of the barrage concept is the tidal lagoon, which depends on a containment structure being built *within* the estuary to retain a small proportion of water in the estuary (in contrast to a barrage that acts on the whole estuary). However the mode of operation is similar – the tidal range is used to create a head of water within the containment structure. Tidal barrages and tidal lagoons are therefore often collectively labelled **tidal range systems**, to distinguish them from systems using tidal currents.

Before looking at the details of, in turn, barrages, lagoons and tidal current schemes it is instructive to consider the physics behind tides and to appreciate what factors are involved in locating tidal energy systems.

The physics of tidal energy

The existence of tides is due primarily to gravitational interaction between the Earth and the Moon. This gravitational force, combined with the rotation of the Earth, produces, at any particular point on the globe, a twice-daily rise and fall in sea level, this being modified in height by the gravitational pull of the Sun and by the topography of land masses and ocean beds. A detailed analysis of this interaction between Earth, Moon and Sun is quite complex, but we will attempt to describe it in simple terms.

The first part of the explanation is relatively straightforward. Starting first with just the Earth and the Moon, the gravitational pull of the Moon draws the seas on the side of the Earth *nearest* to the Moon into a bulge *towards* the Moon. That gives us one tide per day at any one point, as the planet rotates through the bulge. But what about the second tide each day? This is more difficult to explain. Sometimes it is explained in simple terms by saying that the waters that make up the bulge facing the Moon are drawn from the seas at each side of the Earth, but the water at the far side is 'left behind', at its original level. However that does not really explain the fact that the second tide is roughly the same height as the first. Neither does the fact that the water in the seas *furthest* from the Moon experiences slightly less of the lunar pull, being further away, although it may be part of the explanation.

The full explanation of the bulge that forms on the side of the Earth away from the Moon depends on a more complicated analysis, based on understanding the effect of the relative movements of the Earth and Moon (see Box 6.2).

The basic pattern described in Box 6.2 is also modified by the pull of the Sun. Although the Sun is much larger than the Moon, its distance from the Earth is much greater, and the Moon's gravitational influence on the seas is therefore approximately twice that of the Sun. The final impact depends on their relative orientation.

BOX 6.2 **The Earth and the Moon**

A useful mathematical analysis of the generation of tides is given in *Renewable Energy Resources* (Twidell and Weir, 2006). This identifies *two* processes at work in relation to the Earth and the Moon: a rotational effect as well as a gravitational effect.

The first process, the rotational effect, is the result of the fact that the Earth and the Moon rotate around each other, somewhat like a 'dumb-bell' being twirled. This rotation gives rise to an outward force, sometimes (rather loosely) called a centrifugal force, which acts on the water in the seas. However, this giant dumb-bell does not rotate around the halfway point between the Earth and the Moon. Since the Earth is much larger than the Moon, their common centre of rotation is close to the Earth; in fact it is just below its surface (Figure 6.4). The mutual rotation around this point produces a relatively large *outward* centrifugal force acting on the seas on the side of the Earth *furthest* from the Moon, bunching them up into a bulge. There is also a much smaller centrifugal force, directed *towards* the Moon, that acts on the seas *facing* the Moon. (This force is smaller since here the distance from the Earth's surface to the common rotation point, just below the surface, is smaller.)

The second process, the gravitational effect, is more familiar and relates to the gravitational pull of the Moon, which draws the seas on the side of the Earth *nearest* to the Moon into a bulge *towards* the Moon, whilst the seas *furthest* from the Moon, being slightly further away, experience a reduced lunar pull towards the Moon.

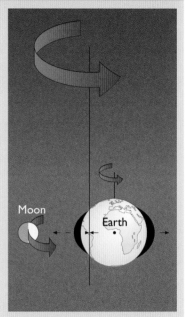

Figure 6.4 Relative rotation of the Earth and the Moon (not to scale)

There is thus, to summarize, a small centrifugal force and an increased lunar pull acting on the seas facing the Moon, and a larger centrifugal force and a decreased lunar pull acting on the seas on the other side of the Earth. The end result, on the basis of this analysis, is essentially a rough symmetry of forces, small and large, on each side of the Earth, producing tidal bulges of roughly the same size on each side of the Earth. In practice, the bulges may differ significantly, due, for example, to the tilt of the Earth's axis in relation to the orbit of the Moon and to local topographic effects.

As the Earth rotates on its axis, the lunar pull will maintain the high-tide patterns, as it were 'under' the Moon. That is, the two high-tide configurations will in effect be drawn around the globe as the Earth rotates, giving, at any particular point, *two* tides per day (or, more accurately, two tides in every 24.8 hour period), occurring approximately 12.4 hours apart. Since the Moon is also moving in orbit around the Earth, the timing of these high tides at any particular point will vary, occurring approximately 50 minutes later each day.

When the Sun and the Moon pull together (in line), whether both pulling on the same side of the Earth or each on opposite sides, the result is the very high **spring tides**; when the Sun and Moon are at 90° to each other (relative to the Earth), the result is the lower **neap tides**. The period between neap and spring tides is approximately 7 days – that is, a quarter of the 29.5-day lunar cycle (Figure 6.5). The ratio of output from a tidal range plant at the maximum spring tide to the output at a minimum neap tide can be more than two to one (Figure 6.6).

However, it is not simply a matter of the Earth's oceans rising and falling under gravitational forces. The pull of the Moon sets the whole sea into a state of oscillation, much as water might slop to and fro in a shallow dish if slightly shaken. The scale and frequency of these oscillations is influenced by the shape and size of the area of water, as defined by the surrounding land masses. The effect is somewhat like the resonance that occurs in musical instruments at various frequencies and volumes. Tidal resonances

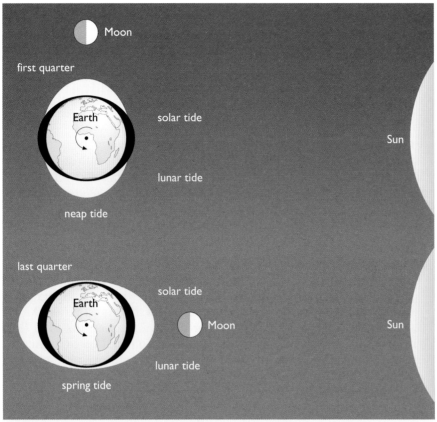

Figure 6.5 Influence of the Sun and the Moon on tidal range (not to scale)

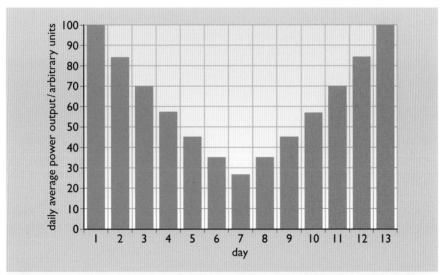

Figure 6.6 Typical variation in daily average output from a tidal range power plant over the spring–neap cycle (source: adapted from SDC, 2007)

can occur in well defined estuaries but can also be significant across entire oceans. For example, while the tidal range in open sea in mid-Atlantic is about 0.5 metres, at the coasts it can be enhanced to about 3 metres. In the case of the much wider Pacific the resonance effects are much smaller, so that the tides can be very small. In addition, the Earth's rotation causes the tidal flows to be deflected slightly, which can create a swirling pattern – in the North Atlantic there is a slow anticlockwise motion.

The rotation of the Earth also means that high tide occurs at each point on the Earth at a different time. Moreover, when the tide reaches the coast, the general topography of the landmass then defines the rate at which the tide moves along a coast. For example it takes about six hours for a tide to progress down the English Channel from Land's End to Dover and there is several hours difference between high tides at other points around the east and west coast of the UK. This has important implications for the integration of the output of tidal power plants into an electricity grid.

Equally important for tidal power generation is the fact that the tidal ranges and tidal stream velocities experienced in practice at coastal sites are also sometimes significantly modified and amplified by *local* topographic variations, for example in shallow coastal waters and in estuaries. As the tide approaches the shore and the water depth decreases, the tidal flow is concentrated and the range can be increased to reach up to, typically, 3 metres. Examples of coastlines giving rise to funnelling and tidal range enhancement include the Severn Estuary in the UK (Figure 6.7), the Gulf of St Malo in France and the Bay of Fundy in Canada.

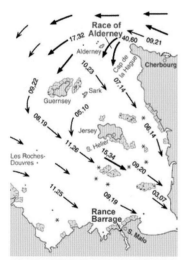

Figure 6.8 Tidal stream map for the Race of Alderney and La Rance Barrage 4 hours after high water at Dover; tidal stream rate in tenths of a knot expressed in the form: neap tide, spring tide (i.e. 10,23 means on a neap tide the stream runs at 1.0 knots and on a spring tide 2.3 knots). Note: 1 knot = 0.51 m s⁻¹ (source: adapted from *Admiralty Tidal Stream Atlas: The English Channel NP250*, 1992)

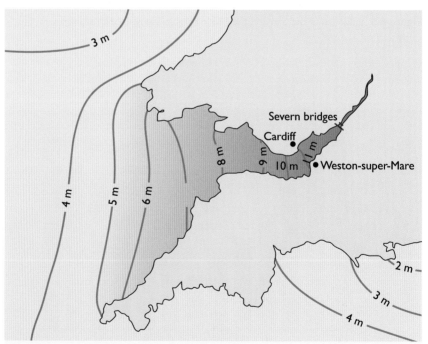

Figure 6.7 The effects of concentration of tidal flow in the Severn Estuary (tidal ranges in metres) (source: Department of Energy, 1981)

The tidal stream velocity can also be amplified if the tidal flow is constrained, for example in the Race of Alderney between the island of Alderney and the French mainland (Figure 6.8).

However, there are also frictional effects to take into account. For example, energy is lost as the tidal flow moves over differing estuary bed materials. Thus, the extent to which the tidal range is magnified at any point depends on the balance between the energy losses and the concentration of the tidal flow by the topography. The frictional effects will usually begin to outweigh the concentration gains at some point upstream when the funnel-like layout of an estuary gives way to a more parallel-sided, flat-bottomed river configuration. On the Severn, this 'natural' optimal point normally occurs around the site of the (first) Severn Bridge, where the tidal range reaches 11 metres. The range decreases further upstream, as do flow rates.

Occasionally, in some long estuaries, dramatic tidal effects can occur upstream. For example, rather than producing a relatively slow rise, as normally happens in the main part of the Severn Estuary, the upstream tidal flow further up the Severn can be concentrated so abruptly that it rises into an almost vertical step or wave: the so-called Severn Bore. A similar effect occurs on other long estuaries, including the Humber in the UK and the Hoogly near Calcutta in India.

The water level actually experienced at a given location is also influenced by the weather – so-called **surge tides** are generated by storms rather than the Moon and Sun. Their effects can also be amplified with devastating effects by topographic features and tidal power plants have to be built to withstand such extreme phenomena.

So, even leaving aside freak effects like the Severn Bore and occasional surge tides, there is a complex range of tidal phenomena. Fortunately for the designers of tidal barrages and other tidal devices, the end results – that is, the normal tidal patterns in estuaries – although very site specific, are predictable and reliable. The tides will continue to ebb and flow, on schedule, indefinitely.

But is the energy in the tides *really* 'renewable'? As we have seen, the primary mechanism in tide generation is the gravitational interaction between the Earth and the Moon, and the forces produced by their relative orbital movements. These forces create bulges of water. The rotation of the Earth draws these resulting tidal bulges across the seas, or, more precisely, the water in the seas rise into a bulge as the water rotates with the planet. The result is that there are horizontal tidal flows, as water is drawn into the moving bulge. The rotation of the Earth is being very gradually slowed by tidal processes (by approximately one-fiftieth of a second every 1000 years), because of frictional effects – it takes energy to drag the water along, especially through areas where there are topographical constrictions. However, the extra frictional effect that would be produced by even the widespread use of tidal barrages would be extremely small. The influence on the Moon's orbital velocity (which is also being very slowly reduced by the tidal interaction) would be even smaller. So overall, tidal energy is renewable on any reasonable interpretation of the concept.

6.2 Power generation from barrages

The basic physics and engineering of tidal barrage power generation are relatively straightforward.

Tidal barrages, built across suitable estuaries, are designed to extract energy from the rise and fall of the tides, using turbines located in water passages in the barrages. The potential energy, due to the difference in water levels across the barrage, is converted into kinetic energy in the form of fast-moving water which passes through the turbines – the spinning turbines then drive generators to produce electricity.

The *average* power output from a tidal barrage is roughly proportional to the square of the tidal range. The mathematical derivation of this is fairly simple, as is demonstrated by the analysis in Box 6.3.

BOX 6.3 Calculation of power output from a tidal barrage

Figure 6.9 Power generation from tides (source: Twidell and Weir, 2006)

Let us assume that we have a rectangular basin with a constant surface area A, behind a barrage, and a high-to-low tidal range R (Figure 6.9). In conventional ebb generation, when the tide comes in, it is freely allowed to flow into the basin, but when the tide goes out, the water in the basin is held there, at the high-tide level. When the sea has retreated to its low-tide level, the surface of the water held behind the barrage will be at a height R above the sea.

Given a rectangular basin, the centre of gravity of the usable mass of water will be at a height $R/2$ above the low-tide level. The total volume of water in the basin will be AR and, if the density of the water is ρ it will have a mass ρAR, i.e. ρ multiplied by the volume of water (A times R). This water could all now be allowed to flow out of the barrage through a turbine to the low-tide level. The maximum potential energy E available per tide if all the water falls through a height of $R/2$ is therefore given by the mass of water (ρAR) times the height ($R/2$) times the acceleration due to gravity (g); that is, $E = \rho ARg(R/2)$. The basin could then be allowed to fill on the next incoming tide and the cycle repeated again and again. If the tidal period is T, then the average potential power that could be extracted becomes E/T or $\rho AR^2g/2T$.

Clearly, even small differences in tidal range, however caused, can make a significant difference to the viability and economics of a barrage. A mean tidal range of at least 5 metres is usually considered to be the minimum for viable power generation, depending on the economic criteria used. As the analysis in Box 6.3 indicates, the energy output is also roughly proportional to the area of the water trapped behind the barrage, so the geography of the site is very important. All of this means that the siting of barrages is a crucial element in their viability.

Many studies have been carried out on tidal power in the UK, dating from the early 1900s onwards (see Box 6.1). This is hardly surprising, as the UK holds about half the total European potential for tidal barrage energy, including one of the world's best potential sites, the Severn Estuary. There is also a range of possible medium- and small-scale sites, including locations on the Mersey and Solway Firth (further details are in Section 6.6). The total UK tidal barrage resource potential has been put at around 53 TWh per year (ETSU, 1990) which was about 14% of UK electricity generation in 2010. In practice, the contribution to electricity consumption that could be achieved in the UK and elsewhere would depend on a range of *technical, environmental* and *economic* factors. Although these factors interact, we can explore each in turn before attempting a synthesis.

Barrage designs

The input energy source for a barrage, the rise and fall of the tides, follows a roughly sinusoidal pattern (see the sea-level curves in Figures 6.10–6.12). The tides have a 12.4 hour cycle, with the actual tidal range varying from site to site as a result of complex resonance and funnelling effects as mentioned earlier.

Given the complexity of estuary configurations, the actual resonances and funnelling effects are very difficult to model accurately, with variations in depth, width and friction over differing estuary bed materials introducing many local variations.

However, it is well worth the effort required to analyse these effects when deciding on the precise siting and orientation of a barrage, since they will have a major effect on its output. Indeed, it may be possible to locate and/or operate a barrage so as to 'tune' the barrage to the estuary tidal pattern, and thus to increase energy output. Certainly, any disturbance that might reduce existing resonance effects should be avoided.

In addition to the basic issues of location and orientation, a second set of factors that influences the likely energy output of a barrage relates to its *operational pattern*.

Energy can be generated from a barrage in three main ways. The most commonly used method is **ebb generation**. Here the incoming tide is allowed to pass through the barrage sluice gates. The water is trapped behind the barrage at high-tide level by closing the sluices. The head of water then passes back through the turbines on the *outgoing* ebb tide in order to generate energy (Figure 6.10). Alternatively, **flood generation** uses the *incoming* tide to generate electricity as it passes through the turbines mounted in the barrage (Figure 6.11). In each case, two bursts of energy are produced in every 24.8 hour period. **Two-way operation**, on the ebb *and* the flood, is also possible (Figure 6.12).

The basic technology for power production is well developed, having much in common with conventional low-head hydro systems (see Chapter 5). Figure 6.13 is an artist's impression of the typical layout of a power generation scheme.

A number of different turbine configurations are possible. At La Rance, a so-called **bulb** system is used, with the turbine generator sealed in a bulb-shaped enclosure mounted in the flow (Figure 6.14). As the water has to flow around the large bulb, access (for maintenance) to the generator involves cutting off the flow of water.

These problems are reduced in the **Straflo** or **rim generator** turbine (as used at Annapolis Royal in Canada), with the generator mounted radially around the rim and only the runner (that is, the turbine blades) in the flow (Figure 6.15). Although it is more efficient than the bulb design, because the water flow is not so constricted, this design introduces extra problems with the sealing between the runner blades and the radial generator.

Alternatively, there is the **tubular** turbine configuration, with the runner set at an angle so that a long (tubular) shaft can take rotational energy out to an external generator (Figure 6.16). This design also avoids constricting

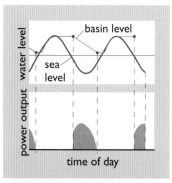

Figure 6.10 Schematic diagram of water levels and power outputs for an ebb generation scheme (source: Department of Energy, 1981)

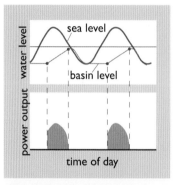

Figure 6.11 Schematic diagram of water levels and power outputs for a flood generation scheme (source: Department of Energy, 1981)

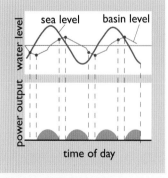

Figure 6.12 Schematic diagram of water levels and power outputs for a two-way generation scheme (source: Department of Energy, 1981)

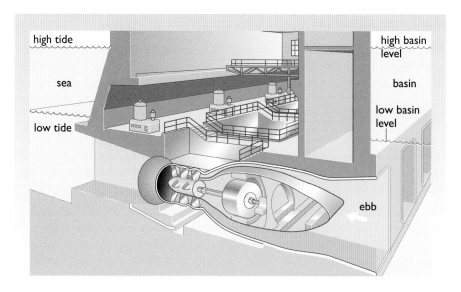

Figure 6.13 Artist's impression of the typical layout of a power generation scheme

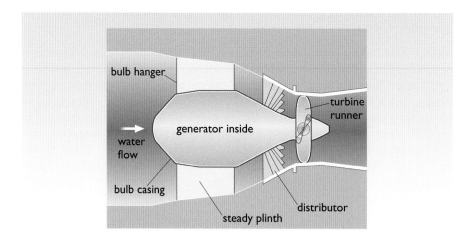

Figure 6.14 Bulb turbine as used at La Rance

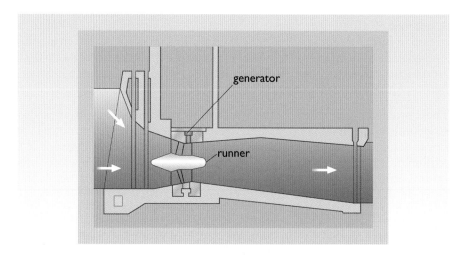

Figure 6.15 Straflo or rim generator turbine as used at Annapolis Royal

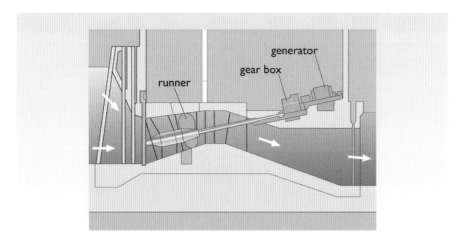

Figure 6.16 Tubular turbine

the flow of the water and, since the generator is not in a confined space, there is room for a gearbox, which can allow for efficient matching to the higher speed generators usually used with hydro plants. Several such units have been used in hydro power plants in the USA, the largest being rated at 25 MW. However, there have been vibration problems in the long drive shaft and, so far, bulb turbines have proved to be the most popular with barrage designers.

As mentioned above, the rotational speeds of the turbines in tidal barrages tend to be lower than those for turbines in hydro plants (50–100 revolutions per minute, in comparison to 200–450 revolutions per minute for hydro generators) and therefore wear is also reduced. Since large volumes of water have to pass through the barrage in a relatively short time, large numbers of turbines are required in a large-scale barrage (see Boxes 6.4 and 6.5 for details of La Rance and the proposed Severn Barrage).

In simple ebb or flood generation, this large installed capacity is used only for a relatively short period (three to six hours at most) in each tidal cycle, producing a large but short burst of power, which may not match the demand for electricity. However, using reversible-pitch turbines it is possible to operate on both the ebb and the flood in a two-way operation. Reversible-pitch turbines are more complex and costly than standard turbines and, although the output will be more evenly distributed over time, there will be a net decrease in electricity output for each phase compared with a simple ebb generation scheme. This is because, in order to be ready for the next cycle, neither the ebb nor the flow generation phases can be taken to completion: it is necessary to open the sluices and reduce water levels ready for the next flood cycle, and vice versa for ebb generation (see Figure 6.12).

Flood pumping is another option for electricity generation. Here, the turbine generators are run in reverse and act as motor-pump sets, powered by electricity from the grid. Additional water is pumped behind the barrage into the basin at around high tide, when there is a low head difference across the barrage. This provides extra water for the subsequent ebb generation phase when there is a high head difference. This is, in effect, a way to store excess off-peak power from the grid, but with a net energy gain.

BOX 6.4 La Rance

The 740-metre long Rance Barrage was constructed between 1961 and 1967. It has a road crossing and a ship lock (Figure 6.17) and was designed for maximum operational flexibility. It contains 24 reversible (that is, two-way) turbines, each of 10 MW capacity, operating in a tidal range of up to 12 metres, with a typical head of approximately 5 metres.

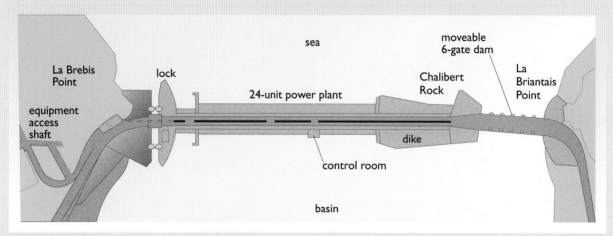

Figure 6.17 Layout of the Rance Barrage (source: Department of Energy, 1981)

The operational pattern initially adopted at La Rance was to optimize the uniformity of the power output by using a combination of two-way generation (which meant running the turbines at less than the maximum possible head of water) and incorporating an element of pumped storage. For spring tides, two-way generation was favoured; for neap tides and some intermediate tides, direct pumping from sea to basin was sometimes carried out to supplement generation on the ebb.

Although some mechanical problems were encountered in 1975, which subsequently led to two-way operation mostly being avoided, overall the barrage has been very successful. Typically the plant has been functional and available for use more than 90% of the time, and net output has been approximately 480 GWh y^{-1} with, in some years, significant energy gains from pumping.

The construction of the barrage involved building two temporary coffer-dams, with the water then being pumped out of the space in between, to allow work to be carried out in dry conditions (see Figure 6.18). River water was allowed to

Figure 6.18 La Rance coffer-dam during construction of the barrage

pass via sluices, but the reduced ebb and flow resulted in effective stagnation of the estuary and the subsequent partial collapse of the ecosystem within it. Since construction, exchange of water between the open sea and the estuary has restored the estuarine ecosystem, but because there was no monitoring it is difficult to establish what changes the barrage caused to the original environment in the estuary.

BOX 6.5 The Proposed Severn Barrage

Basic data for the Cardiff to Weston Barrage is given below (Department of Energy, 1989). The same design specifications were used for the assessment carried out by the Sustainable Development Commission in 2007 (SDC, 2007) and in the Department of Energy and Climate Change/Welsh Assembly Government/SWRDA study in 2010 (DECC, 2010a).

Number of turbine generators	216
Diameter of turbines	9.0 metres
Operating speed of turbines	50 rpm
Turbine generator rating	40 MW
Installed capacity	8640 MW
Number of sluices, various sizes	166
Total clear area of sluice passages	35 000 m²
Average annual energy output	17 TWh
Operational mode	ebb generation with flood pumping
Length of barrage:	
total	15.9 km
including: powerhouse caissons	4.3 km
sluice caissons	4.1 km
other caissons	3.9 km
embankments	3.6 km
Area of enclosed basin at mean sea level	480 km²

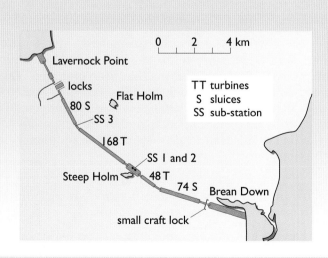

Figure 6.19 Layout of the proposed Cardiff to Weston Severn Barrage (source: adapted from Department of Energy, 1989)

In addition, as will be discussed in more detail later, many different types of **double-basin** system have been proposed (see, for example, Figure 6.20), often using pumping between the basins. During periods of low demand, excess electricity generated by the turbines of the first basin can be used to pump water into the second basin, ready for the latter to use for generation when power is required.

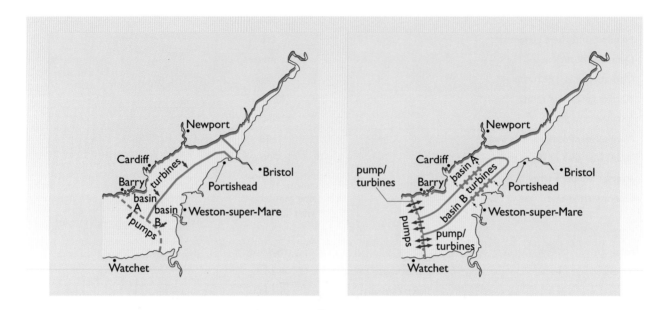

Figure 6.20 Severn Estuary with possible double-basin schemes (source: adapted from Department of Energy, 1981)

Whatever the precise configuration chosen for a barrage, the basic components are the same: *turbines*, *sluice gates* and, usually, *ship locks*, to allow passage of ships, all linked to the shore by *embankments*.

The turbines are usually located in large concrete units. For the Severn Barrage, the use of large concrete, or steel, caissons (containment and support structures) to house the turbines has been proposed. These could be constructed on shore, in dry dock facilities, and then floated onto site and sunk into place. It has been claimed such techniques could cut civil engineering costs by up to 30% as compared to the temporary coffer-dam approach taken at La Rance (although there could be additional difficulties in placing caissons in strong currents).

Sluice gates are another essential operational feature of a barrage, to allow the tide to flow through ready for ebb generation, or back out after flow generation. These can also be mounted in caissons.

The rest of a barrage is relatively straightforward to construct. La Rance, for example, has a rock-filled embankment, whilst one design proposed for the Severn used sand-filled embankments faced with suitable concrete or rock protection.

6.3 Environmental considerations for tidal barrages

The construction of a large barrier across an estuary will clearly have a significant effect on the local ecosystem. Some of the effects will be negative, and some will be positive. The negative local impacts have to be weighed up against the role that barrages could play in helping to resolve some global environmental and energy problems (such as global warming caused by CO_2 emissions from the burning of fossil fuels), and in offering improved energy security through decreased reliance on imported fuel.

In the UK, much research has gone into trying to ascertain the probable final balance of positive and negative impacts, and the overall cost effectiveness, focusing mainly on the proposed Severn Tidal Barrage (Department of Energy, 1989). The most recent review was carried out as part of the UK government's Severn Tidal Power Feasibility Study (DECC, 2010a). This looked at a number of tidal range projects for the Severn, including the Cardiff–Weston Barrage and some smaller barrages and tidal lagoons, and concluded that of all the schemes studied, the Cardiff–Weston Barrage would 'have the greatest impact on habitats and bird populations and the estuary ports', as well as very high capital costs.

Certainly the most obvious potential impact of any barrage would be on local wildlife, that is, fish and birds, many of the latter being migratory. The UK's estuaries play host to approximately 28% of European swans and ducks and to 47% of European geese. There are also large populations of fish: the Severn, for example, is well known for its salmon and eels (elvers). Many of these species rely on the estuaries for food, and access to that supply might be affected by a tidal barrage (Department of Energy, 1989).

The proposed Severn Barrage would decrease a large area (200 km^2 or more) of mud flats exposed each day, since the water level variations behind the barrage would be significantly reduced (Figure 6.21). Some species (for example, mud-wading birds) feed on worms and other invertebrates from the exposed mud flats, and could be adversely affected. Similar issues would apply for salt marshes that might be exposed daily by the tides at other potential barrage sites.

However, the barrage could have a compensating impact on the level of silt and sediment suspended in the water – the action of the tide in churning up silt makes the water in the Severn Estuary impenetrable to sunlight. With the barrage in place and the tidal ebbs and flows reduced, some of this silt would drop out, making the water clearer. Given this change in **turbidity** sunlight would penetrate further down, increasing the biological productivity of the water and therefore increasing the potential food supplies for fish and birds. The net impact is likely to be mixed: some species might not find a niche in the new ecological balance, whilst others previously excluded from the estuary might become established.

This rather simplified example illustrates the general point that there are complex interactions at work making it difficult to predict the outcome.

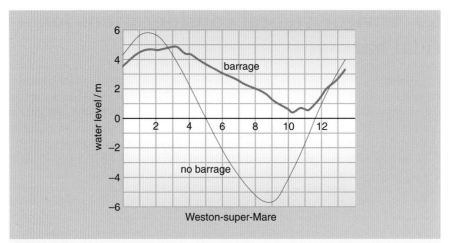

Figure 6.21 Variation in tide levels with and without the Severn Barrage over a 12 hour period (source: Department of Energy, 1989)

Similar interactions and trade-offs occur in relation to other ways in which barrages can impact on their surroundings. Clearly, the construction of a barrage across an estuary will impede any shipping, even though ship locks are likely to be included. The fact that the sea level behind the barrage would, on average, be higher could improve navigational access to ports, the net effect depending on tidal cycles and the precise location of the barrage and of any ports.

Visually, barrages present fewer problems than comparable hydro schemes. Even at low tide, the flank exposed would not be much higher than the maximum tidal range. From a distance, all that would be seen would be a line on the water.

Barrages could also play a useful role in providing protection against floods and storm damage, since they could be operated to control very high tidal surges and limit local wave generation. Conversely, for some sites, due to the change in tidal patterns (with the tide upstream staying above mid-level for longer periods), there might be a need for improved land drainage upstream.

A barrage would have some effect on the local economy both during the construction phase and subsequently, in terms of employment generation and local spending, tourism and, in particular, enhanced opportunities for water sports. Depending on the scale and the site, there could also be the option of providing a new road or rail crossing, as with the Rance Barrage. The incorporation of a public road was part of the plans proposed for the Severn barrage.

Whether these local infrastructural improvement options represent environmental benefits or costs depends on your views on industrial and commercial development (some conservation and wildlife groups, for example, baulk at the prospect of increased tourism), but many people would be likely to welcome local economic growth. Indeed, that was the message from local populations faced with barrage proposals. Local commercial and civic interests and the wider public have on the whole been supportive of such plans while other special interest groups have opposed barrages. For example the Royal Society for the Protection of Birds (RSPB) sees barrages as inherently damaging, reducing habitats for key species, particularly migrant birds. This problem could clearly be compounded if several barrages were to be built. In 2008, when the Severn Barrage idea was moving up the UK political agenda, the National Trust, RSPB and World Wide Fund for Nature (WWF), in a coalition with other groups, came out strongly against the barrage.

6.4 Integration of electrical power from tidal barrages

The electricity produced by barrages must usually be integrated with the electricity produced by the other power plants that feed into a national grid power transmission network.

The key problem in feeding power from a tidal barrage into national grid networks is that with conventional ebb or flood generation schemes the tidal energy inputs come in relatively short bursts at approximately twelve-hour intervals. Typically, power can be produced for five to six hours during spring tides and three hours during neap tides, within a tidal cycle lasting 12.4 hours (Figure 6.22).

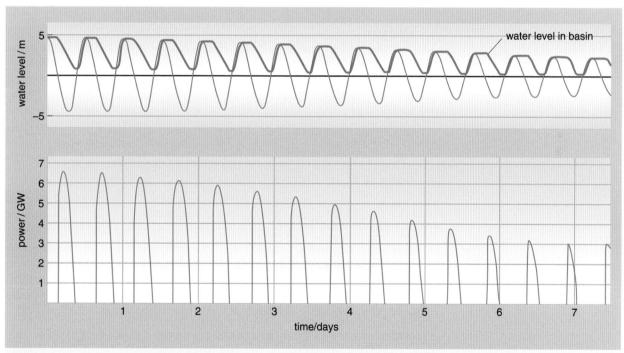

Figure 6.22 Water level and power output of the proposed Severn Barrage over a spring–neap tide cycle (source: Laughton, 1990)

Clearly, such availability of power from a barrage will not always match the pattern of demand for power on the grid. With a large, well-developed grid, as in the UK, with a large load and many other types of power plant connected into it, this problem might not be too severe, depending on the size of the barrage output. The two daily bursts of tidal power could be used to reduce the load on older, less efficient and/or more expensive plants such as older coal-fired plants. The barrage would thus be operating in a 'fuel-saving' mode, the predictability of the tides allowing for the process of substitution to be planned well in advance. Even so, with a barrage the size of that proposed for the Severn, absorbing all the power would clearly represent a significant task. At its peak, it would be generating over 8 GW of power, which represents a sizeable proportion of the UK's peak electricity demand of almost 60 GW (in 2010), and an even larger proportion of the minimum demand level during summer nights (around 20 GW – see Figure 10.13).

Two-way generation, on the incoming flood as well as the ebb, might offer some advantages since that would give *four* bursts of power for each 24 hour cycle, but as noted earlier the more complex turbines required add to the cost. There is also the option of using modified turbines as pumps, running on grid power (e.g. during off-peak periods), to pump extra water behind a barrage during the flood phase. Typically, between 5% and 15% extra output can be gained, with little additional capital cost and with no loss of generating efficiency. This has been the favoured option for all the barrage schemes proposed in the UK. That said, the overall *economic* advantages of pumping may be fairly small (the electricity used has to be paid for), and they depend crucially on tidal timing, which will not necessarily coincide with off-peak grid power availability.

Another option we noted earlier for providing power output more evenly over time is to construct two (or more) basins. For example, the first basin could operate in the normal manner, but any excess power it produced during low-demand periods, could be used to pump water out of the second basin to keep it below tide level, so that power could be generated from the second basin when needed by filling it through its turbines. As these configurations involve building what amounts to two or more barrages costs increase significantly – at the time of writing, no double-barrage systems have been seriously considered, and simple ebb generation is seen as the most economic option.

Whichever system is used, the overall economic viability of tidal power might be enhanced, in theory at least, if several barrages were in operation in a range of sites, given the fact that the tidal maxima occur at slightly different times around the coast. For example, the Solway Firth and Morecambe Bay on the northwest coast are approximately five to six hours out of phase with the Severn, so that the output from these and other possible barrage sites could be fed into the grid to provide a contribution from tidal power over a longer period of time, although at neap tides this input would be low.

It is possible that there might be synergies between tidal and other renewable energy systems, for example via the installation of wind turbines along barrages in the same way as some wind farms have been constructed on causeways in harbours. The wind plant might even be used at times for pumping water behind the barrage, although the energy contribution that could be made, even if wind turbines were located regularly along the entire length of a barrage, would be relatively small compared with the output of a large barrage. For example, if, say, thirty 2 MW wind turbines were installed along parts of the 16 km long proposed Severn Barrage, their total annual electricity output would be just over 1% of that from the barrage.

Finally, the linking of power from barrages to the national grid could present some practical problems. As with many other types of large, new, power plant, extra grid connections would have to be made, and in some circumstances existing local grid lines strengthened to carry the extra power. Fortunately, most potential barrage sites in the UK are reasonably near to existing power lines, so the problems such as those of harnessing deep-sea wave power (much of which would have to be transmitted from the north of Scotland; see Chapter 8) would be avoided. In the case of the proposed Severn Barrage new power lines, perhaps stretching to the major loads in the Midlands, would probably be required. It has been estimated that these grid connections might add 10% to the capital costs of the barrage.

In summary, the key issue to integration is cost, whether this is for additional grid linkages or for systems that allow power to be produced on a more nearly continuous basis. We now move on to look at economic factors more generally.

6.5 The economics of tidal barrages

The overall economics of tidal barrages depends both on their operational performance and their initial capital costs. The latter are high – it was estimated in the late 2000s that the 10 mile Cardiff (Lavernock Point) – Weston-super-Mare (Brean Down) barrage would cost over £30 billion (DECC, 2010a). The civil engineering works are the single largest element in the total cost, closely followed by turbine manufacture and installation. In addition to the construction and equipment costs, there is also the cost of borrowing the money, with interest having to be paid during the course of construction, when of course there is no income since the barrage is not yet operating. This raises a key issue for the economics of large capital-intensive projects like this with long construction times.

Although there are running costs (approximately 1% of the total capital cost per year), tidal barrages, like all renewable energy systems based on natural flows, have no fuel cost. After initial construction, they can generate power for many years without major civil engineering effort with the low-speed turbines needing replacement perhaps every 30 years.

Furthermore, in addition to the significant initial capital costs, the period during which power can be produced, at least for simple single-basin ebb or flood systems, is clearly less than for a conventional power plant. For example, because it would only operate during tidal cycles, the 8.6 GW turbine capacity of the Severn Barrage could only offer the same output, averaged out over a year, as a conventional plant with around 2 GW of generating capacity. In other words, the barrage requires a large investment in expensive capacity which is only used intermittently and can therefore only replace a limited amount of conventional plant output. The value of a barrage as a replacement for conventional plant(s) will depend in practice on the scale and timing of the outputs of the plants they can replace, not all of which will be able to generate continuously either.

As explained in Chapters 1 and 5, a convenient way to compare systems is in terms of their capacity factors (the ratio of a plant's actual energy output over a given period to the theoretical output if the plant operated at its full-rated capacity). The annual capacity factor for the proposed Severn Barrage is estimated at around 23%. By comparison, in typical UK conditions the average annual plant capacity factor for nuclear stations has in recent years been around 77%, and for combined cycle gas turbine power plants around 84%.

Thus, compared with most other types of power plant, tidal projects have a relatively high capital cost in relation to the energy output with, consequently, long capital payback times and low rates of return on the capital invested, the precise figures depending on the price that can be charged for the electricity.

In the 1980s, when the Severn Tidal Power Group analysed the economics of a Severn Barrage, the UK electricity industry was state-owned. As such it was expected to make a return on investment of about 5% per annum. However, the policy of the Conservative government of the time was to privatise the industry (which eventually happened in 1989). Any barrage project would thus be expected to be competitive with existing coal-fired and nuclear plant. It would also be privately financed and required to make commercial rates of return (10–15%). This would require a high electricity price to cover the interest on the large capital outlay. The Severn Barrage was not considered to be economic under these conditions (STPG, 1986). That was one reason why the project was not supported at that particular time.

Although fuel prices have increased dramatically and concerns about climate change have grown, the basic economic assessment has not changed significantly since then. It has, however, been argued that, given wider strategic concerns, such as climate change and security of energy supplies, it might be justifiable to accept a lower rate of return, similar to that used for public projects.

In its 2003 Green Book, the UK Treasury suggests that public sector organizations should make a real return of 6% on investments, though this point is qualified with a long discussion on the proper evaluation of risk and 'optimism bias' (HM Treasury, 2003).

In its 2007 study of tidal options for the Severn, the UK government's Sustainable Development Commission (SDC) noted that 'The high capital cost of a barrage project leads to a very high sensitivity to the discount rate used' (a truism for all projects with significant initial capital expenditure, as exemplified in Chapter 5 where a cost comparison of CCGT and hydro plants was made). At a low discount rate of 2%, which it argued could be justified for a climate change mitigation project, the cost of electricity output 'is highly competitive with other forms of generation'. On that basis they felt that the project should be financed as a public project (SDC, 2007).

In contrast they noted that, at a more 'commercial' discount rate of 8% or more, costs escalated significantly, making private sector investment unlikely without significant market intervention by government. For example they calculated that at a 2% discount rate the Cardiff–Weston Barrage would generate at between 2.27 and 2.31p per kWh depending on how long it took to build, whereas at 10% discount rate the cost would rise to 11.18–12.37p per kWh. These costs were expressed in the money of the day, i.e. £(2006).

In its 2010 study of Severn tidal schemes the UK government compared the impact of 3.5% public sector 'social' discount rates with 10% commercial 'investor' discount rates (see Chapter 10 and Appendix B for a further discussion of discount rates). Table 6.1 illustrates the dramatic difference in energy costs and also the different levels of environmental impact of different options. Table 6.1 also compares the costs of large and smaller barrage schemes on the Severn. It might be expected that the smaller schemes further upstream would be more attractive financially, because they can be built more rapidly – but this is evidently not the case, in part due to the lower energy output from the smaller tidal ranges at these sites. So whether under 'investor' or 'social' financial frameworks, smaller barrages on the Severn are less attractive in terms of the cost of electricity produced. Although in the case of the Severn this seems clear-cut, small barrages in other estuaries, in locations with higher tidal ranges, might do better.

Table 6.1 Comparison of costs and impacts of large and small barrages on the Severn

Scheme	Installed capacity /MW	Annual energy generated /TWh y^{-1}	Levelized energy cost/£ MWh^{-1}		Inter-tidal habitat loss/km^2
			Investor (10% discount rate)	Social (3.5% discount rate)	
Cardiff–Weston Barrage	8640	15.6	312	108	160
Shoots Barrage	1050	2.7	335	121	33
Beachley Barrage	625	1.2	419	151	27

Source: DECC, 2010a

Given the lifetime of the low-speed turbines (replacement needed perhaps only every 30 years), low running costs (approximately 1% of the total capital cost per year) and zero fuel cost, it has been argued that, like hydro projects, barrages make a very good long-term investment. This is unfortunately not reflected in contemporary financial approaches, with the emphasis usually being on short-term returns.

Quite apart from the problems of funding and the vagaries of finance capital, interest rates, etc., on the 'supply' side, there are technical and environmental uncertainties, with trade-offs between operational efficiency and likely impact. To these uncertainties must be added the 'demand side' uncertainties, with the price that can be charged for tidal electricity having to be estimated over the very long term and compared with conventional fuel prices.

At £30 billion or so for cyclically varying power, the Severn barrage might not be the best use of any money available. Indeed a study by Frontier Economics for the National Trust, RSPB, WWF and other environmental groups opposed to the Severn barrage, suggested that large barrages were significantly more expensive on a £ per MWh basis than any other major energy generation option (Frontier Economics, 2008; see Figure 6.23). On this basis, it would seem that other options are likely to be more attractive, a view that now seems to be shared by the government, which in its 2010 study of the Severn tidal options commented: 'The Government believes that other options, such as the expansion of wind energy, carbon capture and storage (CCS) and nuclear power without public subsidy, represent a better deal for taxpayers and consumers at this time' (DECC, 2010a).

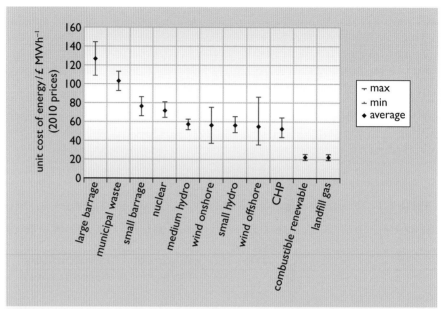

Figure 6.23 Comparison of costs of energy from tidal and other renewable energy projects, at 7% discount rate (source: adapted from Frontier Economics, 2008)

Although it seems likely that fossil fuel (and possibly nuclear) electricity prices will increase over time, so that tidal barrage projects will become more attractive, under the conditions present at the time of writing tidal barrages appear to be relatively unattractive commercial investment options. Nevertheless, with concerns about climate change and energy security, interest in the concept might revive, in part because of the large amount of renewable energy that could be generated.

In carbon balance terms, barrages do not generate CO_2 during their operation, but building them inevitably does. The UK SDC estimated that the carbon dioxide that would be produced by making the materials for and constructing the Cardiff–Weston barrage effectively amounted to 2.42 gCO_2 per kWh generated. This translates into a carbon payback time (based on how long it would take to recoup those emissions by the use of the barrage rather than a fossil-fuelled plant) of around 5–8 months (SDC, 2007).

Once built, a barrage could be seen as either displacing the output of existing fossil plant, or displacing the need for some other form of new capacity, and the associated emissions. The SDC decided to focus on the latter and assume that a Severn barrage would displace the need for new gas-fired combined cycle gas turbine (CCGT) plant, since 'a Severn barrage is unlikely to be operational for at least 10 years, during which time much of the UK's coal capacity will be taken out of service' (SDC, 2007).

On this basis, SDC calculated that the Cardiff–Weston Barrage would avoid the emission of 5.5 $MtCO_2$ per year compared to alternative CCGT gas-fuelled electricity generation, which represents a 0.92% reduction in UK carbon emissions compared to the 1990 baseline level used by the government. That might be seen as a relatively small saving for an estimated outlay of £30 billion or more.

6.6 Tidal barrages: potential projects

United Kingdom

The UK has some of the largest, and most studied, tidal energy resources in the world. In general, the best potential tidal barrage sites in the UK are on the west coasts of England and Wales, where the highest tidal ranges are to be found. Despite its indented coastline, the tidal energy potential of Scotland is very small (1–2 TWh y^{-1}) due to its generally low tidal range.

As we have seen, the practical potential of tidal power depends crucially on economics, as well as on environmental factors. In theory, the exploitable potential, assuming that every practical UK scheme was developed, could rise to approximately 53 TWh per year, or around 14% of UK electricity generation in 2010. Approximately 90% of this potential (48 TWh y^{-1}) lies in eight large sites, each offering between 1 TWh y^{-1} and 17 TWh y^{-1}, while the remaining 10% relates to 34 small sites, each providing somewhere in the range 20–150 GWh y^{-1} (ETSU, 1990).

The Cardiff –Weston Barrage, if built, would make the largest contribution, approximately 17 TWh y^{-1}, but initial estimates have suggested that the Wash, the Mersey, the Solway Firth, Morecambe Bay and possibly the Humber, amongst others, could also make significant contributions (Figure 6.24).

In addition to these larger sites, there are many smaller estuaries and rivers that could be used. Feasibility studies were carried out in the 1980s on the Loughor Estuary (8 MW) and Conwy Estuary (33 MW) in Wales, the Wyre (64 MW) in Lancashire, and the Duddon (100 MW) in Cumbria.

Overall, the total UK potential for small tidal schemes (that is, schemes of up to 300 MW capacity), has been estimated at nearly 2% of UK electricity requirements.

Studies have continued on some of these smaller sites, for example on the Duddon and there have also been proposals for larger schemes, including a 1 GW rated £2 billion barrage 11 miles across the Wash, and a 4.75 km² tidal lagoon in the Thames estuary, coupled, in a £2 – £4 billion project, with a tunnel crossing and tidal surge barrier between Medway and Canvey Island.

In addition there is continuing interest in other locations, where some even larger schemes might be possible. For example, a study led by the North West Regional Development Agency looked at tidal energy options for the Solway Firth, the UK's third largest estuary, on the border of England and Scotland. It identified nine main options including four tidal barrages, the largest scheme running from Workington to Abbey Head.

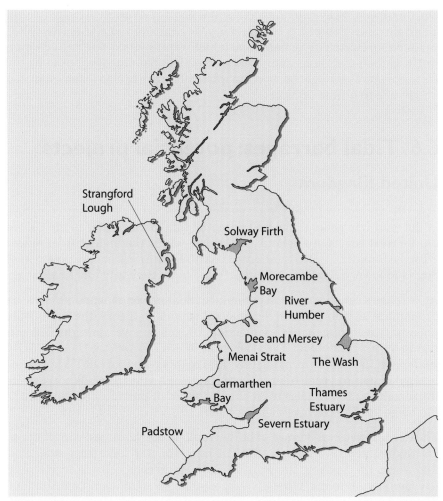

Figure 6.24 Some potential locations for tidal barrages in the UK

The schemes were compared on the basis of their likely environmental impact and the levelized cost of energy generated. The conclusions for the barrage projects were as follows.

- Workington to Abbey Head: £16 billion cost and large environmental impact; 5.9 GW installed, delivering 11.5 GWh y^{-1} – this scheme had a CoE of £183 per MWh (18.3p/kWh).

- Southerness Point to Beckfoot: £6.1 billion cost and substantial environmental impact; 2.7 GW installed – this had the lowest CoE of all the Solway schemes studied at £175 per MWh (17.5 p/kWh).

- Bowness to Annan: £1.2 billion cost; 316 MW installed delivering 320 GWh y^{-1} – CoE of £389 per MWh (38.9p/kWh).

- Morecambe Bay barrage (located out of the main estuary to reduce ecological impact): 113 MW installed delivering 120 GWh y^{-1} – CoE of £553 per MWh (55.3p/kWh).

- For comparison, the 8.6 GW Cardiff–Weston Severn Barrage (using an 8% rate of return, not the 10% discount rate in Table 6.1) was estimated to give a CoE of £160 per MWh (16.0 p/kWh).

The non-barrage tidal projects examined (including lagoons and tidal flow turbine systems) all had similarly high CoEs, but lower environmental impacts (Halcrow Group Ltd et al., 2009).

A 2010 study by the North West Regional Development Agency and renewable energy developer Peel Energy looked at a range of tidal options for the Mersey, including barrages of various sizes but also tidal lagoons and a variety of tidal current turbine concepts. The study reviewed an earlier 700 MW barrage proposal by the Mersey Barrage Company, but suggested that, although viable, the impacts of a barrage on shipping, sedimentation, water quality and the local ecology would need to be very carefully assessed. A smaller 500 MW barrage was also seen as viable, delivering 900 GWh y^{-1}. Moreover, it also found that some the other tidal options, although producing less energy, might have significantly lower environmental impacts (Peel Energy/NWRDA, 2010).

World

The total power in the tides globally is very large – of the order of 3000 GW.

In practice, since there are geographical access and location constraints on siting barrages or other means of extracting tidal energy, and they also have to be reasonably near to major lines of a national grid, the realistically available resource is much smaller. Jackson (1992) has put the realistically recoverable resource at 100 GW. Although that is under 12% of the existing global hydroelectric capacity, it still represents a significant resource. As Table 6.2 indicates, there are a number of potential large-scale sites for tidal barrages around the world, in Russia, Canada, the USA, Argentina, Korea, Australia, France, China and India, with an estimated total potential of perhaps as much as 300 TWh y^{-1}. There are some locations where some very large projects might be possible, (e.g. the potential 87 GW scheme at Penzhinsk in Russia) as well as many smaller ones.

So far, only a few of these potential sites have been exploited, notably, some small projects in Canada, Russia and the medium-scale project of La Rance in France. However, South Korea has a significant tidal barrage programme, which is developing some sites in addition to those identified by the World Energy Council in 2001 (see Table 6.2).

A 2006 study (Lee, 2006) estimated that South Korea had around 2400 MW of barrage potential, including possibly 700–1000 MW at Incheon, 600–800 MW at Cheonsu, and 480–520 MW at Garolim (this estimate being an increase on the 400 MW shown in Table 6.2). A 2008 study reported that a new 430 MW barrage project was planned near a dam at Saemangeum, along with a 520 MW project at Garolim and that an 812 MW Gangwha/Incheon project was planned to be completed in 2014 (Jo, 2008) At the time of writing a 254 MW tidal range project is being completed at Sihwa, in Gyeonggi province, bordering the West Sea. This is designed to generate power from a tidal rise of 5.64 metres and when fully operational will become the world's largest tidal power plant.

There have also been some very ambitious (if very speculative) proposals for very large barrages, such as huge barrages across the Bering Straits and the Irish Sea although the latter proposal would be considered by many engineers to be technologically unrealistic, as well as economically and politically challenging.

Table 6.2 Potential non-UK tidal power sites

Country	Mean tidal range/m	Basin area/km²	Installed capacity/MW	Approx annual output/TWh	Annual plant capacity factor/%
Argentina					
San José	5.8	778	5040	9.4	21
Golfo Nuevo	3.7	2376	6570	16.8	29
Rio Deseado	3.6	73	180	0.45	28
Santa Cruz	7.5	222	2420	6.1	29
Rio Gallegos	7.5	177	1900	4.8	29
Australia					
Secure Bay	7.0	140	1480	2.9	22
Walcott Inlet	7.0	260	2800	5.4	22
Canada					
Cobequid	12.4	240	5338	14.0	30
Cumberland	10.9	90	1400	3.4	28
Shepody	10.0	115	1800	4.8	30
India					
Gulf of Kutch	5.0	170	900	1.6	22
Gulf of Khambat	7.0	1970	7000	15.0	24
South Korea[1]					
Garolim	4.7	100	400	0.836	24
Cheonsu	4.5	–	–	1.2	–
Mexico					
Rio Colorado	6–7	–	–	5.4	–
USA					
Pasamquoddy	5.5	–	–	–	–
Knik Arm	7.5	–	2900	7.4	29
Turnagain Arm	7.5	–	6500	16.6	29
Russian Federation					
Mezen	6.7	2640	15 000	45	34
Tugur[2]	6.8	1080	7800	16.2	24
Penzhinsk	11.4	20 530	87 400	190	25

[1] See text for more recent additions/updates
[2] 7000 MW variant also studied
Source: WEC, 2001

6.7 Tidal lagoons

Somewhat less ambitiously but still speculatively, in addition to large barrages, there have also been proposals for offshore 'bounded reservoir' or 'tidal lagoon' systems, consisting of circular low-head dams in open water, trapping water at high tide, with the water then being released to drive turbines in the usual way (Figure 6.25).

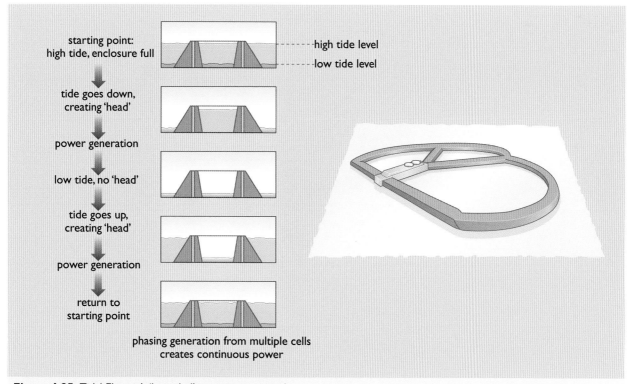

Figure 6.25 Tidal Electric's 'bounded' reservoir proposal

Lagoons would be in relatively shallow water and would be constructed like causeways with rock infill. Lagoons have the key advantage that although, as with conventional barrages, they would involve the creation of a head of water, they would not involve blocking off an estuary and so should have a lower environmental impact and avoid interfering with the passage of ships.

As well as generating power directly for the grid, lagoons might also be used as a low-head offshore pumped storage facility (see Chapter 5). Moreover, as with the double barrage idea, segmented lagoons could enable phased operation and pumping between segments.

The US company Tidal Electric has been investigating potential sites for tidal lagoons in Alaska, Africa, Mexico, India and China. At the time of writing a 1000 MW project is being considered in India and the company has also been in discussion with the Chinese government, which, it says, has expressed interest in a 300 MW offshore tidal lagoon to be built near the mouth of the Yalu River. Three sites are also under consideration in the UK, off the coast of Wales, including a 60 MW lagoon off Swansea. As illustrated in Figure 6.26 many other locations in that area have also been seen as possible lagoon sites.

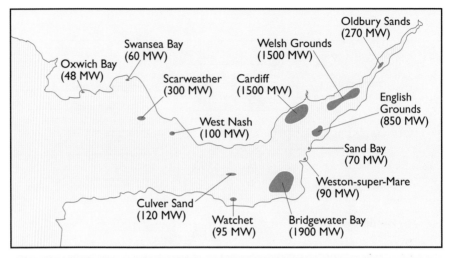

Figure 6.26 Possible tidal lagoon sites in and around the Severn estuary (source: Salter and Walker, 2010)

The disadvantage of offshore lagoons is that the entire containment wall has to be constructed, so increasing the cost. This is in contrast to barrages where the estuary shore provides free containment for the bulk of the water held behind it at high tide. Even so, given that lagoon construction is expected to be easier and quicker than for barrages, it is claimed that the cost of power from offshore lagoons can be competitive (Atkins Consultants Ltd, 2004).

While Tidal Electric have focused on fully offshore lagoons, a compromise option, to reduce cost further, would be to site the lagoon near to the shore, with part of the containment being the shore line. However inshore lagoons of this type are more likely to impact on mud flats, and the wildlife that uses them, than fully offshore schemes.

A 2010 DECC review of Severn tidal options (DECC, 2010a) concluded that although the cost of energy was relatively high, a near-shore lagoon at Bridgwater Bay on the English side could be feasible, with 'lower environmental impacts than barrage options', but that the Welsh Grounds near-shore lagoon was not viable (see Table 6.3). Tidal Electric's fully offshore scheme for Swansea Bay was not reviewed, being deemed to be outside of the geographical area of the study.

Lagoon proposals have also emerged for other locations around the UK, notably the Solway Firth and the Mersey estuary (see Section 6.6 above), but, so far, none have gone ahead.

Table 6.3 Comparison of costs and impacts of lagoons in the Severn (see Table 6.1 for the equivalent information for barrages)

| Scheme | Installed capacity /MW | Annual energy generated /TWh y⁻¹ | Levelized energy cost/£ MWh⁻¹ | | Inter-tidal habitat loss/km² |
			Investor (10% discount rate)	Social (3.5% discount rate)	
Welsh Grounds Lagoon	1000	2.1	515	169	73
Bridgewater Bay Lagoon	3600	6.2	349	126	25

Source: DECC, 2010a

No doubt the debate over the relative merits of barrages, large and small, and lagoons near shore and offshore, will continue. The combined resource is certainly not insignificant – an assessment of the energy resource offered by tidal range projects in the UK carried out in 2010 suggested that, assuming a maximum credible expansion programme, 20 GW of tidal range capacity might be expected by 2050 delivering 40 TWh per year. That included contributions from tidal lagoons, as well as barrages on the Severn, Mersey and Solway (DECC, 2010b).

The 2010 DECC Severn Tidal Feasibility Study concluded that the over £30 billion cost of the Cardiff–Weston Barrage made it high cost and high risk in comparison to other methods of generating electricity and the smaller barrages and lagoons were even less attractive economically (Figure 6.27).

Although the DECC report did say that the results for other locations around the UK might be different, it is hard to see how they could do better than the Severn – the best site by far in terms of tidal range, although other factors may in the future prove to be more important.

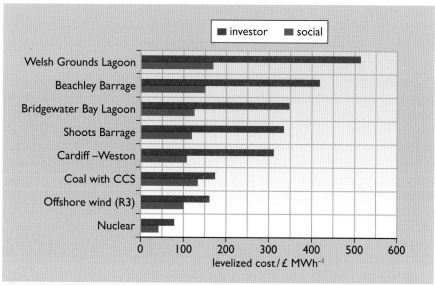

Figure 6.27 DECC estimates of Severn barrage and lagoon electricity prices compared to other options at 'investor' (10%) and 'social' (3.5%) discount rates (source: DECC, 2010a)

6.8 Tidal streams/currents

As noted in the introduction to this chapter, the use of tidal streams or currents is being followed up with some enthusiasm. Instead of using costly and potentially invasive barrages located in estuaries to exploit the *vertical* rise and fall of tides, and the *potential* energy of heads of water trapped behind dams or lagoons, it is possible to harness the energy in the *horizontal* movement of the tides – that is, the *kinetic* energy of the tidal ebbs and flows.

In open sea, the speed of the tidal ebbs and flows is relatively low, but it can be much higher when the tidal movements are concentrated by passing through narrow channels or around islands, headlands, or other topographical constraints. The enhanced velocity of the tidal streams or currents in such locations enables maximum energy extraction, for example by using relatively simple, submerged, wind turbine-like rotors (Figure 6.3).

As was noted in Section 6.1, the terms 'tidal current' and 'tidal stream' are often used interchangeably, and to confuse matters more not all the water flows in oceans are actually 'tidal' in the sense of being driven by the Moon. Some of these flows, especially the larger offshore ones, such as the Gulf Stream, are the result of complex interactions between warm and cold layers of water in the oceans around the world, and the associated effects of varying salinity – they are, in effect, solar-driven. Strictly speaking the correct terms to use for such non-tidal flows are **ocean currents** or **ocean streams**. We will be looking at some projects aiming to tap this energy in a later section.

Whatever type of flow is used, the basic physics of power extraction via turbines is similar to that for wind turbines. As Chapter 7 will explain, the power P in the wind is given by the formula $P = 0.5\ \rho a v^3$, where ρ is the density of air, v the wind speed, and a the area of the circle swept by the rotor (which is πr^2, where r is the rotor blade length). Thus the energy available is proportional to the square of the rotor size and, in the case of a turbine mounted in water, the cube of the water velocity.

The major difference between wind turbines and tidal devices is that with tidal turbines the fluid concerned, water, is much denser than air. Moreover, tidal flows interact with the seabed and other topographical features, and these frictional effects are much more significant (given the density of water and the often narrow and relatively shallow channels involved) than in the case of wind flows. These interactions are especially significant in more complex channels where there are direction changes and varying topographical features, with complex resonances also sometimes playing a role (see Section 6.1 for discussion of resonances). Tidal flows will also be affected by interactions with the interface between air and water at the sea/river surface. Estimating the energy resource available is therefore complex and site specific, requiring detailed local modelling (see Hardisty, 2007; Bryden et al., 2007).

So, while areas where there are high tidal ranges are important (see Figure 6.24), the local topography is crucial in determining whether tidal currents/streams can be utilized effectively – flow rates can sometimes be low at sites with a high tidal range and vice versa.

The resource and its location

Although tidal current resources are site specific, there is a lot of energy in the enhanced tidal currents in appropriate locations. For example, it has been estimated that the power flowing through the north channel of the Irish Sea is equivalent to 3.6 GW, while the flow through Pentland Firth in Scotland is the equivalent of 6.1 GW.

Mean peak spring tidal currents of at least 2–2.5 m s^{-1} (~4–5 knots) and a depth of between 20 and 35 metres are seen as necessary for economic exploitation. On this basis, the UK has some of the world's best sites for producing energy from tidal streams, as shown (see Figure 6.28). Table 6.4 gives details of the expected energy outputs in some of these sites.

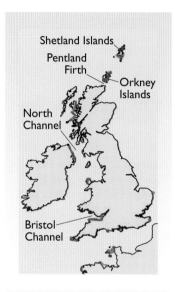

Figure 6.28 Possible tidal current sites around and near the British Isles

Table 6.4 Some major tidal current sites in the vicinity of the British Isles and their associated resource potential

Site name	Location	TWh y^{-1}
Pentland Skerries	Pentland Firth	3.9
Strøm	Pentland Firth	2.8
Duncansby Head	Pentland Firth	2.0
Casquets	Alderney	1.7
South Ronaldsay	Pentland Firth	1.5
Hoy	Pentland Firth	1.4
Race of Alderney	Alderney	1.4
South Ronaldsay	Pentland Firth	1.1
Rathlin Island	North Channel	0.9
Mull of Galloway	North Channel	0.8

Source: SDC, 2007

The 2050 Pathways report produced by Department of Energy and Climate Change in 2010 saw UK tidal current projects as potentially delivering up to 69 TWh per year in total by 2050, from 21.3 GW of installed generating capacity, assuming maximum possible development (DECC, 2010b).

Moreover, a study by an industry group led by the Public Information Research Centre (PIRC), put the practical resource potential for tidal currents much higher, at 116 TWh per year, from 33 GW installed capacity (OVG, 2010).

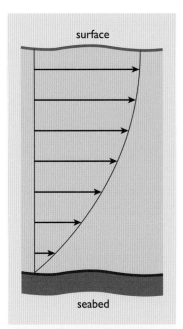

Figure 6.29 The shear effect – water speed (as indicated by arrow length) reduces with depth

Tidal current turbine design constraints

When considering device design and location it is crucial to note that friction with the seabed produces a vertical 'shear' effect in the current: surface water moves fastest, with the flow being slower lower down. Around 75% of the energy in the flow is contained in the top half of the water (Figure 6.29), but since the rotors have to be fully submerged, and to be efficient and avoid breaking the surface, a mid-depth location can make sense, with the size of the device being limited by the water depth. Given that most sites with high velocity water flow are relatively shallow (below 40–50 metres), tidal devices thus have a size constraint that does not apply to wind turbines.

Since the available energy is proportional to the square of the rotor blade length and the cube of the water velocity, even a small increase in blade size or water speed will yield a significantly increased amount of energy production. Figure 6.30 shows how the relationship works out in practice – the faster the flow and the larger rotor (subject to practical considerations) the better.

Submerged free-standing tidal rotors can operate at lower rotational speeds than wind turbines, the rotor speed in both cases being proportional to the flow velocity, with typical sea current velocities only being approximately 3 m s^{-1} compared with, say, 5–15 m s^{-1} for wind machines. Given that the density of the working fluid is much higher, the power output for a tidal stream machine is much larger than for a wind machine of equivalent rotor diameter.

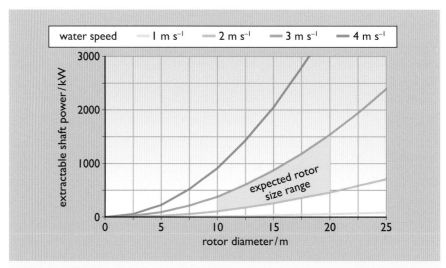

Figure 6.30 Relationship between rotor diameter, water speed and extractable power (source: adapted from Marine Current Turbines, 2010)

In comparison to barrage systems, the output from a tidal stream turbine is likely to be lower than that from an equivalent-sized conventional turbine in a barrage, which has the advantage of the funnelling effect of the barrage and the creation of an enhanced head of water. However, free-standing tidal current systems can have much larger rotors than is practical in barrage-mounted turbines.

An advantage of tidal current devices over simple ebb or flood tidal barrages is that they can operate on both ebb and flood tides and thereby attain a much higher capacity factor. Other advantages are that individual units can be constructed on a modular basis and installed incrementally in gradually expanding arrays giving both economies of scale and a much shorter lead time between investment and gaining revenue.

As with wave energy systems (see Chapter 8) there may be problems of fouling (for example, by seaweed, lost fishing nets, etc.), tethering and power take-off to overcome, but on the other hand, expensive and environmentally intrusive barrages are not required. There will be little visual impact, as in most designs all of the device would be under water, and there is virtually no noise.

There is a range of possible device configurations. Energy in tidal and ocean streams/currents can be captured using wind turbine-like horizontal or vertical axis rotors, supported under floating pontoons tethered to the seabed, fixed directly on the seabed, or mounted on towers. Ducts can be used to accelerate the water flow through the turbine. Alternatively, use can be made of hydroplane arrangements, involving seabed mounted hydrofoils oscillating up and down in the tidal flows. The following sections look at some examples of each approach and at the status, at the time of writing, of development in the UK and elsewhere.

Some of the projects reviewed below are well established, having been fully tested at sea, and some have been deployed at full scale. However, it should be stressed that many of the projects being discussed are at a relatively early stage of development, with the claims about potential energy outputs and generation capacities still unproven – indeed some descriptions are of currently speculative design concepts and proposals. When examining novel proposals, care has to be taken to assess the credibility of the claims being made. For example, generation capacities are sometimes claimed which could only be achieved, given the size of the device, at very high water speeds. While many of the devices and proposals described in the sections below may not eventually prove successful, a wide sample of projects has been included here, to give a feel for the range and vitality of the innovative and experimental processes underway in this field.

6.9 An overview of projects and innvovative tidal stream concepts in the UK

Tidal current/stream development has only begun to be developed on a significant scale in the UK relatively recently, partly because it was previously thought that the cost of generating electricity via such technologies was likely to be high. Such cost considerations meant that work on tidal stream technology was given a low priority in the renewable energy development strategy adopted in the early 1990s by the (then) Department of Trade and Industry (ETSU, 1993).

Subsequently, following a review by the Marine Foresight panel set up by the Office of Science and Technology, support for tidal stream projects has grown, partly because the report argued that economies of scale via the production of large numbers of turbines (albeit only after a substantial development programme) could reduce these costs (OST, 1999) and also because novel devices have begun to emerge.

In 1994, in a pioneering project, a prototype two-bladed 10 kW tidal current turbine was tested in the Corran Narrows in Loch Linnhe, near Fort William in Scotland, by a consortium involving IT Power, the National Engineering Laboratory and Scottish Nuclear. This was the world's first tidal current turbine. The tests involved a rotor of 3.9 metres in diameter submerged under a small catamaran pontoon in the sea at a depth of 5 metres. In a fully operational system it was proposed that the rotor unit would be supported by a cable attached to a floating buoy and also tethered to the seabed. It could therefore swivel around on the change of the tide to absorb power from tidal currents in either direction (Figure 6.31).

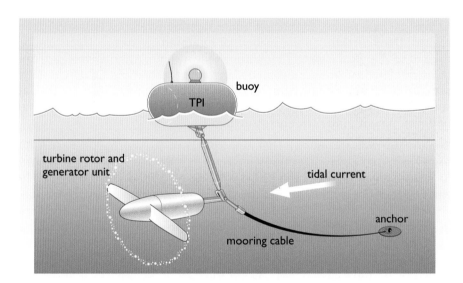

Figure 6.31 The IT Power tidal current turbine concept, developed and tested at Loch Linnhe, Scotland (source: adapted from Marine Current Turbines, 2010)

This free-floating, mid-depth swivelling system concept might in some locations have cost advantages over fixed bottom-mounted devices, e.g. in places where there is sufficient room for the structure to swing around. However, the need for flexible marine cable power take-off arrangements could add to the overall cost, and Marine Current Turbines (MCT) (an offshoot of IT Power) went on to develop a system with a rotor mounted on a fixed steel pile driven into the seabed.

Subsequently, MCT implemented this approach in a two-bladed 300 kW experimental prototype named SeaFlow which was tested 1 km off the coast of north Devon near Lynmouth between 2003 and 2006 before being decommissioned in 2009 after testing had finished.

MCT then developed a larger two-bladed twin-rotor design of 1.2 MW capacity (at 2.4 m s⁻¹), called SeaGen, which they installed in Strangford Narrows, Northern Ireland in 2008 (Figure 6.32). The rotors of the first prototype are 16 metres in diameter and their pitch can be rotated through 180 degrees, to re-orient them to the tide when it changes direction, thus allowing efficient operation on both the ebb and the flood. They can be

Figure 6.32 MCT 1.2 MW SeaGen in Strangford Narrows with the rotor blades raised out of the water for access

raised out of the water for easy access. Although the prototype is rated at a relatively low power level its successors are expected to be capable of up to 2 MW, using rotors of up to 20 m diameter.

SeaGen has proved very successful, delivering power to the grid and, from 2009 onwards, earning Renewable Obligation Certificates for the electricity supplied, under the UK Renewable Obligation scheme – it was the first tidal project to obtain support from this scheme. Following this project, the aim is to build a 10 MW 'tidal farm' array, using several machines off the coast of Wales near Anglesey, with a similar project planned for Kyle Rhea off the Isle of Skye in Scotland. In addition MCT and ESB International are developing the initial phase of a 100 MW project off the Antrim coast, in Northern Ireland, the initial 50 MW phase of which could be in operation by 2018. MCT also has plans for a 100 MW project off the Orkney Islands with a 2020 completion date.

A new development of SeaGen has multiple rotors (3 to 6) to gain further economies of scale. It operates completely submerged but has the facility to be automatically surfaced by remote control to enable safer and easier maintenance; the prototype, a 2 MW triple rotor device, is intended to be installed in 2012 in the Bay of Fundy in Canada.

Conventional axial flow or 'propeller-type' designs like MCT's SeaGen seem the most obvious way forward: they mimic the successful approach used by the wind industry. Although not yet fully developed tested or deployed like MCTs systems, there have been several similar designs, like Swanturbine's seabed mounted Cignet, developed at Swansea University, and Tidal Energy's DeltaStream triangular frame seabed-mounted system, being tested off Wales – although both of these use three-bladed turbines, rather than two-bladed ones as in SeaGen. A double rotor system, with two triple-bladed contra-rotating rotors mounted on a common shaft, has also been developed by Strathclyde University and a 500 kW version has been tested by spin-off company Nautricity.

Novel designs

The turbines we have looked at so far may look like aircraft or ship propellers, but of course they do not propel anything – they absorb energy from the tidal flow. Propeller-type configurations have certainly been popular, but they are not the only option and the tidal turbine field is currently at the innovative stage where many different concepts are being proposed and tested. For example the Engineering Business group developed a novel design - the **Stingray** - with a set of totally submerged hydroplanes, oscillating up and down in the tidal flows and driving a generator mounted on the seabed (Figure 6.33). A 150 kW prototype was installed and tested in 2002 and

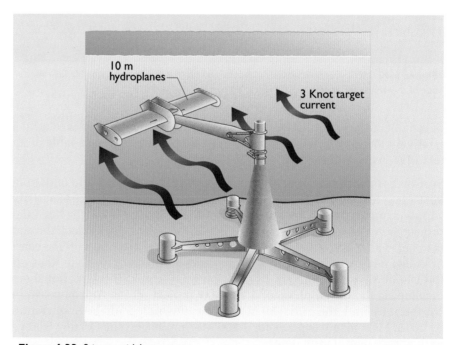

Figure 6.33 Stingray tidal generator

again 2003, in Yell Sound in the Shetlands. However it proved to be very inefficient and was subsequently abandoned.

It seems the energy conversion efficiency was low (around 9.5%) due in part to the relatively slow oscillation rate, but a derivative, with two sets of hydrofoils on a 'see-saw' arrangement oscillating much more rapidly, was subsequently developed by Pulse Tidal. In 2009, a 100 kW test rig was tested in the Humber estuary, with power being delivered to shore. The company has been looking in particular at Kyle Rhea as a possible site for a 1.2 MW unit, possibly to be followed by linking eight tidal power devices together in a 9.6 MW tidal fence (Figure 6.34).

Given the relatively shallow depth of most tidal channels, hydroplane type designs are claimed to have an advantage over propeller-like designs in that they can operate in shallower water, with the blades able to traverse a larger cross-sectional area of tidal flow than the circular area defined by the rotating blades of propeller-like devices. This comparison does need to be treated with some caution, since reciprocating devices operate most efficiently in mid-stroke and lose energy as they slow at the end of each stroke.

Many other designs have been proposed. One approach has been to enhance/accelerate the tidal flow using an annular duct surrounding a conventional horizontal-axis rotor (Figure 6.35).

A very different approach to ducting has been adopted by Neptune Renewable Energy. Their Proteus system has a vertical-axis rotor mounted in a duct system submerged under a floating pontoon (Figure 6.36). The system was tested at the University of Hull and a 150 tonne prototype was then built and installed in the Humber Estuary, the aim being to use the electricity generated to power The Deep, an aquarium centre in Hull.

An issue with ducting is whether the enhancement of the flow justifies the extra cost of the duct. With wind turbines, this does not seem to be the case, especially since wind directions vary, but tidal flows always follow fixed paths, back and forth, so fixed ducting might be economically viable.

One of the most novel designs is the 'open-centred' turbine developed by Irish company OpenHydro, which has rotors mounted around the inner rim of a sealed circular generator system, which also acts as a duct (Figure 6.37). Following tests of a 250 kW rated prototype at the

Figure 6.34 Pulse Tidal's two hydrofoils mounted on a see-saw concept

Figure 6.35 Example of a ducted rotor being developed by Rotech

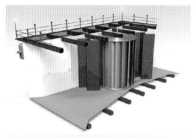

Figure 6. 36 Neptune's Proteus – a vertical-axis ducted rotor

European Marine Energy Centre (EMEC), a 1 MW version was developed and installed for testing in the Bay of Fundy, Canada in 2008, and, in 2010, a 200 MW project was given a site in the Pentland Firth off the northern coast of Scotland. In addition there are plans for a 284 MW project off Alderney.

BOX 6.6 A European test centre

The European Marine Energy Centre (EMEC) in Stromness in the Orkney Islands, provides support for development in tidal and wave power with a site at the Fall of Warness off the island of Eday supporting tidal power research and a site at Billia Croo (Mainland) where wave power devices may be tested.

Apart from offering sites that allow developmental systems to be deployed in the open sea and connected to a grid, EMEC has expertise in device monitoring and with its links to different developers can support best practice throughout this part of the renewables industry. The site is not just restricted to UK devices as the centre is being used by developers from across the world.

One of the key factors in tidal turbine design, as with wave energy devices, is accessibility for maintenance and repair, since this is likely to be a dominant factor in the running costs. Many devices are designed so that the rotor unit and/or generator can be winched to the surface by a boat. Alternatively some have in-built mechanisms for raising the turbine assembly out of the water, as shown in Figure 6.32 in the case of MCT's SeaGen. In a similar vein, TidalStream's Triton design has a series of propeller-type units on a hinged structure fixed to the seabed which can be tilted up so that the turbines are on the surface (Figure 6.38). A one-tenth scale prototype has been tested in the Thames. MCT have also proposed something similar for

Figure 6.37 The OpenHydro open-centred turbine

Figure 6.38 TidalStream's Triton

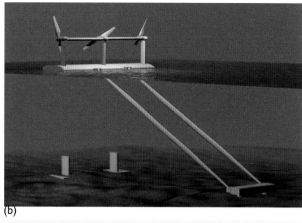

(a)

(b)

Figure 6.39 MCT's next generation SeaGen – the seabed-mounted array can be raised to the surface for access

the follow-up SeaGen deployments – a hinged frame fixed to the seabed (Figure 6.39).

Most of the tidal current schemes and designs discussed above involve single machines or groups of machines in free-standing 'tidal farm' arrays, but there are also proposals for more integrated schemes, with devices mounted in structures such as permeable 'tidal fences' across estuaries (see Box 6.7).

BOX 6.7 Integrated tidal current turbine structures for the Severn

Although some integrated schemes were looked at, along with barrages and lagoons, in the first round of the UK government's Severn Tidal Feasibility Study, they were not included on the final shortlist. An ancillary DECC-supported assessment programme, the Severn Embryonic Technology Scheme (SETS), which started in 2008, followed up some of the ideas described here.

Tidal Fence

As an alternative to the Severn Tidal Barrage, the Severn Tidal Fence group (STF) proposed a 1.3 GW permeable tidal fence in the same location, with a row of large-diameter ducted tidal current turbines, which STF said would be less environmentally invasive than a barrage – since it was not trying to create a head of water, it would not have to block the entire estuary (Figure 6.40).

Figure 6.40 The STF Tidal Fence Proposal

Tidal Reef

The Tidal Reef concept, developed by Evans Engineering, is something of a compromise between a full barrage and the tidal fence, designed to minimize environmental impact. It would be a 15 mile long causeway from Minehead to Aberthaw in Wales, with, in one version, vertical-axis tidal turbines running on both ebb and flood tides, mounted in floating caissons, which would rise, from the supporting causeway, with the tides. The project would comprise over 1000 5 MW turbines of 10 m diameter with an annual generation of around 20 TWh. This compares to the 17 GWh expected from the Cardiff–Weston Barrage.

The basic concept was for the reef to be very permeable, so, unlike the Cardiff–Weston Barrage it would not hold back the full height of the tide or create a large head of water – the low-speed turbines would run on only about a 2 m head, so that the environmental impact should be lower.

Atkins and Rolls Royce subsequently developed a new version of the idea (the Rolls Royce/Atkins Tidal Bar) which was chosen, along with STF's Tidal Fence, for further assessment funded by the government's SETS scheme.

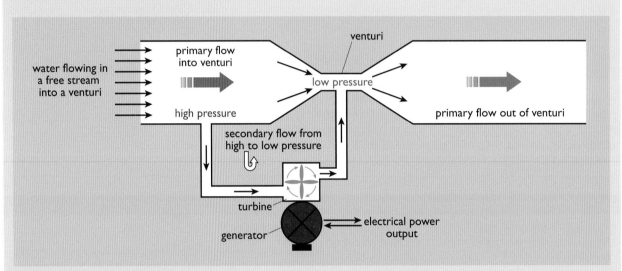

Figure 6.41 SMEC 'venturi' concept – basic principles (source: adapted from VerdErg, 2009)

Spectral Marine Energy Converter (SMEC)

A third project looked at under SETS was the Spectral Marine Energy Converter (SMEC), a novel device being developed by VerdErg using the venturi effect – the reduction in fluid pressure that results when a fluid flows out through a constriction. In this system the tidal flow passes through vanes, creating a pressure drop which is used to drive a turbine and generator via a secondary water flow. The secondary flow will have a higher velocity than the primary flow (Figure 6.41). Furthermore this secondary flow can power a turbine and generator located more conveniently out of the water.

The SMEC concept could be used at a variety of scales – in free-standing mid-stream locations in rivers, or across whole estuaries in a permeable fence-like structure. Vertical vanes would be used in the latter case to produce the pressure

drop, horizontal vanes for shallow mid-channel use in rivers (see Figure 6.42). An estuary-wide tidal fence version (Figure 6.43), would, VerdErg claims, produce nearly as much electrical power as a full conventional barrage, but at two thirds of the cost and with much less environmental impact.

Reporting on the outcome of the SETS project in 2010, DECC's Severn Tidal Feasibility Study noted that the tidal bar/reef and the Spectral Marine Energy Converter 'showed promise for future deployment within the Severn estuary – with potentially lower costs and environmental impacts than either lagoons or barrages.' However they added 'these proposals are a long way from technical maturity and have much higher risks than the more conventional schemes the study has considered. Much more work would be required to develop them to the point where they could be properly assessed.' (DECC, 2010a).

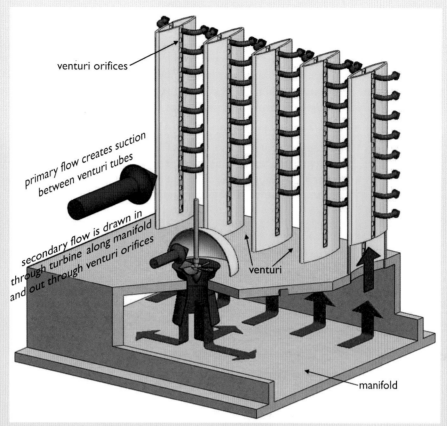

Figure 6.42 Proposed SMEC tidal system for a river

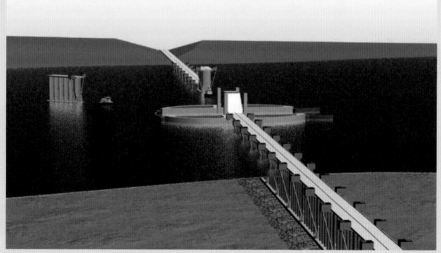

Figure 6.43 Proposed SMEC tidal fence system for an estuary.

The devices looked at in the Severn Embryonic Technology Scheme (SETS) are all at very early stages in their development, as are some of the other devices mentioned above, with only a handful having been tested. There are also many other novel designs emerging. Clearly not all of the designs discussed will prove successful, although MCTs SeaGen has broken through to commercial-scale deployment, and several others seem likely to do so.

In 2010 The Crown Estate announced successful bidders for 10 tidal current and wave energy projects on sites in Scotland's Pentland Firth and Orkney waters. Projects totalling 1.2 GW of installed capacity, 600 MW each from wave and tidal project bids, were given the go-ahead for development, with the target completion date being 2020.

The successful tidal bids included a 100 MW project from MCT, a 200 MW OpenHydro project, a 200 MW project at Westray South led by SSE Renewables Developments (UK); and a 100 MW project led by Scottish Power Renewables UK, using a 1 MW turbine being developed by Norwegian company Hammerfest Strøm (see Figure 6.44).

Figure 6.44 The Hammerfest Strøm tidal turbine

Subsequently the go ahead was also given for a 400 MW project in the Inner Sound area of Pentland Firth, involving the Atlantis Resources Corporation, a Singapore based company that has developed a 1 MW double rotor turbine (see Figure 6.45). In this device one rotor runs on the ebb tide and the other runs on the flood tide (on each tide the non-functioning rotor is fixed).

Figure 6.45 Atlantis AK-1000 1MW double rotor tidal turbine

In all, 1 GW of tidal current projects are now planned for the area and, as there are also plans for other projects elsewhere off the UK coast, in total perhaps 1.2 GW capacity may be in place around the UK by around 2020.

6.10 Tidal current projects and concepts around the world

The idea of using tidal currents is also being explored around the world, although, since UK tidal current resources are amongst the best globally, some of the devices being developed by non-UK companies are being tested in UK waters, with some subsequently being designated for full scale deployment in UK waters, sometimes as joint projects with UK companies. As noted in the previous section both the Norwegian Hammerfest Strøm (see Figure 6.44) and the Singaporean Atlantis Resources Corporation (see Figure 6.45) are deploying devices into UK waters.

Tidal current projects – a world overview

A wide variety of concepts are being developed and tested around the world, for example in Europe (including most notably Norway, but also France, the Netherlands, Germany and Italy), and in the USA, Canada, New Zealand and Australia. In this section a sample of the larger projects underway at the time of writing will be outlined starting with South Korea, which has several medium-scale projects underway.

South Korea

Following tests with a 100 kW tidal current power facility installed in 2003 in Uldolmog on the South coast, in a narrow channel with a 5.5 m s^{-1}

current speed, a 1 MW rated facility is being built and there are plans for a larger project in Daebang Strait, of up to 20 MW. There are also plans for an ocean current power farm in the Sihwa area where there is already a tidal barrage nearing completion . The largest tidal stream project so far is at Wando, where there are plans to have 300 MW of tidal current turbines installed by 2015. Here Voith Hydro is to install up to 600 of its 1 MW Seaturtle tidal current turbines.

Voith have also developed a three-bladed 1 MW seabed mounted propellor-type system, which they plan to test at EMEC, and then to install in an array of 100 MW capacity at Jindo, Jeollanamdo Province. A unique feature is that it has unsealed sea-water compatible bearings, which it is claimed reduce maintenance requirements (Figure 6.46).

Figure 6.46 Voith 1 MW turbine

Norway

In addition to the Hammerfest Strøm project already mentioned, Hydra Tidal are developing a multi-rotor floating propeller-type device with laminated wood composite blades. A 1.5 MW four-rotor two-bladed version, Morild II, with 23m diameter blades, was officially opened offshore from Lofoten, northern Norway, in 2010, with grid links planned.

USA

Six of Verdant Power's small (35 kW) three bladed propeller-type tidal turbines have been tested in New York City's East River near Roosevelt Island. By 2008 they claimed to have clocked up 9000 turbine-hours of operation (Figure 6.47). There are plans for expanding the project to 30 turbines, possibly feeding power to the United Nations building nearby. In addition, the Massachusetts Tidal Energy Company has been looking at the possibility of installing up to 150 2 MW devices at Vineyard Sound on the New England coast.

Meanwhile, following tests with a prototype, the UEK Corporation, based in Annapolis, has plans for a project at the mouth of the river Delaware, using a series of its two-way 45 kW Underwater Electric Kite ducted turbines.

In addition, the Ocean Renewable Power Company (ORPC) is deploying its cross-flow tidal turbine in Eastport, Maine and in 2010 reportedly landed power from its 60 kW prototype (Figure 6.48). The device has a helical turbine design similar to that developed by Gorlov (see below). The company is also looking at sites in Alaska, which has some of the USA's largest tidal resources.

On the US West coast, the Pacific Gas & Electric Company has signed an agreement with the City and the County of San Francisco, as well as the Golden Gate Energy Company, to assess possibilities for harnessing the tides in San Francisco Bay. UK developer Hydroventuri has also been looking at

Figure 6.47 Verdant Power's turbine being lowered into East River, New York City in 2007

Figure 6.48 ORPC cross-flow turbine: a 250 kW version is planned

the possibility of using its innovative venturi turbine device (see Box 6.7) in that location.

Canada

Canada has had a long involvement with tidal barrage technology and is now also looking at tidal current systems. For example the Atlantic Tidal Energy Consortium has outlined plans to install up to 600 MW of tidal systems in the Bay of Fundy. Canadian company Clean Current has developed a version of the ducted rotor design and plans to test it in the Bay of Fundy. In addition, Canada's Minas Basin Pulp and Power Company has an agreement with the UK's MCT to install a 2 MW SeaGen device in the Bay of Fundy and the OpenHydro turbine has also been tested there (a 1 MW device was installed in 2008, although it suffered blade failure).

Australia and New Zealand

In Australia, Tenax Energy has plans for a 200 MW tidal project in fast-flowing waters in Clarence Strait between northern Australia and the Tiwi Islands, while in New Zealand Neptune Power is planning an array of 1 MW floating sub-sea turbines in the tidal currents off the Cook Strait between the North and South Islands, and CREST Energy have plans for a major $400m 300 MW project at Kaipara Harbour near Auckland.

India

The Atlantis Resources Corporation is planning to install a 50 MW tidal farm in the Gulf of Kutch in Gujarat, western India, to be followed up by a 250 MW project if all goes well.

Other novel projects and devices

An EU-funded study has been made of the potential for tidal current generation in the Strait of Messina, between Sicily and mainland Italy. This involves installing 100 vertical-axis turbines on the seabed at a depth of 100 metres (Figure 6.49). In preparation for this, the EU-backed Enemar project has tested a novel vertical-axis turbine in the Strait of Messina. Interest has been shown in this system by China, which is looking at a possible project in the Jintang Strait in the Zhoushan Archipelago.

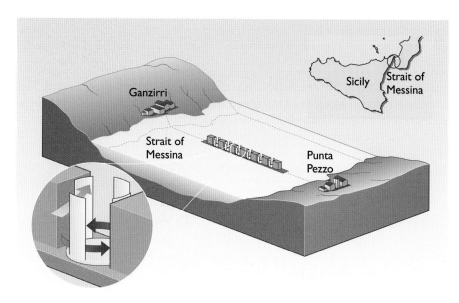

Figure 6.49 The Strait of Messina project

There are many other novel projects under development around the world, some of which are quite exotic, for example:

- tidal kites (aerofoil wings carrying a lightweight rotor and generator that are tethered to the seabed but effectively 'fly' in tidal currents)
- tidal sails (current driven 'sails' mounted on an underwater 'ski-lift' like arrangement).

Some devices are designed for slower water speeds and for use in rivers with smaller rotors: although the basic physics is against them (to get the same output as from fast-moving water, you would need much larger rotors). Although, in general, big is best there may be locations where useful and commercially viable energy can be obtained from such systems.

Large tidal projects and ocean current schemes

While small projects may have a place, there are also proposals for some very large projects. For example, Dr Alexander Gorlov, Professor of Mechanical Engineering at Northeastern University in Boston, has been developing ideas for turbine devices for use with *ocean* currents further out to sea. Dr Gorlov is the inventor and patent holder of the Gorlov Helical Turbine (Figure 6.50) – a variant of the Darrieus vertical-axis wind turbine design

Figure 6.50 Gorlov's helical turbine

(see Chapter 7). Small prototypes have been tested and Gorlov is proposing large-scale applications in offshore locations – in particular, the Gulf Stream. He has noted that the total power of the kinetic energy of the Gulf Stream near Florida is equivalent to approximately 65 000 MW.

Clearly, there exists a large potential energy resource and Gorlov has ambitious plans for exploiting it. His aim is to construct a very large tidal farm of one hundred power modules each with hundreds of triple-helix Gorlov turbines mounted in columns in a lattice array.

Although Gorlov's scheme is still highly-speculative, interest in tidal power is growing in the USA. Florida Atlantic University's Center for Ocean Energy Technology is now looking at the potential of the Gulf Stream more generally, and has suggested that up to 10 GW of capacity could ultimately be installed.

Equally ambitious in conception, the Canadian company Blue Energy has developed a tidal fence concept in which H-shaped vertical-axis turbines are mounted in a modular framework structure (Figure 6.51). Blue Energy has been investigating the idea of a 50 MW rated demonstration tidal power plant in the Philippines, possibly to be followed by a 1000 MW plant, with arrays of vertical-axis turbines mounted in a permeable causeway, or tidal fence, between two islands, extracting power from tidal current flows.

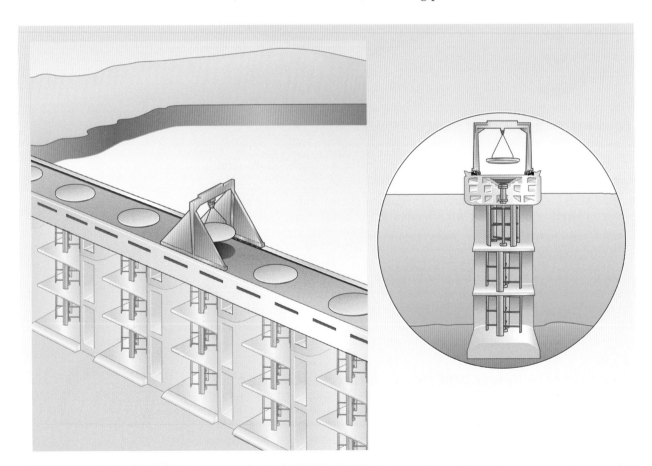

Figure 6.51 Blue Energy's tidal fence concept (source: adapted from Blue Energy, 2011)

Even though projects on that scale may be very speculative, they do indicate the potential for large-scale tidal current systems.

6.11 Tidal current assessment

Although tidal barrages offer larger amounts of energy per project, and tidal current technology is in its infancy, at the time of writing, the prospects for the latter look more promising.

Tidal current systems may also be easier to develop than the wave energy systems described in Chapter 8. Whereas wave energy systems have to operate in a chaotic interface between air and water and have to try to capture energy from multiply-vectored wave motions using complex technology, tidal currents are smoother, laminar flows, in a much more stable, although still turbulent, undersea environment, and the energy can be captured using relatively simple turbines.

Not surprisingly then, given the relatively easier technological challenge, tidal current projects have proliferated – in 2011 there were perhaps 100 different devices under development around the world at various scales. Few are likely to be commercially successful, but those that are may reap large rewards. One study suggested that the annual world market for tidal technology might be £155–444 billion (Westwood, 2005).

The UK is well placed to exploit tidal flows in geographical terms (Atlas of UK Marine Renewable Energy Resources, 2011). A study by ABPmer has suggested that ultimately about 36 GW of tidal current device capacity might in theory be installed in the UK. It estimated that devices in under 40 m depth might generate around a total of up to 94 TWh per year – about a quarter of the 2010 UK electricity requirement (ABPmer, 2007).

However, it is unlikely to be possible to harness all of that resource, and certainly it could take time to do so. The DTI/Carbon Trust Renewables Innovation Review in 2004 put the UK's total practical tidal current resource at around 31 TWh per year, about 10% of UK electricity demand at the time, while a subsequent study by the Carbon Trust, taking economic and other constraints into account, put the tidal figure at only 18 TWh per year (Carbon Trust, 2006).

These estimates may be pessimistic and depend on the assumptions used. For example, Professor Stephen Salter has argued that existing estimates of the tidal stream resource ignore changes in the resource that might be caused by the physical installation of turbines lowering the current velocity (and thus lowering the energy 'lost' to seabed friction). Looking at the Pentland Firth, he claimed that 'Any small reduction in velocity caused by turbine installations will release large amounts of energy. About one third of the present total friction loss could be extracted, giving a possible resource of 10–20 GW, much higher than previous estimates' (Salter, 2008).

On a similar basis, Prof. David Mackay at Cambridge University has claimed that the UK tidal current resource has been seriously underestimated, by perhaps a factor of 10 or more (MacKay, 2007). Similarly, DECC's 2050 Pathways report noted the following.

It has been widely quoted that the total UK tidal stream potential is of the order of 17 TWh/year. This is derived from a method that provides the most conservative estimate. ... However, academic research has highlighted uncertainties surrounding the calculation of practical resource and other methods of estimating the tidal stream resource have resulted in higher technical potentials of up to 197 TWh/year. ...

Industry and academics across a range of disciplines, including oceanography, turbulence, marine energy and physics, need to collaborate to come to a consensus on the appropriate methods for estimating resource and the subsequent predictions that result.

(DECC, 2010b)

Whether or not these views are confirmed, the resource is far from small and significant resources have also been identified elsewhere in the world. (Hardisty 2007, 2009).

Perhaps inevitably, estimates of what might actually be obtained in practice are fluid when a technology is at a relatively early stage of development. It is not clear which systems will be successful in the long run, much less where they can best be sited, how much of the resource they can harvest, and at what cost.

It is reasonable to draw parallels with wind power (see Chapter 7): significant cost-reducing improvements can be expected when and if capacity is established and as economies of volume production are achieved. A 2011 Carbon Trust report claimed that, given targeted and accelerated development, by 2025 electricity costs could be comparable to those from offshore wind projects, and, for the best sites, possibly even to those from land-based wind projects and nuclear plants (Carbon Trust, 2011).

Environmental Impact

In addition to operational and economic issues, if tidal current devices are to be used on a wide scale then a key issue will be their environmental impact. As indicated earlier, this is a major concern for tidal barrages and may also prove to be of some significance for tidal lagoons. In contrast, most studies of tidal current turbines so far have suggested that impacts will be low. Even large arrays will not impede flows significantly and as the rotor blades will turn slowly (slower than wind turbines, and much slower than the turbines in barrages and lagoons) they should not present a hazard to marine life – fish will be unaffected. Certainly experience with MCT's SeaGen has not indicated any problems. At the time of writing, no seal deaths have been attributed to the turbines since their installation in 2008. A sonar system has been used to detect the approach of any marine mammals and shut the turbines down.

All structures put in the sea will have some impact, and this needs to be, and already is, carefully assessed when considering possible locations. Comparing various tidal schemes is difficult – although tidal current turbines may have less impact than barrages or lagoons, many 2 MW tidal turbines are required to equal the output of, say, an 8 GW barrage, but then again a lower-rated network of turbine farms and arrays around the UK coast may be able to give as much *useful* output – the power available when needed – as a single large barrage.

6.12 **Summary**

In terms of actual developments, while there is continuing interest in tidal barrages and lagoons around the world, and there are some significant barrage projects in Korea, at the time of writing it is tidal current turbines that have caught the imagination of many engineers, and some large developments seem likely to go ahead.

As this chapter has shown, the energy potential is significant, especially in the UK: an expansion programme (DECC, 2010b) could theoretically deliver by 2050:

- up to 21.3 GW of tidal current generation capacity delivering 69 TWh per year
- 20 GW of tidal range capacity (barrages and lagoons) delivering 40 TWh per year.

The key issues facing this technology area are cost and environmental impact.

Barrages have some similarity to low-head hydro, and thus their economics are likely to be relatively challenging. Once built, tidal barrages may be able to generate at reasonably competitive electricity costs for long periods, but capital costs remain stubbornly high. Thus it is hard to see large-scale barrages as viable except as public-sector led projects. By contrast, smaller, easier and quicker to install, tidal lagoons might conceivably be more viable as private investments.

However, at the time of writing, overall it seems that tidal range projects are not economically attractive in a UK context, given the financial climate.

As regards the environment, some of the impacts will be positive and of global significance. For example, the carbon dioxide emissions resulting from power generated via fossil fuel plants could be avoided if this power could be provided through tidal barrages instead. The negative impacts are more localized and further research might show how these can be reduced. One of the key factors will probably be how people perceive barrages, Given that so far most large tidal barrage projects have been opposed by environmental groups in the UK, it may be hard to win approval, and, at the very least, design compromises might be needed in response to public concerns. Overall, although there are disagreements, it seems likely that schemes that block entire estuaries are generally likely to have the largest impact per MW installed and per useful MWh delivered.

The outlook for tidal current devices is more encouraging, since they are modular with relatively short installation lead times and lower environmental impact. At present, generation costs for tidal current turbines are relatively high, but this is for prototype devices. Some manufacturers have claimed prices may fall relatively quickly, given the right support, and, with an expanding commercial market, 'mass' production of turbines.

The large potential of tidal energy has been talked about for many years, but the economics have often been seen as challenging. Now, with increasing concerns about climate change and energy security, it seems that priorities may be changing and more effort is being put into developing new technologies. While interest in barrages and lagoons remains, the rapid development of tidal current technology could well mean that it will become the leading approach in this field.

References

ABPmer (2007) *Quantification of Exploitable Tidal Energy Resources in UK waters*, ABPmer, R.1349; available at http://www.abpmer.co.uk/files/report.pdf (accessed 7 March 2011).

Admiralty Charts and Publications (1992) *Admiralty Tidal Stream Atlas: The English Channel NP250*, Taunton, UK Hydrographic Office.

Atkins Consultants Ltd (2004) *Feasibility Study for a Tidal Lagoon in Swansea Bay* [online], http://www.tidalelectric.com/resources-feasibility.shtml (accessed 7 March 2011).

Atlas of UK Marine Renewable Energy Resources (2011) *Atlas of UK Marine Renewable Energy Resources* [online], http://www.renewables-atlas.info/ (accessed 11 August 2011).

Blue Energy (2011) *Blue Energy* [online], http://www.bluenergy.com (accessed 21 May 2012).

Bryden, I. G., Couch, S. J., Owen, A. and Melville, G. (2007) 'Tidal current resource assessment', *Proc IMechE, Part A: Journal of Power and Energy*, vol. 221, no. 2, pp. 125–135.

Carbon Trust (2006) *Future Marine Energy: Results of the Marine Energy Challenge: Cost competitiveness and growth of wave and tidal stream energy*, Carbon Trust, CTC601; available at http://www.carbontrust.co.uk/Publications/pages/PublicationDetail.aspx?id=CTC601 (accessed 7 March 2011).

Carbon Trust (2011) *Accelerating Marine Energy*, Carbon Trust, CTC797; available at http://www.carbontrust.co.uk/publications/pages/publicationdetail.aspx?id=CTC797 (accessed 25 August 2011).

DECC (2010a) *Severn Tidal Power: Feasibility Study Conclusions and Summary Report*, Department of Energy and Climate Change, URN 10D/808; available at http://www.decc.gov.uk/assets/decc/What%20we%20do/UK%20energy%20supply/Energy%20mix/Renewable%20energy/severn-tp/621-severn-tidal-power-feasibility-study-conclusions-a.pdf (accessed 8 March 2011).

DECC (2010b) *2050 Pathways Analysis*, Department of Energy and Climate Change, URN 10D/764; available at http://www.decc.gov.uk/assets/decc/What%20we%20do/A%20low%20carbon%20UK/2050/216-2050-pathways-analysis-report.pdf (accessed 8 March 2011).

Department of Energy (1981) *Tidal Power from the Severn Estuary, Vol. I*, Department of Energy, Energy Paper 46, HMSO.

Department of Energy (1989) *The Severn Barrage Project: General Report*, Department of Energy, Energy Paper 57, HMSO.

ETSU (1990) *Renewable Energy Research and Development Programme*, Report R56, Energy Technology Support Unit.

ETSU (1993) *Tidal Stream Energy Review*, Energy Technology Support Unit, ETSU-T-05/00155/REP.

Frontier Economics (2008) *Analysis of a Severn Barrage, A report prepared for the NGO steering group*, London, Frontier Economics; available at http://assets.wwf.org.uk/downloads/frontier_economics_barrage_repo.pdf (accessed 8 March 2011).

Halcrow Group Ltd, Mott MacDonald and RSK Group plc (2009) *Solway Energy Gateway Feasibility Study*, Halcrow Group Ltd, Mott MacDonald and RSK Group plc; available at http://www.solwayenergygateway.co.uk/ (accessed 11 August 2011).

Hardisty, J. (2007) 'Assessment of tidal current resources: case studies of estuarine and coastal sites', *Energy & Environment*, vol. 18, no. 2, pp. 233–49.

Hardisty, J. (2009) *The Analysis of Tidal Stream Power*, London, Wiley.

HM Treasury (2003) *The Green Book: Appraisal and Evaluation in Central Government*, London, TSO; available at http://www.hm-treasury.gov.uk/data_greenbook_index.htm (accessed 12 March 2011). See especially Annex 4 on risk and uncertainty and Annex 6 on discount rate.

Jackson, T. (1992) 'Renewable energy: Summary paper for the renewables series', *Energy Policy,* vol. 20, no. 9, pp. 861–83.

Jo, C. H. (2008) 'Recent Development of Ocean Energy in Korea', *BWEA Marine '08*, Edinburgh, February 28, 2008 [online], (accessed 25 August 2011).

Laughton, M. A. (1990) *Renewable Energy Sources, Report 22*, London, Watt Committee on Energy/Elsevier Applied Science.

Lee, K-S. (2006) *Tidal and Tidal Current Power Study in Korea*, OREG 2006 Spring Symposium, Victoria, May 4–5, 2006 [online], www.oreg.ca/docs/May%20Symposium/KOREA.pdf (accessed 25 August 2011).

MacKay, D. J. C. (2007) *Under-estimation of the UK Tidal Resource* [online], Cavendish Laboratory, University of Cambridge http://www.inference.phy.cam.ac.uk/sustainable/book/tex/TideEstimate.pdf (accessed 8 March 2011).

Marine Current Turbines (2010) 'Presentation to 4th International Tidal Energy Summit', *4th International Tidal Energy Summit*, London, November 23–24, 2010.

OST (1999) *Energies From the Sea: Towards 2020*, Marine Foresight Panel Report.

OVG (2010) *The Offshore Valuation – A valuation of the UK's offshore renewable energy resource*, Machynlleth, Public Interest Research Centre; available at http://www.offshorevaluation.org/downloads/offshore_vaulation_full.pdf (accessed 11 August 2011).

Peel Energy/NWRDA (2010) *Mersey Tidal Power Feasibility Study: Stage 1 Options Report*, Peel Energy/North west Regional Development Agency; available at http://www.merseytidalpower.co.uk/content/project

Salter, S. (2008) *Renewable electricity generation technologies, Fifth Report of Session 2007–08, Volume II, HC216-II*, page Ev 106, House of Commons, Innovation, Universities, Science and Skills Committee, London, The

Stationery Office; available at http://www.publications.parliament.uk/pa/cm200708/cmselect/cmdius/216/216ii.pdf (accessed 11 August 2011).

Salter, M. and Walker, C. (2010) *The Severn Barrage Digest*, report by two Members of Parliament, London, All Party Parliamentary Group on Angling; available at http://www.martinsalter.com/wp-content/uploads/2010/04/severn-barrage-digest-chapter-6-post-short-list-conclusions.pdf (accessed 11 August 2011).

SDC (2007) *Turning the Tide: Tidal Power in the UK*, Sustainable Development Commission; available at http://www.sd-commission.org.uk/publications/downloads/Tidal_Power_in_the_UK_Oct07.pdf (accessed 8 March 2011).

STPG (1986) *Tidal Power from the Severn*, Severn Tidal Power Group.

Twidell, J. W. and Weir, A. J. (2006) *Renewable Energy Resources* (2nd edition), Oxford, Taylor & Francis.

VerdErg (2009) *'SMEC' Spectral marine energy converter* [online], http://www.verderg.com/attachments/-01_SMEC_Doc_Oct%2009.pdf (accessed 8 March 2011).

WEC (2001) *2001 Survey of Energy Resources*, London, World Energy Council; available at http://www.worldenergy.org/documents/ser_sept2001.pdf (accessed 12 August 2011).

Westwood, A. (2005) *Refocus Marine Renewable Energy Report: Global Markets, Forecasts and Analysis 2005–2009*, London, Elsevier Science.

Chapter 7

Wind energy

By Derek Taylor

7.1 Introduction

Wind energy has been used for thousands of years for milling grain, pumping water and other mechanical power applications. Today, there are many thousands of windmills in operation around the world, a proportion of which are used for water pumping. But it is the use of wind energy as a pollution-free means of generating electricity on a significant scale that is attracting most current interest in the subject. Strictly speaking, a wind*mill* is used for milling grain, so modern technology for electricity generation is generally differentiated by use of the term **wind turbines**, partly because of their functional similarity to the steam and gas turbines that are used to generate electricity, and partly to distinguish them from their traditional forebears. Wind turbines are also sometimes referred to as **wind energy conversion systems (WECS)** and sometimes described as **wind generators** or **aerogenerators**.

Attempts to generate electricity from wind energy have been made (with various degrees of success) since the late nineteenth century when Professor James Blyth of the Royal College of Science and Technology, now Strathclyde University, built a range of wind energy devices to generate electricity, his first being in 1887. A later design built at Marykirk in Scotland continued to generate electricity for over 20 years.

For many years, small-scale wind turbines have been manufactured to provide electricity for remote houses, farms and remote communities, and for charging batteries on boats, caravans and holiday cabins (thousands of small turbines similar to that shown in Figure 7.1 are in use worldwide). More recently they have been used to provide electricity for cellular telephone masts and remote telephone boxes.

However, it is only since the 1980s that the technology has become sufficiently mature to enable rapid growth of the sector. Between the early 1980s and the late 2000s the cost of wind turbines fell steadily and the rated capacity of typical machines increased significantly. Now, on reasonably windy and accessible sites, wind turbines are one of the most cost-effective methods of electricity generation. Given continuing improvements in cost, capacity and reliability, it can be expected that wind energy will become even more economically competitive over the coming decades. Moreover, as wind turbines are increasingly deployed offshore, where wind speeds are generally higher and planning constraints perhaps less demanding, the technically accessible wind resource is massively increased. Of course, as will be seen later, there are significant additional technical challenges associated with offshore wind and the cost of generation is inevitably higher.

Figure 7.1 The 70 watt rated Marlec 'Rutland Wind Charger'

The improvements in wind power technology have made it one of the fastest growing renewable energy technologies worldwide in terms of installed rated capacity. A total of over 194 GW of wind generating capacity had been installed by the end of 2010, with almost 36 GW added in that year. This is about 11 times the capacity that had been installed by the end of 2000, and, at the time of writing, the current average growth rate is around 22% per annum.

To understand the machines and systems that extract energy from the wind involves an appreciation of many fields of knowledge, from meteorology, aerodynamics and planning to electrical, structural, civil and mechanical engineering. Hence, this chapter begins with a description of the atmospheric processes that give rise to wind energy. Wind turbines and their aerodynamics are then described, together with various ways of calculating their power and energy production. This is followed by discussions of the environmental impact and economics of wind energy, together with an examination of recent commercial developments and a discussion of its future potential. The final section looks at offshore wind power, which seems likely to be one of the most important areas of wind energy development in coming decades, especially for the UK and northern Europe.

7.2 **The wind**

As mentioned in Chapter 2, one square metre of the Earth's surface on or near the equator receives more solar radiation per year than one square metre at higher latitudes. The curvature of the Earth means that its surface becomes more oblique to the Sun's rays with increasing latitude. In addition, the Sun's rays have further to travel through the atmosphere as latitude increases, so more of the Sun's energy is absorbed *en route* before it reaches the surface. As a result of these effects, the tropics are considerably warmer than higher latitude regions.

This differential solar heating of the Earth's surface causes variations in atmospheric pressure, which in turn give rise to the movement of atmospheric air masses which are the principal cause of the Earth's wind systems (see Box 7.1 for more details).

BOX 7.1 The Earth's wind systems

Like all gases, air expands when heated, and contracts when cooled. Thus warm air is less dense than cold air and will rise to high altitudes when strongly heated by solar radiation.

A low pressure belt (with cloudy and rainy weather patterns) is created at the equator due to warm humid air rising in the atmosphere until it reaches the tropopause (the top of the troposphere). At the surface the equatorial region is called the 'doldrums' (from an old English word meaning dull) by early sailors who were fearful about becoming becalmed.

At the tropopause in the northern hemisphere the air moves northwards and in the southern Hemisphere it moves southwards. This air gradually cools until it

reaches latitudes of about 30 degrees, where it sinks back to the surface, creating a belt of high pressure at these latitudes (with dry clear weather patterns). The majority of the world's deserts also occur in these high-pressure regions.

Some of the air that reaches the surface at these latitudes is forced back towards the low-pressure zone at the equator. These air movements are known as the '**trade winds**'. On reaching the equator these air movements complete the circulation of what is known as the **Hadley cell** – named after the scientist (George Hadley) who first described them in 1753.

However, not all of the air that sinks at the 30 degree latitudes moves toward the equator. Some of it moves poleward until it reaches the 60 degree latitudes, where it meets cold air coming from the poles at what are known as the 'polar fronts'. The interaction of the two bodies of air causes the warmer air to rise and most of this air cycles back to the 30-degree latitude regions where it sinks to the surface, contributing to the high-pressure belt. This completes the circulation of what is known as the **Ferrel cell** (named after William Ferrel who first identified it in 1856).

The remaining air that rises at the polar fronts moves poleward and sinks to the surface at the poles as it cools. It then returns to the 60-degree latitude region completing the circulation of what is known as the **polar Hadley cell** or **polar cell**.

There is a further complication in that, because the Earth itself rotates, winds moving across the Earth's surface are subject to a phenomenon known as the Coriolis Effect. The net result of this effect, given the Earth rotates in an eastwards direction, is that in the northern hemisphere 'north bound' winds are caused to veer 'right'. Such winds are known as 'westerlies' as, whilst they are veering toward an easterly direction, it is the convention when referring to wind direction to use the direction *from* which winds blow. In the southern hemisphere 'north bound' winds veer to the 'left' ('trade winds'). Likewise, 'south bound' winds veer 'right' ('trade winds') in the northern hemisphere and 'left' ('westerlies') in the southern hemisphere. Figure 7.2 shows the overall pattern of global wind circulation.

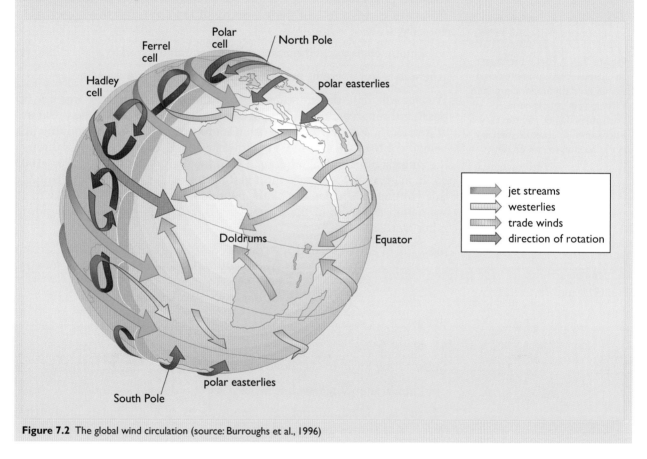

Figure 7.2 The global wind circulation (source: Burroughs et al., 1996)

Atmospheric pressure is the pressure resulting from the weight of the column of air above a specified surface area, with the unit of atmospheric pressure being known as the bar. Atmospheric pressure is measured by means of a barometer (Figure 7.3). These devices are usually calibrated in millibars (mbar), that is, thousandths of a bar. The average atmospheric pressure at sea level is about 1013.2 mbar (approximately 1 bar). The SI unit of pressure, the pascal (Pa), is defined as one newton per square metre and 1 bar is equivalent to 100 kPa.

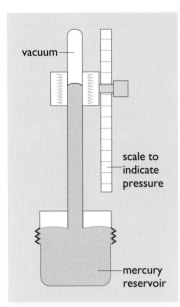

Figure 7.3 A Fortin barometer, an example of a barometer used to measure atmospheric pressure. Variations in atmospheric pressure acting on the mercury in the reservoir cause the mercury in the column to rise or fall

On the weather maps featured in television weather forecasts or in newspapers, there are regions marked 'high' and 'low', surrounded by contours (Figure 7.4). The regions marked 'high' and 'low' relate to the atmospheric pressure and the contours represent lines of equal pressure called **isobars**. The high-pressure regions tend to indicate fine weather with little wind, whereas the low-pressure regions tend to indicate changeable windy weather and precipitation.

In addition to the main global wind systems shown in Box 7.1 there are also local wind patterns, such as sea breezes (Figure 7.5) and mountain-valley winds (Figure 7.6).

Energy and power in the wind

The energy contained in the wind is its kinetic energy, and as we saw in Chapter 1 the kinetic energy of any particular moving mass (moving air in this case) is equal to half the mass, m, (of the air) times the square of its velocity, V:

$$\text{kinetic energy} = \text{half mass} \times \text{velocity squared} = \tfrac{1}{2}mV^2 \tag{1}$$

where m is in kilograms and V is in metres per second (m s^{-1}).

We can calculate the kinetic energy in the wind if, first, we imagine air passing through a circular ring or hoop enclosing an area A (say 100 m^2) at a velocity V (say 10 m s^{-1}) (see Figure 7.7). As the air is moving at a velocity of 10 m s^{-1}, a cylinder of air with a length of 10 m will pass through the ring each second. Therefore, a volume of air equal to $100 \times 10 = 1000$ cubic metres (m^3) will pass through the ring each second. By multiplying this volume by the density of air, ρ (which at sea level is 1.2256 kg m^{-3}), we obtain the mass of the air moving through the ring each second. In other words:

mass (m) of air per second = air density × volume of air passing per second
$\qquad\qquad\qquad\qquad$ = air density × area × length of cylinder of air passing per second
$\qquad\qquad\qquad\qquad$ = air density × area × velocity

that is:

$$m = \rho AV$$

Substituting for m in (1) above gives:

$$\text{kinetic energy } per\ second = 0.5\ \rho AV^3 \text{ (joules per second)}$$

where ρ is in kilograms per cubic metre (kg m^{-3}), A is in square metres (m^2) and V is in metres per second (m s^{-1}).

If we recall that energy per unit of time is equal to power, then the power in the wind is P (watts) = kinetic energy in the wind traversing the circular ring per second (joules per second), that is:

$$P = 0.5\ \rho AV^3 \tag{2}$$

The main relationships that are apparent from the above calculations are that the power in the wind is proportional to:

- the density of the air
- the area through which the wind is passing (i.e. through a wind turbine rotor), and
- the cube of the wind velocity.

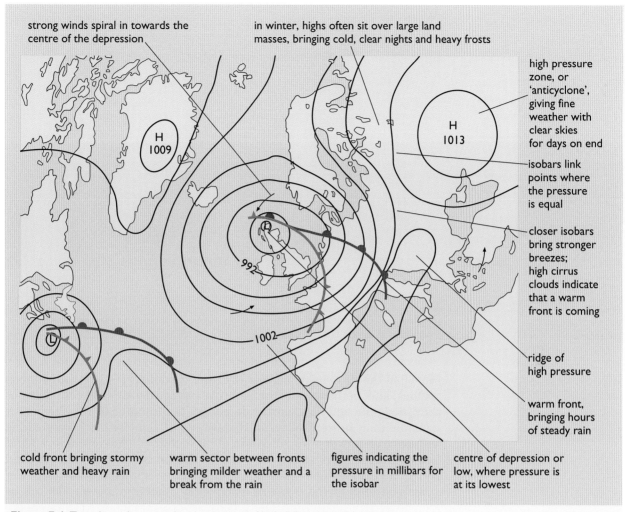

strong winds spiral in towards the centre of the depression

in winter, highs often sit over large land masses, bringing cold, clear nights and heavy frosts

high pressure zone, or 'anticyclone', giving fine weather with clear skies for days on end

isobars link points where the pressure is equal

closer isobars bring stronger breezes; high cirrus clouds indicate that a warm front is coming

ridge of high pressure

warm front, bringing hours of steady rain

cold front bringing stormy weather and heavy rain

warm sector between fronts bringing milder weather and a break from the rain

figures indicating the pressure in millibars for the isobar

centre of depression or low, where pressure is at its lowest

H 1009

H 1013

992

1002

Figure 7.4 Typical weather map showing regions of high (H) and low (L) pressure

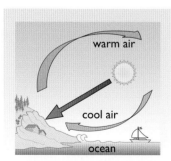

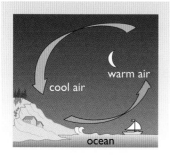

warm air

cool air

ocean

cool air

warm air

ocean

Figure 7.5 Sea breezes are generated in coastal areas as a result of the different heat capacities of sea and land, which give rise to different rates of heating and cooling. The land has a lower heat capacity than the sea and heats up quickly during the day, but at night it cools more quickly than the sea. During the day, the sea is therefore cooler than the land and this causes the cooler air to flow shoreward to replace the rising warm air on the land. During the night the direction of air flow is reversed

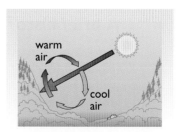

warm air

cool air

warm air

cool air

warm air

cool air

Figure 7.6 Mountain-valley winds are created when cool mountain air warms up in the morning and, as it becomes lighter, begins to rise: cool air from the valley below then moves up the slope to replace it. During the night the flow reverses, with cool mountain air sinking into the valley

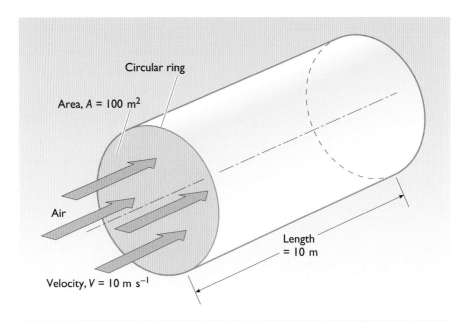

Figure 7.7 Cylindrical volume of air passing at velocity V (10 m s⁻¹) through a ring enclosing an area, A (100 m²), each second

Note that the air density is lower at higher elevations (e.g. in mountainous regions) and, perhaps more importantly, average densities in cold climates may be significantly higher than in hot regions (more than 10% higher for example than in tropical regions). Also, wind velocity has a very strong influence on power output because of the 'cube law'. For example, a wind velocity increase from 6 m s⁻¹ to 8 m s⁻¹ will increase the power in the wind by a factor of more than two. It is also important to appreciate that the power contained in the wind is not in practice the amount of power that can be extracted by a wind turbine. This is because losses are incurred in the energy extraction/conversion process (see Section 7.4 on aerodynamics). Moreover there are additional mechanical-to-electrical power conversion losses.

7.3 Wind turbines

A brief history of wind energy

Wind energy was one of the first non-animal sources of energy to be exploited by early civilizations. It is thought that wind was first used to propel sailing boats, but the static exploitation of wind energy by means of windmills is believed to have been taking place for about 4000 years.

Windmills have traditionally been used for milling grain, grinding spices, dyes and paint stuffs, making paper and sawing wood. Traditional wind pumps were used for pumping water in Holland and East Anglia in the UK, and, because they often used identical forms of sails and support structures, they were (and are) often also referred to as windmills.

Many early windmills were of the *vertical-axis* type and, unlike modern wind turbines which are driven by lift forces (see below), these were drag-driven devices and relied on differences in drag on either side of the

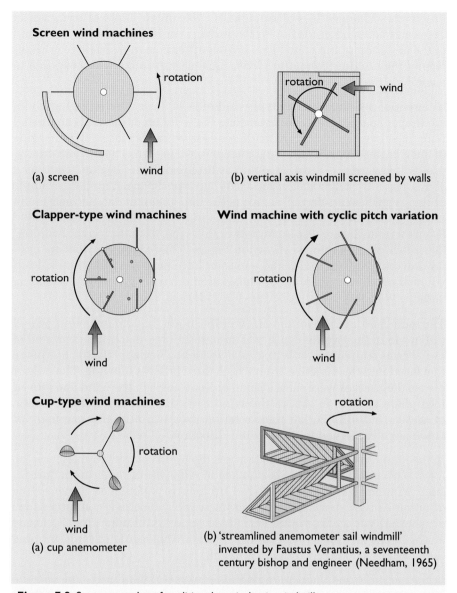

Screen wind machines

(a) screen

(b) vertical axis windmill screened by walls

Clapper-type wind machines

Wind machine with cyclic pitch variation

Cup-type wind machines

(a) cup anemometer

(b) 'streamlined anemometer sail windmill' invented by Faustus Verantius, a seventeenth century bishop and engineer (Needham, 1965)

Figure 7.8 Some examples of traditional vertical-axis windmills

vertical shaft in order to function. Some examples are shown in Figure 7.8 and include the following.

Screened windmills. These windmills employ screens or partial walls around the windmill, which are positioned to screen the windmill sails from the wind during the 'backward' part of the cycle, when the sails are moving towards the wind.

'Clapper' windmills. These windmills are so called because the moveable sails 'clap' against stops as the rotor turns with the wind (forwards), maximizing their air resistance, but align themselves with the wind (like a weather vane) when on the part of their cycle in which they are moving into the wind (backwards), so reducing their air resistance.

Cyclically pivoting sail windmills. These windmills are similar to the 'clapper' windmills, but use a more complex mechanism to achieve progressive changes in sail orientation. The pitch angle of each sail is cyclically adjusted according to its position during its rotation cycle and to the direction of the wind. This gives a difference in resistance on either side of the windmill's rotation axis, causing it to rotate when exposed to a wind stream.

Differential resistance or cup type windmills. In these windmills, the blades are shaped to offer greater resistance to the wind on one surface compared with the other. This results in a difference in wind resistance on either side of the windmill axis, so allowing the windmill to turn. The first electricity-producing wind generator, invented by Professor James Blythe, mentioned above, was a 10 m diameter vertical axis cup type device. A modern example of this type of wind-driven device is the cup anemometer, an instrument used for measuring wind speed. The simple 'S' type and multi-bladed S-type windmills are also examples of this type, as is the 'Savonius rotor' (a 'split-S' shaped rotor as shown in Figure 7.9). Savonius rotors are used for powering fans in trucks and vans and have been used for simple do-it-yourself windmills. They are produced as micro wind generators, including variants with helically twisted semi-cylindrical 'cups'.

The more familiar *horizontal-axis* windmills are thought to have appeared in Europe in the twelfth century. These traditional machines consisted of radial arms supporting sails that rotated about a horizontal axis, in a plane that faced into the direction from which the wind was blowing. The sails or blades themselves were set at a small oblique angle to the wind and moved in a plane at right angles to the wind direction. Another characteristic of these windmills is that their rotation axes were usually manually or automatically aligned with the wind direction.

In the Mediterranean regions of Europe, the traditional windmills took the form of triangular canvas sails attached to radial arms. In northern Europe, such windmills were characterized by long rectangular sails consisting of either canvas sheets on lattice frameworks, so-called 'common sails', or 'shutter-type sails', which resembled venetian blinds. The shuttering arrangement gave a degree of control over starting, regulating and stopping the windmill according to wind strength.

In northern Europe there were two main forms of windmill. One was the less common 'post mill', in which the whole windmill was moved about a large upright post when the wind direction changed; the other was the more common 'tower mill' (Figure 7.10), in which the rotor and cap were supported by a relatively tall tower, usually of masonry. In the tower mill, only the cap (in combination with the rotor and its shaft) were moved in response to changes in wind direction. The sails turned fairly slowly and provided mechanical power.

At their zenith, before the Industrial Revolution, it is estimated that there were some 10 000 of these windmills in Britain (Golding, 1955) and they formed a familiar feature of the countryside.

Wind turbine types

The variety of machines that has been devised or proposed to harness wind energy is considerable and includes many unusual devices. Figure 7.9 shows a small selection of the various types of machines that have been proposed over the years.

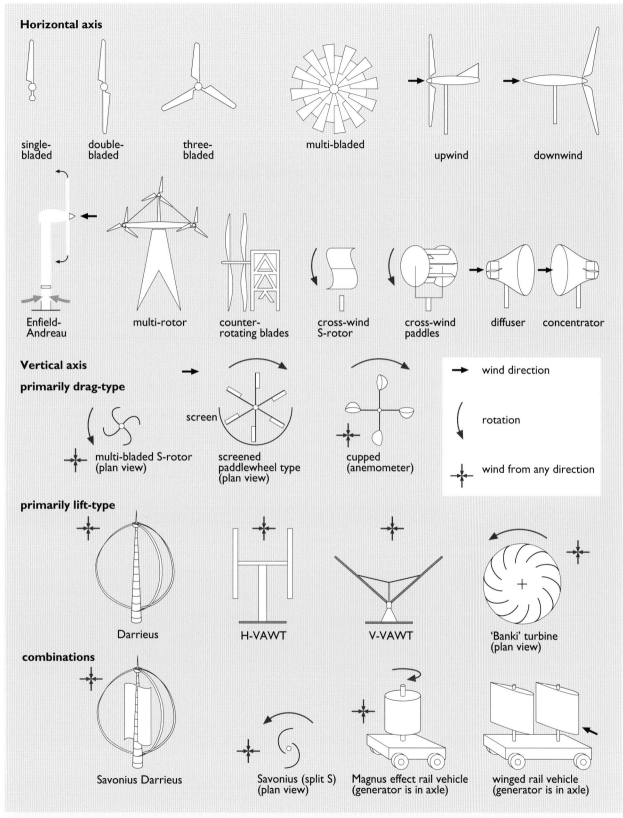

Horizontal axis

single-
bladed   double-
bladed   three-
bladed   multi-bladed   upwind   downwind

Enfield-
Andreau   multi-rotor   counter-
rotating blades   cross-wind
S-rotor   cross-wind
paddles   diffuser   concentrator

Vertical axis

primarily drag-type

screen

multi-bladed S-rotor
(plan view)   screened
paddlewheel type
(plan view)   cupped
(anemometer)

→ wind direction

↳ rotation

✳ wind from any direction

primarily lift-type

Darrieus   H-VAWT   V-VAWT   'Banki' turbine
(plan view)

combinations

Savonius Darrieus   Savonius (split S)
(plan view)   Magnus effect rail vehicle
(generator is in axle)   winged rail vehicle
(generator is in axle)

Figure 7.9 Some examples of the machines that have been proposed for wind energy conversion. (Source: partly based on Eldridge, 1975. For further information on these machines see Eldridge, 1975 and Golding, 1955). The figure is divided into two sections, Horizontal axis and Vertical axis machines.

Figure 7.10 Traditional north European tower windmill

Figure 7.11 Multi-bladed wind pump

Most modern wind turbines come in one of two basic configurations: horizontal axis and vertical axis. Horizontal axis turbines are predominantly of the 'axial flow' type (i.e., the rotation axis is in line with the wind direction), whereas vertical axis turbines are generally of the 'cross flow' type (i.e., the rotation axis is perpendicular to the wind direction). They range in size from very small machines that produce a few tens or hundreds of watts to very large turbines producing as much as 7.5 MW. Larger turbines rated at 10 to 15 MW are now being considered and 20 MW designs are being investigated.

Horizontal axis wind turbines

Horizontal axis wind turbines (HAWTs) generally have either two or three blades, but can have many more. Multi-bladed wind turbines have what appears to be virtually a solid disc covered by many solid blades (usually of slightly cambered sheet metal construction). They have been used since the nineteenth century for water pumping on farms (Figure 7.11). Appropriately for their application they produce high torque at low rotor speeds.

The term '**solidity**' is used to describe the fraction of the swept area that is solid. Wind turbines with large numbers of blades, such as these multi-bladed devices, have highly solid swept areas and are referred to as **high-solidity** wind turbines. Wind turbines with small numbers of narrow blades have a swept area that is largely void: only a very small fraction of the area appears to be 'solid' – such devices are referred to as **low-solidity** wind turbines. Multi-blade wind pumps have **high-solidity rotors** and modern electricity-generating wind turbines (with one, two or three blades) have **low-solidity rotors**.

Low-solidity devices work effectively at much higher rotational speeds making them attractive for electricity generation (Box 7.2 discusses the effect of blade number on turbine characteristics).

BOX 7.2 **Effect of the number of blades**

The speed of rotation of a wind turbine is usually measured in either revolutions per minute (rpm) or radians per second (rad s^{-1}). The **rotation speed** in revolutions per minute (rpm) is usually symbolized by N and the **angular velocity** in radians per second is usually symbolized by Ω. The relationship between the two is given by:

$$1 \text{ rpm} = \frac{2\pi}{60} \text{ rad s}^{-1} = 0.10472 \text{ rad s}^{-1}$$

A useful alternative measure of wind turbine rotor speed is **tip speed**, U, which is the **tangential velocity** of the rotor at the tip of the blades, measured in metres per second. It is the product of the **angular velocity**, Ω, of the rotor and the **tip radius**, R (in metres):

$$U = \Omega R$$

Alternatively, U can be defined as:

$$U = \frac{2\pi R N}{60}$$

By dividing the **tip speed**, U, by the **undisturbed wind velocity**, V_0, upstream of the rotor, we obtain a non-dimensional ratio known as the **tip speed ratio**, usually symbolized by λ. This ratio provides a useful measure against which aerodynamic efficiency can be plotted. The aerodynamic efficiency of a wind turbine is usually described as its **power coefficient** (effectively the ratio of power output from the turbine to the theoretical power in the wind). This quantity is symbolized by C_P and given as a fraction, such that 1 equates to 100% efficiency. When the power coefficient is plotted against tip speed ratio, such $C_P - \lambda$ curves provide an effective way to present the performance of a rotor and to compare wind turbines with differing characteristics.

A wind turbine of a particular design can operate over a range of tip speed ratios, but will usually operate with its best (maximum) efficiency at a particular tip speed ratio, i.e. when the velocity of its blade tips is a particular multiple of the wind velocity. This optimum tip speed ratio (λ_{opt}) is also commonly denoted as λ_{max} with the corresponding efficiency (i.e. power coefficient) being C_{Pmax}. The optimum tip speed ratio for a given wind turbine rotor will depend upon both the number of blades and the width of each blade.

In order to extract energy as efficiently as possible, the blades have to interact with as much as possible of the wind passing through the rotor's **swept area**. The blades of a high-solidity, multi-blade wind turbine interact with all the wind at very low tip speed ratios, whereas the blades of a low-solidity turbine have to travel much faster to 'virtually fill up' the swept area, in order to interact with all the wind passing through. If the tip speed ratio is too low, some of the wind travels through the rotor swept area without interacting with the blades; whereas if the tip speed ratio is too high, the turbine offers too much resistance to the wind, so that some of the wind goes around it. A two-bladed wind turbine rotor with each blade the same width as those of a three-bladed rotor will have an optimum tip speed ratio *one-third higher* than that of a three-bladed rotor. Optimum tip speed ratios for modern low-solidity wind turbines range between about 6 and 20.

In theory, the more blades a wind turbine rotor has, the more efficient it is. However, when there are large numbers of blades in a rotor, the flow becomes more disturbed, so that they aerodynamically interfere with each other. Thus high-solidity wind turbines tend to be less efficient overall than low-solidity turbines. Of low-solidity machines, three-bladed rotors tend to be the most energy efficient; two-bladed rotors are slightly less efficient and one-bladed rotors slightly less efficient still. Wind turbines with more blades can be generally expected to generate less aerodynamic noise as they operate at lower tip speeds (see Section 7.6) than wind turbines with fewer blades.

The mechanical power that a wind turbine extracts from the wind is the product of its angular velocity and the torque imparted by the wind. **Torque** is the moment about the centre of rotation due to the driving force imparted by the wind to the rotor blades. Torque is usually measured in newton metres (N m) (see Box 7.4). For a given amount of power, the *lower* the angular velocity the *higher* the torque; and conversely, the *higher* the angular velocity the *lower* the torque.

The pumps that are used with water pumping wind turbines require a high starting torque to function. Multi-bladed turbines are therefore generally used here because of their low tip speed ratios and resulting high torque characteristics.

Conventional electrical generators run at speeds many times greater than most wind turbine rotors so they generally require some form of gearing when used with wind turbines. Low-solidity wind turbines are better suited to electricity generation because they operate at high tip speed ratios and therefore do not require as high a gear ratio to match the speed of the rotor to that of the generator. In addition, many low-solidity small wind turbines (and even certain very large wind turbines) have avoided using gearboxes by using directly coupled low-speed multi-pole generators.

Figure 7.12 Two-bladed HAWT (WEG MS400 turbine)

Figure 7.13 Three-bladed HAWT (Vestas V52 850 kW turbine)

Figure 7.14 Single bladed HAWT (MBB 600 kW turbine)

Modern *low-solidity* HAWT rotors evolved from traditional windmills and superficially resemble aircraft propellers. Wind turbines with such rotors are by far the most common design manufactured today. They have a clean streamlined appearance, due in part to their design being driven by aerodynamic considerations derived largely from developments in aircraft wing and propeller design. HAWT rotors generally have two or three wing-like blades (Figures 7.12 and 7.13). They are almost universally employed to generate electricity. Some experimental single-bladed HAWTs have also been produced (Figure 7.14) and continue to be researched.

Vertical axis wind turbine

Vertical axis wind turbines (VAWTs), unlike their horizontal axis counterparts, can harness winds from any direction without the need to reposition the rotor when the wind direction changes. However, despite this advantage, they have found little commercial success to date, in part due to issues with power quality, cyclic loads on the tower systems and the lower efficiency of some VAWT designs. A technical description of how VAWTs operate is given in Section 7.4.

The modern VAWT evolved from the ideas of the French engineer, Georges Darrieus, whose name is used to describe one of the VAWTs that he invented in 1925 – such devices were independently re-invented in Canada by South and Rangi at the National Aeronautical Establishment of the National Research Council in the 1960s (South and Rangi, 1972). This device, which resembles a large eggbeater, has curved blades (each with a symmetrical aerofoil cross-section) the ends of which are attached to the top and bottom of a vertical shaft (see Figure 7.15). Several hundred were manufactured in the USA and installed in wind farms in California in the 1980s. A small number were produced in Canada, including 'Eole', the largest VAWT yet built, a 100 m tall 60 m diameter turbine. Eole operated for six years from 1988 in Quebec and achieved a 94% availability (see Figure 7.15(b)).

These Darrieus VAWTs were guyed structures, which added complexity and limited their height, but a new design of Darrieus VAWT, mounted on a free-standing cantilevered tower that avoids the difficulties of guys, is under development in New Mexico (VPM, 2011). There is also a floating enhanced Darrieus VAWT design under development (FWC, 2011).

The blades of a Darrieus VAWT take the form of a 'troposkien' (the curved, arch-like shape taken by a spinning skipping rope). This shape is a structurally efficient one, well suited to coping with the relatively high centrifugal forces acting on VAWT blades. However, they can be difficult to manufacture, transport and install, though the advent of modern composites and manufacturing methods may help to address some of the difficulties. In order to overcome these problems, straight-bladed VAWTs have been developed: these include the 'H'-type vertical axis wind turbine (H-VAWT) and the 'V'-type vertical axis wind turbine (V-VAWT).

The H-VAWT (Figure 7.16) consists of a tower (which may house a vertical shaft), capped by a hub to which is attached two or more horizontal cross arms that support the straight, upright, aerofoil blades. In the UK, this type of turbine was developed by VAWT Ltd which built 125 kW and 500 kW prototypes at Carmarthen Bay and a 100 kW turbine on the Isles of Scilly in the 1980s. There continues to be interest in H-VAWTs both at small scale and for large offshore applications, (VertAx, 2011).

The V-VAWT consists of straight aerofoil blades attached at one end to a hub on a vertical shaft and inclined in the form of a letter 'V'. Its main features include a shorter tower, ground/water level-mounted generator options, ground/water level blade installation and the ability to self start without needing complex variable pitch blades or the electrical starting required by other types of VAWTs. Experimental prototypes were tested at the Open University (Figure 7.17(a)) as was the *Sycamore Rotor*, a single bladed version. New generation *V2 Turbine* variants (Figure 7.17(b)) suited to large scale and offshore fixed/floating applications are being researched by the author.

At the present time, VAWTs are not generally economically competitive with HAWTs. However, they continue to attract research as they should, in principle, offer significant advantages over HAWTs in terms of blade loading and fatigue, if they can be built in very large sizes (such as are becoming desired for offshore applications). Whilst VAWTs are subject to wind-induced cyclic loads (which do not progressively increase with increasing size of turbine) they are not subject to the major gravitational cyclic loadings (which do progressively increase with rotor diameter) that large diameter HAWTs experience. As the Canadian Eole demonstrated in the 1980s/90s, large VAWTs can be operated with very high reliability.

More recent variants in vertical axis wind turbines are VAWTs with helically shaped blades. These were first advocated in the 1990s by the Swedish engineer Olle Ljungstrom and additionally by US engineer Alexander Gorlov. Gorlov also suggested such turbines could be used in hydro and, as discussed in Chapter 6, tidal current applications (Gorlov, 1998). Stimulated by perceived concerns about cyclic torque (due to the cyclic variation in the position of a VAWTs blade, relative to the wind direction – see Section 7.4), in the 1990s Ljungstrom produced designs of Darrieus VAWTs that employed helically shaped blades. A number of more recent designs have employed such blades. In practice, however, simply employing three blades should usually be sufficient to even out the torque variation for VAWTs, without the extra complexity and cost of manufacturing helical blades (Musgrove, 1990).

As was mentioned above, most types of VAWT (with aerofoil blades) are not able to self-start without some extra mechanism (as they are generally unable to produce sufficient aerodynamic starting torque). Examples of such additions are drag-driven 'vertical axis starter rotors' (which can reduce aerodynamic efficiency), complex variable pitch blades, or some form of electrical starting mechanism. Electrically assisted starting is not a major issue for the medium/large scale VAWTs employed in wind farms, but it is a major shortcoming for small/micro scale VAWTs, especially for off-grid applications or on relatively low wind speed sites, when electrically started VAWTs can consume large amounts of electricity, thus greatly reducing their net productivity (Day et al., 2010).

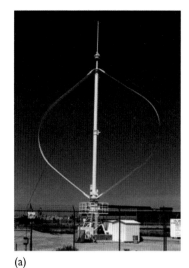

(a)

(b)

Figure 7.15 (a) Seventeen metre diameter Darrieus-type VAWT at Sandia National Laboratories, New Mexico. (b) Sixty metre diameter Eole VAWT in Quebec, Canada

Figure 7.16 500 kW 'H'-type VAWT at Carmarthen Bay, Wales

(a)

(b)

Figure 7.17 (a) V-VAWT prototype developed and tested at the Open University in Milton Keynes in the 1980s (b) Multi-megawatt scale V-Turbine concept in offshore configuration

7.4 **Aerodynamics of wind turbines**

Aerodynamic forces

When a force is transferred by a moving solid object to another solid object, the second object will generally move in either the same direction or in a direction at a small angle (less than 90 degrees) to the direction of motion of the first object, unless subjected to another force. However, the method by which forces are transferred from a fluid to a solid object is very different.

Wind turbines are operating in an unconstrained fluid, in this case air. To understand how they work, two terms from the field of aerodynamics will be introduced. These are 'drag' and 'lift'.

An object in an air stream experiences a force that is imparted from the air stream to that object (Figure 7.18). We can consider this force to be equivalent to two component forces acting in perpendicular directions, known as the *drag* force and the *lift* force. The magnitude of these drag and lift forces depends on the shape of the object, its orientation to the direction of the air stream, and the velocity of the air stream.

The **drag force** is the component that is in line with the direction of the air stream. A flat plate in an air stream, for example, experiences maximum drag forces when the direction of the air flow is perpendicular (that is, at right angles) to the flat side of the plate; when the direction of the air stream is in line with the flat side of the plate, the drag forces are at a minimum. Traditional vertical axis windmills and undershot water wheels (see Chapter 5) are driven largely by drag forces.

Objects designed to minimize the drag forces experienced in an air stream are described as streamlined, because the lines of flow around them follow smooth, stream-like lines. Examples of streamlined shapes are teardrops, the shapes of fish such as sharks and trout, and aeroplane wing sections (aerofoils) (Figure 7.19).

The **lift force** is the component that is at right angles to the direction of the air stream. It is termed 'lift' force because it is the force that enables aeroplanes to *lift* off the ground and fly, though in other applications it may induce a *sideward* (as in a sailboat) or *downward* force (as in the downforce aerofoil used in some racing cars). Lift forces acting on a flat plate are smallest when the direction of the air stream is at a zero angle to the flat surface of the plate. At small angles relative to the direction of the air stream – that is, when the so-called *angle of attack* (see below for more detail) is small – a low pressure region is created on the 'downstream' (or 'leeward') side of the plate as a result of an increase in the air velocity on that side (Figures 7.20 and 7.21 show this effect on aerofoil sections).

In this situation, there is a direct relationship between air speed and pressure: the faster the airflow, the lower the pressure (i.e. the greater the 'suction effect'). This phenomenon is known as the **Bernoulli effect** after Daniel Bernoulli, the Swiss mathematician who first explained it. The lift force thus acts as a 'suction' or 'pulling' force on the object, in a direction at right angles to the airflow.

As well as enabling aeroplanes and gliders to fly, it is the lift force that propels modern sailing yachts, and supports and propels helicopters. Lift is also the principal force that drives a modern wind turbine rotor and thus allows it to produce power.

Aerofoils

Arching or cambering a flat plate will cause it to induce higher lift forces for a given angle of attack, but the use of so-called **aerofoil sections** is even more effective. There are two main types of aerofoil section that are conventionally distinguished: asymmetrical and symmetrical (Figure 7.22). Both have a markedly convex upper surface, a rounded end called the 'leading edge' (which faces the direction from which the air stream is coming), and a pointed or sharp end called the 'trailing edge'. It is the shape of the 'under surface' or high pressure side of the sections that identifies the type. Asymmetrical aerofoils are optimized to produce most lift when the underside of the aerofoil is closest to the direction from which the air is flowing. Symmetrical aerofoils are able to induce lift equally well (although in opposite directions) when the air flow is approaching from either side of the **chord line** (the 'length', from the tip of its leading edge to the tip of its trailing edge, of an aerofoil section).

The angle which an aerofoil (or flat or cambered plate profile) makes with the direction of an airflow, measured against a reference line (usually

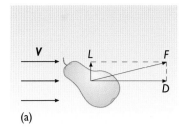

(a)

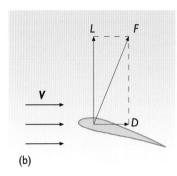

(b)

Figure 7.18 (a) and (b) An object in an air stream is subjected to a force, *F*, from the air stream. This is composed of two component forces: the drag force, *D*, acting in line with the direction of air flow and the lift force, *L*, acting at 90° to the direction of air flow

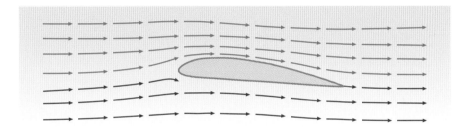

Figure 7.20 Streamlined flow around an aerofoil section

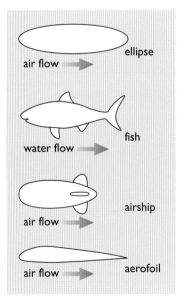

Figure 7.19 Some examples of streamlined shapes

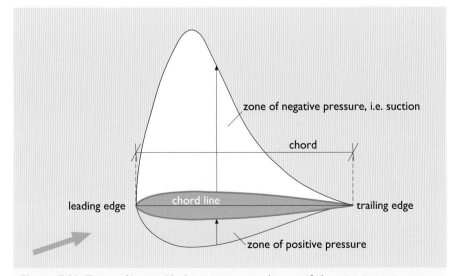

Figure 7.21 Zones of low and high pressure around an aerofoil section in an air stream

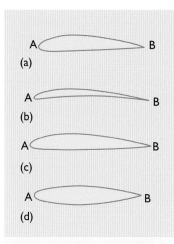

Figure 7.22 Types of aerofoil section: (a), (b) and (c) are various forms of asymmetrical aerofoil section and (d) is a symmetrical aerofoil section

the chord line of the aerofoil), is called the **angle of attack** α *(alpha)* (Figure 7.23). When airflow is directed towards the underside of the aerofoil, the angle of attack is usually referred to as positive.

When employed as a wing profile, asymmetrical aerofoil sections will, subject to a net incident airflow velocity (in aircraft this is due to forward flight), tend to accelerate the airflow over the more convex 'upper' surface. The high air speed thus induced results in a large reduction in pressure over the upper surface relative to the lower surface. This results in a 'suction' effect which 'lifts' the aerofoil-shaped wing, although it should be noted that this lift can only be sustained if the airflow leaves the aerofoil at the downstream edge (known as the trailing edge) in a smooth manner that prevents the high pressure air recirculating around the trailing edge and cancelling out the reduced pressure. The strength of the lift force induced by an aerofoil section is well demonstrated by its ability to support the entire mass of a large aircraft such as the Airbus A380.

The lift and drag characteristics of many different aerofoil shapes, for a range of angles of attack, have been determined by measurements taken in wind tunnel tests, and catalogued (e.g. in Abbott and von Doenhoff, 1958). The lift and drag characteristics measured at each angle of attack can be described using non-dimensional **lift** and **drag coefficients** (C_L and C_D) or as **lift to drag ratios** (C_L/C_D). These are defined in Box 7.3.

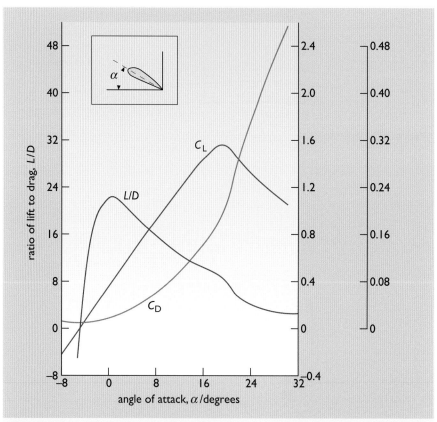

Figure 7.23 Lift coefficient, C_L, drag coefficient, C_D, and lift to drag ratio (L/D) versus angle of attack, α (shown inset), for a Clark Y aerofoil section. The region just to the right of the peak in the C_L curve corresponds to the angle of attack at which stall occurs

BOX 7.3 Aerofoil sections and lift and drag coefficients

Note that the chord of an aerofoil section is also the same as the *width* of the blade in a wind turbine at a given position along the blade.

Drag coefficient (C_D)

The drag coefficient of an aerofoil is given by the following expression:

$$C_D = \frac{D}{0.5pV^2A_b}$$

where:

  D is the drag force in newtons (N)

  ρ is the air density in kilograms per cubic metre (kg m^{-3})

  V is the velocity of the air approaching the aerofoil in metres per second (m s^{-1})

  A_b is the blade area (i.e. chord × length) in square metres (m^2).

In the case of a blade element, the area is equal to the mean chord × length of the blade element.

Lift coefficient (C_L)

The lift coefficient of an aerofoil is given by the following expression:

$$C_L = \frac{L}{0.5pV^2A_b}$$

where L is the lift force in newtons.

The lift and drag coefficients of an aerofoil can be measured in a wind tunnel at different angles of attack and wind velocities. The results of such measurements can be presented in either tabular or graphical form as in Figure 7.23.

Each aerofoil has an angle of attack at which the lift to drag ratio (C_L/C_D) is at a maximum. This angle of attack results in the maximum force and is thus the most efficient setting of the blades of a HAWT. Consequently, plots of this ratio against angle of attack can be useful to turbine designers (Figure 7.23).

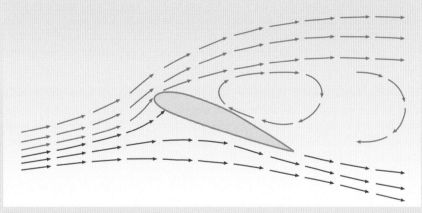

Figure 7.24 Aerofoil section in stall

Another important characteristic relationship of an aerofoil is its **stall angle**. This is the angle of attack at which the aerofoil exhibits stall behaviour. Stall occurs when the flow suddenly leaves the suction side of the aerofoil (when the angle of attack becomes too large), resulting in a dramatic loss in lift and an increase in drag (Figures 7.23 and 7.24). When this happens during the flight of an aeroplane, it can be extremely dangerous unless the pilot can make the plane recover. One of the methods used by wind turbines to limit the power extracted by the rotor in high winds takes advantage of this phenomenon; such turbines are known as stall regulated (see below for more details).

Aerofoils can also now be designed with the aid of specially developed software, and new aerofoils are being designed and optimized to be more efficient in the aerodynamic conditions experienced by wind turbines. Examples of wind turbine specific aerofoils include the DU airfoils (Rooij and Timmer, 2004) and NREL airfoils (Tangler and Somers, 1995). Figure 7.23 shows typical lift and drag coefficients, and lift to drag ratios, for one aerofoil section. Knowledge of these coefficients is essential when selecting appropriate aerofoil sections in wind turbine blade design. Lift and drag forces are both proportional to the energy in the wind.

Relative wind velocity

When a wind turbine is stationary, the direction of the wind as 'seen' from a wind turbine blade is the same as the undisturbed wind direction. However, once the blade is moving, the direction from which it 'sees' the wind approaching effectively changes in proportion to the blade's velocity. (In the case of a moving vertical axis wind turbine blade, the direction from which the blade 'sees' the wind is also affected by its position during its rotation cycle – see Figure 7.27). Two-dimensional **vectors** are used to represent this effect graphically. A two-dimensional vector is a quantity that has both magnitude and direction. A velocity vector can be represented graphically in the form of an arrow, the length of which is proportional to speed, and the angular position of which indicates the direction of flow.

The wind as seen from a point on a moving blade is known as the **relative wind**, and its velocity is known as the **relative wind velocity** (usually symbolized by W). This is a vector which is the *resultant* (i.e. the vector sum) of the **wind velocity at the rotor**, V_1 (i.e. the **undisturbed wind velocity vector**, V_0, reduced by a factor known as the axial interference factor) and the tangential velocity vector of the blade at that point on the blade, u (see Box 7.4). Note that the tangential velocity, measured in metres per second (m s^{-1}), is distinct from the angular velocity, which is measured in radians per second or in revolutions per minute (Box 7.2).

The angle from which the point on the moving blade sees the relative wind is known as the **relative wind angle** (usually symbolized by ϕ) and is measured from the tangential velocity vector, u. The **blade pitch angle**, β, at this point on the blade is the relative wind angle minus the angle of attack, α, at that point on the blade (see Box 7.4).

Harnessing aerodynamic forces

Modern horizontal and vertical axis wind turbines make use of the aerodynamic forces generated by aerofoils in order to extract power from the wind, but each harnesses these forces in a different way.

In the case of a HAWT with fixed-pitch blades with its rotor axis assumed to be in constant alignment with the undisturbed wind direction, for a given wind speed and constant rotation speed, the angle of attack at a given position on the rotor blade *stays constant throughout its rotation cycle* (Box 7.4).

BOX 7.4 **HAWT** rotor blades wind forces and velocities

Figure 7.25 shows a section through a moving rotor blade of a HAWT. Also shown is a vector diagram of the forces and velocities at a position along the blade at an instant in time.

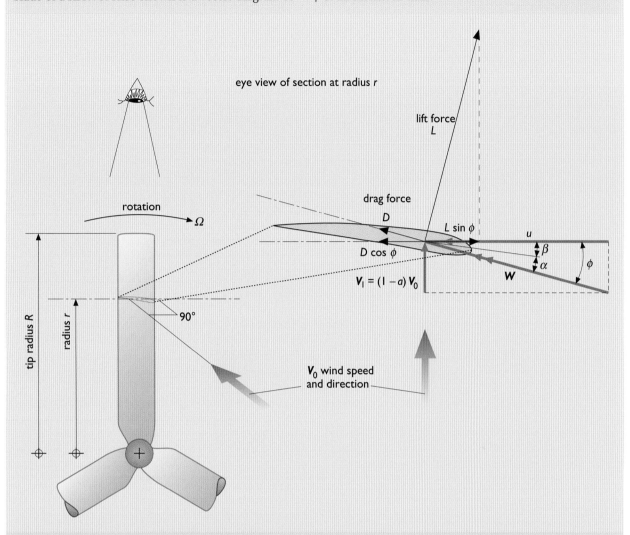

Figure 7.25 Vector diagram showing a section through a moving HAWT rotor blade. Notice that the drag force, D, at the point shown is acting in line with the direction of the relative wind, W, and the lift force, L, is acting at 90° to it

Because the blade is in motion, the direction from which the blade 'sees' the relative wind velocity, W, is the resultant of the tangential velocity, u, of the blade at that position and the wind velocity at the rotor, V_1. Note that in this diagram, the direction of u is shown in the opposite direction to the direction of motion of the rotating blade, in the same manner that a flag on a motor boat moving in calm weather points to the stern of the boat, showing the air to be 'flowing' from the opposite direction to the boat's forward motion.

The wind velocity at the rotor, V_1, is the undisturbed wind velocity upstream of the rotor, V_0, reduced by a factor that takes account of the wind being slowed down as a result of power extraction. This factor is often referred to as the **axial interference factor**, and is represented by a.

The tangential velocity, u, (in metres per second) at a point along the blade is the product of the angular velocity, Ω (in radians per second) of the rotor and the local radius, r, (in metres), at that point, that is:

$$u = \Omega r \qquad (3)$$

Albert Betz showed in 1928 that the maximum fraction of the power in the wind that can theoretically be extracted is 16/27 (59.3%). This occurs when the undisturbed wind velocity is reduced by one-third, in other words, when the axial interference factor, a, is equal to one-third. The value of 59.3% is often referred to as the **Betz limit**.

The relative wind angle, ϕ, is the angle that the relative wind makes with the blade (at a particular point with local radius, r, along the blade) and is measured from the plane of rotation. (Note: if it were not for the fact that the wind is slowed down as a result of the wind turbine extracting energy – in other words if V_0 was not reduced to V_1 at the rotor – the tip speed ratio would be equal to the reciprocal of the tangent of the relative wind angle at the blade.) The angle of attack, α, at this point on the blade can be measured against the relative wind angle, ϕ. The blade pitch angle (usually represented by β) is then *equal* to the relative wind angle *minus* the angle of attack. Since the rotor is constrained to rotate in a plane at right angles to the undisturbed wind, the driving force at a given point on the blade is that component of the aerofoil lift force that *acts in the plane of rotation*. This is given by the product of the lift force, L, and the sine of the relative wind angle, ϕ (that is, $L \sin \phi$). The component of the drag force in the rotor plane at this point is the product of the drag force, D, and the cosine of the relative wind angle, ϕ (that is, $D \cos \phi$).

The torque, q (that is, the moment about the centre of rotation of the rotor in the plane of the rotor), in newton metres (N m) at this point on the blade is equal to the *product* of the *net driving force in the plane of rotation* (that is, the component of lift force in the plane of rotation *minus* the component of the drag force in the rotor plane) and the local radius, r. The total torque, Q, acting on the rotor can be calculated by summing the torque at all points along the length of the blade and multiplying by the *number* of blades. The power from the rotor is the *product* of the total torque, Q and the rotor's angular velocity, Ω.

Why are rotor blades twisted?

The magnitude and direction of the relative wind angle, ϕ, varies along the length of the blade according to the local radius, r. Equation (3) shows that the tangential velocity varies with radius, so as the tangential speed *decreases* towards the hub, the relative wind angle, ϕ, *progressively increases* (see Figure 7.26). A HAWT rotor designed for optimum performance will have a tapered blade, and to have a constant angle of attack along its length (assuming the same aerofoil section is used throughout its length), it will have to have a built-in twist. The amount of twist will vary (as the relative wind angle varies) progressively from tip to root. Figure 7.26 demonstrates the progressive twist of such a HAWT

rotor blade. Most manufacturers of HAWT blades use tapered and twisted blades, although it is possible to build functional HAWT rotor blades that are not twisted. These are cheaper, but less efficient and how well they function depends in part on both the aerofoil characteristics and the overall blade pitch angle.

Blade pitch

As well as the blade pitch angle defined above the term **blade pitch** also refers to the whole blade's angular position about the blade's longitudinal axis (also known as the pitch axis) such that in the case of a **variable pitch** rotor blade, the whole blade is able to be rotated about its pitch axis. In most cases all of the blades of a variable pitch HAWT rotor change pitch at the same time in order to:

optimise the turbine's power production across a range of wind speeds (in order to maintain the angle of attack at or near to the optimal angle across a range of wind speeds)

reduce its output at high wind speeds

to stop the rotor during very high wind speeds or to 'park' the rotor (such that the blade pitch is in its 'feathered' position, e.g. each blade pitch is at or near 90 degrees, relative to the plane of rotation), when it is necessary to prevent the rotor from operating for any reason.

In the case of a **fixed pitch** rotor the blade pitch angle remains unchanged, which makes them less productive and less controllable. Most large wind turbines employ variable pitch blades, but most micro wind turbines and small wind turbines and some medium-scale wind turbines use fixed pitch blades.

Stall control of wind turbines

Let us assume that a wind turbine is rotating at a constant rotation speed, regardless of wind speed, and that the blade pitch angle is fixed. As the wind speed *increases* the tip speed ratio *decreases*. At the same time, the relative wind angle *increases*, causing an increase in the angle of attack.

It is possible to take advantage of this characteristic to control a turbine in high winds, if the rotor blades are designed so that above the rated wind speed they become less efficient because the angle of attack approaches the stall angle. This results in a loss of lift, and thus torque, on the regions of the blade that are in 'stall'.

This method of so-called **stall regulation** has been employed successfully on numerous fixed-pitch HAWT rotors and has also been employed on most modern lift-driven VAWT rotors. However, as turbines have increased in size this approach to power regulation has generally been discontinued in HAWTs in favour of variable speed turbines with variable pitch regulation. The main reasons for this are that it is very hard for designers to predict exactly when a given rotor will stall and that variable pitch rotors are more efficient over a range of wind speeds.

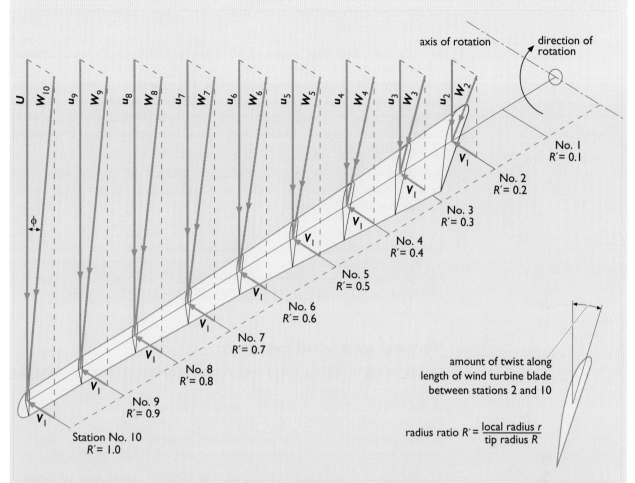

Figure 7.26 Three-dimensional view of an optimally tapered and twisted HAWT rotor blade design (the blade is shown in a horizontal position and moving through the upward part of its cycle about its axis of rotation). The figure shows how the relative wind angle, ϕ, changes along the blade span (length). Note that the blade aerofoil section and the angle of attack are assumed to be constant along the length of the blade. The diagram (lower right) of the view along the blade indicating the amount of built-in twist along the length of the blade shows the blade cross-section at station 2: the cross section of the blade at station 10 has been omitted for clarity

By contrast, in a VAWT with fixed pitch blades, under the same conditions, the angle of attack at a given position on the rotor blade is *constantly varying throughout its rotation cycle*.

During the normal operation of a horizontal axis rotor, the direction from which the aerofoil 'sees' the wind is such that the angle of attack remains positive throughout.

In the case of a vertical axis rotor, however, the angle of attack changes from positive to negative and back again over each rotation cycle. This means that the 'suction' side reverses during each cycle, so a symmetrical aerofoil has to be employed to ensure that power can be produced irrespective of whether the angle of attack is positive or negative.

Horizontal axis wind turbines

Most horizontal axis wind turbines are axial flow devices – the rotation axis is maintained in line with wind direction by a 'yawing' mechanism, which constantly realigns the wind turbine rotor in response to changes in wind direction.

In addition to its swept area and rotor diameter, the performance (power output, torque and rotation speed) of an axial flow horizontal axis wind turbine rotor is dependent on numerous other factors. These include the number and shape of the blades and the choice of aerofoil section, the length of the blade chord, the tip speed ratio, the blade pitch angle, the relative wind angle and angle of attack at positions along the blade, and the amount of blade twist between the hub and tip (see Box 7.4).

Box 7.4 explains how the relative wind velocity, W, and relative wind angle, ϕ, both vary along the blade, and describes their influence on the optimum blade pitch angle.

A few examples of cross-flow horizontal axis windmills have also been proposed historically (see Figure 7.9) and cross-flow horizontal axis turbines are under development for use in building integrated wind energy systems (see Section 7.8 and Taylor, 1998). In this context, cross-flow horizontal axis turbines function in a similar manner to the vertical axis wind turbines described below.

Vertical axis wind turbines

Modern VAWTs, unlike HAWTs, are 'cross-flow devices'. This means that the direction from which the undisturbed wind flow comes is at right angles to the axis of rotation; that is, the wind flows across the axis. As the rotor blades turn, they sweep a three-dimensional surface (a cylindrical surface in the case of an H-VAWT or a conical surface in the case of a V-VAWT), as distinct from the single circular plane swept by a HAWT's rotor blades.

In contrast to traditional vertical-axis windmills, the blades of modern vertical axis wind turbines extract most of the power from the wind as they pass across the front and rear – as distinct from one side (relative to the undisturbed wind direction) of the swept surface.

A vertical axis wind turbine will function with the wind blowing from any compass direction, but let us assume initially that it is blowing from one particular direction and also that the setting angle of the blade is such that its chord is in line with a tangent to the circular path of rotation (that is, it has 'zero set pitch'). Clearly, the angle of the blade, in relation to the direction of the undisturbed wind, changes from zero to 360 degrees over each cycle of rotation. It might appear that the angle of attack of the wind to the blade would vary by the same amount, and so it might seem impossible for a VAWT to operate at all. However, we have to take into account the fact that when the blade is moving, the relative wind angle 'seen' by the blade is the resultant (W) of the wind velocity V_1 at the rotor and the blade velocity u (see Box 7.4). Provided that the blade is moving sufficiently fast relative to the wind velocity (in practice, this means at a tip speed ratio of three or more), the angle of attack that the blade makes with the relative wind velocity W will only vary within a small range (see Figure 7.27).

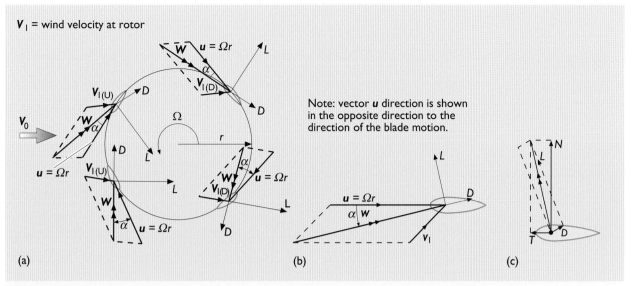

Figure 7.27 The lift and drag forces acting on VAWT rotor blades can be resolved into two components: 'normal', N, (that is, in line with the radius) and 'tangential', T, (that is perpendicular to the radius). The magnitude of both components varies as the angle of attack varies: (a) blade forces and relative velocities for a VAWT, showing angles of attack at different positions; (b) detail of aerodynamic forces on a blade element of a VAWT rotor blade; (c) normal (radial) and tangential (chord-wise) components of force on a VAWT blade. Note $V_{1(U)}$ is the wind velocity at the rotor on the upwind side; $V_{1(D)}$ is the wind velocity at the rotor on the downwind side

7.5 Power and energy from wind turbines

How much power does a wind turbine produce?

The power output of a wind turbine varies with wind speed: every turbine has a characteristic wind speed–power curve, often simply called the **power curve**. The shape of a wind speed-power curve is influenced by the:

- rotor swept area
- choice of aerofoil
- number of blades
- blade shape
- optimum tip speed ratio
- speed of rotation
- cut-in wind speed (the wind speed at which a turbine begins to generate power)
- rated wind speed (the wind speed at which a turbine generates its rated power)
- shut-down or cut-out wind speed (the wind speed at which a turbine is shut down and stops generating – also known as the furling wind speed)
- aerodynamic efficiency (power coefficient)
- gearing efficiency, and
- generator efficiency.

An example of such a curve is shown in Figure 7.28.

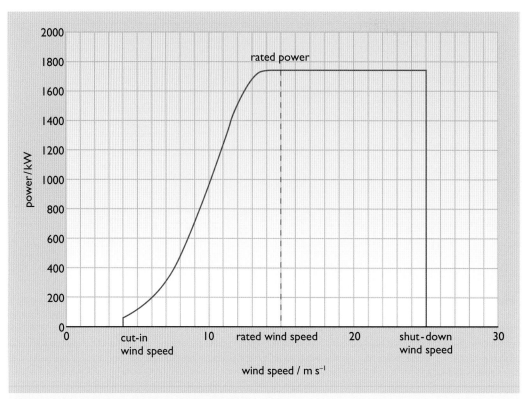

Figure 7.28 Typical wind turbine wind speed–power curve

How much energy will a wind turbine produce?

The energy that a wind turbine will produce depends on both its wind speed–power curve and the **wind speed frequency distribution** at the site. The latter is essentially a graph or histogram showing the number of hours for which the wind blows at different wind speeds during a given period of time. Figure 7.29 shows a typical wind speed frequency distribution.

For each incremental wind speed within the operating range of the turbine (that is, between the cut-in wind speed and the shut-down wind speed), the energy produced at that wind speed can be obtained by multiplying the number of hours of its duration by the corresponding turbine power at this wind speed (given by the turbine's wind speed–power curve). This data can then be used to plot a **wind energy distribution** such as that shown in Figure 7.30. The total energy produced in a given period is then calculated by summing the energy produced at all the wind speeds within the operating range of the turbine.

The best way to determine the wind speed distribution at a site is to carry out wind speed measurements with equipment that records the number of hours for which the wind speed lies within each given 1 m s^{-1} wide speed band, e.g. 0–1 m s^{-1}, 1–2 m s^{-1}, 2–3 m s^{-1}, etc.

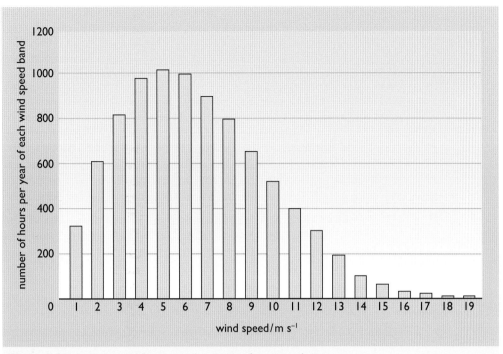

Figure 7.29 A wind speed frequency distribution for a typical site

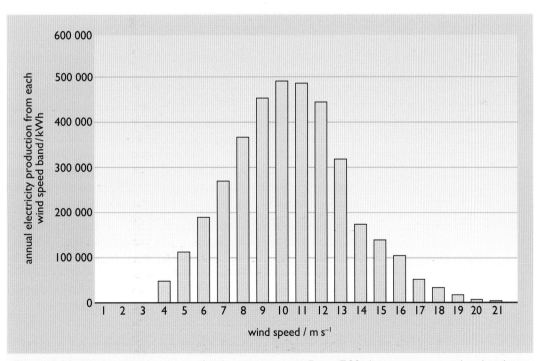

Figure 7.30 Wind energy distribution for the same site as in Figure 7.29, showing energy produced at this site by a wind turbine with the wind speed-power curve shown in Figure 7.28

The longer the period over which measurements are taken, the more accurate is the estimate of the wind speed frequency distribution. As the power in the wind is proportional to the cube of the wind velocity – see Equation (2) – a small error in estimating the wind speeds can produce a large error in the estimate of the energy yield.

Additional factors that affect the total energy generated include transmission losses and the availability of the turbine. **Availability** is an indication of the reliability of the turbine installation and is the fraction or percentage of a given period of time for which a wind turbine is available to generate, when the wind is blowing within the turbine's operating range. Current commercial wind turbines typically have annual availabilities in excess of 90%, many have operated at over 95% and some are achieving 98%.

If the mean annual wind speed at a site is known, or can be estimated, the following formula (Beurskens and Jensen, 2001) can be used to make a rough *initial estimate* of the electricity production (in kilowatt-hours per year) from a number of wind turbines:

Annual electricity production $= K \, V_m^3 \, A_t \, T$

where:

$K = 3.2$ and is a factor based on typical turbine performance characteristics and an approximate relationship between mean wind speed and wind speed frequency distribution (see below)

V_m is the annual mean wind speed at the site in metres per second

A_t is the swept area of the turbine in square metres

T is the number of turbines.

This formula should be used with caution, however, because it is based on an average of the characteristics of the medium- to large-scale wind turbines currently available and assumes an approximate relationship between annual mean wind speed and the frequency distribution of wind speeds that may not be accurate for an individual site. It also does not allow for the different power curves of wind turbines that have been optimized either for low or high wind speed sites. The K factor of 3.2 given above assumes a well designed turbine suited to its site (Beurskens and Jensen, 2001), but it should not be used with small-scale wind turbines as their performance varies greatly and they are more likely to be located at sites with lower annual mean wind speeds and potentially in suburban and urban areas.

For a small wind turbine (less than 200 m² swept area) estimates of potential electricity production can be derived if the supplier provides a British Wind Energy Association Reference Annual Energy (RAE) value (BWEA, 2008) or an American Wind Energy Association Rated Annual Energy (RAE) value together with an AEP (Annual Energy Production) curve (AWEA, 2009), although there will still be a high level of uncertainty in urban areas.

Estimating the wind speed characteristics of a site

It is expensive to carry out detailed measurements at a site and wind speed measurements are often not carried out for small wind turbine installations. However the use of remote sensing methods such as SODAR (SOnic Detection And Ranging) and Doppler LIDAR (LIght Detection And

Ranging) makes it feasible to monitor wind speeds without the need for tall towers. In addition, lower cost instrumentation is becoming available for monitoring small wind turbine sites.

It is preferable to record the wind speed and direction as close as possible to the proposed site for at least a year. However this will only give information for a particular time period, and weather patterns change. In order to ascertain the longer term wind speed characteristics, it is useful to correlate the measured data with data measured at one or more nearby meteorological stations or other wind recording sites. Then by statistical analysis of the two data sets, and extrapolating over the long-term data from the meteorological station, an estimate of the longer term wind speed characteristics at the site can be made. This technique is referred to as the **Measure–Correlate–Predict** or **MCP** method.

There are a number of different ways of implementing the MCP methodology all based on different statistical analysis techniques (Rogers et al, 2005, 2006a). These methods are embedded into different software packages, but their application requires careful judgement – consistency in the use of the methods is important and more than one algorithm should be employed in order to avoid bias.

If it is not possible to carry out wind speed and direction measurements at a proposed site, or where a preliminary analysis is required prior to installing instrumentation, there are a number of techniques that can be employed to give an approximate estimate of the wind speed characteristics of a site.

Using wind speed measurements from a nearby location

This involves making use of existing wind speed measurements from one or more locations nearby and deriving the data for the proposed site by interpolation or extrapolation, taking into account differences between the proposed site and the sites for which measurements are available.

Using wind speed maps and atlases

Maps are available that give estimates of the mean wind speeds over the UK and many other countries. However, most of these maps were made using data from meteorological stations, which tend to be located in places that are often not appropriate for wind energy, so wind speed maps and atlases specifically for wind energy purposes have also been developed for many countries.

Using long-term wind measurements and the WAsP model mentioned below, a *European Wind Atlas* (Troen and Petersen, 1989) has been produced by the Risø Laboratory in Denmark for the European Commission. This document includes maps of various areas within the European Union (for example, Figure 7.31), which show the annual mean wind speed at 50 m above ground level for five different topographic conditions: sheltered terrain, open plain, sea coast, open sea, hills and ridges. The atlas includes a series of procedures for taking account of site characteristics to estimate the wind energy likely to be available. These procedures work quite well on sites with a gentle topography but are not so good for very hilly terrain or urban areas. A similar atlas (included in Figure 7.31) has

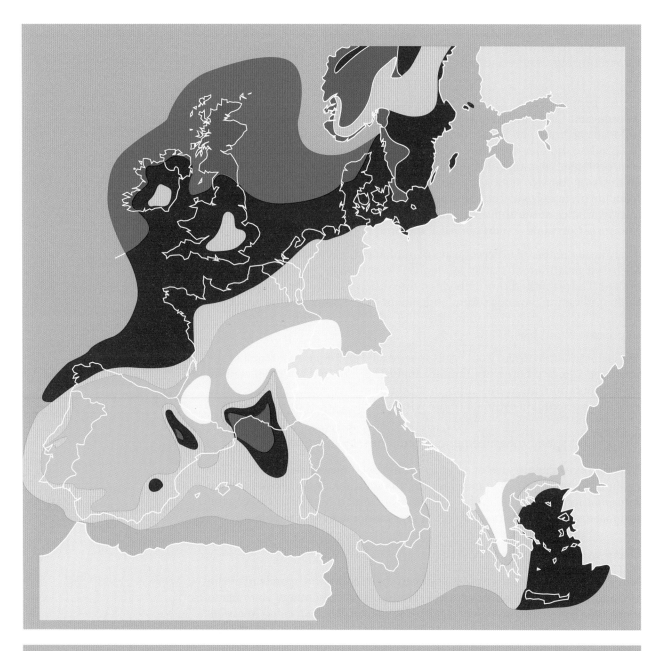

Wind resources at 50 m above ground level for five different topographic conditions										
		Sheltered terrain		Open plain		At a sea coast		Open sea		Hills and ridges	
		m s⁻¹	W m⁻²	m s⁻¹	W m⁻²	m s⁻¹	W m⁻²	m s⁻¹	W m⁻²	m s⁻¹	W m⁻²

Wind class		m s^{-1}	W m^{-2}	m s^{-1}	W m^{-2}	m s^{-1}	W m^{-2}	m s^{-1}	W m^{-2}	m s^{-1}	W m^{-2}
	5	>6.0	>250	>7.5	>500	>8.5	>700	>9.0	>800	>11.5	>1800
	4	5.0–6.0	150–250	6.5–7.5	300–500	7.0–8.5	400–700	8.0–9.0	600–800	10.0–11.5	1200–1800
	3	4.5–5.0	100–150	5.5–6.5	200–300	6.0–7.0	250–400	7.0–8.0	400–600	8.5–10.0	700–1200
	2	3.5–4.5	50–100	4.5–5.5	100–200	5.0–6.0	150–250	5.5–7.0	200–400	7.0–8.5	400–700
	1	<3.5	<50	<4.5	<100	<5.0	<150	<5.5	<200	<7.0	<400

Figure 7.31 Annual mean wind speeds and wind energy resources over Europe (EU Countries) combining land-based and offshore wind atlases (source: Troen and Petersen, 1989)

also been produced to cover the *offshore* wind energy resource in the European Union (Risø, 2009). Similar wind atlases based on the same approach have also been produced for Russia, South Africa and parts of North Africa. Also wind speed/energy atlases have been produced for Ireland, USA and Canada.

The Energy Technology Support Unit (ETSU) also prepared a wind atlas and database of the UK. Using wind speed data from meteorological stations, a digital terrain model of the UK and a wind speed prediction computer model known as NOABL (Numerical Objective Analysis of Boundary Layer), ETSU estimated an annual mean wind speed (AMWS) value for each 1 km × 1 km Ordnance Survey grid square in the UK (Burch and Ravenscroft, 1992). Whilst this is a useful atlas/database for rural areas, it consistently overpredicts the AMWS in urban and suburban areas and (like most wind atlases developed for wind energy development in windy areas) it should not be used for that purpose. The UK Microgeneration Installation Standard, MIS 3003, (MIS, 2011 and MCS, 2011a), includes some adjustment factors that try to take account of this for small or micro wind turbines, but it is still not very reliable when used in those situations. This atlas is no longer being updated, but, at the time of writing, can still be accessed via the Renewable UK and Department for Climate Change (DECC) websites (Renewable UK, 2011a; DECC, 2011a).

Because of the unreliability of the NOABL database, the UK Energy Savings Trust has made available the wind speed data that it accumulated whilst carrying out field trials of domestic small wind turbines. This database of AMWS can be accessed by entering a UK post code together with the rural, suburban or urban site classification (EST, 2011).

For similar reasons the UK Carbon Trust commissioned the UK Meteorological Office to develop a new wind speed database derived from its own extensive wind speed data set together with a number of adjustments for urban and suburban locations. This can be accessed via the Carbon Trust's website (Carbon Trust, 2011) by entering the relevant OS grid reference or postcode together with a number of site classification parameters. It seems to be an improvement on the NOABL database for suburban and urban areas, but installing productive wind turbines in such locations still requires care.

Wind flow simulation computer models

A number of computer models have been developed that aim to predict the effects of topography on wind speed. Data from the nearest wind speed measurement station, together with a description of its site, is required and local effects are taken into account to arrive at estimated wind data for the proposed wind turbine site. Examples include NOABL and WAsP. As described above, NOABL was used in the development of the UK wind atlas/database and WAsP was used in the development of the *European Wind Atlas* (Figure 7.31). WAsP also forms the basis of at least two proprietary wind speed assessment computer software models. There are also improved versions of WAsP, for example for offshore use, and some that are better suited to modelling wind speeds in complex terrain.

CFD (computational fluid dynamics) is increasingly used to model wind flows in complex terrain, over and around forests and for designing wind farms. Berge et al., (2006) compares two CFD models with WAsP in modelling complex terrain.

Used with care, CFD models can be useful for carrying out initial assessments though they are complex, computer intensive, require training and need to be well understood. They tend to be used for wind farm projects in difficult terrain. There is also a need to have access to high quality digital terrain maps (special types of files containing three-dimensional map data), although some models are able to utilize/synchronize with Google Earth.

7.6 Environmental impact

Wind energy development has both positive and negative environmental impacts. The scale of its future implementation will rely on successfully maximizing the positive impacts whilst keeping the negative impacts to the minimum (see for example the US National Research Council's report for Congress, (NRC, 2007) which gives a comprehensive overview of the positive and negative environmental impacts of wind energy).

The generation of electricity by wind turbines does not involve the release of carbon dioxide or pollutants that cause acid rain or smog, that are radioactive, or that contaminate land, sea or water courses. Large-scale implementation of wind energy within the UK would probably be one of the most economic and rapid means of reducing carbon dioxide emissions. Over its working lifetime, a wind turbine can generate approximately 40 to 80 times the energy required to produce it (see Everett et al., 2012).

In addition, wind turbines do not require the consumption of water, unlike many conventional (and some renewable) energy sources. This benefit could be of growing importance if water shortages occur with increasing frequency in the future.

Of course wind power is not without certain negative (or perceived negative) repercussions and the following subsections will look into the following issues:

- noise
- electromagnetic interference
- aviation related issues
- wildlife
- public attitudes and planning.

Wind turbine noise

Whilst wind turbines are often described as noisy by opponents of wind energy, in general they are not especially noisy compared with other machines of similar power rating (see Table 7.1 and Figure 7.32). However, there have been incidents where wind turbine noise has been cited as a nuisance. Currently available modern wind turbines are generally much quieter than their predecessors and conform to noise immision level requirements (see below). **Noise immision** is a measure of the cumulative noise energy to which an individual is exposed over time. It is equal to the average noise level to which the person has been exposed, in decibels, plus 10 times the logarithm ($\log_{10}$) of the number of years for which the individual is exposed,

Table 7.1 Noise of different activities compared with wind turbines

Source/activity	Noise level in dB(A)*
Threshold of pain	140
Jet aircraft at 250 m	105
Pneumatic drill at 7 m	95
Truck at 48 km h⁻¹ (30 mph) at 100 m	65
Busy general office	60
Car at 64 km h⁻¹ (40 mph) at 100 m	55
Wind farm at 350 m	35–45
Quiet bedroom	20
Rural night-time background	20–40
Threshold of hearing	0

*dB(A): decibels (acoustically weighted to take into account that the human ear is not equally sensitive to all frequencies)
Source: ODPM, 2004b

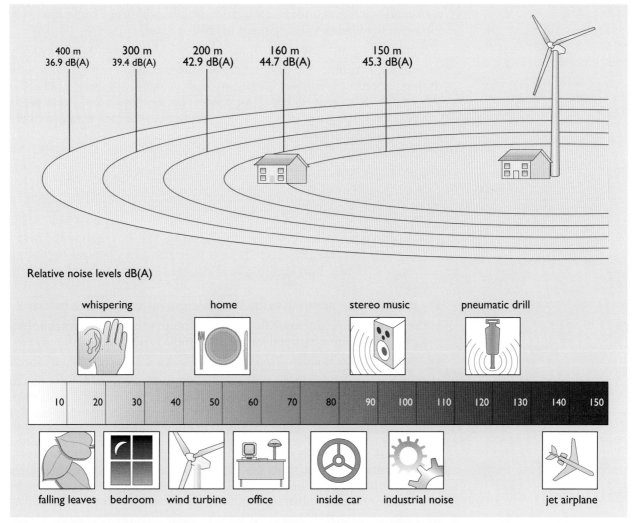

Figure 7.32 Wind turbine noise pattern from a typical wind turbine (source: EWEA, 1991)

(Note: the figure shows indicative sound levels interpolated from the pictorial positions along the horizontal line. For more precise values see Table 7.1 and its source reference.)

There are two main sources of wind turbine noise. One is that produced by mechanical or electrical equipment, such as the gearbox and the generator, known as **mechanical noise**; the other is due to the interaction of the air flow with the blade, referred to as **aerodynamic noise**.

If noise does occur, then mechanical noise is usually the main problem, but it can be remedied fairly easily by the use of quieter gears, mounting equipment on resilient mounts, and using acoustic enclosures – or eliminating the gearbox altogether by opting for a direct drive low speed generator.

Aerodynamic noise

The aerodynamic noise produced by wind turbines can perhaps best be described as a 'swishing' sound. It is affected by: the shape of the blades; the interaction of the airflow with the blades and the tower; the shape of the blade trailing edge; the tip shape; whether or not the blade is operating in stall conditions; and turbulent wind conditions, which can cause unsteady forces on the blades, causing them to radiate noise.

Aerodynamic noise will tend to increase with the speed of rotation. For this reason, some turbines are designed to be operated at lower rotation speeds during periods of low wind. Noise nuisance is usually more of a potential risk in light winds: at higher wind speeds, background wind noise tends to mask wind turbine noise. Increasing the numbers of blades is also likely to reduce aerodynamic noise.

Noise regulations, standards, controls and reduction

Most commercial wind turbines undergo noise measurement tests in accordance with one of the following:

- the recommended procedure developed by the International Energy Agency (Ljungren 1994, 1997). This procedure lies behind an online computer model provided by the National Physical Laboratory (NPL, 2011)
- a procedure conforming to the Danish noise regulations (see below)
- the method documented in the IEC (International Electrotechnical Commission) international standard 61400-11 (IEC, 2006).

The measured noise level values from such tests provide information that enables the turbines to be sited at a sufficient distance from habitations to minimize (or avoid) the risks of noise nuisance. This standard procedure also allows manufacturers to identify any noise problem and take remedial action before the commercial launch of the machine.

In Denmark, in order to control the effects of noise from wind turbines, there is a standard that specifies that the maximum wind turbine noise level permitted at the nearest dwelling in open countryside should be 45 dB(A). At habitations in residential areas a noise level of only 40 dB(A) is permitted. This noise limit has been demonstrated to be achievable with commercially available turbines.

In the UK, the current guidance (principally for medium- and large-scale turbines) is that noise limits should be set relative to background noise with different limits for daytime and night-time (35 – 40 dB(A) for daytime and 43 dB(A) for night-time). The decision of which daytime limit to use – either 35 or 40 dB(A) – depends on the following:

- the number of dwellings in the neighbourhood of the wind farm
- the effect of noise limits on the number of kWh generated
- the duration and level of exposure.

For micro/small wind turbines the noise limit is slightly higher at 45dB LAEQ 5 min (i.e. the equivalent continuous A-weighted sound level over a 5 minute period) at 1 metre from the window of a habitable room in the façade of any neighbouring residential property (DCLG, 2007, 2009).

Turbines that comply with the BWEA standard (BWEA, 2008), must provide a noise immision map – drawn up using specified methods (BSI, 2003) – that shows the noise at different distances from the hub for different wind speeds across the range of operation of the turbine. At the time of writing, further information on wind turbine siting is provided in Planning Policy Statement 22 (PPS22) (ODPM, 2004a) and its accompanying guidance (ODPM, 2004b). Fiumicelli and Triner (2011) provide extensive information together with a wind farm noise complaint methodology. The Microgeneration Certification Scheme (MCS) discussed in the section on small-scale wind turbines (Section 7.8) also addresses this topic.

Noise is a sensitive issue. Unless it is given careful consideration at both the turbine design and the project planning stages, taking into account the concerns of people who may be affected, opposition to wind energy development is likely.

By eliminating the gearbox, mechanical turbine noise can be considerably reduced – as mentioned previously, some manufacturers have developed gearbox free turbines with low speed generators directly coupled to the rotor. This makes the turbines very quiet and able to be more comfortably sited close to buildings.

There is much ongoing research to reduce wind turbine noise. A useful overview of this work is included in Legerton (1992) and refinements in methods of measurement and noise propagation modelling are described in Kragh et al. (1999), Rogers et al. (2006b) and IEC (2006).

Electromagnetic interference

When a wind turbine is positioned between a radio, television or microwave transmitter and receiver (Figure 7.33) it can sometimes reflect some of the electromagnetic radiation in such a way that the reflected wave interferes with the original signal as it arrives at the receiver. This can cause the received signal to be distorted significantly.

The extent of electromagnetic interference caused by a wind turbine depends mainly on the materials used to make the blades and on the surface shape of

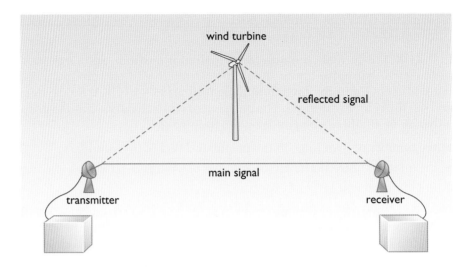

Figure 7.33 Scattering of radio signals by a wind turbine

the tower. If the turbine has metal blades (or glass-reinforced plastic blades containing metal components), electromagnetic interference may occur if it is located close to a radio communications service. The laminated timber blades used in some turbines absorb rather than reflect radio waves so do not generally present a problem. There has also been research into applying radar absorbing materials (RAMs) into blades. Faceted towers reflect more than smooth rounded towers, due to their flat surfaces.

Chignell (1987) and ODPM (2004b) give some simplified guidance about how to prevent/minimize electromagnetic interference, and Ofcom have examined the impact of tall structures, including wind turbines, on broadcasting and other wireless services (Ofcom, 2009).

The most likely form of electromagnetic interference is to television reception. This is relatively easily dealt with by the installation of relay transmitters or by connecting cable television services to the affected viewers. In the UK, the BBC provides an online assessment tool, to evaluate the impact of wind turbines on TV broadcast services (BBC, 2011).

Microwave links, very high frequency (VHF) omni-directional ranging (VOR) and instrument landing systems (ILS) can also be affected by wind turbines. A method of determining an acceptable exclusion zone around radio transmission links has been developed which takes account of the characteristics of antennae (Bacon, 2002). In the UK, Ofcom commissioned detailed research into potential wind farm interference to fixed link and scanning telemetry devices (Randhawa and Rudd, 2009); at the time of writing Ofcom maintains a website that provides guidance on wind farms and electromagnetic interference (Ofcom, 2011).

Wind turbines and aviation

According to Renewable UK (one of the main UK renewable energy trade bodies) over half the planned wind energy projects in the UK are likely to be affected by concerns about the possible impact of wind turbines on aviation.

The UK Ministry of Defence (MOD) has voiced concern about the interference with military radar that could be caused by wind turbines. In addition, the MOD is concerned that wind turbines (particularly those with large diameters and tall towers), when located in certain areas, will penetrate the lower portion of the low flying zones used by military aircraft. These various concerns have led to the MOD's intervention which has impeded the development of a number of wind farms in the UK. NATS (National Air Traffic Services Ltd) has produced a series of maps that indicate the areas around the UK where radar interference may be considered a potential hazard to aviation. The maps cover different wind turbine tip heights (20 m to 200 m).

Renewable UK maintains a website (Renewable UK, 2011a) giving information about wind turbines and aviation, including a series of maps from NATS, MOD and RESTATS (RESTATS, 2011) that show the consultation zone areas in the UK for which NATS requires notification of wind turbine planning applications.

There have been a number of studies (for example Jago and Taylor, 2002) aimed at clarifying the precise nature of the effects of wind turbines on radar – it seems the experience of a number of European countries is that the effect of wind turbines on military aviation was not a major problem.

The UK Government and the wind industry have funded a number of projects to address these issues and these seem to be arriving at solutions acceptable to the MOD and CAA. One involves adapting the design of wind turbine blades to include RAMs (radar absorbing materials). A joint project between QinetiQ and Vestas (Appleton, 2010) has tested a turbine equipped with a set of RAM blades in Norfolk and this 'stealthy' turbine has demonstrated a substantially reduced impact on radar. Another approach is the development of filtering systems that can reliably filter out the interference from wind turbines. One such system is BAE's ADT (Advanced Digital Tracking) system (Butler, 2007) and this appears to be another viable solution which may be able to be deployed at affected radar sites around the UK.

Impact on wildlife

In the UK, a collaboration between English Nature, the Royal Society for the Protection of Birds (RSPB), the World Wide Fund for Nature (WWF) and the British Wind Energy Association (BWEA) yielded a guidance document (English Nature et al., 2001) for nature conservation organizations and developers when consulting over wind farm proposals in England. This covers nature conservation, environmental impact assessments, the planning process and a checklist of possible impacts of relevance to nature conservation (both flora and fauna). Since this report was published Natural England, the Countryside Council for Wales and Scottish Natural Heritage have all developed extensive guidance on wind energy and wildlife.

In the case of offshore wind turbines, there are concerns about the possible impact on fish, crustaceans, marine mammals, marine birds and migratory birds and these are the subject of ongoing research by a number of organizations including Natural England, Scottish Natural Heritage, COWRIE (Collaborative Offshore Wind Research Into the Environment) and CEFAS (Centre for Environment, Fisheries and Aquaculture Science) amongst others.

Elsewhere, the US Fish and Wildlife Service has produced draft voluntary guidelines (USFWS, 2011) providing advice and recommendations for land-based wind energy projects to follow to avoid or minimise impacts on wildlife and habitats.

The impact of wind turbines on bats has been of particular concern and is discussed further below.

Wind turbines and birds

In addition to potential disturbance, barriers and potential habitat loss, the main potential hazard to birds presented by wind turbines is that they could be killed by flying into the rotating blades (Drewitt and Langston, 2006).

So far the worst location for bird strikes has been the Altamont Pass in California, where raptor species have been killed. However, apart from bird strikes on wind turbines at Tarifa and Navarra in Spain, this raptor mortality does not seem to have been duplicated elsewhere, so it may be due to special circumstances.

The American Bird Conservancy (ABC) reports that 100 000–440 000 bird collisions occur per year with wind turbines, 4–50 million with towers, 10–154 million with power lines, 10.7–380 million with roads/vehicles, over 31 million with urban lights and 100 million–1 billion with glass (ABC, 2011).

The US FWS guidelines (USFWS, 2010), mentioned above, includes guidance to avoid and minimize the impacts of wind turbines on birds. In addition, the US National Wind Coordinating Collaborative has developed a guide for studying wind turbines and wildlife interactions (NWCC, 2011), a mitigation tool box (NWCC, 2007) and a birds and bat factsheet (NWCC, 2010).

According to Natural England (2010), there is little evidence that wind farms in England have a significant impact on birds, but nonetheless Natural England and Scottish Natural Heritage provide guidance about wind turbines and birds, and post-construction monitoring of bird impacts.

Studies were carried out by Denmark's National Environmental Research Institute on the offshore wind farm at Tunø Knob (which was deliberately located where there was a large marine bird population, in order to monitor the interaction of birds – mainly eiders – and wind turbines). The institute's conclusions were that the eiders keep a safe distance from the turbines but are not scared away from their foraging areas – it was felt that the offshore wind turbines had no significant impact on water birds. (NERI, 1998).

Nonetheless there are concerns that a substantial increase in the number of wind turbines could result in an increase in bird strikes, so projects will need to take account of bird sensitivity areas and to take particular care when locating turbines in bird migratory routes. The RSPB has produced a map of wildlife sensitive areas in Scotland that can be used to inform developers about the appropriate siting of wind turbines. It is also now possible to install bird control radar systems which automatically detect approaching birds and, if there is a likelihood of collisions, bird deterrent devices can be activated, or the turbines shut down until after the birds have passed.

Wind turbines and bats

There is growing concern that wind turbines may have an impact on bats – particularly along migration routes.

Natural England has produced interim guidance (Natural England, 2009a and 2009b) to help planners and wind turbine operators take account of potential impacts to bats when developing or assessing wind turbine developments. At the time of writing, the UK Department of Agriculture, Food and Rural Affairs (DEFRA) is funding research at Exeter University into such impacts.

The UK Bat Conservation Trust (BCT) has produced a scoping report (BCT, 2009) with regard to bats and wind farms, but has also raised specific concerns about bats and micro wind turbines. BCT have released a Position Statement (BCT, 2010) to raise awareness of the potential hazards.

Public attitudes to wind power/planning considerations

Additional environmental factors that should be considered in assessing the impact of a wind turbine installation include safety and shadow flicker (ODPM, 2004b), however there are also concerns about the visual aspect of turbines.

Visual impact and attitudes

The visual perception of a wind turbine or a wind farm is determined by a variety of factors. These will include physical parameters such as turbine size, turbine design, number of blades, colour, the number of turbines in a wind farm, the layout of the wind farm and the extent to which moving rotor blades attract attention. Figure 7.34 compares wind turbines and other large constructions in the UK.

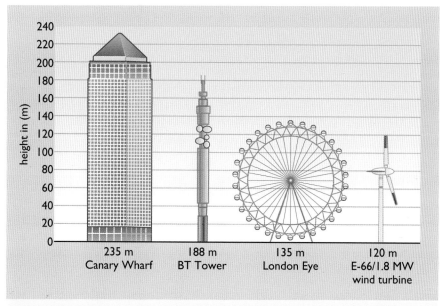

Figure 7.34 Comparison of wind turbines and other structures in the landscape

'Flicker' may be caused by sunlight interacting with rotating blades on sunny days. A comprehensive peer-reviewed study commissioned by DECC into wind-turbine related flicker showed that 'the frequency of flickering caused by wind turbines is such that it should not cause a significant risk to health' (PB, 2011).

An individual's overall perception of a wind energy project will also depend on a variety of less easily defined psychological and sociological parameters. These may include the individual's level of understanding of the technology, opinions on what sources of energy are desirable, and his or her level of involvement with the project. Newspaper and television reports are often the only source of information to which many people have access about wind energy, and these may well influence their opinions on the subject.

Much of the controversy about wind energy development has been due to opposition to changes to the *visual* appearance of the landscape. However, whether this is due to a visual dislike of wind turbines specifically, or simply to a general dislike of changes in the appearance of the landscape is often unclear. Resistance to the visual appearance of new structures or buildings is not a new phenomenon and opinions often change once structures become familiar.

Since the 1990s, over 60 independent survey projects have been carried out to monitor public attitudes and they have consistently shown that on average 70% to 80% support the development of wind farms in the UK (see for example NOP, 2005 and YouGov, 2010). However, there is still opposition to change and it is important for projects to be well designed and planned, and also to engage with local communities to provide trusted and reliable information together with meaningful community benefits.

Planning

Planning controls have a major influence on the deployment of wind turbines and some local authorities have developed policy guidelines about the planning aspects of wind energy. However, the planning aspects of wind energy have been treated differently in different locations.

The UK Government included planning guidance for wind energy in Planning Policy Statement 22 (PPS 22) (ODPM, 2004a), however, with the proposed National Planning Policy Framework, (DCLG, 2011) – in draft form at the time of writing – PPS 22 will probably be superseded, though it and its companion guide (ODPM, 2004b) do provide useful guidance.

Guidelines for developers and planners have also been prepared by Natural England, Scottish Natural Heritage and the Countryside Council for Wales.

A useful wind energy planning conditions guidance note was produced for the UK Department of Business, Enterprise and Regulatory Reform (BERR) by TNEI Services (TNEI, 2007). This summarizes the various factors and conditions that are likely to affect the granting of planning permission for a wind energy project.

Small and micro wind turbines can be installed under 'permitted development' regulations (see MCS, 2011a).

7.7 **Economics**

Calculating the costs of wind energy

The economic appraisal of wind energy involves a number of specific factors. These include:

- the annual energy production from the wind turbine installation;
- the capital cost of the installation;
- the discount rate being applied to the capital cost of the project (see Chatper 10 and Appendix B);
- the length of the contract with the purchaser of the electricity being produced;
- the number of years over which the investment in the project is to be recovered (or any loan repaid), which may be the same as the length of the contract;
- the operation and maintenance costs, including maintenance of the wind turbines, insurance, land leasing, offshore leasing etc.

A simple procedure for calculating the cost of wind energy is given in Appendix B. More information on costing and investing in energy projects can be found in Everett et al. (2012).

The estimates for land-based operation and maintenance costs are quite varied but seem to be equivalent to 3 to 4.5% of capital cost per year, e.g. approximately £40/kW to £50/kW per year (Renewable UK, 2010a). It is extremely difficult to predict the operation and maintenance costs for offshore wind energy projects although one source (McMillan and Ault, 2007) estimates that they could be of the order of 3 to 5 times land-based costs, but this has yet to be confirmed.

Greenacre et al. (2010) gives some indication of the difficulty in predicting offshore O&M costs as these may be influenced by a range of factors that include location, distance offshore, water depth, turbine redundancy and reliability, remote condition monitoring, uncertain 'weather windows' (especially in winter), material supply chains, currency exchange rates and vessel availability.

As we have seen, the annual energy produced by a wind turbine installation depends principally on the wind speed–power curve of the turbine, the (hub height) wind speed frequency distribution at the site, and the availability of the turbine.

As already noted in earlier chapters *capacity factor* is widely used to describe the productivity of a power plant over a given period of time.

If a wind turbine were able to operate at full rated power throughout the year, it would have an annual capacity factor of 100%. However, in reality, the wind does not blow constantly at the full-rated wind speed throughout the year, so in practice a wind turbine will have a much lower capacity factor. On moderate land-based wind sites in the UK, with annual mean wind speeds equivalent to half the rated wind speed, a turbine capacity factor of 0.25 (i.e. 25%) is typical. However, on better land-based wind sites, such as Carmarthen Bay in Wales, St Austell in Cornwall or the Orkney Islands, capacity factors of 0.35–0.40 or more are achievable.

The capital cost of large and medium scaled land-based wind turbines currently ranges from approximately £1300 to £1600 per kilowatt of output (Renewable UK, 2010a, but see also Renewable UK, 2010b). With 15 to 20 year contracts and on sufficiently windy sites, wind energy may be competitive with conventional forms of electricity generation, if the costs of the latter are calculated on a comparable basis. The cost of wind-generated electricity is very dependent on the way the plant is financed and this can strongly affect the price of the electricity produced.

As the cost of wind energy does not include the cost of fuel, it is relatively straightforward to determine, compared with the cost of energy from fuel-consuming power plants which are dependent on estimates of future fuel costs. High or escalating fuel prices tend to favour zero (or low) fuel cost systems such as wind energy, but steady or falling fuel prices are less favourable to them.

Wind turbines (on land) are very quick to install, so they can be generating before they incur significant levels of interest on the capital expended during construction – in contrast to many other highly capital-intensive electricity generating plant (e.g. large hydro stations, tidal barrages and nuclear power stations).

7.8 Commercial development and wind energy potential

Wind energy developments worldwide

The present healthy state of the wind energy industry is due largely to developments in Denmark and California in the 1970s and 1980s, and Germany in the 1990s and 2000s.

In Denmark, unlike most other European countries that historically employed traditional windmills, the use of wind energy never ceased completely, largely because of the country's lack of fossil fuel reserves and because windmills for electricity generation were researched and manufactured from the nineteenth century until the late 1960s. Interest in wind energy took on a new impetus in the 1970s, as a result of the 1973 'oil crisis'. Small Danish agricultural engineering companies then undertook the development of a new generation of wind turbines for farm-scale operation.

It was California, however, that gave wind energy the push needed to take it from a small, relatively insignificant industry to one with the potential for generating significant amounts of electricity. A rapid flowering of wind energy development took place there in the mid-1980s, when wind farms began to be installed in large numbers. As a result of generous environmental tax credits, an environment was created in which it was possible for companies to earn revenue both from the sale of wind-generated electricity to Californian utilities and from the manufacture of wind turbines. The new Californian market gave Danish manufacturers an opportunity to develop a successful export industry, taking advantage of the experience acquired within their home market.

Since the 1980s, Europe has taken the lead in wind energy, with over 86 000 MW of wind generating capacity (over 44% of the world total) installed by the end of 2010. Germany in particular has been in the vanguard of deployment in Europe to date and by the end of 2010 had installed over 27 000 MW. By the end of 2010, China had achieved the world's largest wind energy capacity, with over 42 000 MW installed. The USA has the next largest with over 40 000 MW installed.

In the UK, by contrast, progress has been more modest. Over 3100 grid connected wind turbines had been installed by the end of 2010, representing a combined capacity of over 5200 MW, generating enough electricity for 2.9 million homes and offsetting some 5.8 million tonnes of CO_2 per year (Renewable UK, 2011c). Figure 7.35 shows the location of wind energy projects in the UK at the time of writing.

Details of levels of wind generating capacity installed in various countries and continents in 2010 are given in Table 7.2.

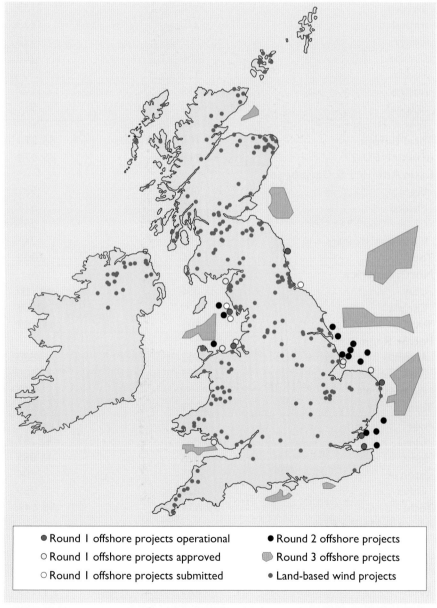

● Round 1 offshore projects operational   ● Round 2 offshore projects
○ Round 1 offshore projects approved   ◖ Round 3 offshore projects
○ Round 1 offshore projects submitted   • Land-based wind projects

Figure 7.35 Location of wind energy projects in the UK (source: Renewable UK, 2011c)

Table 7.2 Breakdown of Global Wind Power Capacity in 2010

	Power/MW	
	Start 2010	**End 2010**
Europe		
Germany	25 777	27 214
Spain	19 160	20 676
Italy	4849	5797
France	4574	5660
UK	4245	5204
Denmark	3465	3752
Portugal	3357	3702
Netherlands	2223	2237
Sweden	1560	2163
Ireland	1310	1428
Turkey	801	1329
Greece	1086	1208
Poland	725	1107
Austria	995	1107
Belgium	563	911
Rest of Europe	1611	2677
Total Europe	**76 300**	**86 075**
of which EU-27	**74 919**	**85 074**
Latin America & Caribbean		
Brazil	606	931
Mexico	202	517
Chile + Costa Rica	291	295
Caribbean + others	207	265
Total	**1306**	**2008**
North America		
USA	35 086	40 180
Canada	3319	4009
Total	**38 405**	**44 189**
Pacific Region		
Australia	1712	1880
New Zealand	497	506
Pacific Islands	12	12
Total	**2221**	**2397**
Asia		
China	25 805	42 287
India	10 926	13 065
Japan	2085	2304
Other	823	985
Total	**39 639**	**58 641**
Africa & Middle East		
Egypt	430	550
Morocco + Tunisia	307	400
Others	129	129
Total	**866**	**1,079**
World Total	**158 738**	**194 390**

Source: GWEC, 2011

Small-scale wind turbines

Small-scale wind turbines are more expensive per kilowatt of capacity than medium-scale wind turbines. In most cases, the cost of the power they produce is not competitive with mains electricity (except in remote areas) without support schemes. The need for batteries also tends to greatly increase the cost of such systems. However, with the wider deployment of feed-in tariffs, such as those in place in certain European countries and introduced in the UK in 2010, there may be fresh opportunities for carefully sited small wind systems that are grid-linked, avoiding the need for batteries.

The steady demand from people interested in obtaining electricity from pollution-free sources, or who are in locations where conventional supplies are not available, already provides enough support to sustain a significant number of manufacturers of small-scale wind turbines throughout the world (Figures 7.1 and 7.36).

Figure 7.36 Small-scale wind turbine

Planning controls are being relaxed in the UK for so-called micro wind turbines and, subject to a number of constraints in terms of maximum size, maximum height and maximum noise etc, these micro wind turbines are to be considered 'permitted development' (that is they do not in general require planning permission). However the maximum heights permitted are unfortunately too low for satisfactory output from small wind turbines, as these turbines would invariably require relatively tall towers to be productive. There is therefore a risk that inappropriate installations of micro/small wind turbines will occur as a result of the current permitted development proposals.

Recent experience of ill-conceived micro and small wind turbines (both horizontal axis and vertical axis designs) and inappropriate use of the

NOABL wind speed database (principally in urban and suburban locations) and inappropriate siting has, perhaps not surprisingly, damaged the reputation for micro/small scale turbines in the UK.

To improve the situation Renewable UK has introduced a small wind turbine standard (BWEA, 2008). This requires manufacturers of micro/ small wind turbines with swept areas of 200 m^2 or less to provide consistent, independently verified data including the BWEA Reference Power output at 11 m s^{-1} and a standard BWEA RAE (discussed in Section 7.5). In addition there is also a UK Microgeneration Scheme standard (and website, MCS, 2011b) for micro and small wind turbines (MCS 006), (see MCS, 2009) and a Microgeneration Installation Standard (MIS 3003) (see MIS, 2011) for installing wind turbines that utilize the UK feed-in tariff.

However there is also a need for special care when assessing the wind characteristics of the site for a micro/small wind turbine. If possible, wind speed measurements should be carried out, together with a survey and site visit, to filter out obviously inappropriate sites, e.g. those surrounded by trees, large numbers of buildings or other obstacles to wind flow.

Local community and co-operatively owned wind turbines

Another type of wind energy development that is gaining support is that of local community wind turbines. This can take a variety of forms. In Denmark it usually involves a group of people from a local community buying a turbine or group of turbines. The local community benefits from the sale of the electricity produced, or makes use of it for its own purposes.

This approach can encourage a positive attitude towards wind energy in communities that might be opposed to commercial wind energy developers from outside the area. A number of organizations have attempted to develop such projects in the UK, and there are now a number of innovative projects and several more are being planned. The first community-owned wind farm (Baywind) has been operating successfully in Cumbria since the 1990s. There are a number of other community wind farms in the UK, with most in Scotland – including some which have brought useful income to Scottish island communities such as those on the Isle of Skye and the Isle of Gigha.

One recent innovative project is the Westmill Wind Farm Cooperative (Westmill Wind Farm Cooperative, 2011) in Oxfordshire that has five 1.3 MW wind turbines (Figure 7.37). In 2008 this was the UK's largest community-owned wind farm.

It is being increasingly recognized that local people are likely to be more supportive of community wind turbines, and interest in the concept of the 'village wind turbine' or 'town wind farm' appears to be increasing.

Numerous single medium scale and large scale wind turbines are currently operating in the UK, either as community wind turbines, or in supplying factories, hospitals, supermarkets or housing projects.

(a)   (b)

Figure 7.37 (a) Westmill Wind Farm Co-operative, (b) opening ceremony.

Wind energy and buildings

The established wisdom of experienced designers and installers of small wind turbines for many years has been that 'wind turbines and buildings do not generally mix'. A small wind turbine should be located at a distance from the nearest building equivalent to around ten times the height of the building, and supported on a tower with a hub height equivalent to at least twice the height of the building. This is still good advice, but unfortunately it is not consistent with the UK conditions of permitted development for micro/small wind turbines. An example of what happens if this good advice is not followed is demonstrated by the poor performance of the building-mounted wind turbines monitored in the Energy Savings Trust's field trial of domestic wind turbines (EST, 2009).

In spite of the established wisdom, the UK government promoted the installation of building-mounted micro/small wind turbines, in part because of over-optimistic estimates from new manufacturers of building-mounted micro wind turbines, over-predicted wind speeds from NOABL in urban and suburban areas, and because a number of studies and reports made predictions of the likely electricity production from building-mounted wind turbines in these areas that were very optimistic. Whilst estimates for tower mounted wind turbines in these areas were also over optimistic, building mounted wind turbines experience higher levels of turbulence which further degrades productivity. Some building mounted-wind turbines are attached to the wall of a gable end of a building, which is similar to mounting a wind turbine near a cliff edge: a type of location discouraged by established manufacturers of small turbines. This raises structural integrity issues, and damage to the house is quite possible depending on the size of turbine. Other mounting approaches are possible, and some manufacturers give guidance (Udell et al. (2010) and suggest low cost mounting arrangements), but great care is needed.

If the building is designed to integrate the wind turbines, the turbines may be less affected by turbulence and damaging flows, but will still be dependent on the local wind speed conditions – although taller buildings may be able to take advantage of higher wind speeds one such example being the Strata 3 Tower in London. Such approaches are likely to be bespoke and not easily replicated on other buildings.

In addition there may be opportunities for using buildings (in appropriate locations) as a means of enhancing wind energy production by carefully designing the building form to accelerate wind velocities. In doing so it may be possible to reduce the size of wind turbine required for a given power output and to offset the topographical roughness effects which slow down local winds in urban and suburban environments.

The author of this chapter has patented an approach using one or more wing-like 'planar concentrators'. These can be used as free-standing systems, or as building-augmented wind energy systems when used in combination with a building surface such as a roof (as in the 'Aeolian Roof', Figure 7.38). Axial flow or cross-flow wind turbines are located between the planes, or between the planes and the building surface (e.g. roof or wall). For a given wind speed, this approach has demonstrated a significant increase in power output compared to the same wind turbine in normal non-augmented operation. Such systems can be deployed on both new and existing buildings and could generate a high proportion of the electricity requirements of appropriately oriented energy-efficient buildings.

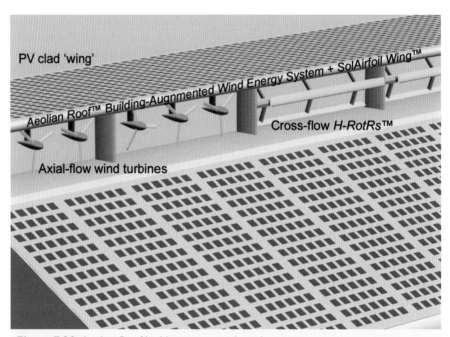

Figure 7.38 Aeolian Roof building-augmented wind energy systems

Wind energy potential

In an extensive study, Greenblatt (2005) scaled a previous estimate of the world land-based wind energy potential (Grubb and Meyer, 1993) at Class 4+ sites, to take account of taller towers and increasing hub heights (Class 4+ meaning Class 4 sites and above – see Figure 7.31). They arrived at a figure of 185 000 TWh per year.

With the advent of turbines capable of electricity generation from lower wind speed sites, Greenblatt (2005) also explored the wind energy potential if Class 3+ (i.e. including Class 3 sites and above) were included in the analysis, concluding that the global annual wind generated electricity resource could be around 335 000 TWh per year (Table 7.3).

Table 7.3 Available world 'land-based' wind resources and future electricity demand

Region of the world	Electricity demand by 2025 TWh y⁻¹	Installed capacity GW	Wind resource TWh y⁻¹ (Class 4+ sites)	Wind resource TWh y⁻¹ (Class 3+ sites)
North America	6700	18 700	62 400	93 500
Latin America	1800	6100	20 400	36 300
Europe	6200	15 200	50 500	92 500
Western	3100	4400	14 700	21 000
Eastern and Former Soviet Union	3100	10 800	35 800	71 300
Africa/Middle East	2200	10 400	34 700	71 300
Asia	8700	1900	6400	21 500
India	1300			
China	4300			
Other Asia	3100			
Australia/Oceania	400	3200	10 700	20 200
World Total	**26 000**	**70 400**	**185 000**	**335 400**

Note: Class 4+ = Class 4 and above; Class 3+ = Class 3 and above
Source: Greenblatt, 2005

The Global Climate and Energy Project at Stanford University carried out an evaluation of global potential of wind energy using five years of data from the US National Climatic Data Center and the Forecasts Systems Laboratory (Archer and Jacobson, 2005). This study estimated the global wind speeds at 80 metres above ground level (based on wind speed data for the year 2000) and found that using only 20% (e.g. 14.4 TW or 10 800 Mtoe) of the potential viable land-based resource wind energy could satisfy the global electricity demand (given at that time as 1.6 to 1.8 TW) *seven times over*.

In 2010 the GWEC (Global Wind Energy Council) produced a series of global wind energy outlook scenarios (GWEC, 2010) to examine the future potential for wind energy up to 2020, 2030 and 2050. These were based on three scenario assumptions:

(1) a reference scenario (RS) based on the projections in the International Energy Agency 2009 World Energy Outlook (IEA, 2009)

(2) a moderate scenario (MS) which takes into account policy measures to support renewable energy and targets either enacted or in the planning stages around the world

(3) an advanced scenario (AS) which has more ambitious assumptions based on an estimate of the extent to which the wind industry could grow in a best case 'wind energy vision'.

Figure 7.39 shows the predicted increases in global cumulative wind power capacity based on these scenarios up to 2030.

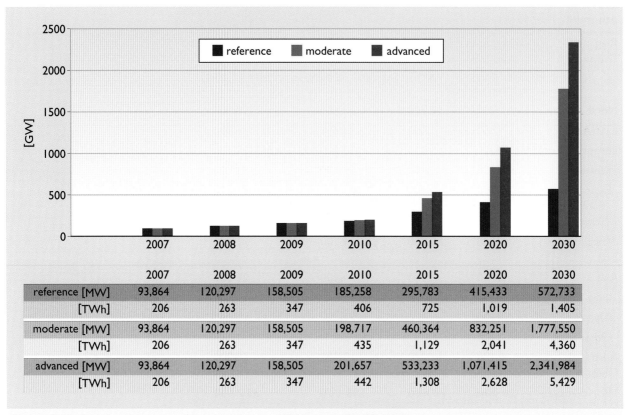

	2007	2008	2009	2010	2015	2020	2030
reference [MW]	93,864	120,297	158,505	185,258	295,783	415,433	572,733
[TWh]	206	263	347	406	725	1,019	1,405
moderate [MW]	93,864	120,297	158,505	198,717	460,364	832,251	1,777,550
[TWh]	206	263	347	435	1,129	2,041	4,360
advanced [MW]	93,864	120,297	158,505	201,657	533,233	1,071,415	2,341,984
[TWh]	206	263	347	442	1,308	2,628	5,429

Figure 7.39 Global cumulative wind power capacity to 2030 (source: GWEC, 2010)

European potential

The EEA (European Environment Agency) carried out a detailed land/sea use analysis of the land-based and offshore wind energy potential within Europe (EEA, 2009). This analysis estimated the technical potential (TP) that could be generated in 2020 and 2030 assuming 80 m hub height on land and 120 m hub height offshore. This was then filtered to exclude environmentally sensitive areas plus a number of offshore constraints, such as shipping lanes etc, and zoning to yield the 'constrained potential' (CP). The 'economically competitive potential' (ECP) was calculated on the basis of projected costs of developing and running wind farms in 2020 and 2030 derived from EC, 2009. The 2009 EEA report estimates the Economically Competitive Potential for 2030 as 27 000 TWh per year from land based and 3 400 TWh per year from offshore projects. Table 7.4 gives a summary of the TP, CP and ECP estimated on this basis for 2020 and 2030.

The EWEA (European Wind Energy Association) commissioned a study (Zervos and Kjaer, 2009) to establish wind energy targets for Europe for

Table 7.4 Projected technical, constrained and economically competitive potential for European wind energy development in 2020 and 2030

		Year	TWh	Share of 2020 and 2030 demand (*)
Technical potential	Onshore	2020	45 000	11–13
		2030	45 000	10–11
	Offshore	2020	25 000	6–7
		2030	30 000	7
	Total	2020	70 000	17–20
		2030	75 000	17–18
Constrained potential	Onshore	2020	39 000	10–11
		2030	39 000	9
	Offshore	2020	2 800	0.7–0.8
		2030	3 500	0.8
	Total	2020	41 800	10–12
		2030	42 500	10
Economically competitive potential	Onshore (*)	2020	9 600	2–3
		2030	27 000	6
	Offshore	2020	2600	0.6–0.7
		2030	3400	0.8
	Total	2020	12 200	3
		2030	30 400	7

Note: (*) European Commission projections for energy demand in 2020 and 2030 (EC2008 a, b)are based on two scenarios: 'business as usual' (4 078 TWh in 2020 – 4 408 TWh in 2030) and 'EC Proposal with RES trading' (3 537 TWh in 2020 – 4 279 TWh in 2030). The figures here represent the wind capacity relative to these two scenarios, e.g. onshore capacity of 45 000 TWh in 2020 is 11–12.7 times the size of projected demand.
(*) These figures do not exclude *Natura* 2000 areas (which form an EU-wide network of nature protection areas).
Source: EEA, 2008

2020 and 2030. The study assumed low and high scenarios for 2020. The low scenario for EU27 countries is predicted to achieve installed capacities of 190 GW (land-based) and 40 GW (offshore), with 210 GW (land-based) and 55 GW (offshore) capacity for the high scenario. The total 2020 low scenario capacity is predicted to generate 589.1 TWh per year and the total for the 2020 high scenario capacity is predicted to be 683 TWh per year. The EWEA also projects that 50% of electricity in the EU could eventually come from wind energy (EWEA, 2010).

A European Commission report (EC, 2009) which refers to investing in technologies for the Strategic Energy Technologies plan (SET-Plan (EU, 2010)) states that 'With additional research efforts, and crucially, significant progress in building the necessary grid structure over the next ten years, wind energy could meet one fifth of the EU's electricity demand in 2020, one third in 2030 and half by 2050', (see also Zervos and Kjaer, 2009). Achieving this would require achieving 400 GW wind energy capacity in 2030 and 600 GW capacity in 2050 with the majority (350 GW) of the 2050 capacity coming from offshore turbines. Table 7.5 gives a summary of the wind energy capacity needed to meet the European Commission's SET-Plan targets.

Table 7.5 Wind energy capacity needed to meet the European Commission's SET-plan targets

	Onshore wind (GW)	Offshore wind (GW)	Total wind energy capacity (GW)	Average capacity factor onshore	Average capacity factor offshore	TWh onshore	TWh offshore	TWh Total	EU-27 gross electricity consumption*	Wind power's share of electricity demand
2020**	210	55	265	26.0%	42.3%	479	204	683	3494	20%
2030	250	150	400	27.0%	42.8%	592	563	1155	3368	34%
2050	250	350	600	29.0%	45.0%	635	1380	2015	4000	50%

* Electricity demand assumes the European Commission's New Energy Policy $100 oil/barrel scenario until 2020 and High Renewables/Energy Efficiency scenario for 2030. Demand in 2050 is assumed to be 4000 TWh.
** Assuming 265 GW by 2020 in accordance with EWEA's 'high' scenario combined with the European Commission's 'New Energy Policy' Assumption for demand.
Source: Zervos and Kjaer, 2009

7.9 Offshore wind energy

The capital costs of energy from offshore wind farms are generally higher than those of onshore installations because of the extra costs of civil engineering for substructure, higher electrical connection costs and the higher specification materials needed to resist the corrosive marine environment.

However, offshore wind speeds are generally higher and more consistent than on land (apart from certain mountain and hill tops) and test results from the Tunø Knob offshore wind farm in Denmark indicate that actual output is 20–30% higher than estimated from wind speed prediction models. Availability was also higher than expected with an average of 98% being achieved, though this may not necessarily be typical. These wind energy characteristics, together with likely reductions in offshore costs as experience is gained in this environment, are expected to make offshore wind energy costs competitive in the medium to long term (although it should also be noted that capital costs doubled in the five years up to 2009 (Willow and Valpy, 2011)). In deeper water, further offshore, capital and operational costs will be higher, but it is anticipated that the increased energy yield, particularly from larger rotors (offshore, it is more feasible to utilize very large-scale wind turbines than it is on land, see Box 7.5) will more than compensate for these additional costs (Willow and Valpy, 2011).

Europe is the world's leader in offshore wind, having installed 1136 offshore turbines (over 2946 MW in 45 wind farms) with grid connections to nine European countries by the end of 2010, in shallow waters – depths mainly up to 30–45 m (EWEA, 2011a). EWEA forecasts that between 1000 and

BOX 7.5 Very large turbines

It is more feasible to utilize very large-scale wind turbines offshore than on land. This may improve economic viability, as more energy can be captured from a single platform and this can have benefits in terms of reduced maintenance costs. However, working against these benefits is the tendency to increasing capital costs associated with turbines above approximately 2 MW rated capacity. The latter effect is due to the scaling laws for strength and weight of materials with increasing rotor swept area, together with the much higher torque loading experienced by gearboxes.

5 MW offshore wind turbines have been installed by Repower, who are also developing 6 MW variants, Enercon is demonstrating 6 and 7.5 MW direct drive turbines without gear boxes (although at the time of writing these have only been used on land), Alstom Wind and Siemens are each developing 6 MW turbines, Vestas is trialling a 7 MW turbine and designs for 15 MW turbines are being considered for offshore applications by both GE and Gamesa.

Figure 7.40 E126 7.5 MW 126 m diameter Enercon wind turbine, an example of very large turbine (source: Enercon, 2011)

It is uncertain how much larger horizontal axis turbines can be scaled. This will involve developing blades which are lighter and have improved fatigue resistance, because when HAWTs are built to the sizes necessary to achieve very high power ratings, the reversing gravity loads on the rotor (as the blades move up and downwards during their rotation cycles) become a significant structural limitation.

Similarly there may be an increase in the cyclic impact on large HAWT wind turbine blades due to higher levels of 'wind shear' that occur with large rotor diameters. As wind speed increases with height, the blade tip of a large diameter HAWT will move through wind velocities that vary in magnitude between the upper and lower parts of the rotation cycle and the larger the diameter the greater is the difference in wind speed experienced. Design feasibility studies are underway to explore whether larger HAWTs up to 20 MW are possible and to explore the benefits of material substitution together with improvements to the structural design of blades.

Vertical axis wind turbines do not experience the gravity-driven fatigue loading encountered in large HAWTs. VAWT blades are also not as greatly affected by wind shear generated cyclic loading, and towers for some large swept area VAWT designs do not have to be built as tall in order to provide sufficient clearance above the sea level – thus reducing overturning moments. So in principle, VAWTs could be scaled-up to potentially very large sizes.

1500 MW of new offshore wind energy capacity will be fully connected in Europe in 2011. Ten new wind farms totalling 3 GW are currently under construction and when completed, Europe's installed offshore wind capacity will have grown to 6200 MW. A further 19 000 MW are currently fully consented (EWEA, 2011b).

Over 140 GW of offshore wind energy projects have been proposed or are being developed by European developers, which seems to indicate that the EWEA's targets of 40 GW by 2020 and 150 GW by 2030 could be realistic (EWEA, 2011b).

Figure 7.41 The Hywind floating 2.3 MW wind turbine

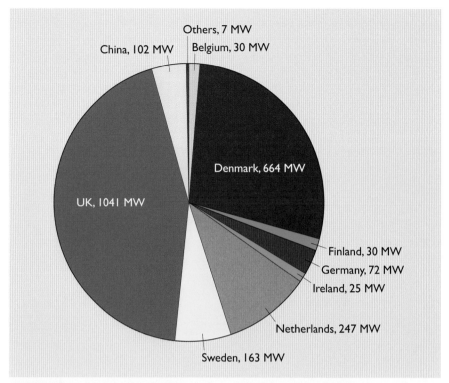

Figure 7.42 Global installed offshore wind energy capacity by country (January 2010) (sources: 4C Offshore Ltd, 2010; Musial and Ram, 2010)

The 2009 EEA study (EEA, 2009) mentioned previously estimates that the 'economically competitive' European offshore wind energy potential 'in 2020 is 2600 TWh y^{-1}, equal to between 60% and 70% of projected electricity demand, rising to 3400 TWh y^{-1} in 2030 equal to 80% of the projected EU electricity demand.' EEA, 2009, also estimates the technical potential of offshore wind energy to be 'seven times greater than the projected EU electricity demand in 2030'.

When realized, the ambitious programmes for substantial offshore projects established by several European countries will mean that wind power will ultimately become a major provider of electricity in those countries.

One interesting development which has taken place in Norway was an innovative floating 2.3 MW wind turbine (Figure 7.41), known as the Hywind project, that was successfully operated in a sea water depth of 220 m. This has significant implications for expanding the potential offshore wind energy beyond the shallow continental shelf sites so far developed.

Whilst the USA has the second largest installed land-based wind energy capacity, it has no offshore wind energy capacity at the time of writing. However a major study of the offshore wind energy potential around the USA was completed by the National Renewable Energy Laboratory (NREL). The US offshore wind energy resource gross capacity is estimated to be around 4000 GW (assuming one 5 MW turbine at 1 km × 1 km spacing, at sites which have 7 m s^{-1} annual mean wind speeds or more – no account being taken of potential constraints on turbine location). This is 'roughly

BOX 7.6 Thanet Offshore Wind Farm

The world's largest single offshore wind farm at the time of writing (Figure 7.45) began officially operating in September 2010 at a site 12 km off Foreness Point in Thanet, Kent and forms the second wind farm (of potentially a total of five) in the Thames Estuary after Kentish Flats Wind Farm. The Thanet wind farm cost £780 million, took two years to construct and was completed in June 2010.

The project consists of 100 Vestas V90-3 turbines each rated at 3 MW, giving a total installed capacity of 300 MW. The 90 metre diameter turbines are spaced 500 m apart in one direction and 800 m apart in the other, over an area of 35 km², in a water depth of 20 to 25 m. Each turbine has a hub height above sea level of approximately 70 m. The wind farm is estimated to generate an electricity output equivalent to the consumption of more than 200 000 UK homes.

Figure 7.43 The Thanet Offshore Wind Farm (a) location, (b) view of the wind farm and (c) view of turbines (source: Vattenfall, 2011)

equivalent to four times the generating capacity currently carried on the US grid' (Musial and Ram, 2010).

About 20 US offshore wind energy projects (with a combined capacity of over 2000 MW) were in the planning and permitting process as of September 2010, with the majority of these being planned for the north east

Table 7.6 Global offshore wind energy development in permitting and under construction stages

Country	Permitting, approved or under construction (MW)	In operation (MW)
Belgium	1194	30
Canada	1826	0
China	201	102
Denmark	653	664
Estonia	1000	0
Finland	1306	30
France	1455	0
Germany	25411	72
Greece	1101	0
Ireland	1530	25
Italy	2526	0
Japan	0	1
Maldives	75	0
Netherlands	3969	247
Norway	565	2
Romania	500	0
Spain	70	0
Sweden	3346	163
United Kingdom	6085	1041
United States	~2000	0
Total	54813	2377

Sources: Musial and Ram, 2010; 4C Offshore, 2010

and mid-Atlantic regions of the US coastline, with some being considered at the Great Lakes, the Gulf of Mexico and along the Pacific Coast (Musial and Ram, 2010).

At the time of writing the total current global offshore wind energy capacity is 2377 MW installed, with another 54 813 MW in permitting, approval or construction stages and, according to the Global Offshore Wind Farm Database (4C Offshore, 2011), over 900 offshore wind farms are being planned in 36 countries around the world. Table 7.6 and Figure 7.42 summarize the global offshore wind projects at the end of 2009.

Offshore wind energy in the UK

With over 1300 MW offshore wind energy capacity installed (from 436 offshore wind turbines) by the end of 2010, the UK has the world's largest offshore wind energy capacity at the time of writing.

In 2008, the UK BERR Department published an update to its marine renewable energy atlas (BERR, 2008). This includes a map of UK offshore annual mean wind speeds at 100 m height above sea level (Figure 7.44) within the 'UK Renewable Energy Zone'.

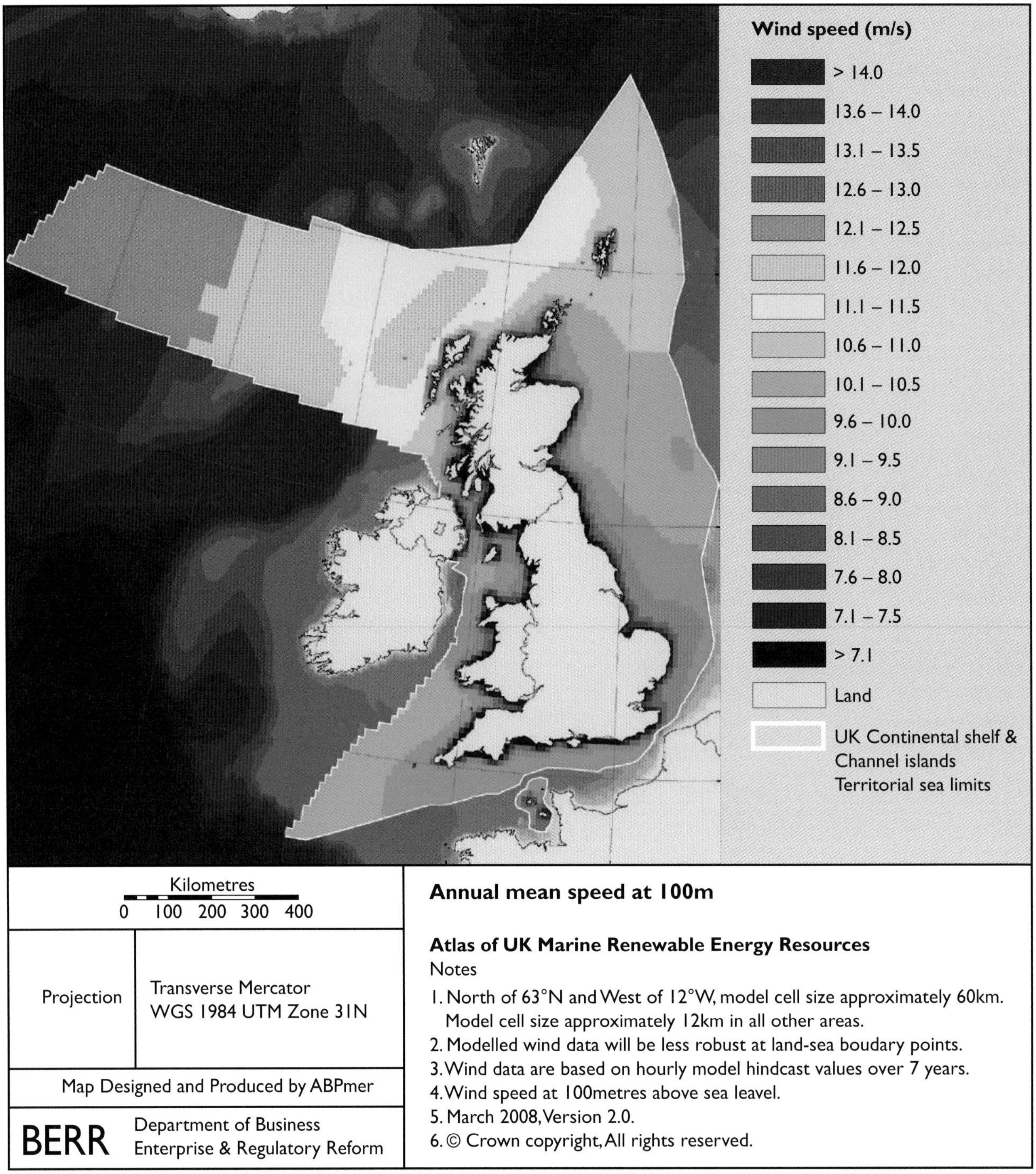
Wind speed (m/s)
> 14.0
13.6 – 14.0
13.1 – 13.5
12.6 – 13.0
12.1 – 12.5
11.6 – 12.0
11.1 – 11.5
10.6 – 11.0
10.1 – 10.5
9.6 – 10.0
9.1 – 9.5
8.6 – 9.0
8.1 – 8.5
7.6 – 8.0
7.1 – 7.5
> 7.1
Land
UK Continental shelf & Channel islands Territorial sea limits
Kilometres
0 100 200 300 400
Projection
Transverse Mercator WGS 1984 UTM Zone 31N
Map Designed and Produced by ABPmer
BERR Department of Business Enterprise & Regulatory Reform
Annual mean speed at 100m
Atlas of UK Marine Renewable Energy Resources
Notes
1. North of 63°N and West of 12°W, model cell size approximately 60km. Model cell size approximately 12km in all other areas.
2. Modelled wind data will be less robust at land-sea boudary points.
3. Wind data are based on hourly model hindcast values over 7 years.
4. Wind speed at 100metres above sea leavel.
5. March 2008, Version 2.0.
6. © Crown copyright, All rights reserved.

Figure 7.44 Annual mean wind speeds offshore at 100 m above sea level from the *Atlas of UK Marine Renewable Energy* (source: BERR, 2008)

The atlas resulted in a range of more detailed assessments of the UK offshore wind energy potential, including one from the Offshore Valuation Group (OVG, 2010). This evaluated all of the marine renewable energy resources available in the UK (including tidal and wave energy as well as offshore wind). It estimated that the total practical resource from offshore fixed wind turbines (i.e. turbines fixed to the seabed) was some 406 TWh per year (including those already installed). It also estimated the total practical resource from offshore floating wind turbines that could be deployed in deeper water to be 1533 TWh per year indicating a total combined output (fixed and floating wind turbines) of over 1900 TWh per year from offshore wind energy.

Since the first UK offshore wind energy project was installed at Blyth in 2000, many more offshore wind farms have been installed both in the UK and in other parts of Europe and much has been learnt, important experience and confidence gained to encourage a substantial growth both in successfully operating offshore turbines and attracting investment into substantial planned projects.

The UK government supported the initial offshore wind farms with capital grants and, through the Crown Estates Office (which is responsible for the area of seabed that surrounds the UK), has allocated licences for three rounds of offshore wind energy development (see Figure 7.35).

- Round One (announced in 2001). This was effectively a demonstration phase to allow offshore developers and wind turbine manufacturers to gain experience and to transfer expertise from the UK's offshore oil and gas industries. Round One sites had water depths less than 20 metres and were situated within 12 km of the coast. Of the seventeen Round One projects that were initially awarded, eleven are now generating electricity (with a total installed capacity of 962 MW).

- Round Two (announced in 2003). Fifteen sites were announced in this round giving a combined capacity of 7000 MW in 2003. Of these the first two (Gunfleet 2 and Thanet) are now fully operational bringing the UK's offshore wind energy capacity up to 1330 MW. Five further Round Two sites (Greater Gabbard, London Array, Sheringham Shoal and Walney 1) are under construction.

- Round Three (announced in 2010). This round is aimed at facilitating the development of 33 GW of generating capacity. The nine areas allocated for development are shown in Figure 7.36 and listed as follows in Table 7.7. Many of these are well advanced.

Table 7.7 Details of Round Three sites

Round 3 zone	Name	Area /km^2	Depth /m	Capacity /GW
Zone 1	Moray Firth	520	30 to 57	1.3
Zone 2	Firth of Forth	2852	30 to 80	3.5
Zone 3	Dogger Bank	8660	18 to 63	9
Zone 4	Hornsea	4735	30 to 40	4
Zone 5	Norfolk	6036	5 to 70	7.2
Zone 6	Hastings	270	19 to 62	0.6
Zone 7	West of Isle of Wight	723	22 to 56	0.9
Zone 8	Bristol Channel	949	19.5 to 60.9	1.5
Zone 9	Irish Sea	2200	28 to 78	4.2

Source: Derived from Crown Estate, 2008a

In 2009, the *UK Renewable Energy Strategy* (DECC, 2009), was published. This included the target of obtaining more than 30% of electricity from renewables by 2020. It projected that most of this would come from a combination of land-based and offshore wind energy projects.

This has the potential to be a step change in serious support for wind energy development in the UK. If and when these projects are successfully delivered, the UK would then have started to become a major generator of electricity from wind energy.

The UK Government 2050 Pathways Analysis (DECC, 2010) mentioned above, assumed a number of scenario trajectories and four levels of offshore wind energy deployment. Level four represents the most ambitious of these. As a result of evidence submitted in response to consultation (DECC, 2011b), DECC doubled its estimates, increasing the potential contribution from offshore wind from 430 TWh y^{-1} to 929 TWh y^{-1} of electricity by 2050 (assuming a 45% annual capacity factor). It also included the possibility of including floating wind turbines. In the December 2011 update of its Carbon Plan, DECC also expanded the proposed contribution from wind energy – both from land and offshore – envisaged in 2050. In this plan, it is not only envisaged that wind energy would be making a major contribution to electricity generation by 2050, but also contributing to low/zero carbon heating of buildings (via heat pumps) and also providing a major proportion of the electricity to power vehicles (in parallel to electrifying the car fleet) in 2050 (DECC, 2011c).

7.10 Summary

This chapter covers a wide range of aspects of wind energy, including the causes of winds, the energy and power contained within winds, and a historical review of wind energy technology from traditional windmills through to modern wind turbines, including how vertical axis and horizontal axis wind turbines work. The importance of aerodynamics in the design of efficient wind turbines is discussed.

The chapter includes a section on wind turbine power curves and the techniques employed for estimating the electricity production of wind turbines in particular locations.

The environmental impacts of wind turbines are also discussed, including potential issues related to noise, electromagnetic interference, aviation, wildlife, public attitudes and planning aspects. The economics of wind energy are reviewed, although Appendix B covers this in more detail.

There is an overview of the commercial development of wind energy and its future potential. This includes a review of the current global wind power installed capacity and current estimates of the European and global potential energy contribution from land-based wind turbines. Although the chapter mainly focuses on large scale projects, it also discusses small and medium scale wind turbines, as well as community wind energy projects and building-integrated wind turbines.

The final section provides a brief overview of offshore wind energy, offshore wind turbine technology (fixed and floating) and very large wind turbines. It also reviews the current status of offshore wind energy and summarizes some of the ambitious current proposals for offshore wind energy and its future potential in the UK, Europe, the USA and worldwide.

Wind energy continues to be one of the fastest growing energy technologies and is set to become a major generator of electricity throughout the world. Particularly in Europe, the offshore exploitation of wind energy is likely to become one of the most important means of reducing carbon dioxide emissions from the electricity sector. However, to achieve the potential of the EWEA's predicted 30% of 2030 EU electricity demand (or 50% of 2050 demand) will require considerable investment in the electricity grids, interconnection and in other infrastructure, but it does appear that there is strong motivation from governments and industry to facilitate this expansion.

Clearly there are still risks and uncertainties with the approach, but the growth of offshore wind energy developments in 20 years, from the first project at Vindeby, in Denmark, to the present, with over 900 offshore wind projects worldwide, is an impressive achievement – especially given the extra difficulties of deployment in wind-swept seas.

References

4C Offshore (2010) website (updated weekly) http://www.4coffshore.com (accessed 8 January 2010).

4C Offshore (2011) *Global Offshore Wind Farm Database, 4C Offshore Ltd.* (accessed March 2011) http://www.4coffshore.com/offshorewind/.

Abbott, I. H. and von Doenhoff, A. E. (1958) *Theory of Wing Sections*, New York, Dover Publications Inc.

ABC (2011) *American Bird Conservancy (ABC) policy on bird collisions*, http://www.abcbirds.org/abcprograms/policy/collisions/index.html (accessed 8 November 2011).

Appleton, S. (2010) *Stealth blades – a progress report*, http://www.all-energy. co.uk/UserFiles/File/25Appleton.pdf (accessed 8 November 2011).

Archer, C. L. and Jacobson, M. Z. (2005) 'Evaluation of global wind power', *J. Geophys Res.*, 110, D12110, http://www.agu.org/journals/jd/ jd0512/2004JD005462/ (accessed 8 November 2011).

AWEA (2009) *AWEA Small Wind Turbine Performance and Safety Standard*, AWEA 9.1, American Wind Energy Association.

Bacon, D. F. (2002) *Fixed-link wind turbine exclusion zone method, A proposed method for establishing an exclusion zone around a terrestrial fixed radio link outside of which a wind turbine will cause negligible degradation of the radio link performance*, Ofcom. http://licensing. ofcom.org.uk/binaries/spectrum/fixed-terrestrial-links/wind-farms/ windfarmdavidbacon.pdf (accessed 8 November 2011).

BBC (2011) *Wind Farm Assessment Tool*, http://www.bbc.co.uk/reception/ info/windfarm_tool.shtml (accessed 8 November 2011).

BCT (2009) *Determining the potential ecological impact of wind turbines on bat populations in Britain, Phase 1 report Final Report*, Scoping & Method Development Report, May. www.bats.org.uk/pages/wind_turbines.html and http://www.bats.org.uk/data/files/determining_the_impact_of_wind_ turbines_on_british_bats_final_report_29.5.09_website.pdf (accessed 9 December 2011).

BCT (2010) Position Statement: *Microgeneration Schemes: Risks, Evidence and Recommendations*, Bat Conservation Trust. http://www. bats.org.uk/publications_download.php/1042/Position_Statement_ Microturbines_2010.pdf (accessed 9 December 2011).

Berge, E., Gravdahl, A. R., Schelling, J., Tallhaug, L. and Undeheim, O (2006) *A comparison of WAsP and two CFD-models*, Presentation for EWEC 2006 http://www.windsim.com/documentation/papers_presentations/0602_ ewec/ewec_berge.pdf (accessed 9 December 2011).

BERR (2008) *Atlas of UK Marine Renewable Energy Resources*, Department for Business & Regulatory Reform, April http://www.renewables-atlas.info/ downloads/documents/Renewable_Atlas_Pages_A4_April08.pdf (accessed 9 December 2011).

Beurskens, J. and Jensen, P. H. (2001) 'Economics of wind energy – Prospects and directions', *Renewable Energy World*, July–Aug.

BSI (2003) *Wind turbine generator systems Part 11: Acoustic noise measurement techniques*, (BS) EN 64100-11:2003, London, British Standards Institution.

Burch, S. F. and Ravenscroft, F. (1992) *Computer Modelling of the UK Wind Energy Resource: Final Overview Report*, ETSU WN7055, ETSU.

Burroughs, W. J., Crowder, B., Robertson, E., Vallier-Talbot, E. and Whitaker, R. (1996) *Weather – the ultimate guide to the elements*, London, HarperCollins.

Butler, G. (2007) 'Advanced Digital Tracker – Update' at *All Energy '07 Aviation and Technical Workshop*, May http://www.all-energy.co.uk/UserFiles/File/2007GeoffButler.pdf (accessed 9 December 2011).

BWEA (2008) *BWEA Small Wind Turbine Performance and Safety Standard*, 29 February.

Carbon Trust (2011) *Wind Power Estimator*, Carbon Trust, http://www.carbontrust.co.uk/cut-carbon-reduce-costs/products-services/technology-advice/renewables/Pages/small-scale-wind-energy.aspx (accessed 8 November 2011).

Chignell, R. J. (1987) *Electromagnetic Interference from Wind Turbines – A Simplified Guide to Avoiding Problems*, National Wind Turbine Centre, National Engineering Laboratory, East Kilbride.

Crown Estate (2008a) *Round 3 offshore wind farms table*, The Crown Estate, June, http://www.thecrownestate.co.uk/interactive_map_round3_table.

Crown Estate (2008b) *Round 3 Offshore wind farms map*, The Crown Estate, http://www.thecrownestate.co.uk/media/158181/round3_map.pdf (accessed 8 November 2011).

Day, A., Dance, S., Moseley, T. and Dunlop, B. (2010) *Ashenden Wind Turbine Trial: Phase II Progress Report*, London, South Bank University.

DCLG (2007) *Domestic Installation of Microgeneration Equipment – Final report from a Review of the related Permitted Development Regulations*, London, Department for Communities and Local Government.

DCLG (2009) *Permitted development rights for small scale renewable and low carbon energy technologies, and electric vehicle charging infrastructure – Consultation,* London, Department for Communities and Local Government.

DCLG (2011) *Draft National Planning Policy Framework*, London, Department for Communities and Local Government.

DECC (2009) *The UK Renewable Energy Strategy*, London, Department of Energy and Climate Change, HMSO.

DECC (2010) *2050 Pathways Analysis*, London, Department of Energy and Climate Change.

DECC (2011a) Department of Energy and Climate Change's wind speed database website URL: http://www.decc.gov.uk/en/content/cms/meeting_energy/wind/windsp_databas/windsp_databas.aspx (accessed 9 December 2011) or http://www.bwea.com/noabl/index.html (accessed 9 December 2011).

DECC (2011b) *2050 Pathways Analysis – Response to Call for Evidence*, London, Department of Energy and Climate Change.

DECC (2011c) *The Carbon Plan: Delivering our low carbon future*, London, Department of Energy and Climate Change.

Drewitt, A. L. and Langston, R. H. W. (2006) 'Assessing the impacts of wind farms on birds', *British Ornithologists' Union*, vol. 148, pp. 29 to 42.

EC (2009) *Investing in the Development of Low Carbon Technologies (SET-Plan)*, European Commission.

EEA (2009) *Europe's on shore and offshore wind energy potential – An assessment of environmental and economic constraints*, EEA Technical Report No. 6/2009, European Environment Agency.

Eldridge, F. R. (1975) *Wind Machines*, Mitre Corporation.

English Nature, RSPB, WWF-UK, BWEA (2001) *Wind farm development and nature conservation*, WWF-UK.

EST (2009) *Location, location, location: Domestic small-scale wind field trial*, Energy Savings Trust.

EST (2011) *Domestic Wind Speed Prediction Tool*, Energy Savings Trust, http://www.energysavingtrust.org.uk/Generate-your-own-energy/Can-I-generate-electricity-from-the-wind-at-my-home (accessed 8 November 2011).

EU (2010) *The European Strategic Energy Technology Plan – SET-Plan – Towards a low carbon future*, European Union.

Everett, B., Boyle, G. A., Peake S. and Ramage, J. (eds) (2012) *Energy Systems and Sustainability: Power for a Sustainable Future* (2nd edn), Oxford, Oxford University Press/Milton Keynes, The Open University.

EWEA (1991) *Time for Action: Wind Energy in Europe,* European Wind Energy Association.

EWEA (2010) *2050: Facilitating 50% Wind Energy – Recommendations on transmission infrastructure, system operation and electricity market integration*, European Wind Energy Association.

EWEA (2011a) *List of Operational Wind Farms end 2010*, European Wind Energy Association.

EWEA (2011b) *The European offshore wind industry – key trends and statistics*, European Wind Energy Association, January.

EWEA (2011c) *Offshore wind,* www.ewea.org/index.php?id=203 (accessed 8 November 2011).

Fiumicelli, D. and Triner, N. (eds.) (2011) *Wind Farm Noise Statutory Nuisance Complaint Methodology* (NANR 277) AECOM for DEFRA.

FWC (2011) http://floatingwindfarms.com/index.html) (accessed 8 November 2011).

Golding, E. W. (1955) *Generation of Electricity by Wind Power*, London, E. and F. N. Spon.

Gorlov, A. (1998) *Development of the helical reaction hydraulic turbine*, Final Technical Report (DE-FG01-96EE 15669), Northeastern University for US Department of Energy.

Greenacre, P., Gross, R. and Heptonstall, P. (2010) *Great Expectations: The cost of offshore wind in UK waters – understanding the past and projecting the future*, London, UK Energy Research Centre.

Greenblatt, J. B. (2005) *Wind as a source of energy, now and in the future*, Amsterdam, Environment Defense for InterAcademy Council.

Grubb, M. and Meyer, N. (1993) 'Wind energy resources, systems and regional strategies' in Johansson, T. B. et al (eds) *Renewable Energy – Sources for Fuels and Electricity*, Earthscan, pp. 157–212.

GWEC (2010) *The Global Wind Energy Outlook 2010*, The Global Wind Energy Council.

GWEC (2011) *Global Wind Statistics 2010*, Global Wind Energy Council.

IEA (2009) *World Energy Outlook*, Paris, International Energy Agency.

IEC (2006) *Wind turbine generator systems – Part 11: Acoustic noise measurement techniques, IEC Standard 61400-11,* Geneva, International Electrotechnical Commission.

IPCC (2011) *Special Report of Renewable Energy Sources and Climate Mitigation (SEREN) - Summary Report*, Abu Dhabi, Inter-government Panel on Climate Change.

Jago, P. and Taylor, N. (2002) *Wind turbines and aviation interests – European Experience and Practice*, STAYSIS Ltd for the DTI, ETSU W/14/00624/REP (DTI PUB URN No 03/5151).

Kragh, J. et al. (1999) *Noise imission from wind turbines*, National Engineering Laboratory for the Energy Technology Support Unit (ETSU), ETSU W/13/00503/REP.

Legerton, M. (ed.) (1992) *Wind Turbine Noise Workshop Proceedings,* ETSU-N-123, Department of Trade and Industry/British Wind Energy Association.

Ljungren, S. (ed.) (1994) *Recommended practices for Wind Turbine Testing. 4. Acoustics Measurement of Noise Emission from Wind Turbines* (3rd edn), submitted to the Executive Committee of the International Energy Agency Programme for Research and Development on Wind Energy Conversion Systems.

Ljungren, S. (ed.) (1997) *Recommended practices for Wind Turbine Testing. 10. Measurement of Noise Emission from Wind Turbines*, submitted to the Executive Committee of the International Energy Agency Programme for Research and Development on Wind Energy Conversion Systems.

McMillan, D. and Ault, G. (2007) 'Quantification of Condition Monitoring Benefit for Offshore Wind Turbines' in *Wind Engineering*, vol. 31, no. 4, pp. 267–285.

MCS (2009) *MCS 006 Issue 1.5 – Product certification scheme requirements: Micro and Small Wind Turbines*, MCS Working Group 3 'Micro and Wind Systems' for DECC, http://www.microgenerationcertification.org/admin/documents/MCS%20006%20-%20Issue%201.5%20Product%20Certification%20Scheme%20Requirements%20-%20Micro%20and%20Small%20Wind%20Turbines%2010%20July%202009.pdf (accessed 9 December 2011).

MCS (2011a) *MCS 020 Planning standards for permitted development installations of wind turbines and air source heat pumps on domestic premises*, MCS for DECC, http://www.microgenerationcertification.org/admin/documents/MCS%20020%20Planning%20Standards%20Issue%201.0.pdf (accessed 9 December 2011).

MCS (2011b) *Microgeneration Certification Scheme* website: URL www.microgenerationcertification.org.

MIS (2011) *MIS 3003:Version 3 Microgeneration Installation Standard: Requirement for contractors undertaking the supply, design, installation, set to work commissioning, and handover of micro and small wind turbines.* For DECC http://www.microgenerationcertification.org/admin/documents/MCS%20006%20-%20Issue%201.5%20Product%20Certification%20Scheme%20Requirements%20-%20Micro%20and%20Small%20Wind%20Turbines%2010%20July%202009.pdf (accessed 9 December 2011).

Musgrove, P. J. (1990) 'Vertical axis WECS design' in Freris, L.L. (ed.) *Wind Energy Conversion Systems*, Prentice Hall.

Musial, W. and Ram, B. (2010) *Large Scale Wind Power in the United States – Assessment of opportunities and barriers*, National Laboratory for Renewable Energy.

Natural England (2009a) Technical Information Note TIN051: *Bats and onshore wind turbines – Interim guidance*, First Edition, February.

Natural England (2009b) Technical Information Note TIN059: *Bats and single large wind turbines: Joint Agencies interim guidance*, First Edition, September.

Natural England (2010) Technical Information Note TIN069: *Assessing the effects of onshore wind farms on birds*, First Edition, January.

Needham, J. (1965) *Science and Civilisation of China*, Cambridge University Press.

NERI (1998) *Impact Assessment of an Off-shore Wind Park on Sea Ducks*, National Environmental Research Institute (NERI) Technical Report No. 227, Denmark.

NOP (2005) *Survey of Public Opinion of Wind Farms*, NOP for BWEA, UK.

NPL (2011) *Wind Turbine Noise Model*, National Physical Laboratory, http://resource.npl.co.uk/acoustics/techguides/wtnm/ (accessed 9 November 2011).

NRC (2007) *Environmental Impacts of Wind Energy Projects*, Committee on Environmental Impacts on Wind Energy, Washington DC, National Academies Press.

NWCC (2007) *NWCC Mitigation Toolbox*, National Wind Coordinating Collaborative.

NWCC (2010) *Wind Turbine Interactions with Birds, Bats and their Habitats: A Summary of Research Results and Priority Questions*, National Wind Coordinating Collaborative.

NWCC (2011) *Comprehensive Guide to Studying Wind Energy/Wildlife Interactions*, National Wind Coordinating Collaborative.

ODPM (2004a) *Planning Policy Statement 22 (PPS22): Renewable Energy*, Office of the Deputy Prime Minister.

ODPM (2004b) *Planning for Renewable Energy – A Companion Guide to PPS22*, Office of the Deputy Prime Minister.

Ofcom (2009) *Tall structures and their impact on broadcast and other wireless services*, http://licensing.ofcom.org.uk/radiocommunication-licences/fixed-terrestrial-links/guidance-for-licensees/wind-farms/tall_structures/ (accessed 9 November 2011).

Ofcom (2011) *Ofcom website guidance on wind farms and electromagnetic interference*, http://licensing.ofcom.org.uk/radiocommunication-licences/fixed-terrestrial-links/guidance-for-licensees/wind-farms (accessed 9 March 2012).

OVG (2010) *The Offshore Valuation – A valuation of the UK's offshore renewable energy resource*, Offshore Valuation Group and Public Interest Research Centre, 108pp. Downloadable from http://offshorevaluation.org/downloads/offshore-valuation-full.pdf (accessed 26th June 2012).

PB (2011) *Update of UK Shadow Flicker Evidence Base – Final Report*, Parsons Brinckerhoff for Department of Energy and Climate Change.

Randhawa, B.S. and Rudd, R. (2009) *RF Measurement Assessment of Potential Wind Farm Interference to Fixed Links and Scanning Telemetry Devices*, ERA Report Number 2008-0568 (issue 3) Final Report, ERA Technology Ltd for Ofcom.

RESTATS (2011) *Aviation Safeguarding Maps*, https://restats.decc.gov.uk/cms/aviation-safeguarding-maps/&sa=U&ei=k_eaTb7SBoi7hAe49cDQBg&ved=0CBsQFjAC&usg=AFQjCNEYtrYHSkp2NDen0qGJsAVExTLupw (accessed 9 March 2012).

Renewable UK (2010a) *Wind Energy Generation Costs*, Fact Sheet 04, June, http://www.bwea.com/pdf/briefings/Wind-Energy-Generation-Costs.pdf (accessed 9 December 2011).

Renewable UK (2010b) *Onshore Costs/Benefit Study*, November, http://www.bwea.com/pdf/publications/RenewableUK_Onshore_costs.pdf (accessed 9 December 2011).

Renewable UK (2011a) *UK wind speed database* http://www.bwea.com/noabl/ (accessed 9 November 2011).

Renewable UK (2011b) *Aviation resources* http://www.bwea.com/aviation/aviation_resources.html (accessed 9 November 2011).

Renewable UK (2011c) *UKWED the UK Wind Energy Database*, www.bwea.com/ukwed/index.asp (accessed 9 November 2011).

Reynolds, J. (1970) *Windmills and Watermills*, London, Hugh Evelyn Ltd Publishers.

Risø (2009) *European wind resources over open sea*, http://www.windatlas.dk/europe/oceanmap.html (accessed 9 November 2011).

Rogers, A. L., Rogers, J. W. and Manwell, J. F. (2005) 'Comparison of the Performance of Four Measure-Correlate-Predict Algorithms' in *Journal of Wind Engineering and Industrial Aerodynamics*, vol. 93, no. 3, pp. 243–264.

Rogers, A. L., Rogers, J. W., Manwell, J. F. (2006a) 'Uncertainties in Results of Measure-Correlate-Predict Analysis', *European Wind Energy Conference*, February/March.

Rogers, A. L., Manwell, J. F., Wright, S. (2006b) *Wind Turbine Acoustic Noise – A White Paper*, Renewable Energy Research Laboratory, University of Massachusetts at Amherst, 2003, amended January 2006.

Rooij, R. van and Timmer, N. (2004) *Design of Airfoil for Wind Turbine Blades*, DUWIND, Delft University of Technology, The Netherlands.

South, P. and Rangi, R. (1972) *Wind tunnel investigation of a 14 feet diameter vertical axis windmill*, Report Number LTR-LA-105, National Aeronautical Establishment, National Research Council, Canada.

Tangler, J. L and Somers, D. M. (1995) NREL *Airfoil Families for HAWTs*, Golden, USA, National Renewable Energy Laboratory.

Taylor, D.A. (1998) 'Using buildings to harvest wind energy' in *Building Research and Information*, E & FN Spon.

TNEI (2007) *Onshore Wind Energy Planning Conditions Guidance Note, A report for the Renewables Advisory Board and BERR*, TNEI Services Ltd.

Troen, I. and Petersen, E. L. (1989) *European Wind Atlas*, Risø, Denmark for the Commission of the European Communities.

Udell, D., Infield, D. and Watson, S. (2010) 'Low-cost mounting arrangements for building-integrated wind turbines', *Wind Energy*, vol 13, no. 7, pp. 657–669.

USFWS (2010) *Wind Turbine Guidelines Advisory Committee – Recommended Guidelines*, US Fish and Wildlife Service.

USFWS (2011) *US Fish and Wildlife Service Land-Based Wind Energy Guidelines*, Draft, July, US Fish and Wildlife Service.

VertAx (2011) *VertAx Wind Offshore Wind Energy H-Type VAWT project*, http://vertaxwind.com/ (accessed 9 March 2012).

VPM (2011) Information from website: http://www.vawtpower.blogspot.com/ (accessed 9 March 2012).

Westmill Wind Farm Co-operative (2011) *Westmill Wind Farm Co-operative Website*, http://www.westmill.coop/westmill_home.asp (accessed 9 March 2012).

Willow, C. and Valpy, B. (2011) *Offshore Wind – Forecasts of future costs*, BVG Associates for Renewable UK, April.

Wolfram (2011) The *Wolframs Demonstrations Project demonstration of 2D Vector Addition*, http://demonstrations.wolfram.com/2DVectorAddition/ (accessed 9 March 2012).

YouGov (2010) *Public Attitudes to Wind Farm YouGov Survey*, YouGov plc for Scottish Renewables.

Zervos, A and Kjaer, C. (2009) *Pure Power – Wind energy targets for 2020 and 2030*. 2009 update. European Wind Energy Association.

Chapter 8

Wave energy

By Les Duckers

8.1 Introduction

The possibility of extracting energy from ocean waves has intrigued people for centuries. However, although there are concepts over 200 years old, it was only in the latter half of the twentieth century that viable schemes began to emerge. In general, these modern wave energy conversion schemes have few environmental drawbacks, and the prospects that some of them may make a significant energy contribution are promising. In fact, in areas of the world where the wave climate is energetic and where conventional energy sources are expensive, such as remote islands, some of these schemes are already competitive. By 2020 it is expected that a number of commercial schemes will be in operation around the world, and as experience is gained costs are expected to fall.

The total wave energy resource is large, however estimates of its magnitude vary widely as the oceans have not been fully monitored over a long enough time period to establish a reasonable value. Moreover, even if a reasonable value could be determined there are difficult questions about how much of the resource is economically viable given the limitations of location and technology.

Nonetheless, it is clear that wave energy could make a major contribution to meeting world energy needs in a sustainable way. The World Energy Council has estimated the exploitable worldwide resource to be 2000 TWh per year if all potential improvements to existing devices are realized (WEC, 2010). Therefore for some countries – the UK is one example – wave energy offers a very large potential resource to be tapped. Thorpe estimates that between 15 and 20% of current UK electrical demand could be met from marine energy sources (Thorpe, 2003).

Technological developments could enable wave energy to fulfil this promise. A number of shore-mounted, near-shore and offshore prototypes are already planned or in operation. Refinements of these prototype designs could open up the possibility of harvesting vast quantities of energy from the oceans, particularly from floating offshore wave farms, consisting of tens or hundreds of devices.

The European Ocean Energy Association's ocean energy roadmap (EU-OEA, 2010) provides a set of steps which, once implemented, would facilitate exploitation of the vast European ocean energy resource and enable the realization of 3.6 GW of installed capacity by 2020, and close to 188 GW by 2050 (these figures encompassing both the wave devices of this chapter and the tidal technologies of Chapter 6).

History

Following the UK 'energy crisis' of 1973 and Stephen Salter's landmark paper (Salter, 1974) a large number of device concepts were invented, mathematically modelled and experimentally tested, with support from commercial sponsors and the former UK Department of Energy. Unfortunately, insufficient time and money was allocated to bring the various concepts and associated technologies to maturity, and in 1982, the Department of Energy scaled down the UK wave energy programme (Ross, 1995).

Some of the research teams involved were, however, able to sustain a minimal effort on wave energy projects. In 1989 a 75 kW prototype oscillating water column (OWC) wave energy converter was installed on Islay in Scotland. This had been fully funded by the Department of Energy following a recommendation that small-scale devices should be investigated as a source of energy for islands and remote communities where diesel normally provided the main energy source (ETSU, 1985).

Meanwhile, during this period of reduced funding in the UK, a number of other countries, notably Norway and Japan, increased their research and development wave energy programmes. With hydroelectric schemes supplying virtually all of its electricity, Norway had little immediate domestic need for wave energy, but it was keen to develop an export market for wave energy technology. In contrast, Japan did require more energy sources, but its wave climate is very modest.

Japan has conducted a substantial wave energy research programme, with many teams working on a variety of projects (see Section 8.5 for details of the Whale, BBDB and Pendulor). The late wave energy pioneer Yoshio Masuda, who began work in Japan in the 1940s, is generally considered to be the inventor of the oscillating water column (see below) and is often described as the 'father' of wave energy. He was the inspiration behind the Kaimei, the first large floating wave power station. Further details can be found in Miyazaki (1991); Miyazaki and Hotta (1991); Kondo (1993) and ISOPE (2002).

In the 1990s there was a revival of awareness of the potential of wave energy amongst politicians and others in a number of countries. In particular, a European Union initiative was launched (Caratti et al., 1993), which provided funding for a small number of projects and led to the formation of a European Wave Energy Thematic Network. Between 1992 and 2002, Thorpe carried out several surveys for the UK Department of Trade and Industry (DTI) and, by looking at the main types of device, he estimated the electricity generation costs to be around 5p per kWh (in 2001 prices), based on an annual practical wave resource of 30 TWh (Thorpe, 1992; 1998; 2001).

Over the years 2001–2011 there has been substantial UK investment in wave energy technology, both from commercial investors and from the government. Two marine energy centres have been established, in Scotland and Cornwall, and these centres offer facilities for developers to test their pilot schemes. There has been considerable growth in interest in wave energy, especially in Scotland and Ireland, where the most significant wave energy resources in the British Isles exist.

International interest in wave energy has also grown in the last 10 years with developments in India, China, the United States, Japan, Korea, South Africa, Australia, and the EU amongst others. An extensive list of wave devices and developments can be found in Chapter 14 by Thorpe in the 2010 World Energy Council survey (WEC, 2010).

8.2 Introductory case studies

The case studies presented in this section provide some insight into the nature of waves and of schemes designed to harness wave energy.

The TAPCHAN device is particularly valuable in that it incorporates an element of storage, and the oscillating water column (OWC) concept is important because it has been deployed in a number of countries and represents the most common form of wave energy converter deployed to date, probably because of its simplicity and robustness.

Shoreline prototypes are generally of the oscillating water column type. They are often regarded as easier to construct and maintain than offshore devices, but, in practice, each shore-mounted device may have to be purpose-built to exactly fit the specified location, whereas offshore devices can potentially be built in production facilities in large numbers. This should eventually lead to high-volume, high-quality, low-cost fabrication and assembled structures could be towed out in calm conditions for deployment in wave farms. By using an area of ocean a few kilometres or more offshore as a wave farm, it should be possible to deploy a large array of wave energy converters and hence capture large quantities of energy, which could then be transmitted back to shore via subsea electrical cables. Offshore devices can harvest greater amounts of energy, as the waves in deep water have a greater power density than those in the shallower water near to land.

TAPCHAN

In 1985 a 350 kW prototype TAPCHAN wave energy converter, built by the company Norwave, commenced operation on a small Norwegian island some 40 km north-west of Bergen.

The name 'TAPCHAN' comes from the 'TAPered CHANnel' design of the scheme (Figure 8.1). The first, and so far only, example had a channel with a 40 metre wide horn-shaped collector. Waves entering the collector fed into the wide end of the tapered, upward sloping channel, where they then propagated towards the narrow end with increasing wave height. The channel walls on the prototype were 10 m high (from 7 m below sea level to 3 m above) and 170 m long. Because the waves were forced into an ever-narrowing channel, their height became amplified until the crests spilled over the walls into the reservoir at a level of 3 m above the mean sea level. The kinetic energy in the waves was thus converted into potential energy, and this was subsequently converted into electricity by allowing the water in the reservoir to return to the sea via a low-head Kaplan turbine system (see Chapter 5 for details of the Kaplan turbine). This powered a 350 kW generator that delivered electricity into the Norwegian grid.

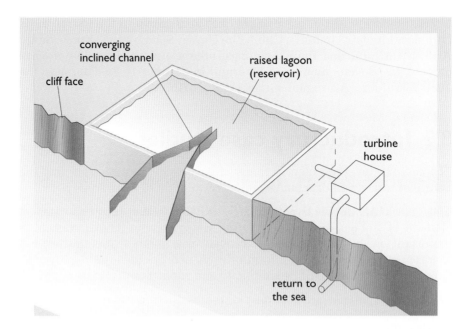

Figure 8.1 (a) The tapered channel (TAPCHAN) wave energy conversion device. (b) Aerial photograph of the Norwegian TAPCHAN: water is entering the reservoir in the centre of the image, where the channel device can just be seen between the cliff walls

The TAPCHAN concept is simple. With very few moving parts, its maintenance costs should be low and its reliability high. The storage reservoir also helps to smooth the electrical output. We shall see in Section 8.3 that ocean waves have a random nature and so most wave energy converters produce a fluctuating power output. In contrast, a TAPCHAN device 'collects' waves in the reservoir, and so the output from the Kaplan turbine is dependent on the relatively steady difference in water levels between the reservoir and the sea. A TAPCHAN device therefore has an integral storage capacity which is generally not found in other wave energy converters.

In the 1990s Norwave considered methods for reducing the cost of construction of future TAPCHANs. Among those methods is a scheme for wave prediction, to allow the Kaplan turbine to run at a greater output for some short time before the arrival of a number of large waves. This reduces the level of water in the reservoir and so makes room for those large waves. This technique may permit the designers to build schemes with smaller reservoirs and hence reduce the construction costs. A second cost-reduction method that has been proposed is to fabricate a shorter channel, and this was tried out on the existing prototype at Bergen by reducing the length of the existing channel. Unfortunately, there were some technical difficulties with the dynamiting of the concrete channel, and the ensuing commercial problems have meant that this prototype is no longer in operation (Petroncini and Yemm, 2000).

Perhaps more than other wave energy systems, the TAPCHAN approach can only be used in very particular places: to be effective it requires a good wave climate (i.e. high average wave energy, with persistent waves); deep water close to shore; a small tidal range (less than 1.0 m), otherwise the low-head hydro system cannot function properly for 24 hours a day (this therefore excludes most of the UK south of the Shetland Isles); and a convenient and cheap means of constructing the reservoir, which usually requires a natural feature of the coastline.

The Islay shoreline oscillating water columns

In the mid-1980s Queen's University, Belfast, (QUB) worked on the development of a shoreline oscillating water column (OWC) device. After surveying several Scottish islands, the island of Islay was chosen for their first OWC scheme, which was installed in a natural gully in 1989. It supplied the local grid with electricity on an intermittent basis from a 75 kW generator from 1991 until it was decommissioned in 1999.

The approach was to develop a device which could be built cheaply on islands using locally available technology and plant. The main part of the OWC consisted of a closed wedge-shaped chamber, made from modular pre-fabricated concrete components, which was placed above the natural gulley. This chamber sloped down to below the sea surface and wave motion drove the water column thus trapped within the lower part of the chamber up and down like a huge piston. A cylindrical tube connected to the atmosphere allowed air to be thus expelled and drawn into the chamber. On its way to and from the atmosphere the air drove a Wells turbine (see Box 8.3), which was directly coupled to an electrical generator.

With experience gained from this natural gully project the team from QUB then collaborated with Wavegen of Inverness to develop what they referred to as a 'designer gully' OWC to overcome some of the limitations of the first Islay project. The principal modifications applied to the second Islay OWC, called LIMPET (Land Installed Marine Powered Energy Transformer), were in the construction method and the shape of the oscillating column.

The designer gully was excavated behind a natural rock wall, which was only removed at the end of the installation (Figure 8.2). As regards the water column, the chamber in the first Islay OWC had a horizontal floor at right angles to the back wall, causing turbulence and consequent loss of energy

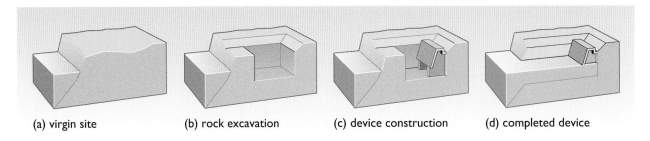

(a) virgin site   (b) rock excavation   (c) device construction   (d) completed device

Figure 8.2 Construction sequence used by Queen's University and Wavegen for the 'designer gully' LIMPET OWC

within the chamber itself so, in order to improve the flow of water into and out of the oscillating chamber, the main part of the chamber of the designer gully OWC was built at a slope so as to efficiently change the water motion from horizontal to vertical and vice versa (Figure 8.3).

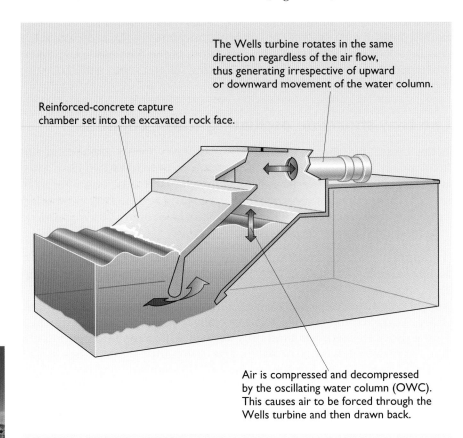

The Wells turbine rotates in the same direction regardless of the air flow, thus generating irrespective of upward or downward movement of the water column.

Reinforced-concrete capture chamber set into the excavated rock face.

Air is compressed and decompressed by the oscillating water column (OWC). This causes air to be forced through the Wells turbine and then drawn back.

Figure 8.3 Outline of the LIMPET device on Islay

Figure 8.4 Photograph of the LIMPET OWC device

Construction of LIMPET was completed in September 2000, and Figure 8.4 shows the finished structure, consisting of a rectangular sloping chamber which ducts the airflow through two contra-rotating Wells turbines. Each turbine is coupled to a 250 kW induction generator, giving the device a 500 kW maximum power output. Details of the first year of operation of LIMPET can be found in Boake et al. (2002). LIMPET has now accumulated many thousands of hours of operation, in more than ten years, and is used by Wavegen as a test bed for air turbines.

8.3 Physical principles of wave energy

Ocean waves are generated by wind passing over long stretches of water (known as 'fetches'). The precise mechanisms involved in the interaction between the wind and the surface of the sea are complex and not yet completely understood, but three main processes appear to be involved.

(1) Initially, air flowing over the sea exerts a tangential stress on the water surface, resulting in the formation and growth of waves.

(2) Turbulent air flow close to the water surface creates rapidly varying shear stresses and pressure fluctuations. Where these oscillations are in phase with existing waves, further wave development occurs.

(3) Finally, when waves have reached a certain size, the wind can exert a stronger force on the upwind face of the wave, causing additional wave growth.

Because the wind is originally derived from solar energy we may consider the energy in ocean waves to be a stored, moderately high-density form of solar energy. Solar power levels, which are typically of the order of 100 W m^{-2} (mean value), can be eventually transformed into waves with power levels of over 100 kW per metre of crest length. Note that these power levels are reported in different units, one *per square metre*, and the other *per metre of crest length*. This is because the nature of waves makes it impossible to refer to power per unit area: we have to consider that the wave action takes place throughout the depth of water, and so we consider the power passing through a *1 metre wide slice of water*. We cannot therefore directly compare solar and wave power densities, but can state that the solar to wind to wave sequence gradually concentrates the power density.

A simple, 'regular' wave can be characterized by its wavelength, λ; height, H; and period, T (see Box 8.1). Waves of greater height contain more energy per metre of crest length than small waves. It is usual to quantify the power of waves rather than their energy content.

BOX 8.1 Wave characteristics and wave power

The shape of a typical wave is described as **sinusoidal** (that is, it has the form of a mathematical sine function). The difference in height between peaks and troughs is known as the **height**, H, and the distance between successive peaks (or troughs) of the wave is known as the **wavelength**, λ.

Suppose that the peaks and troughs of the wave move across the surface of the sea with a velocity, v. The time in seconds taken for successive peaks (or troughs) to pass a given fixed point is known as the **period**, T. The **frequency**, f, of the wave describes the number of peak-to-peak (or trough-to-trough) oscillations of the wave surface per second, as seen by a fixed observer, and is the reciprocal of the period. That is, $f = 1/T$.

If a wave is travelling at velocity v past a given fixed point, it will travel a distance equal to its wavelength λ in a time equal to the wave period T. So the velocity v is equal to the wavelength λ divided by the period T, i.e.:

$$v = \lambda/T$$

The power, P, of an idealized ocean wave is approximately equal to the square of the height, H (metres), multiplied by the wave period, T (seconds). The expression is:

$$P = \frac{\rho g^2 H^2 T}{32\pi} \text{ W m}^{-1}$$

or (approximately) in kW m^{-1}, $P \approx H^2 T$ kW m^{-1}

where ρ is the density of water and g is the acceleration due to gravity.

Deep water waves

In terms of wave propagation, water is considered to be 'deep' when the water depth is greater than about half of the wavelength λ. The velocity of a deep-water ocean wave can be shown to be proportional to the period as follows:

$$v = gT/2\pi$$

This leads to the useful approximation that the velocity in metres per second is about 1.5 times the wave period in seconds.

An interesting consequence of this result is that in the deep ocean the long waves travel faster than the shorter waves. This is referred to as 'dispersion', a unique feature of deep water waves which can lead to dangerous and hard to predict combinations of wave crests. Long waves can catch up with preceding shorter waves creating large combined waves where they meet.

If both the above relationships hold, we can find the deep water wavelength, λ, for any given wave period:

$$\lambda = gT^2/2\pi$$

Intermediate depth waves

As the water becomes shallower, the properties of the waves become increasingly dominated by water depth. When waves reach shallow water, their properties are completely governed by the water depth, but in intermediate depths (i.e. between $d = \lambda/2$ and $d = \lambda/4$) the properties of the waves will be influenced by both the water depth d and wave period T.

Shallow water waves

As waves approach the shore, the seabed starts to have an effect on their velocity, and it can be shown that if the water depth d is less than a quarter of the wavelength, the velocity is given by:

$$v = \sqrt{g(d)}$$

In other words, the velocity under these conditions is equal to roughly three times the square root of the water depth d – it no longer depends on the wave period.

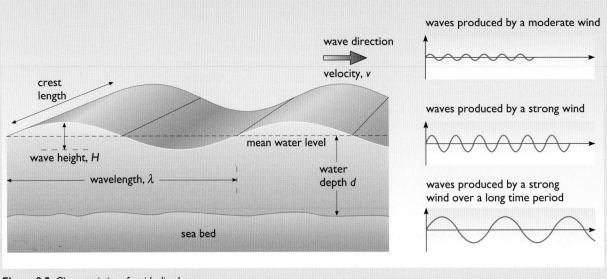

Figure 8.5 Characteristics of an idealized wave

Waves located within, or close to, the areas where they are generated are sometimes referred to as a 'wind sea'. When winds change in strength and direction, the resulting wave patterns can become quite complex. Waves can travel out of these areas with minimal loss of energy to produce 'swell waves' at great distances from their area of origin. As waves travel, there is a systematic tendency for periods and wavelengths to increase, and so swell seas are typically made up of relatively long period waves. The height and steepness of the waves generated by any wind field depends upon three factors: the wind speed; its duration; and the fetch, i.e. the distance over which wind energy is transferred into the ocean to form waves. When a steady wind has blown for a sufficient time over a long enough fetch, the waves are referred to as constituting a fully developed sea.

The UK is well situated to make use of wave energy because it lies at the end of a long fetch (the Atlantic Ocean) with the prevailing wind blowing towards it. The UK's western approaches have therefore been of most interest to wave energy engineers and developers. Although often stormy, the fetches in the North Sea are relatively short and so the wave power densities are generally smaller than those of the Atlantic Ocean.

Typical sea state

A typical sea state is actually composed of many individual components, each of which is like the idealized wave described in Box 8.1. Each wave has its own properties, i.e. its own period, height and direction. It is the combination of these waves that we observe when we view the surface of the sea, and the total power in each metre of wave front of this irregular sea is of course the sum of the powers of all the components. It is obviously impossible to measure all the heights and periods independently, so an averaging process is used to estimate the total power, as follows.

(1) By deploying a wave-rider buoy it is possible to record the variation in surface level during some chosen period of time. (Satellites are also used to detect wave heights.)

(2) The average water height will always be zero, since the average value also defines the zero value, but we can obtain a meaningful figure by calculating the significant wave height, H_s. This is defined as $4 \times$ the *root mean square* of the water elevation – i.e. the instantaneous elevations are first squared, making all of the values positive, then the mean over a number of waves is calculated, then the significant wave height is calculated as four times the square root of the mean. The **significant wave height** is approximately equal to the average of the highest one-third of the waves (which generally corresponds to the estimation of height made by eye, since the smaller waves tend not to be noticed).

(3) The **zero-up-crossing period** T_z (or in abbreviated form, the zero-crossing period) is defined as the average time – counted over ten crossings or more to get a reasonable average – between upward movements of the surface through the mean level. (Note that including the downward movements would give the half-period.)

The **energy period**, T_e, is more useful to us as it characterizes the energy in the waves, whereas T_z represents all of the content of the waves including very small components which have negligible energy. The energy period is slightly longer than the zero-up-crossing period: $T_e \sim 1.12 T_z$.

(4) For a typical irregular sea, it can then be shown (the derivation lies outside the scope of this book) that the average total power in one metre of wave crest can be approximated by $P = 0.5 H_s^2 T_e$ where P is in units of kW per metre length of wave crest. The approximation $P = 0.5 H_s^2 T_e$ is commonly used.

Figure 8.6(a) illustrates a typical wave record and shows the significant wave height and zero-crossing period. Figure 8.6(b) shows two further wave records for the same location, recorded on different days.

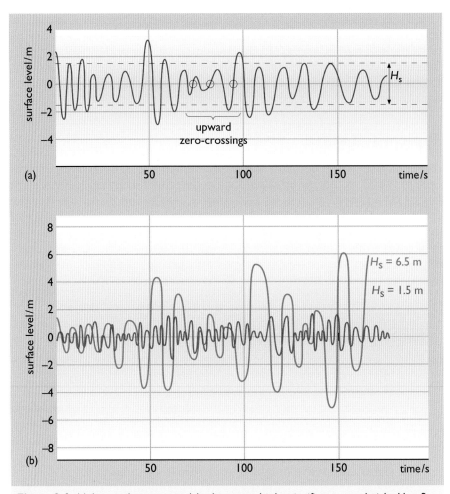

Figure 8.6 (a) A typical wave record. In this example the significant wave height H_s = 3 m (from −1.5m to +1.5m). Successive upward movements of the surface are indicated with small circles. In this case there are 15 crossings in 150 seconds, so T_z = 10 seconds. Then T_e = 11.2 seconds. From this, P = 0.5 (3² × 11.2) kW m⁻¹ = 50.4 kW m⁻¹ (b) Two further wave records are shown here for the same location but represent recordings taken on different days

Variations in the wave power at a given location

Sea level recordings made at different times or dates will of course differ, leading to different values of H_s and T_e. Suppose that each recording represents a time period of one-thousandth of a year or 8.76 hours. If we record the sea states at our chosen location over a whole year, characterizing each of them by their values of H_s and T_e, we can build up a statistical picture of the distribution of wave conditions at our chosen location. This picture, or scatter diagram, gives the relative occurrences in parts per 1000 of the contributions of H_s and T_e. The example of a scatter diagram shown in Figure 8.7 is for the north Atlantic and shows that the waves at this location have a high average power density. In water 100 m deep at South Uist (Hebrides, Scotland), for example, the annual average might typically

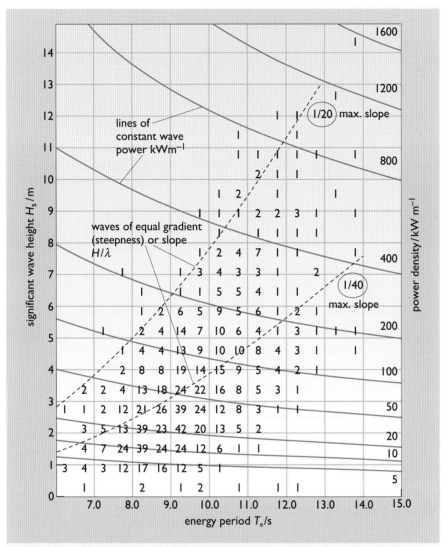

Figure 8.7 Scatter diagram of significant wave height (H_s) against energy period (T_e) for 58°N 19°W in the middle of the north Atlantic. The numbers on the graph denote the average number of occurrences of each combination of H_s and T_e in each set of one thousand 8.76 hour-long measurements made over one year. The most frequent occurrences are at $H_s \sim 2$ m, $T_e \sim 9$ s

be around 70 kW m^{-1} (or 613 000 kWh m^{-1} per year), whereas closer to shore where the depth is 40 m, the corresponding figure might be around 50 kW m^{-1} (or 438 000 kWh m^{-1} per year). These figures indicate that the north Atlantic is indeed a valuable wave energy resource.

Figure 8.8 shows estimates of the average wave power density at various locations around the world. The areas of the world which are subjected to regular wind fluxes are those with the largest wave energy resource. South westerly winds are common in the Atlantic Ocean, and often travel substantial distances, transferring energy into the water to form the large waves which arrive off the European coastline.

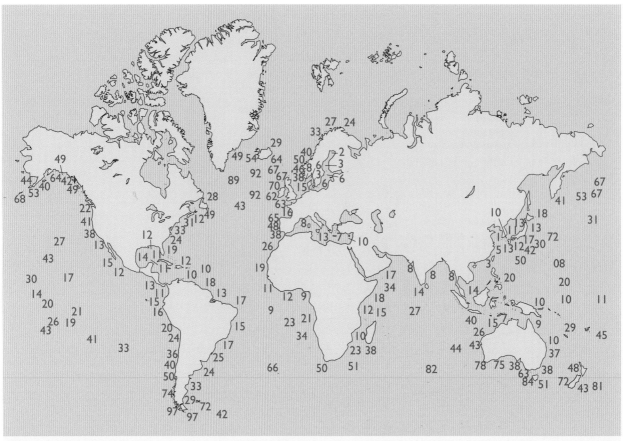

Figure 8.8 Annual average wave power in kilowatts per metre (kW m⁻¹) of crest length, for various locations around the world. (source: adapted from Claeson, 1987)

Wave pattern and direction

The direction of waves travelling in deep water is obviously dictated by the direction of the wind generating them. Waves can travel vast distances across open water without much loss of energy. At any given location we can therefore expect to observe waves arriving from different sources, and hence different directions. For example, we might see waves approaching us from the south-west which were produced by the winds over the mid-Atlantic, but at the same time find that some waves have been generated by storm conditions to the south and east of our position. It is easy to imagine that the resulting wave pattern will be complex, and indeed such patterns are commonly observed. (Figure 8.9a).

A representation of the annual average power as a function of direction at a given location can be given by a 'directional rose' (Figure 8.9b).

What happens beneath the surface?

The surface profile of the ocean is the obvious evidence for the existence of waves, but we also need to understand the sub-surface nature of waves if we are to design schemes to capture energy from them (Figure 8.10).

Waves are composed of orbiting particles of water. Near to the surface, these orbits are the same size as the wave height, but the orbits decrease

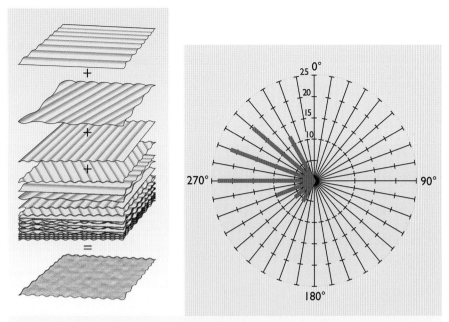

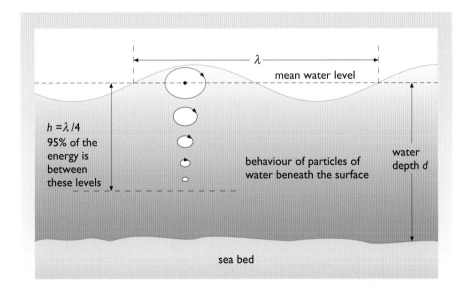

Figure 8.9 (a) The development of a complex sea – wave components produced by different weather conditions in different distant geographical areas arrive and combine in the local area (source: adapted from Claeson, 1987) (b) A directional rose for waves. The length of the line in each sector represents the average annual power in that sector. In this case most of the waves are coming from the west (source: Thorpe, 1999)

Figure 8.10 Behaviour of particles of water beneath the surface

in size as we go deeper below the surface. The size of orbits decreases exponentially with depth.

To capture the maximum energy from a wave we could try to construct a device that was deep enough to intercept all of the orbiting parts of that wave. But this would be impractical and uneconomic, since the lowest orbits actually contain very little energy. In deciding how deep a structure to extract wave energy should be, it is useful to know that 95% of the wave

Table 8.1 North Atlantic offshore wave conditions

	Period/ s	Height/ m	Power density/ kW m⁻¹	Velocity/ m s⁻¹	Wavelength/ m
Storm	14	14	1700	23	320
Average	9	3.5	60	15	150
Calm	5.5	0.5	1	9	50

energy is contained in the layer between the surface and a depth h equal to a quarter of the wavelength λ (i.e. $h = \lambda/4$). From Table 8.1 we can see that a typical north Atlantic wavelength might be 150 m, which tells us that a device would have to be 38 m deep to intercept 95% of the incident wave energy. Such a device would be too expensive, and the optimum depth for most practical devices will be between 4 and 10 m.

Moving into shallow water

There are a few areas in the world where the shoreline is formed by a steep cliff which drops into reasonably deep water. These are the areas most suitable for shore-mounted wave energy converters because the incident waves have a high power density. However, for most of the coastlines around the world the near-shore water is quite shallow. Due to the frictional coupling between the water particles at the greatest depths with the seabed, deep water waves gradually give up their energy as they move into shallower water and eventually run up the shore to the beach. The frictional effect becomes significant when the water depth is less than quarter of a wavelength, and the power loss can be tens of watts per metre of crest length for every metre travelled in-shore.

This power loss is very important, because it obviously reduces the total wave energy resource. Typically, waves with a power density of 50 kW m⁻¹ in deep water might contain 20 kW m⁻¹ or less when they are closer to shore in shallow water (depending on the distance travelled in shallow water and the roughness of the seabed). However, storm waves are also attenuated in the same way and are therefore less likely to destroy shoreline devices.

A further mechanism for energy loss as waves run up the beach is the formation of breaking waves, which are turbulent and energy-dissipating. Although such waves ('breakers') are potentially desirable for leisure activities such as swimming and surfing, they can be very damaging to structures such as wave energy converters, and so should be avoided when choosing suitable locations.

Refraction

As ocean waves approach the shore they will usually be entering shallower water and, as we saw in Box 8.1, the velocity of the waves then becomes governed by the water depth. Shallower water means lower wave velocity. This in turn leads to **refraction** (change of wave direction due to change of wave velocity).

Imagine that a wave crest approaches shallow water at an angle, so that one end of the crest reaches the shallow water first. This part then moves more

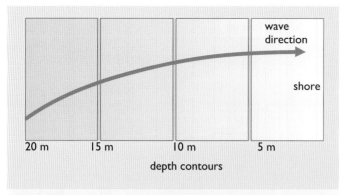

Figure 8.11 As waves travel from deep water into shallow water their velocity reduces and the resulting refraction generally causes waves to approach a beach at right-angles to the shore

slowly than the rest, changing its direction. The remainder of the crest progressively adopts this new direction as its velocity is also reduced on entering the shallow water. The effect of refraction, caused by the reducing depth and hence velocity, is gradually to change the direction of the crest to be roughly parallel with the shore (Figure 8.11).

Knowledge of the depth contours around a coast allows the identification of areas where waves will be concentrated, and are thus the most cost-effective for wave energy developments. Consider a shoreline with headlands (Figure 8.12) – notice how the varying water depth, as shown by the contours (white lines), causes refraction to occur. This concentrates the waves onto the headlands, and leaves the other areas with reduced wave density.

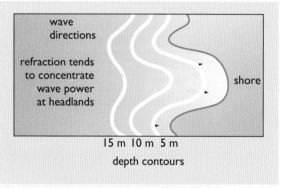

Figure 8.12 Concentration effects of refraction around a shoreline with headlands

8.4 **Wave energy resources**

Wave power resources (the annual average kW per metre of crest length) around the world were shown in Figure 8.8. As already mentioned, the World Energy Council has estimated a total worldwide resource of 2 000 TWh per year (Thorpe, 1999), although this estimate might in fact prove to be rather conservative as others have suggested higher figures.

Regarding the UK, Thorpe (2003) estimated that the total annual average wave energy along the north and west side of the United Kingdom (i.e. from the south-west approaches to Shetland) ranges from 100 to 140 TWh per year at the near shore to about 600 to 700 TWh per year in deep water. Figure 8.13 shows the seasonal variation in significant wave height, a measure of the available resource, in British waters (BERR, 2008). The variation is quite marked, with a greater resource being available, on average, during the winter months, coinciding with the highest demand for energy. The proportion of this resource that could actually be harnessed to produce electrical power depends, of course, on various practical and technical constraints.

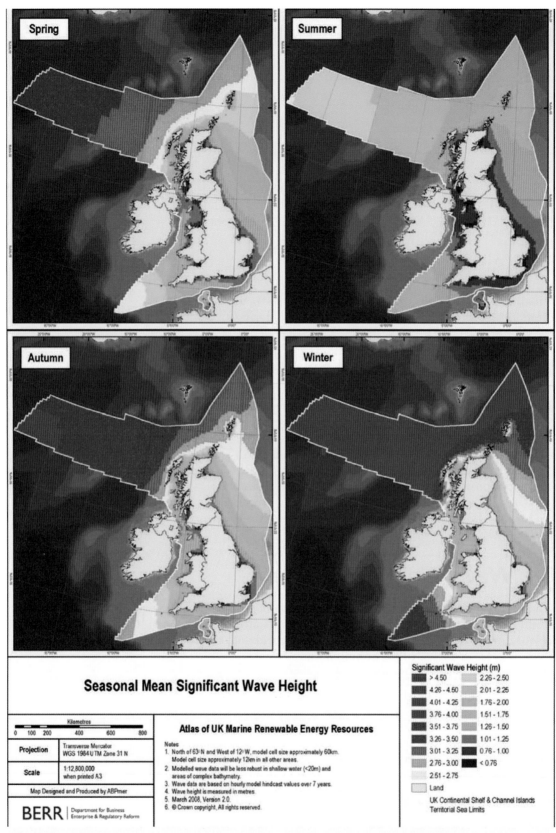

Figure 8.13 Seasonal variation in significant wave height around the UK (source: BERR, 2008)

Thorpe (1992) estimated that the technical resource (i.e. the resource technically available regardless of cost, see Chpater 10) is between 7 GW and 10 GW annual average power, which is equivalent to 61 to 87 TWh per year, depending on the water depth. In comparison, total UK annual electricity production in 2010 was approximately 360 TWh. Table 8.2 gives a breakdown of the resource at different water depths.

Table 8.2 The natural and technical wave energy resource for the north and west side of the UK

Water depth/m	Average natural resource		Average technical resource	
	GW	TWh per year	GW	TWh per year
100	80	700	10	87
40	45	394	10	87
20	36	315	7	61
Shoreline	30	262	0.2[1]	1.75

[1] The technical shoreline resource is very dependent on details of the local shoreline structure, for example the nature and shape of the rock formations and of gullies and beaches.
Source: Thorpe 1992, 2001

8.5 Wave energy technology

In order to capture energy from sea waves it is necessary to intercept the waves with a structure that will react in an appropriate manner to the forces applied to it by the waves. In the case of a shore-mounted device, like TAPCHAN and most OWC devices, the structure is firmly fixed to the seabed, and the waves make water move in a useful way. For other types of device some part of the structure may be fixed, perhaps anchored to the seabed, but another part may be a float which moves in response to the waves by pulling against the anchor. In this case the relative motion between the anchor and the float provides the opportunity to extract energy.

Very loosely tethered floating structures can also be employed, but a stable frame of reference must still be established so that the 'active' part of the device moves relative to the main structure. This can be achieved by taking advantage of inertia, or by making the main structure so large that it spans several wave crests and hence remains reasonably stable in most sea states.

A body in the sea subject to waves can respond via six types of movement. These are sway, roll and yaw, which are not generally harnessed in wave energy conversion technology, and heave, surge and pitch (Figure 8.14), which are harnessed to varying degrees in most wave energy converters. these three modes of movement can be defined as follows:

- pitch – waves cause the device, or part of it to rotate about its axis
- heave – waves cause the device to rise and fall vertically
- surge – waves cause the device to move horizontally backwards and forwards.

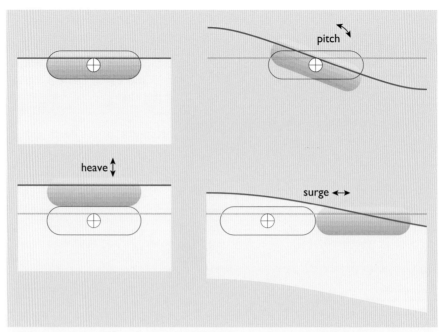

Figure 8.14 The pitch, heave and surge responses of a floating object to incident waves

It is easy to imagine a heaving device, rising and falling with the waves, but such devices have too high a natural frequency to be particularly effective (see the subsection on multi-resonant OWCs for more details). Surge devices tend to have low natural frequency and so can respond to a wide range of wave frequencies. Theoretically, surging motions are twice as energetic as heaving ones, thus making it preferable to harness the surge component of waves (see for example Salter and Lin (1995)).

Economics demands that a device should survive at sea for at least five years. During that time some of its components will have to execute 15 to 30 million cycles, placing severe constraints on material selection and strain levels. A structure designed to operate at a particular design wave power density will also have to endure storms with power densities ten to thirty times higher than the design value.

The physical size of the structure of a wave energy converter is a critical factor in determining its performance. The appropriate size can be estimated by considering the volume of water involved in the upper particle orbits in a wave. In most circumstances a wave energy converter will have to have a swept volume which is similar to this volume of water in order to capture all of the energy contained in the wave. A variety of wave energy concepts are discussed below. The precise physical size and shape of each device will be governed by its mode of operation, but as a rough guide the swept volume must be of the order of several tens of cubic metres per metre of device width. A device with a swept volume much smaller than this would have a limitation on the total energy that it could capture from each typical wave cycle: although it might still be capable of capturing most of the energy from small waves it would be restricted in its response to larger waves, thereby reducing its overall efficiency.

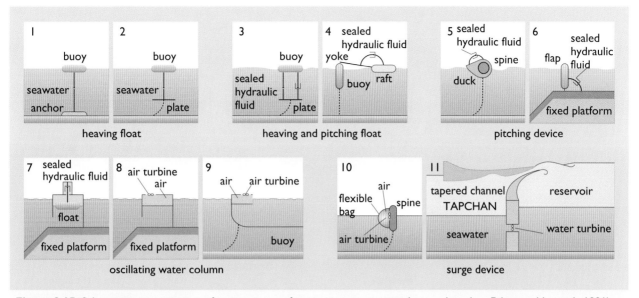

Figure 8.15 Schematic representation of various types of wave energy converter (source: based on Falnes and Løvseth, 1991)

There are many different configurations of wave energy converter, and a number of ways of classifying them have been proposed. One approach is to classify by mode of operation (Figure 8.15).

Another approach is to consider the device location (Figure 8.16) – here the three general classifications are:

- fixed to the seabed, generally in shallow water (eg TAPCHAN)
- tethered in intermediate depths (eg Oyster)
- floating offshore in deep water (eg AWS-III).

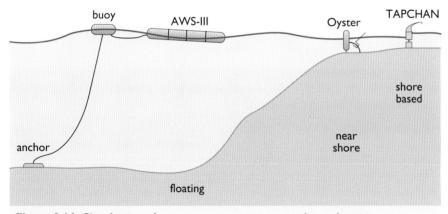

Figure 8.16 Classification of wave energy converters according to location

Alternatively the geometry and orientation of the wave energy converter may be used (Figure 8.17). Here the options are:

- terminators
- attenuators
- point absorbers.

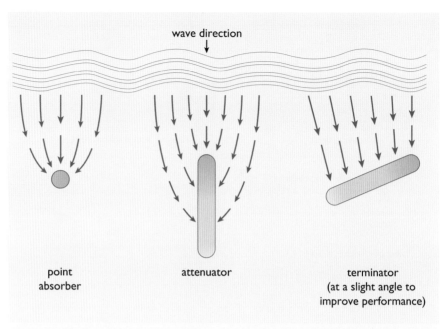

Figure 8.17 Classification of wave energy converters according to size and orientation

Terminator devices have their principal axis parallel (or almost parallel, since a slight angle helps to increase the stability of the whole body) to the incident wave front, and they physically intercept the waves. Depending on the depth, a 300 m long device in a wave climate of 60 kW per metre thus intercepts up to 300 × 60 kW = 18 000 kW. Typically 1/3 of this might be converted into electricity (i.e. 6 000 kW). The terminator might be moored to a leading buoy and anchor, which allows the terminator to pivot as the wave direction changes.

Attenuators have their principal axis perpendicular to the wave front, so that wave energy is gradually drawn in towards the device as the waves move along it. A leading buoy and anchor permit the device to swing round and orientate to the principal wave direction.

An array of terminators or attenuators will require sufficient spacing between individual devices to allow them to swing round on their moorings without fouling their neighbours.

Point absorbers have small dimensions relative to the incident wavelength and work by drawing wave energy from the water beyond their physical dimensions. Typically, being sensitive to waves from all directions, they do not require freedom to orientate to the incoming waves, and so can usually be tightly moored. In principle, they could be extremely slim vertical cylinders which execute large vertical excursions in response to incident waves, but in practice the hardware involved tends to mean that they are a few metres in diameter and absorb energy from perhaps twice their own width. Tethered buoy systems, for example, act as point absorbers.

Point absorber theory (Budal and Falnes, 1975), which is derived from radio antenna theory, suggests that if the body is assumed to be small with respect to the wavelength, the power it may absorb is related, not to the size of the body, but to the wavelength. Therefore, point absorber theory offers the wave energy converter designer the opportunity to create a small device capable of absorbing large amounts of energy from outside its physical bounds. Theoretically the maximum power, P_{max}, absorbed by a perfect point absorber is determined by the mode (heaving or surging) in which it oscillates, and is given by:

$$P_{max} = \varepsilon\left(\frac{\lambda}{2\pi}\right)P_i$$

Where P_i is the incident wave power per metre crest length, and ε is 1 for a heaving device and 2 for a surging device. It is important to note that, from this equation, a body oscillating in surge is theoretically capable of absorbing twice the power of that of a body oscillating in heave, and that for both this could be several times the power incident upon the width of the device.

BOX 8.2 Point absorbers in practice

In theory a perfect wave energy point absorber responding to a heaving motion could capture the energy from a wave front with a length equal to $\lambda/2\pi$ metres. For example, a wave with a period of 6 seconds would have a wavelength of between 56 m and 72 m, depending on whether the water is deep or shallow, and so a perfect (1 m wide) point absorber would absorb the energy from a width of between about 9 m and 12 m). In reality, however, the capture width is much less due to the limitations of the vertical amplitude of the motion of the absorber.

The mathematical explanation of this is outlined by Nielsen and Plum (2000) who go on to report experimental results. These devices are said to have a **capture width ratio**, i.e. the ratio of apparent diameter to physical diameter, greater than 1. Figures 8.25, 8.26 and 8.27 later in the chapter show examples of float systems working off a heaving motion. The concept of **latching**, or holding the float under water for a second or so before allowing it to follow the wave, has been developed to maximize the energy capture by permitting a large amplitude of motion of the float, which is needed for optimal performance (Falnes and Lillebekken, 2003; Falnes, 2011).

Because full-scale devices are generally very large, as a result of needing to be on a similar scale to the wavelengths that they capture, most developments of wave energy technology involve the testing of scale models. Indoor, purpose-built tanks, lakes and reservoirs have been used by research teams. In 1977 a team at the University of Edinburgh constructed a wide wave tank in which it was possible to create repeatable multi-directional mixed-sea conditions, and so test the effects of varying specific design parameters of wave energy converters. The Edinburgh tank was rebuilt in 2002 and is shown in Figure 8.18 with its array of wave-makers.

Figure 8.18 The Edinburgh wave tank (28 m long)

Fixed devices

Fixed seabed and shore-mounted devices are usually terminators, and these have been the most common types of wave energy converter to have been tested as prototypes at sea. Having a fixed frame of reference, being closer to a grid, and with good access for maintenance purposes, they have obvious advantages over floating devices. In addition, the seabed attenuates storm waves which could otherwise destroy the device and turbine.

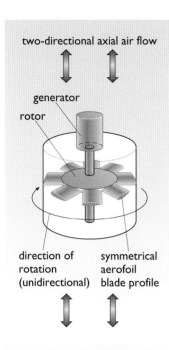

two-directional axial air flow

generator

rotor

direction of rotation (unidirectional)

symmetrical aerofoil blade profile

Figure 8.19 The Wells turbine

However, such devices have the disadvantage that they generally operate in shallow water and hence at lower wave power levels. Another drawback of shore-mounted devices is that of geographical location – only a limited number of sites are suitable for deployment: to optimize output, they need to be positioned in an area of small tidal range, otherwise their performance may be adversely affected. It is also worth noting that mass production techniques are unlikely to be totally applicable to shore-mounted schemes, as site-specific requirements will demand a tailored design for each device, thus adding to the production costs.

The majority of devices tested and planned are of the oscillating water column (OWC) type. In these devices, an air chamber pierces the surface of the water and the contained air is forced out of and then into the chamber by the approaching wave crests and troughs. On its passage from and to the chamber, the air passes through an air turbine (generally of the Wells type) which drives a generator and so produces electricity. The Wells turbine (Figure 8.19) is an axial-flow device which rotates in the same direction irrespective of the direction of airflow, and has aerodynamic characteristics particularly suitable for wave applications (Box 8.3). A very attractive feature of OWCs is that the cross-sectional area of the air turbine can be much smaller than the area of the moving water surface. This reduction in available area for airflow acts like gearing to increase the air velocity through the turbine such that it can operate at high speed.

BOX 8.3 **The Wells turbine**

The Wells air turbine, invented by Professor Alan Wells, is self-rectifying – that is to say, it can accept airflow in either axial direction. To achieve this, the aerofoil-shaped blade profile must be symmetric about the plane of rotation, untwisted and with zero pitch, i.e. the chord line must be in line with the plane of rotation. The vector diagrams in Figure 8.20 show how this occurs. Note that the diagrams are very much like those used in Chapter 7 to explain the properties of wind aerofoils. As the blade moves forward, the angle of attack, which is the angle between the relative airflow velocity and the blade velocity, is small and this produces a large lift force (L). The forward component of L provides the thrust which drives the blade forwards.

The Wells turbine operates in much the same way as would a horizontal-axis wind turbine with symmetrical, untwisted blades and with zero pitch angle. Consider the nearest blade with air flowing in an upward direction (Figure 8.20(a)). If we now work in the frame of reference of the blade – we do this by making the blade appear to be stationary to us (even though it is moving) by considering the blade velocity vector to be in the opposite direction to the blade's actual direction of movement – we get Figure 8.20(b). Note that because the blade chord is in line with the plane of rotation, the angle of attack α is the same as the relative wind angle ϕ referred to in Section 7.4 of Chapter 7.

If we now resolve these vectors we get Figure 8.20(c). From this diagram we can see that there will be a net forward force on the blade acting in the plane of rotation if $(L \sin \alpha) - (D \cos \alpha)$ is greater than zero. The reaction components are of little interest but the rotor bearings must be capable of carrying these forces with minimal friction losses. If the net forward thrust is greater than zero, then the blade will be driven forwards and can usefully extract energy from the airflow. The shape of the blade is extremely important here since it will dictate the values of the lift and drag coefficients C_L and C_D (defined in Chapter 7, Section 7.4) and hence the magnitude of the forward thrust.

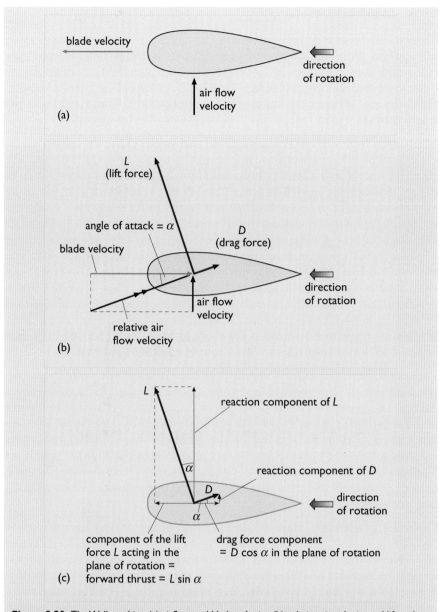

Figure 8.20 The Wells turbine (a) airflow and blade velocity; (b) relative air velocity and lift and drag forces; (c) forces in the plane of rotation

There is a linear relationship between airflow and pressure drop for the Wells turbine rotating at a constant speed. This means that the Wells turbine has a constant impedance to airflow. Careful choice of the design parameters ensures that this impedance matches the requirements of the wave climate at the chosen location. Impedance matching like this maximizes the transfer of energy from the wave to the generator. The Wells turbine is ideally suited to wave energy applications because of its constant impedance, whereas the impedance of a conventional air turbine varies with airflow. A conventional air turbine may have a superior peak efficiency, but a Wells turbine will perform well over a range of air flows giving it a better wave cycle efficiency. A further beneficial characteristic of the Wells turbine is that, at the sizes typically employed, it can rotate at high speed (1500–3000 rpm) so that the electrical generator can be attached directly to the shaft of the turbine, obviating the need for a gear box to raise the generator speed.

OWC-based devices

Fixed oscillating water column devices have been investigated in a number of locations with slight variations being made to the basic concept – this subsection considers bespoke OWC installations, then OWCs combined into other civil engineering projects, before finally considering more advanced OWC designs. For an excellent review of oscillating water columns see Heath (2011).

In addition to the Islay OWC described earlier, an OWC was installed on the island of Pico, part of the Azores, in the north Atlantic. The 500 kW capacity unit is in a slightly sheltered bay which helps to protect it from the excesses of the largest storms. The scheme consists of a 12 metre by 12 metre concrete structure built *in situ* on the rocky sea floor close to the shoreline (it spans a natural harbour), and is equipped with a horizontal-axis Wells turbine-generator capable of producing an average annual output of 124 kW, but rated for a maximum instantaneous value of 525 kW. The Wells air turbine is designed to operate between 750 and 1500 rpm, has eight blades of constant chord and a diameter of 2.3 m (Falcão, 2000). A second turbine with pitch-controlled blades , which permit the blades to be set at the optimum pitch angle for air flow in one direction and then set in the opposite sense when the flow reverses, is to be operated as a contrast to the fixed-pitch one (Le Crom et al., 2009, 2010).

Energetech (now Oceanlinx) in Australia have developed a novel, variable-pitch turbine to improve efficiency and a parabolic wall to focus the wave energy. Designed for use in harbours or rocky outcrops where the water at the coastline is deep, as wave fronts approach they are amplified up to three times at their focal point by the parabola-shaped collector, before entering a 10 metre by 8 metre OWC structure. Oceanlinx (2011) now have 5000 operating hours experience of generating electricity and are planning to deploy their focusing oscillating water column scheme in Hawaii (Gill and Rocheleau, 2009).

Wave energy activity is developing in China – much of the Chinese work is linked to Japan, either in concept or by the exchange of ideas and staff. Some of the work concentrates on OWC powered navigation buoys (developments in floating devices are discussed below), some on theoretical modelling. A shoreline mounted 3 kW OWC, in a natural gully on Dawanshan Island in the Pearl River, has been successful enough to warrant upgrading with a 20 kW turbine.

Harbour and breakwater schemes

When OWCs are incorporated into breakwaters, the double benefits of the breakwater – the provision of a harbour and the generation of electricity – mean that the function of electrical production does not have to justify the total capital cost, and so the generated cost of electricity may be more economically attractive.

Japanese developers have incorporated an OWC into a 20 m section of an extension to the harbour wall at Sakata, on the north-west coast of Japan. The unit was installed in 1989 and contains a Wells turbine rated at 60 kW.

In India, trials of a multi-resonant OWC device (see below), installed in a breakwater and employing a Wells turbine of 2 m in diameter, driving a

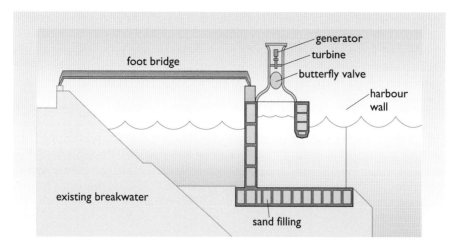

Figure 8.21 Cross-section through the Indian breakwater OWC

150 kW generator (Figure 8.21), commenced off the Trivandrum coast in 1991 (Ravindran et al., 1995).

The device is estimated to be capable of delivering an average of 75 kW during the monsoon period from April to November, and 25 kW from December to March. A power conversion system incorporating two units has been installed so that in the higher power period both units can be operational. Experiments have been conducted to produce desalinated water from the plant, and to test an impulse turbine (Jayashankara et al., 2009).

Since the annual average wave power density along the Indian coast is only between 5 and 10 kW per metre, it is perhaps surprising to see such research and development activity. However, many more harbours are planned on the Indian coastline, hence the consideration of the potential to utilize OWC wave energy converters.

Elsewhere, in July 2011 a 16 chamber, 16 Wells turbine OWC wave energy scheme with a total installed capacity of 300 kW was inaugurated in a breakwater in Mutriku in the Basque Country by the Basque utility Ente Vasco de la Energia. (Renewable Energy Focus, 2011). A Chinese team have also described a breakwater OWC design (Liu et al., 2009).

The multi-resonant OWC (MOWC)

Most wave energy devices are resonant systems – they have a natural resonance period or time over which they repeat their motion. This is just like the motion of a child on a swing who moves with a period of a few seconds. The period is dictated by the length of the swing's ropes – longer ropes give a longer period.

In the case of OWCs the natural period is governed by the length of water from the water or wave surface outside to the water surface inside the device. This can be termed the coupling length (l) and the arrangement behaves like a 'U' tube containing water. The period of oscillation (T) is given by:

$$T = \pi\sqrt{\frac{2l}{g}}$$

For example if $l = 25$ m then T is about 7 s. To put this into a wave energy context: if an OWC has a coupling length of 25 m then it will resonate with a period of 7 s, and it will be very responsive to incoming waves which have an energy period of 7 s. If the incoming waves have greater or smaller periods the response of the OWC will be reduced. This is true of most wave energy devices – the resonance they exhibit makes it difficult to extract energy from a wide range of wave conditions or wave periods.

The MOWC (Figure 8.22) responds more actively across the wave spectrum because a range of coupling lengths can be automatically established due to the harbour walls protruding forwards and so the incoming waves stimulate resonance at longer coupling lengths – hence the term multi-resonant.

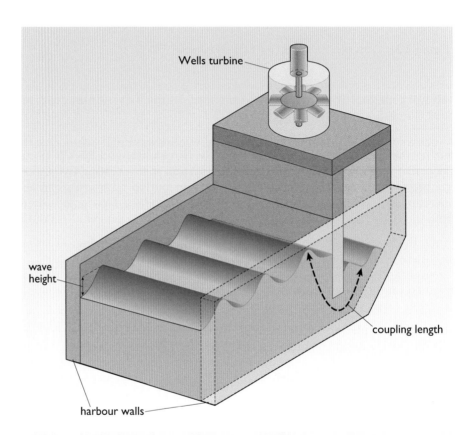

Figure 8.22 A multi-resonant oscillating water column

A multi-resonant oscillating water column (MOWC) was designed and manufactured in 1985 by Kvaerner Brug, and located on the same small Norwegian island as the TAPCHAN prototype. The chamber was set back into a cliff face, which falls vertically to a water depth of 60 m. This created two harbour walls at the entrance to the device, which (as detailed above) allowed the system to absorb energy over a wider range of wave periods. The oscillating airflow was fed through a 2 m diameter Wells turbine rotating within the speed range 1000–1500 rpm. The turbine was directly coupled to a 600 kW generator, and the output passed through a frequency converter before being fed to the grid. The performance exceeded predictions and provided electricity at relatively low cost.

Unfortunately, this device didn't benefit from a shallow water location that often protects fixed OWCs and two severe storms in December 1988 tore the column from the cliff; to date the system has not been replaced. Future designs would therefore have to be much more robust, and would probably involve setting the column into the body of the cliff, as was done with the small OWC built on the island of Islay in Scotland (see Section 8.2 – the first Islay OWC was set into a natural narrowing gully in the rocks and so was able to benefit from the tapered channel effect; the second 'LIMPET' OWC is installed in a 'designer' gully).

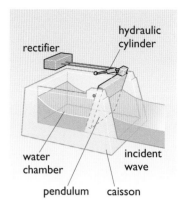

Figure 8.23 The Japanese Pendulor device

Non-OWC fixed devices

Other, non-OWC fixed device prototypes have also been tested in Japan and Scotland. A number have used a mechanical linkage between a moving component, such as a hinged flap, and the fixed part of the device. An example of this approach is the **Pendulor** (Figure 8.23).

The Pendulor is a paddle, hinged at the top, which is fitted one-quarter of a wavelength from the back wall of a caisson. This is at the first anti-node (i.e. the point of maximum amplitude of a series of waves) and so the paddle is subjected to the maximum possible wave movement (note that the paddle can be located at the anti-node for only one particular wavelength). In the regions of Japan where Pendulor devices have been tested, the seas generally have wavelengths close to the design wavelength for much of the year.

A push–pull hydraulic system converts the mechanical energy from the movement of the Pendulor paddle into electrical energy. Two prototypes with nominal power outputs of 5 kW have been operational on Hokkaido, Japan since the early 1980s. The Korean wave energy conversion study group (KORDI) started working on a Floating Pendulor Study in 2010 under a collaborative agreement with a Japanese research group with the intention of accelerating the development of this concept.

In Scotland a working prototype of the Oyster hinged scheme is being trialled. This device is designed to respond to surge forces. The first full-scale Oyster 1 wave power device (Figure 8.24) was installed at the European Marine Energy Centre (EMEC) in Orkney, Scotland in the summer of 2009 and was connected to the national grid in November 2009. At the time of writing Oyster 1 is undergoing sea trials to gather data to finalize the design of the next-generation devices, which are intended to consist of three linked wave power devices with an installed capacity of 2.3 MW. (Aquamarine Power, 2011a). Deployment of these commenced in 2011.

Figure 8.24 The 315 kW Oyster 1 in the factory

Tethered devices

Float systems, with the main body of the structure floating on the surface, but moored to the seabed via a pump, have attracted attention over the years, and are the subject of activity in some current projects. These can act as point absorbers which draw in energy from a greater sea width than their own physical diameter (see Box 8.2).

The European Marine Energy Centre (EMEC) mentioned in Chapter 6, Box 6.6, helps to solve many of the problems that have dogged the construction of prototype wave-power generators in the past. For example, storm-proof moorings and armoured cables facilitate the simple and cheap installation and testing of new devices. Also EMEC can accommodate up to four machines at a time, so designers can directly compare devices under identical conditions, giving them a chance to spot small design 'tweaks' that will improve the efficiency of energy generation.

Another facility providing an infrastructure for the demonstration and proving of arrays of wave energy generation devices over a sustained period of time is the Wave Hub, a grid-connected offshore facility 16 kilometres off the north coast of Cornwall. It holds a 25 year lease on 8 sq km of seabed and can accommodate four separate wave energy schemes each with a capacity of 4–5 MW.

PowerBuoys

The wave energy converter termed 'PowerBuoy', developed by Ocean Power Technology (OPT), consists of a modular buoy-based system which drives generators using mechanical force developed by the vertical movement of the device. Each module is relatively small, permitting low-cost regular maintenance, leading to expected lifetimes of at least 30 years.

An early version of the PowerBuoy was deployed in December 2009 approximately 1.2 km off the coast of Oahu, Hawaii in a water depth of 30 m. To date, this device has operated and produced power from over 3 million power take-off cycles (one take-off cycle is the term for one operational cycle of one significant wave cycle) and 4400 hours of operation. This project is part of OPT's on-going programme with the US Navy to develop and test the PowerBuoy technology, and has contributed to the roll-out of OPT's next generation PB150 system, with a generating capacity of 150 kW, for the electrical utility markets. At the time of writing, the first PB150 device is being prepared for transit to a location off the coast of Inverness, Scotland, for planned ocean testing (OPT, 2011).

OPT has further plans for its devices. It is proposing to develop a commercial wave park in North America at Coos Bay, Oregon, and has signed an agreement to develop a wave-power station in Japan.

The Coos Bay park will be the largest wave energy project in the world on completion and may have a capacity of up to 100 MW. It is intended to be located approximately 4.3 km miles off the coast and will utilize another of OPT's next generation devices – PB500 PowerBuoys (which have a maximum sustained generating capacity of 500 kW). The wave park will consist of up to 200 of these PowerBuoys, 20 undersea substations, and a submarine cable to deliver the electricity generated from this wave park into the grid served by the Pacific Northwest Power Grid (OPT, 2011).

The Japanese development is expected to provide the groundwork for a larger commercial-scale wave-power plant with a capacity of 10 MW or more. The initial phase of the contract will see OPT work with the Japanese consortium of Idemitsu Kosan, Mitsui Engineering and Shipbuilding and

Japan Wind Development to identify ideal sites for the demonstration power station, based on commercial potential. The consortium will then work with OPT to build up to three of its PowerBuoys for the demonstration plant (OPT, 2010).

Hose pump wave energy converter

The average wave power along the Swedish coast is about 0.57–1.1 GW, representing 3–7% of demand. In addition, however, the potential along the neighbouring Norwegian coast is put at around 3.0–3.5 GW, which could contribute some 12–15% of Sweden's electricity demand via the Nordic grid – hence Swedish interest in wave power. Both countries could exploit the Norwegian coastline wave resource.

The Hose Pump Wave Energy Converter was developed by Technocean in Sweden and was intended to pump sea water from an array of hose pumps fixed to the seabed (Figure 8.25). The hose pump consists of a reinforced vertical rubber cylinder, anchored to the seabed and attached to a float on the surface (thus it acts as both tether and pumping mechanism).

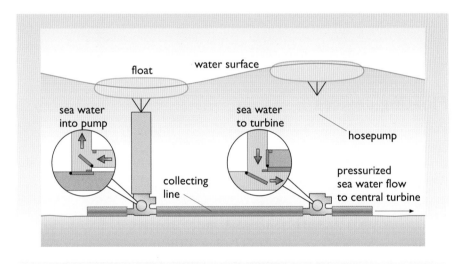

Figure 8.25 Swedish hose pump wave energy converter

The reinforcing cords in the hose are arranged at a carefully chosen angle to the main axis of the hose. As the buoy rises with a wave, the hose is stretched and the cord angle changes in such a way that it causes the internal volume of the hose to be reduced. This raises the pressure of the sea water contained within the hose. The sea water is then pumped out of the tube and up to a storage reservoir on land. A Pelton wheel (see Chapter 5, Section 5.8) extracts energy from the water as it is released from the reservoir back to sea. The tube is refilled with water via a non-return valve. The concept has not been exploited.

Danish research has been conducted in this area on a tethered buoy system where each buoy device contains a generator (Figure 8.26). The floating buoy responds to wave activity by pulling a piston in a seabed-mounted unit. This piston pumps water through a submerged turbine. An array of

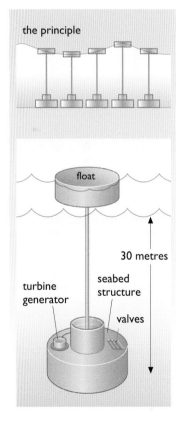

Figure 8.26 Tethered buoy wave energy converter

Figure 8.27 The AquaBuoy system, based on the IPS and Hose Pump concepts

these buoys could be deployed and arranged to have an integrated output (Nielsen and Plum, 2000).

Interproject Service Convertor

The Interproject Service (IPS) Converter was developed by Interproject Service AB, Sweden, in the early 1980s and tested at a 1:10 scale in a lake and as a full-scale prototype at sea. It consists of a long buoy with a tube open at both ends, attached underneath. A piston inside the tube is linked to the buoy and power is extracted by the interaction of the buoy and the water in the tube. A new configuration developed at the University of Edinburgh, the sloped IPS, embodies a number of attractive features, particularly the additional energy capture yielded by operating at an angle to the vertical. By employing a hydraulic accumulator the device can effectively provide reasonably smooth output and so offers the prospect of a reliable source of power for small, isolated electricity networks or small islands (Salter and Lin, 1995).

The AquaBuoy system shown in Figure 8.27 is a development based on both the IPS and the hose pump wave energy converter. An experimental programme aimed at installing a 1 MW scheme consisting of units rated at about 100 kW off the coast of Washington in the US is being conducted by the AquaEnergy Group (AquaEnergy Group, 2010; Wacher and Neilson, 2010).

Floating devices

Floating wave energy conversion devices should be able to harvest more energy than fixed, on-shore devices, since the wave power density is greater offshore than in shallow water and there is little restriction to the deployment of large arrays of such devices.

There is a wide variety of such devices including:

- terminator devices such as the Duck, floating OWCs such as the Whale, OWEL and Backward Bent Duct Buoy (BBDB), and floating tapered channels (described as Floating Wave Power Vessels, FWPV)
- point absorbers such as the AWS-III
- attenuator devices such as the Pelamis.

Terminator devices

OWC-based devices

A number of floating OWC-based devices have been developed – the Kaimei, a converted barge fitted with a number of floating OWC devices and the first sea-going Wells turbine, was first tested in Japan in 1977 – as mentioned previously, this was the first large floating wave power station.

As mentioned earlier, China has an expanding interest in wave energy – numerous OWC-powered navigation buoys have been deployed and the BBDB concept, originally proposed by Masuda, has been tested at model scale (Masuda et al., 2000). Experiments on the BBDB concept (illustrated

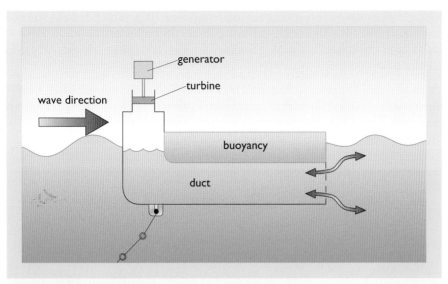

Figure 8.28 The Backward Bent Duct Buoy (BBDB)

in Figure 8.28) have been conducted at the Guangzhou Institute of Energy Conversion (Masuda et al., 2000; Xianguang et al., 2000).

Ocean Energy Ltd has further developed the Backward Bent Duct Buoy and a quarter scale hull has achieved over 20 000 hours of operation in live sea trials at the wave energy test site at Spiddal in Galway in Ireland (Figure 8.29). There it was subjected to a wide range of wave conditions including a severe storm when wind speeds reached 30–35 metres per second and a wave height of 8.2 m was recorded. During the testing the airflows and power output from the tests scaled up predictably and the hull behaviour was also consistent and reliable. A three-quarter scale unit is planned for deployment in 2011–12 (Heath, 2011).

Figure 8.29 Ocean Energy buoy with Wells turbine (courtesy of Ocean Energy)

Other variations on this theme include the Swan DK3, which is based on the L-shaped Backward Bent Duct Buoy (Meyer and Nielsen, 2000; IT Power, 2011) and the Whale, a floating forward facing OWC-based device (Washio et al., 2001).

A multi-cell version of the BBDB, the OWEL, (Figure 8.30) is under development by a consortium led by OWEL (Offshore Wave Energy Limited) and IT Power (2011) which should see a 45 m long pilot version with a rating of 500 kW tested at the Wave Hub site in Cornwall in 2013.

Figure 8.30 Artist's impression of OWEL (source: IT Power)

An alternative design of floating OWC has been developed by Oceanlinx. Their most recent OWC deployment involved its Mk3 floating device. The unit was a one-third scale demonstration version of a full scale 2.5 MW device. It was installed offshore from the eastern breakwater of Port Kembla Harbour from February to May 2010. The unit supplied electrical power directly into the local grid. It broke free of its moorings during a severe storm in May 2010, and ran up onto the beach.

Oceanlinx has now finalized the full-scale development of new shallow and deep water OWC designs, named greenWAVE and blueWAVE respectively. The greenWAVE unit is a single OWC chamber, fixed to the seabed in shallow water (Figure 8.31). In a good wave climate greenWAVE is rated at around 1 MW. The blueWAVE unit is a cluster of six OWC chambers, moored as a floating device in deeper water. Versions of both are expected to be rolled out in commercial operation over the next two years (Oceanlinx, 2011).

Floating wave power vessel (FWPV)

In 1992 Sea Power International installed a floating wave power vessel (FWPV) for testing off the west coast of Sweden. This steel vessel resembles a floating TAPCHAN in that waves run up a sloping ramp and are collected in a raised internal basin. The water flows from the basin back into the sea via low-head turbines. This device is not sensitive to tidal range, and by varying its ballast the device can be tuned to different wave heights.

Figure 8.31 Artist's impression of an Oceanlinx greenWAVE unit

Subsequently, a pilot version based on ship construction was anchored in 50–80 m depth of water 500 m from the Shetland coast in 2002. This device was developed by the Swedish company Sea Power International as part of a Scottish government-backed scheme. It was designed to have a maximum power output of 1.5 MW, producing about 5.2 GWh per year (Lagström, 2000). The device functions by capturing the water from waves that run up its sloping front face. The captured water is returned to the sea via a standard Kaplan hydroelectric turbine (see Chapter 5). In many respects this device may be compared to a floating version of the TAPCHAN.

A Danish version of this floating TAPCHAN concept is the Wave Dragon. Again waves run up a tapered channel and a head of water collects in a reservoir – the water then returns to the sea via a set of simplified Kaplan hydroelectric turbines. A full-scale version was launched near to the University of Aalborg in March 2003, and this delivered its first power to the grid in June 2003. More turbines were installed at sea in September 2003, demonstrating that maintenance at sea is possible (Sorensen et al., 2003). A full scale 7 MW test scheme was designed for deployment in South Wales in 2011.

Another Danish device also uses water to directly drive turbines – the Waveplane is a wedge-shaped device which directs incoming waves of varying frequency into a trough in a spiral configuration, creating a vortex which is used to drive a turbine. Tests using a 1:5 scale model were conducted in Mariager Fjord in Jutland in 1999.

Duck

The Edinburgh Duck concept (Figure 8.32), conceived by Professor Stephen Salter at Edinburgh University in the 1970s, was originally envisaged as many cam-shaped bodies linked together on a long flexible floating spine which was to span several kilometres of the sea. The spine would be oriented almost parallel to the principal wave front, making the Duck largely a terminator. The Duck was designed to extract energy by pitching to match the orbital motion of the water particles, as discussed earlier. To generate power each cam-shaped body – or duck – either moves relative to the spine, producing high pressure in a hydraulic system, or drives a

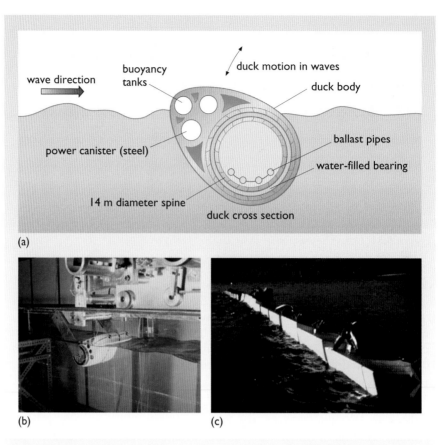

Figure 8.32 (a) The Edinburgh Duck wave energy converter; (b) Duck model being tested in a wave tank; (c) A scale model of the Duck being tested in Loch Ness, Scotland by Coventry University

set of gyroscopes mounted in the noses of the 'ducks'). This matching can be nearly 'perfect' at one wave frequency and the efficiency in long waves can be improved by control of the flexure of the spine through its joints. The concept is theoretically one of the most efficient of all wave energy schemes, but it is likely to take some years to fully develop the engineering necessary to utilize this concept at full scale. The control and engineering challenges of optimizing power conversion efficiency whilst producing steady electrical energy from a 'randomly' rocking body have been studied in great detail by Salter and his team.

Point absorber devices

AWS-III/Clam

AWS Ocean Energy have recently refined the key aspects of this technology, which initially appeared as the Circular Clam, developed at Coventry University in the UK in the 1980s.

AWS-III consists of twelve interconnected air chambers, or cells, arranged around the circumference of a toroid with Wells turbines in each cell. At full scale this would be 60 metres or so in diameter, and would be deployed in deep water (40–100 m). Each cell is sealed against the sea by a flexible reinforced rubber diaphragm. Waves cause the movement of air between cells. Air, pushed from one cell by the incident wave, passes through at least one of the twelve Wells turbines on its way to fill other cells. As the air system is sealed, this flow of air will be reversed as the positions of wave crest and trough on the circle change. Figure 8.33 shows a 1:9 scale test conducted on Loch Ness, Scotland in 2010. After large-scale single cell proving tests, a pilot scheme is scheduled for 2011–12 (AWS, 2011).

Figure 8.33 AWS-III 1:9 scale test on Loch Ness

The important points about AWS-III are its omni-directional energy absorption and the highly sensitive diaphragms which accept energy from the surge motion of the waves. As pointed out earlier in this chapter, the surge component of waves can be twice as energetic as the heave.

Attenuator devices

Pelamis

The Pelamis, or 'sea snake', can trace some of its ancestry to the Edinburgh Duck; it consists of a number of floating cylindrical tubes, connected to each other by active joints. The tubes are arranged at a slight angle off the down-wave direction and so act as attenuator devices. The wave-induced heaving and swaying of the tubes is resisted by hydraulic rams that pump high-pressure oil through hydraulic motors via smoothing accumulators, and the hydraulic motors drive electrical generators to produce electricity.

The ability to withstand high wave power densities has been a key goal of the designers; the Pelamis is capable of inherent load shedding, which means that the spine is not subjected to the full structural loadings that would otherwise be imposed on it during a storm. As an attenuator it sits down the waves rather than across them – see Figure 8.34 – and so becomes detuned in long storm waves, where the waves are much longer than the device.

The prototype 750 kW devices are 150 m long, 3.5 m in diameter, and composed of five modular sections. To reduce risk, wherever possible use

Figure 8.34 The Pelamis P2

is made of existing technology that has been proven offshore. Three 750 kW devices were deployed 5 km into the Atlantic off northern Portugal in 2008. Pelamis P2 has been tested at the European Marine Energy Centre (EMEC), Orkney, Scotland, and the developers have secured the rights to develop several wave farms around Scotland which might give a total installed capacity of 50 MW (Yemm et al., 2012).

The McCabe wave pump

A 40 m long prototype of a device called the McCabe wave pump was deployed off the coast of Kilbaha in Ireland in 2004 (Figure 8.35). This device consists of three narrow rectangular steel pontoons, which are hinged together across their beam, which points into the incoming waves. The pontoons move relative to each other and energy is extracted from this motion by linear hydraulic rams mounted between the pontoons near the hinges. The output is intended for use in applications such as desalination, as well as electrical generation. Experimental and theoretical results of the motion of a McCabe wave pump have been reported by Kraemer et al. (2000) while Nolan et al. (2003) have modelled the power take-off system.

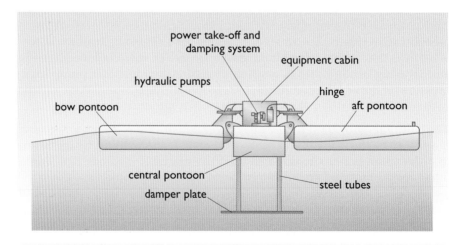

Figure 8.35 The McCabe wave pump – schematic showing operation

The Pitching and Surging FROG

The Pitching and Surging FROG, developed by Professor French and his colleagues at Lancaster University, is a reaction wave energy converter which achieves energy-absorbing behaviour by the movement of internal inertial mass (McCabe et al., 2003). Hence, it is a compact structure which does not require a large spine to provide a stable frame of reference (Figure 8.36).

8.6 Economics

Reducing operation and maintenance costs is the key to the successful economic implementation of wave energy stations. The *capital cost* per kW of establishing a wave-energy run power station is likely to be at least twice that of a conventional station running on fossil fuels; and the *capacity factor* is likely to be much lower than a conventional station due to the variability of the wave climate. Therefore, wave energy costs can only be competitive if the *running costs* are significantly below those for a conventional station. Naturally the 'fuel' or wave energy costs are zero, leaving the operation and maintenance costs as the determining factor. Schemes will therefore have to be reliable in their energy conversion and robust enough to survive the wave climate for many years. This means schemes designed for long lifetimes and with small numbers of moving parts (to minimize failures). The oscillating water columns and TAPCHAN schemes are good examples of what is required.

On the other hand the wave energy pioneer Stephen Salter has long argued against 'simplicity' and in favour, for instance, of sophisticated control and power conversion systems to maximize the useful energy that can be captured from the large and expensive structures that had traditionally been required to intercept the waves (Salter et al., 2002).

The UK Committee on Climate Change (CCC, 2011) has calculated the cost of electricity from a possible future (2030) 50 MW array of shoreline wave energy converters, with a capital cost of £2200 per kW and life expectancy of 40 years. Using a 10% discount rate and expressed in £(2010):

- for a low capacity factor (15%) the cost of electricity is 29.1p kWh^{-1}
- for a high capacity factor (22%) the cost of electricity is 19.9p kWh^{-1}.

As with the hydro and tidal plants discussed in Chapters 5 and 6 the discount rate used is an important factor, and is discussed further in Chapter 10 and Appendix B.

This illustrates the importance of achieving a high capacity factor – making as much use of the device as possible. The cost figures reported here are high and reflect the fact that the fixed devices cannot usually benefit from mass production as they would be purpose built for a specific location, and also that such devices would generally operate in shallow water where there is a much reduced wave energy climate.

The total capital investment required for wave energy schemes is dependent on overall average efficiencies and on location. Many of the devices detailed in this chapter have average efficiencies of around 30%. Frequency response characteristics and limitations of swept volume and survival when operating in very energetic seas are responsible for the generally low overall efficiency. At the time of writing. the capital cost is typically around £3000–£4500 per

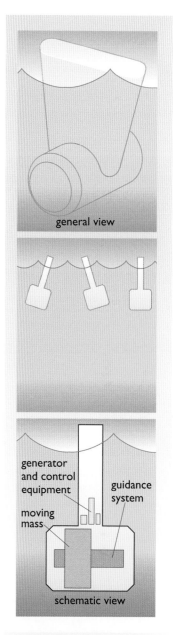

Figure 8.36 The Pitching and Surging FROG wave energy converter

installed kW, although the cost of particular schemes may vary markedly from this.

Large schemes are technically demanding because of the high structural loads imposed by the north Atlantic wave climate. As time has passed there have been improvements in design, performance and construction techniques, together with rationalization of some of the problems and a move to smaller schemes. Smaller schemes are technically simpler and less financially risky, and hence the capital costs and insurance costs are reduced, with commensurate reduction in produced energy costs.

Wave energy technology has moved into the commercial world and several developers are already deploying prototypes and, in some cases, executing schemes generating electricity or desalinating sea water at favourable prices. Coupled with incentives for avoided carbon dioxide emissions, the economic prospects for commercial wave energy exploitation appear to be good.

As wave energy is considered to be environmentally benign (see below), if the technology can be successfully developed it is likely to become an attractive commercial and political proposition, and this should result in an extensive installation programme with wave farms deployed in many locations. The cost of such installations will certainly fall from current levels due to refinement to designs, making them more efficient, and to lower production costs when they are (where possible) mass produced.

8.7 Environmental impact

Wave energy converters may be among the most environmentally benign of energy technologies for the following reasons.

- They have little potential for chemical pollution. At most, they may contain some lubricating or hydraulic oil, which will be carefully sealed from the environment.
- They have little visual impact except where shore-mounted.
- Noise generation is likely to be low – generally lower than the noise of crashing waves (there might be low-frequency noise effects on cetaceans, but this has yet to be confirmed).
- They should present a small (though not insignificant) hazard to shipping.
- They should present no difficulties to migrating fish.
- Floating schemes, since they are incapable of extracting more than a small fraction of the energy of storms, will not significantly influence the coastal environment. Of course, a scheme such as a new breakwater incorporating a wave energy device will provide coastal protection, and may result in changes to the coastline. Concrete structures will need to be removed at the end of their operating life.
- Near-shore wave energy schemes will release (from, for example, construction and material transport) an estimated 11 g of CO_2, 0.03 g of SO_2 and 0.05 g of NO_x for each kWh of electricity generated (Thorpe, 1999), making them very attractive in comparison to the conventional UK electrical generating mix of coal, gas and nuclear plants (see Chapter 10). Thus wave energy can make a significant contribution in meeting climate change and acid rain targets.

8.8 Integration

The electrical output from a wave energy scheme can be used directly but it is much more likely that the electricity will be fed into a grid. The electrical issues associated with such schemes include variability of supply of electricity due to the nature of waves, phasing, power factor correction and transmission losses (Freris and Infield, 2008).

Wave energy for isolated communities

If the grid is small and serves a small, remote community, great care must be taken in integrating the electrical output from a wave energy scheme: the output from a wave energy scheme will vary with time (except in the case of units such as the TAPCHAN) and may cause voltage or frequency fluctuations. Energy storage or other methods of smoothing may thus be necessary.

Many small communities currently depend on diesel generators for their electricity. A diesel generator is best run at a constant output close to its design capacity – say 50 kW. Therefore, if a diesel unit is the sole source of electricity, the load from the grid should always be matched to 50 kW. Clearly, the consumers will cause the load to vary as they switch appliances on and off, so to allow for this, a 'dump' load can be incorporated into the system. The diesel output is directed to this dump load when the load on the grid falls – for instance at night when the demand is low, energy can be used for space heaters, freezers, water heaters or to drive washing machines, etc.

Similarly, the incorporation of a varying electrical output from a wave energy scheme into the grid can partially be accommodated by the use of such dump loads located in houses, schools, factories etc. to stabilize the grid (Figure 8.37).

By careful overall design of an integrated scheme, a remote community could enjoy significant gains in electricity supply from a wave energy scheme. If this produces most of the energy the reduction in diesel oil consumption would be substantial, and since it is costly to transport diesel oil to remote locations, the cost savings could be large. Ideally the diesel generator would be held in reserve and only brought into use when the wave activity is too low to meet demand. An energy storage system such

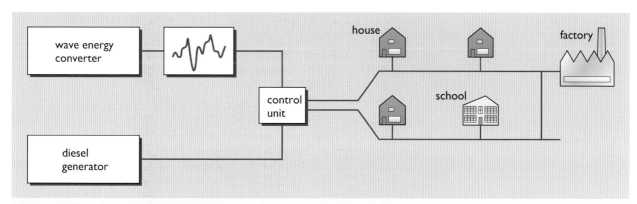

Figure 8.37 An integrated wave energy/diesel system

as batteries would be very useful in smoothing the supply and minimizing the call on the diesel generator.

Wave energy for large electricity grids

When the electrical outputs of several wave energy units are added together, the total output will be generally smoother than for a single unit. If we extend this to an array of several hundred floating devices, then the summed output will be smoother still. In addition, any fluctuations in output will be less important if the electricity is to be delivered to large national systems like those of the UK, where in most locations the grid is 'strong' enough to absorb contributions from a fluctuating source. Figure 8.38 illustrates a typical scheme.

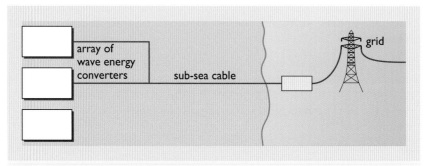

Figure 8.38 Electrical connections for an array of wave energy devices

Finally, although we have dwelt upon short-term fluctuations of seconds or minutes, the wave resource also varies on a day-to-day and season-by-season basis. For those countries in the north-east Atlantic such as the UK, Ireland, Spain and Norway, the seasonal variation shown in Figure 8.39 demonstrates that wave power output reaches a maximum in the bad weather of winter when the electrical demand is greatest (see for example, Rugbjerg et al., 2000 and Petroncini and Yemm, 2000).

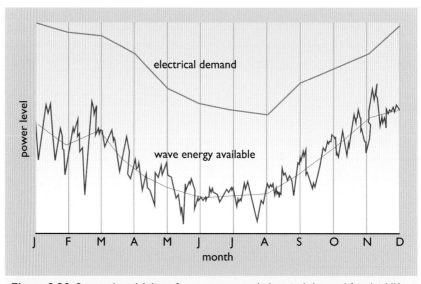

Figure 8.39 Seasonal availability of wave energy and electrical demand for the UK

8.9 Summary

The concept of extracting energy from ocean waves has intrigued people for centuries. The detailed understanding of waves, and of the energy that they contain, is complex, as is the description of a typical sea state which is composed of many waves, with different frequencies moving in different directions at different velocities. However, by developing wave theory even in a simple form we can build up a working knowledge which is suitable for designing wave energy converters.

There are some important outcomes from even a basic appreciation of waves – firstly that the world wave energy resource is extremely large (and even now has not yet been fully assessed). If only a small proportion of this could be harnessed it would make a major contribution to meeting the world's energy requirements.

The waves in deep water are very energy rich, but of course the conditions are difficult to operate in and so wave energy converters for this environment must be highly robust. As waves move towards the shore they lose some of their energy, and while the operating conditions are less strenuous, different technological challenges are raised in designing economic devices for capturing the wave energy. All told, there are over 1000 patents on wave energy devices, resulting in well over 100 wave energy projects – in this chapter a few have been reviewed as being representative of some of the more significant categories of wave energy converter shown in Figure 8.15.

In the UK, while the wave climate is conducive to wave power developments, the political climate has not always been so favourable. In the 1970s and 1980s the UK government's brief to wave energy teams was to design 2 GW schemes. In hindsight we can appreciate that this was remarkably ambitious – analogous to expecting someone to design a Boeing 747 in the early days of aviation without going through the evolution of the biplane, single seater monoplane, jet engine, etc. Attitudes are changing very quickly, however, prompted by the need to address global climate change, by the issue of long-term resource security of fossil fuel supplies and by the increasingly competitive economics of wave energy. Supporting this effort are the EMEC and Wave Hub test sites which have been established in Cornwall and the Orkneys respectively.

UK teams conducted much of the early work in the development of wave energy systems, but as this chapter records many other countries are now very active. Commercial involvement has had an important impact and we can now see private enterprise schemes and concepts emerging from countries such as Norway, Australia, Sweden, Denmark and the USA, as well as the UK. Across the globe, several companies have already commissioned their first wave stations, and more are obtaining permission to exploit sea areas and deploy sea trials.

It should be remembered that the generation of electricity is not the only option for the delivered energy. Desalination, coastal protection, water pumping, mariculture, mineral recovery from sea water and hydrogen generation are among the benefits being developed.

The cost of energy from the current generation of wave energy converters is high, but wave energy developers are confident that with time, experience and technological improvements costs will reduce, rendering wave energy an environmentally attractive and sustainable industry. Indeed some commentators have made the point that no other energy technology (coal, oil, nuclear, wind) has ever started commercial production from such a low unit cost as was initially expected of wave energy. If wave energy costs follow the example of wind power then rapid cost reductions can be expected (Aquamarine Power, 2011b).

The development of wave energy technologies has been a long process, but the economics of current designs are potentially attractive. Proving the long-term survival and cost effectiveness of designs as technologies mature should make the prospect of wave energy stations on a large scale a real possibility – we may well see such converters deployed in large numbers to harvest the considerable wave energy that is present at some locations.

References

AquaEnergy Group (2010) http://aquaenergygroup.com (accessed 18 November 2011).

Aquamarine Power (2011a) Technical Paper 'The Danish wind industry 1980–2010', *International Journal of the Society for Underwater Technology,* vol. 30, no. 1, pp. 27–31.

Aquamarine Power (2011b) http://www.aquamarinepower.com/ (accessed 18 November 2011).

AWS Ocean Energy Ltd (2011) http://www.awsocean.com (accessed 18 November 2011).

BERR (2008) *Atlas of UK Marine Renewable Energy Resources* http://www.renewables-atlas.info/ (accessed 18 November 2011).

Boake, C. B., Whittaker, T. J. T., Folley, M. and Ellen, H. (2002) 'Overview and initial operational experience of the LIMPET wave energy plant', *The Proceedings of the Twelfth (2002) International Offshore and Polar Engineering Conference*, Volume I, Kitakyushu, May 26–31, Cupertino, CA, ISOPE, pp. 586–594.

Budal, K. and Falnes, J. (1975) 'Power generation from ocean waves using a resonant oscillating system', *Marine Science Communications*, vol. 1, pp. 269–288.

Caratti, G., Lewis, A. and Howett, D. (eds) (1993) 'Wave energy R&D', *Workshop Proceedings,* Cork, 1–2 October, 1992, Commission of the European Communities, EUR 15079 EN.

Carbon Trust (2010) http://www.carbontrust.co.uk (accessed 18 November 2011).

CCC (2011) *The Renewable Energy Review: May 2011*, Committee on Climate Change [online], http:/www.theccc.org.uk/reports/renewable-energy-review (accessed 16 July 2012).

Claeson, L (1987) 'Energi från havets vågor', (in Swedish) published by Energiforskningsnämnden Efn-rapport nr 21.

ETSU (1985) *Wave Energy: The Department of Energy's R&D Programme 1974–1983*, Energy Technology Support Unit, Report R26.

EU-OEA (2010) *Oceans of energy: European Ocean Energy Roadmap 2010–2050*, Brussels, European Ocean Energy Association; also available at http://www.eu-oea.com/euoea/files/ccLibraryFiles/Filename/000000000827/ocean%20vectorise%2010%20mai%202010.pdf (accessed 27 May 2011).

Everett, B., Boyle, G. A., Peake S. and Ramage, J. (eds) (2012) *Energy Systems and Sustainability: Power for a Sustainable Future* (2nd edn), Oxford, Oxford University Press/Milton Keynes, The Open University.

Falcão, A. F. de O. (2000) 'The shoreline OWC wave power plant at the Azores', *Fourth European Wave Energy Conference*, Aalborg, Denmark, pp. 42–48.

Falnes, J. (2011) *Heaving buoys, point absorbers and arrays*, Paper given at the meeting on Wave Energy at the Royal Society 12–15th October 2010, to be published October 2011.

Falnes, J. and Lillebekken, P. M. (2003) 'Budal's latching controlled buoy type wave power plant', *Fifth European Wave Energy Conference*, Cork, Ireland.

Falnes, J. and Løvseth, J. (1991) 'Ocean wave energy', *Energy Policy*, vol. 19, pp. 768–775.

Freris, L. and Infield, D. D. (2008) *Renewable Energy in Power Systems*, Wiley.

Gill, A. T. and Rocheleau, R. E. (2009) *Advances in Hawaii's Ocean Energy RD&D*, The eighth European Wave and Tidal Energy Conference EWTEC 2009, Uppsala, Sweden, September 2009.

Heath, T. (2011) *A Review of Oscillating Water Columns*, Paper given at the meeting on Wave Energy at the Royal Society 12–15th October 2010, published October 2011.

ISOPE (2002) *The Proceedings of the Twelfth (2002) International Offshore and Polar Engineering Conference, Volume I: Ocean Resources and Engineering*, Kitakyushu, Japan, ISOPE.

IT Power (2011) http://www.itpower.co.uk (accessed 18 November 2011).

Jayashankara V. S., Anand A. T., Geetha A. S., Santhakumar B. V., Jagadeesh Kumar A. M., Ravindran C. T., Setoguchi D. M., Takao E. K., Toyota F. S. and Nagata F. (2009) 'A twin unidirectional impulse turbine topology for OWC based wave energy plants', *Renewable Energy*, vol. 34, pp. 692–698.

Kondo, M. (1993) *Proceedings of International Symposium on Ocean Energy Development*, Muroran Institute of Technology, Japan.

Kraemer, D. R. B., Ohl, C. O. G. and McCormick, M. E. (2000) 'Comparisons of experimental and theoretical results of the motions of a McCabe wave pump', *Fourth European Wave Energy Conference*, Aalborg, Denmark, pp. 211–218.

Lagström, G. (2000) 'Sea Power International – Floating Wave Power Vessel (FWPV)', *Fourth European Wave Energy Conference*, Aalborg, Denmark, pp. 141–146.

Le Crom, I., Brito-Melo, F., Neumann, A. and Sarmento A. J. N. A. (2009) 'Numerical Estimation of Incident Wave Parameters Based on the Air Pressure Measurements in Pico OWC Plant' EWTEC 2009, *The eighth European Wave and Tidal Energy Conference EWTEC 2009*, Uppsala, Sweden September 2009.

Le Crom, I., Brito-Melo, A., Neumann, F. and Sarmento, A. J. N. A. (2010) 'Portuguese Grid Connected OWC Power Plant: Monitoring Report', *Proceedings of the 20th Int. Offshore and Polar Eng Conf, ISOPE*, Beijing, China, June 2010.

Liu, Z., Hongda Shi N. S. and Beom-Soo Hyan (2009) 'Practical design and investigation of the breakwater OWC facility in China', EWTEC 2009. *The eighth European Wave and Tidal Energy Conference EWTEC 2009*, Uppsala, Sweden, September 2009.

Masuda, Y., Kuboki, T., Xianguang, L. and Peiya, S. (2000) 'Development of terminator type BBDB', *Fourth European Wave Energy Conference*, Aalborg, Denmark, pp. 147–152.

McCabe, A. P., Bradshaw, A., Widden, M. B., Chaplin, R. V., French M. J. and Meadowcroft, J. A. C. (2003) 'PS FROG MK 5: an offshore point absorber wave energy converter', *Fifth European Wave Energy Conference*, Cork, Ireland.

Meyer, N. I. and Nielsen, K. (2000) 'The Danish wave energy programme second year status', *Fourth European Wave Energy Conference*, Aalborg, Denmark, pp. 10–18.

Miyazaki, T. (1991) 'Wave energy research and development in Japan' *Oceans 91 Symposium,* Honolulu, Hawaii.

Miyazaki, T. and Hotta, H. (eds) (1991) *Third Symposium on Ocean Wave Energy Utilization*, (in Japanese) Tokyo, Japan.

Nielsen, K. and Plum, C. (2000) 'Point absorber – numerical and experimental results', *Fourth European Wave Energy Conference*, Aalborg, Denmark, pp. 219–226.

Nolan, G., O'Cathain, M., Ringwood, J. V. and Murtagh, J. (2003) 'Modelling, simulation and validation of the power take-off system for a hinge barge wave energy converter', *Fifth European Wave Energy Conference*, Cork, Ireland.

Oceanlinx (2011) http://www.oceanlinx.com/ (accessed 18 November 2011).

OPT (2010) http://www.oceanpowertechnologies.com/ (accessed 18 November 2011).

OPT (2011) http://www.oceanpowertechnologies.com/projects.htm (accessed 18 November 2011).

Petroncini, S. and Yemm, R. W. (2000) 'Introducing wave energy into the renewable energy marketplace', *Fourth European Wave Energy Conference*, Aalborg, Denmark, pp. 33–41.

Ravindran, M., Jayashankar, P. Jalihal and Pathak, A. G. (1995) 'Indian wave energy programme: progress and future plans', *Proceedings of the Second European Wave Power Conference*, Lisbon, November 1995.

Renewable Energy Focus (2011) http://www.Renewableenergyfocus.com (accessed 4 September 2011).

Ross, D. (1995) *Power from the Waves*, Oxford, Oxford University Press.

Rugbjerg, M., Nielsen, K., Christensen, J. H. and Jacobsen, V. (2000) 'Wave energy in the Danish part of the North Sea', *Fourth European Wave Energy Conference*, Aalborg, Denmark, pp. 64–71.

Salter, S. H. (1974) 'Wave power', *Nature*, vol. 249, pp. 720–724.

Salter, S. H., Taylor, J. R. M. and Caldswell, N. J. (2002) 'Power conversion mechanisms for wave energy', Proc. Inst. Mech. Engineers, vol. 216, Part M: *Engineering for the Maritime Environment*.

Salter, S. and Lin, C. P. (1995) 'The sloped IPS wave energy converter', *Second European Wave Power Conference*, Lisbon.

Sorensen, H. C., Christensen, L. and Hansen, L. K. (2003) 'Development of Wave Dragon from scale 1:50 to prototype', *Fifth European Wave Energy Conference*, Cork, Ireland.

Thorpe, T. W. (1992) *A Review of Wave Energy*, Energy Technology Support Unit, ETSU Report R-72.

Thorpe, T. W. (1998) *Overview of Wave Energy Technologies*, The Marine Foresight Panel, AEAT-3615.

Thorpe, T. W. (1999) *A Brief Review of Wave Energy*, Energy Technology Support Unit, ETSU Report R-122.

Thorpe, T. W. (2001) 'The UK market for marine renewables', *All-Energy Futures Conference*, Aberdeen; also available at http://www.wave-energy.net/Library/The%20UK%20Market%20for%20Marine%20Renewables.pdf (accessed 18 November 2011).

Thorpe, T. W. (2003) *A Brief Overview of Wave & Tidal Energy*, Supergen Future Networks http://oxfordoceanics.co.uk/supergen_future_networks.pdf (accessed 18 November 2011).

Wacher, A. and Neilsen, K. (2010) 'Mathematical and Numerical Modeling of the AquaBuOY Wave Energy Converter', *Mathematics-in-Industry Case Studies Journal*, vol. 2, pp. 16–33.

Washio, Y., Osawa, H. and Ogata, T. (2001) 'The open sea tests of the offshore floating type wave power device "Mighty Whale" – characteristics of wave energy absorption and power generation', *Ocean 2001 Symposium*.

WEC (2010) *Survey Of Energy Resources*, London, World Energy Council; also available online at http://www.worldenergy.org/publications/ser2010_report_1.pdf (accessed 27 May 2011).

Xianguang, L., Peiya, S., Wei, W. and Niandong, J. (2000) 'The experimental study of BBDB model with multipoint mooring', *Fourth European Wave Energy Conference*, Aalborg, Denmark, pp. 173–185.

Yemm, R., Pizer, D., Retzler, C. and Henderson, R. (2012) 'Pelamis: Experience from concept to connection', *Philosophical Transactions of the Royal Society*: A, vol. 370, no. 1959, pp. 365–380.

Chapter 9

Geothermal energy

By John Garnish and Geoff Brown

9.1 Introduction

Geothermal energy – heat from the Earth – is one of the less well recognized forms of renewable energy. This chapter explains the nature of geothermal energy, why it is treated as renewable even though locally heat is being mined, its usage over the past hundred years and its probable future. It is one of the few renewables that is constantly available. Although it used to be thought the preserve of regions prone to volcanic activity, that has now changed; new technologies are making this energy source available almost everywhere and it is realistic to think that geothermal energy could meet a significant fraction of energy demand within the twenty-first century. This chapter documents its current usage and outlines the steps that are being taken to make this forecast a reality.

Geothermal energy – The mining of geothermal heat

In the continuing search to find cost-effective forms of energy that neither contribute to global warming nor threaten national security, geothermal energy has become a significant player. This is the only form of 'renewable' energy that is independent of the Sun, having its ultimate source within the Earth. It is a comparatively diffuse resource; the amount of heat flowing through the Earth's surface, 10^{21} J y^{-1}, is tiny in comparison with the massive 5.4×10^{24} J y^{-1} solar heating of the Earth which also drives the atmospheric and hydrological cycles. Fortunately, there are many places where the Earth's heat flow is sufficiently concentrated to have generated natural resources in the form of steam and hot water (180–250 °C), available in rocks within feasible drilling distance of the ground surface and suitable for electricity generation. These are the so-called 'high-enthalpy' resources (see Box 9.1).

BOX 9.1 Enthalpy

Enthalpy is defined as the heat content of a substance per unit mass, and is a function of pressure and volume as well as temperature.

Geothermal resources are usually classified as 'high enthalpy' (water and steam at temperatures above about 180–200 °C), 'medium enthalpy' (about 100–180 °C) and 'low enthalpy' (<100 °C). The term 'enthalpy' is used because temperature alone is not sufficient to define the useful energy content of a steam/water mixture. A mass of steam at a given temperature and pressure can provide much more energy than the same mass of water under the same conditions. The distinction is important to geothermal practitioners so the term has entered into general use.

For the purposes of this chapter, however, it is usually sufficient to think of temperature and enthalpy as going hand in hand.

The techniques for exploiting the resources are very simple in principle, and are analogous to the well-established techniques for extracting oil and gas. One or more boreholes are drilled into the reservoir, the hot fluid flows or is pumped to the surface and it is then used in conventional steam turbines or heating equipment. Typically, geothermal wells are drilled to depths from 700 to 3000 m.

Obviously, electricity is a more valuable and versatile end-product than hot water, so most attention tends to be focused on those resources capable of supporting power generation, i.e. hot enough to make electricity generation economic. By 2010 world electrical power-generating capacity from geothermal resources had reached 10.7 gigawatts electrical (10.7 GW_e), a small but significant contribution to energy needs in some areas (Table 9.1). It will be seen also that a further 11 countries intend to have geothermal generation plant in service by 2015, with total world capacity forecast to increase to 18.5 GW_e.

About a further 50 gigawatts thermal (50 GW_t) is also being harnessed in non-electrical 'direct use' applications, principally for space heating, agriculture, aquaculture and a variety of industrial processes. Many of these applications occur outside high-enthalpy regions where geological conditions are nevertheless suitable to allow warm water (less than 100 °C) to be pumped to the surface. In many cases, the heat from the water can be used directly, and in other cases the temperature may be increased by the use of heat pumps (see Chapter 2). These are 'low-enthalpy' resources; in 2010, these installations supplied a total of 122 000 GWh (Table 9.2), equivalent to nearly 10 million tonnes of oil.

In almost all these situations heat is being removed faster than it is replaced, so the concept of 'heat mining' is appropriate. Although geothermal resources are non-renewable on the scale of human lifetimes and, strictly, fall outside the remit of this book, they are included because they share many features with true renewable resources. For example, geothermal energy is a natural energy flow rather than a store of energy like fossil or nuclear fuels. At one time it was thought that many high-enthalpy resources were indeed renewable, in the sense that they could be exploited indefinitely, but experience of declining temperatures in some steam producing fields and simple calculations of heat supply and demand show that – locally - heat is being mined on a non-sustainable basis (see Box 9.2). Nevertheless, and especially as techniques become available for extracting energy from rocks that do not support natural water flows (the so-called Hot Dry Rock concept, now more correctly termed Enhanced – or Engineered – Geothermal Systems (EGS): see Section 9.4), the volumes of rock that are potentially exploitable are so large in comparison to an individual reservoir that – to use an agricultural analogy – the resource could be 'cropped' on a

Table 9.1 Geothermal electricity generation: installed capacity and energy produced in 2005 and 2010 and forecast capacity in 2015

Country	Installed in 2005 /MW	Energy in 2005 /GWh	Installed in 2010 /MW	Energy in 2010 /GWh	2015 forecast /MW
Argentina	–	–	–	–	30
Australia	0.2	0.5	1.1	0.5	40
Austria	1.1	3.2	1.4	3.8	5
Canada	–	–	–	–	20
Chile	–	–	–	–	150
China	28	96	24	150	60
Costa Rica	163	1 145	166	1 131	200
El Salvador	151	967	204	1 422	290
Ethiopia	7.3	0	7.3	10	45
France	15	102	16	95	35
Germany	0.2	1.5	6.6	50	15
Greece	–	–	–	–	30
Guatemala	33	212	52	289	120
Honduras	–	–	–	–	35
Hungary	–	–	–	–	5
Iceland	202	1 483	575	4 597	800
Indonesia	797	6 085	1 197	9 600	3 500
Italy	791	5 340	843	5 520	920
Japan	535	3 467	536	3 064	535
Kenya	129	1 088	167	1 430	530
Mexico	953	6 282	958	7 047	1 140
Nevis	–	–	–	–	35
New Zealand	435	2 774	628	4 055	1 240
Nicaragua	77	271	88	310	240
Papua New Guinea	6.0	17	56	450	75
Philippines	1 930	9 253	1 904	10 311	2 500
Portugal	16	90	29	175	60
Romania	–	–	–	–	5
Russia	79	85	82	441	190
Spain	–	–	–	–	40
Slovakia	–	–	–	–	5
Thailand	0.3	1.8	0.3	2.0	1
The Netherlands	–	–	–	–	5
Turkey	20	105	82	490	200
USA	2 564	16 840	3 093	16 603	5 400
TOTAL	**8 933**	**55 709**	**10 717**	**67 246**	**18 500**

Source: Bertani, 2010

Table 9.2 Summary of geothermal direct-use data worldwide, 2010 (included heat from ground source heat pumps)

Country	Capacity /MWt	Annual Use /GWh y⁻¹	Capacity factor	Country	Capacity /MWt	Annual Use /GWh y⁻¹	Capacity factor
Albania	11.5	11.2	0.11	Italy	867	2 761.6	0.36
Algeria	55.6	478.7	0.98	Japan	2 100	7 138.9	0.39
Argentina	307.5	1 085.3	0.40	Jordan	153.3	427.8	0.32
Armenia	1	4.2	0.48	Kenya	16	35.2	0.25
Australia	33.3	65.3	0.22	Korea (South)	229.3	543.0	0.27
Austria	662.9	1 035.6	0.18	Latvia	1.6	8.8	0.62
Belarus	3.42	9.4	0.31	Lithuania	48.1	114.3	0.27
Belgium	117.9	151.9	0.15	Macedonia	47.2	167.1	0.40
Bosnia & Herzegovina	21.7	70.9	0.37	Mexico	155.8	1 117.5	0.82
				Mongolia	6.8	59.2	0.99
Brazil	360.1	1 839.7	0.58	Morocco	5.0	22.0	0.50
Bulgaria	98.3	380.6	0.44	Nepal	2.72	20.5	0.86
Canada	1126	2 464.9	0.25	The Netherlands	1 410	2 972.3	0.24
Caribbean Islands	0.1	0.8	0.85	New Zealand	393	2 653.5	0.77
Chile	9.1	36.6	0.46	Norway	3 300	7 000.6	0.24
China	8 900	20 931.8	0.27	Papua New Guinea	0.1	0.3	0.32
Columbia	14.4	79.7	0.63	Peru	2.4	13.6	0.65
Costa Rica	1	5.8	0.67	Philippines	3.3	11.0	0.38
Croatia	67.5	130.3	0.22	Poland	281	417.0	0.17
Czech Republic	151.5	256.1	0.19	Portugal	28.1	107.3	0.44
Denmark	200	694.5	0.40	Romania	153.2	351.5	0.26
Ecuador	5.2	28.4	0.63	Russia	308.2	1 706.7	0.63
Egypt	1	4.2	0.48	Serbia	100.8	391.7	0.44
El Salvador	2	11.1	0.63	Slovak Republic	132.2	852.1	0.74
Estonia	63	98.9	0.18	Slovenia	104.2	315.7	0.35
Ethiopia	2.2	11.6	0.60	South Africa	6.0	31.9	0.61
Finland	857.9	2 325.2	0.31	Spain	141.0	190.0	0.15
France	1 345	3591.7	0.30	Sweden	4 460	12 584.6	0.32
Georgia	24.5	183.1	0.85	Switzerland	1 061	2 143.1	0.23
Germany	2 485.4	3 546.0	0.16	Tajikistan	2.9	15.4	0.60
Greece	134.6	260.5	0.22	Thailand	2.5	22.0	0.99
Guatemala	2.31	15.7	0.78	Tunisia	43.8	101.1	0.26
Honduras	1.93	12.5	0.74	Turkey	2 084	10 246.9	0.56
Hungary	654.6	2 713.3	0.47	Ukraine	10.9	33.0	0.35
Iceland	1 826	6 767.5	0.42	United Kingdom	186.6	236.1	0.14
India	265	707.0	0.30	United States	12 611	15 710.1	0.14
Indonesia	2.3	11.8	0.59	Venezuela	0.7	3.9	0.63
Iran	41.6	295.6	0.81	Vietnam	31.2	25.6	0.09
Ireland	152.9	212.2	0.16	Yemen	1	4.2	0.48
Israel	82.4	609.2	0.84	**Total**	**50 583**	**121 696**	**0.27**

Source: Lund et al. (2010)

rotational basis. Once a particular zone has been depleted, further boreholes could be drilled to a deeper layer, or a few kilometres further away, and production continued. After a few such operations, each with a lifetime of 20–30 years, the original zone would have regenerated to economically exploitable levels.

BOX 9.2 Deep geothermal extraction and recharge rates

The following simple calculations show the order of magnitude discrepancy between commercial geothermal extraction rates and thermal recharge by the Earth's natural heat flow. They demonstrate that it is *not the heat flow but the heat store that is being exploited*, even in high-enthalpy areas.

(a) The currently exploited area of Tuscany, northern Italy, totals about 2500 km^2. This is a generous estimate that ignores the fact that the active fields are only a small subset of this. The average heat flow is about 200 mW m^{-2} so the total heat flow through this surface is:

$$2500 \times 10^6 \ (m^2) \times 200 \times 10^{-3} \ (W \ m^{-2}) = 500 \ MW_t$$

The region currently supports generating capacity of >700 MW$_e$; at a mean generating efficiency of 15%, this would require 4600 MW$_t$. Moreover, steam supplies have been proved to support an electricity-generating capacity of 1500 MW$_e$, which would require an input of approximately 10 000 MW$_t$. So the ratio of forecast commercial production rate to thermal recharge is at least:

$$\frac{10\,000 \ MW_t}{500 \ MW_t} \ \text{or 20:1.}$$

(b) In the Imperial Valley of California, USA, a commercial lease of 4 km^2 is expected to support generating capacity of some 40 MW$_e$ (which will require at least 250 MW$_t$). Assuming an average heat flow of 200 mW m^{-2}, the thermal recharge is less than 1 MW$_t$, giving a ratio of:

$$\frac{250 \ MW_t}{1 \ MW_t} \ \text{or 250:1.}$$

(c) The well field in Krafla, Iceland, covers an area of some 50 km^2. It currently generates some 60 MW$_e$, implying heat extraction of some 400 MW$_t$. To recharge the field on a sustainable basis, the heat flow would have to be 8 W m^{-2}. Again, this is much higher than could be expected even in this very active area. Of course, there is recharge by the Earth's natural heat flow but a thermally depleted reservoir may take many tens or even hundreds of years to regenerate. However, in regions of large geothermal resources this is generally a minor problem.

Note: MW$_t$ denotes megawatts of thermal energy; MW$_e$ denotes megawatts of electrical energy.

The source of heat

Heat flows out of the Earth because of the massive temperature difference between the surface and the interior: the temperature at the centre is around 7000 °C. So why is the Earth hot? There are two reasons: first, when the Earth formed around 4600 million years ago the interior was heated rapidly as the kinetic and gravitational energy of accreting material was converted into heat. If this were all, however, the Earth would have cooled within 100 million years. Of much greater importance today is the second mechanism: the Earth contains tiny quantities of long-lived radioactive isotopes, principally thorium 232, uranium 238 and potassium 40, all of which liberate heat as they decay. These radiogenic elements are concentrated in the upper crustal rocks. Cumulative heat production from these radioactive isotopes (approximately 5×10^{20} J y^{-1} today) accounts for about half the surface heat flow, though the exponential decay laws for radioactivity imply that heat production was about five times greater soon after the Earth formed. Heat is transferred through the main body of the Earth principally by convection, involving motion of material mainly by creep processes in hot deformable solids. This is a very efficient heat transport process resulting in rather small variations of temperature across the depth of the convecting layer. Closer to the surface, across the outer 100 km or so of the Earth, the material is too rigid to convect because it is colder, so heat is transported by conduction and there are much larger increases of temperature with depth (i.e. larger 'thermal gradients': see Section 9.2). This rigid outer boundary layer, or shell, is broken into a number of fragments, the **lithospheric plates**, which move around the surface at speeds of a few centimetres a year in concert with the convective motions beneath (Figure 9.1). Only this last point is of direct relevance to geothermal exploitation; our ability to drill into the Earth is restricted to the upper few kilometres, so we have to look for mechanisms and locations where the Earth's interior heat is brought within our reach.

From our point of view, it is mainly at the boundaries between plates, particularly where they are in relative extension or compression, that heat flow reaches a maximum. Here, the heat energy flowing through the surface averages around 300 milliwatts per square metre (300 mW m^{-2}) as compared with a global mean of 60 mW m^{-2}.

However, along plate margins heat flow can be even more concentrated locally because rock material reaches the surface in a molten form, resulting in volcanic activity that is often spectacular. Storage of molten, or partially molten, rock at about 1000 °C just a few kilometres beneath the surface strongly augments the heat flow around even dormant volcanoes. These heat flows result in high thermal gradients, really the hallmark of high-enthalpy areas, which are further enhanced in the upper regions by induced convection of hot water. Over geological periods of time, this high heat flow has resulted in large quantities of heat being stored in the rocks at shallow depth, and it is these resources that are mined by geothermal exploitation and commonly used for electricity generation. In areas of lower heat flow, where convection of molten rock or water is reduced or absent, temperatures in the shallow rocks remain much lower, so any resources are likely to be suitable only for direct use applications. So we see that high-enthalpy resources, including all those currently exploited for geothermal electric power (see Figure 9.1), are associated with volcanically active plate margins or localized hot spots like the Hawaiian islands. Boiling mud pools, geysers and volcanic vents with hot steam are characteristic features of such geothermal areas.

Figure 9.1 Map of the earth's lithospheric plates indicating the relative speeds of motions by the lengths of the arrows (generally 1–10 cm per annum). Large dots indicate major high-enthalpy geothermal energy-producing areas

Historical perspective

The historical exploitation of geothermal resources dates back to Greek and Roman times, with early efforts made to harness hot water for medicinal, domestic and leisure applications. Roman spa towns in Britain generally sought to exploit natural warm water springs with crude but reliable plumbing technology. The early Polynesian settlers in New Zealand, who lived undisturbed by European influence for several hundred years until the eighteenth century, depended on geothermal steam for cooking and warmth, and hot water for bathing, washing and healing. Indeed, the healing properties of geothermal waters are renowned throughout the modern world and have important medical benefits.

By the nineteenth century, progress in engineering techniques made it possible to observe the thermal properties of underground rocks and fluids, and to exploit these with rudimentary drills. In Tuscany, where the indigenous geothermal fluids were exploited as a source of boron from the eighteenth century onwards, natural thermal energy was used in place of wood for concentrating and processing the solutions. An ingenious steam collection device, the *lagone coperto* ('covered pool'), sparked a rapid growth in the Italian chemical industry, resulting in flourishing international trade. The generation of electrical power started in 1904, fostered by Prince Piero Ginori Conti, and 1913 saw the arrival of the first 250 kW_e power plant at Larderello, marking the start of new industrial activity. Today the Larderello power station complex has a capacity approaching 800 MW_e and a rebuilding programme is in progress that will take the capacity to some 1000 MW_e (Figure 9.2).

(a)

(b)

Figure 9.2 (a) The old geothermal power station at Larderello (b) the new power station

The Wairakei field in New Zealand was the second to be developed for commercial power generation, though not until the early 1950s. It was followed closely by the Geysers field in northern California where electricity was first generated in 1960. With an installed capacity that peaked at 2800 MW_e in the early 1990s, the Geysers field is still the most extensively developed in the world, though it will be overtaken soon by the Philippines. The Geysers field, however, illustrated the dangers of uncoordinated exploitation; the field was exploited independently by several different companies. In particular, little of the extracted water was reinjected into the reservoir after use. The result was a decline in steam pressure and a reduction in output capacity by several hundred megawatts. It was this decline that accounted for the very low net growth in world capacity during the second half of the 1990s. Fortunately, the problem was recognized in time, and reinjection is now practised widely; steam pressures and volumes are now recovering.

With the notable exceptions of Italy, the most volcanically active country in mainland Europe, and Iceland, which lies on the volcanic ridge of the central Atlantic, the chief geothermal nations are clustered around the Pacific rim. Japan, the Philippines, Indonesia and Mexico have shared in recent technological developments; the installations in El Salvador and Nicaragua are strategically vital to the economies of those nations, and several other countries, notably Costa Rica, Ecuador and Chile, also produce geothermal electricity.

Meanwhile, schemes making direct use of geothermal heat for district heating and agricultural purposes have advanced, with the major producers being China, the United States of America, Sweden, Turkey, Japan, Norway and Iceland. France developed substantial heating systems in the 1970s and 1980s, but the most significant event within continental Europe has been the opening up of the formerly centrally planned economies of eastern Europe. Many of these countries – notably the former East Germany, Poland, Romania, Hungary and Slovenia – possess good low-enthalpy aquifers and, with the increasing drive to minimize the environmental effects of their outdated and dirty solid fuel fired heating systems, began to develop district heating schemes along the lines of those in Paris. (The technology had evolved independently in eastern Europe and a few schemes were already in operation before the borders were opened, but progress was hampered by the lack of high-grade materials for pipelines, pumps and heat exchangers that could withstand the corrosive effects of brines in poorly maintained systems.)

Finally, there has been a quiet revolution in the use of 'shallow' geothermal energy, especially in Europe and the USA, with the widespread adoption of ground source heat pumps (GSHPs). Whereas 'deep' geothermal installations are mainly large scale schemes, the scale for GSHPs can be that of the single dwelling. Larger installations use multiple shallow boreholes feeding large heat pumps. They extract heat at only 12–15 °C from depths of 100–150 m and do not depend on the presence of water-bearing rocks at that depth. Consequently, they can be exploited almost anywhere. Taking a typical unit as having a 12 kW$_t$ output, the most recent survey estimates that there is now the equivalent of nearly 3 million units installed worldwide (Lund et al., 2010). They are particularly valuable when a building requires cooling or air conditioning in summer as well as heating in winter, as they can be designed to reverse the cycle, extracting heat from a building and depositing it in the ground (and thereby helping also to regenerate the resource).

9.2 The physics of deep geothermal resources

Primary ingredients

Geothermal resources of most types must have three important characteristics, as shown in Figure 9.3: an aquifer containing water that can be accessed by drilling; a cap rock to retain the geothermal fluid; and a heat source.

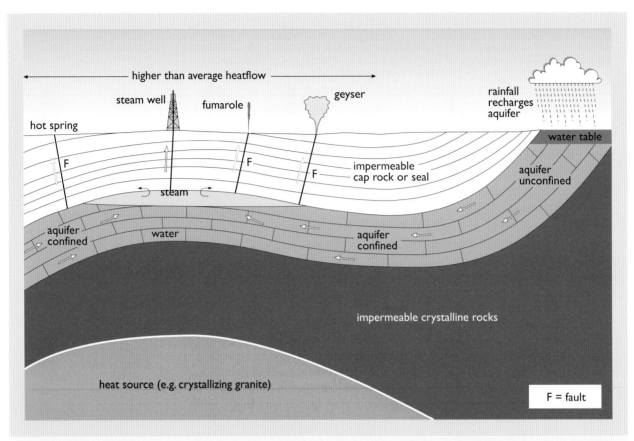

Figure 9.3 Simplified schematic cross-section to show the three essential characteristics of a geothermal site: an aquifer (e.g. fractured limestone with solution cavities); an impermeable cap rock to seal the aquifer (e.g. clays or shales); and a heat source (e.g. crystallizing granite). Steam and hot water escape naturally through faults (F) in the cap rock, forming fumaroles (steam only), geysers (hot water and steam), or hot springs (hot water only). The aquifer is unconfined where it is open to the surface in the recharge area, where rainfall infiltrates to keep the aquifer full, as indicated by the water table just below the surface. The aquifer is confined where it is beneath the cap rock. Impermeable crystalline rocks prevent downward loss of water from the aquifer

First, what is an aquifer? **Natural aquifers** are porous rocks that can store water and through which water will flow. **Porosity** refers to the cavities present in the rock, whereas the ability to transmit water is known as **permeability**. A geothermal aquifer must be able to sustain a flow of geothermal fluid, so even highly porous rocks will only be suitable as geothermal aquifers if the pores are interconnected. In Figure 9.4, rocks (a) and (c) are porous and likely to be highly permeable, whilst (b) and (d) have low porosity and permeability. Example (e), however, has low permeability despite its high porosity whereas cavities developed in (f) by dissolution of more soluble components give high porosity and permeability. Permeability due to fracturing ('fracture permeability'), as in (g), is particularly important in many geothermal fields.

A good measure of the permeability of a rock is its hydraulic conductivity (K_w). Darcy's Law states that the speed (v) of a fluid moving through a porous medium is proportional to the hydraulic pressure gradient causing the flow:

$$v = K_w \frac{H}{L}$$

(1)

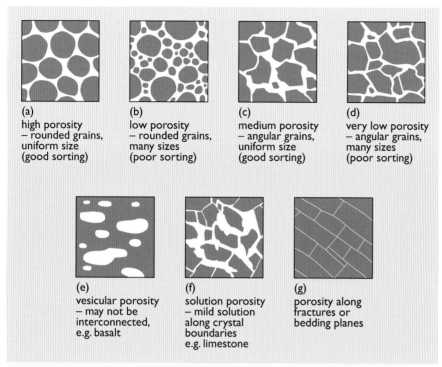

Figure 9.4 The relationship between grain size, shape and porosity in sedimentary rocks, especially sandstones (a–d); vesicular porosity in crystallized lava flows due to gas bubbles (e); and solution porosity resulting from rock dissolution, especially where acid groundwaters attack limestone (f). Porosity also develops in rocks along original planes of weakness, especially bedding planes and fractures (joints and faults) (g)

Here, H is the effective head of water driving the flow, and is measured in metres of water. The pressure gradient, or hydraulic gradient (H/L) is the change in this head per metre of distance L along the flow direction.

The volume of water (Q) flowing in unit time through a cross-sectional area A m^2 is v times A. So Darcy's Law may also be written:

$$Q = A K_w \frac{H}{L} \tag{2}$$

and K_w (the hydraulic conductivity) may be interpreted as the volume flowing through one square metre in unit time under unit hydraulic gradient. Some values of hydraulic conductivity for different rocks are given in Table 9.3.

Notice that the highest values of K_w occur in coarse-grained unconsolidated rocks, such as the ash layers, which are particularly common in volcanic areas, but that values are also quite high in some limestones and sandstones. These are aquifer rocks, with high permeability. It should be remembered also that fracture permeability is often important in geothermal aquifers (see Figure 9.4(g)), and is central to the Hot Dry Rock/Enhanced Geothermal System (HDR/EGS) concept.

In a confined aquifer, such as in Figure 9.3, the fluid pressure beneath the extraction point is high because there is a **cap rock**, a relatively impermeable rock, or seal, to prevent fluid escaping upwards. A cap rock is essential if a

Table 9.3 Typical porosities and hydraulic conductivities

Material	Porosity/%	Hydraulic conductivity/m day⁻¹
Unconsolidated sediments		
Clay	45–60	$<10^{-2}$
Silt	40–50	10^{-2}–1
Sand, volcanic ash	30–40	1–500
Gravel	25–35	500–10 000
Consolidated sedimentary rocks		
Mudrock	5–15	10^{-8}–10^{-6}
Sandstone[1]	5–30	10^{-4}–10
Limestone[1]	0.1–30	10^{-5}–10
Crystalline rocks		
Solidified lava[1]	0.001–1	0.0003–3
Granite[2]	0.0001–1	0.003–0.03
Slate	0.001–1	10^{-8}–10^{-5}

[1] The larger values of porosity and hydraulic conductivity apply to heavily fractured rocks and, for limestones, may also reflect the presence of solution cavities (see Figure 9.4(f)).
[2] Granite is a coarsely crystalline rock that has cooled down slowly from a melt at depth in the Earth. Such rocks are generally non-porous and impermeable, but contain many natural fractures and acquire limited permeability.

steam field is to develop. Mudrocks, clays and unfractured lavas are ideal. The importance of cap rocks was demonstrated in the early 1980s during exploration for geothermal resources in a very obvious place, the flanks of the volcano Vesuvius. Only small amounts of low-pressure fluid were discovered because the volcanic ashes that form its flanks are apparently quite permeable throughout. Given time, alteration of the uppermost deposits or overlying sediments by hot water and steam can create clays or deposit salts in pore spaces, so producing a seal over the aquifer, and in this way many geothermal fields eventually develop their own cap rocks. For this reason, however, the youngest volcanic areas, like Vesuvius, are not necessarily the most productive from a geothermal viewpoint.

The third prerequisite for exploitable geothermal resources is the presence of a heat source. In high-enthalpy regions, abundant volcanic heat is available, but in low-enthalpy areas the heat source is less obvious. In such regions there are two main types of resource: (a) those located in deep sedimentary basins where aquifers carry water to depths where it becomes warm enough to exploit, and (b) those located in 'hot dry rocks' where natural heat production is high but an artificial aquifer must be created by enhancing the rock fractures in order that the geothermal resource may be exploited. Let us now look at each type of resource in more detail.

Volcano-related heat sources and fluids

The heat supply for a high-enthalpy field is usually derived from a cooling and solidifying body of magma (partially molten rock), which need not necessarily be centred directly beneath the geothermal field (Figure 9.5). It may seem surprising that much of the magma rising beneath a volcano is not erupted but instead reaches only a level of neutral buoyancy at which its density is the same as that of the surrounding rocks. Two factors conspire to halt the rising magma: first the pressure of overlying rocks reduces as the magma ascends; this promotes the separation of liquid magma from its dissolved gases, which are lost, increasing the density of the remaining magma; second, shallower rocks are inherently less dense than rocks at greater depth, usually because they are less compressed. So, whereas volcanic eruptions are driven by exceptionally high gas pressures, many magmas form 'intrusions', coming to rest and crystallizing beneath the surface at 1–5 km depth.

In the 1980s, experiments were undertaken in the USA with the ultimate aim of drilling directly into or very close to magma bodies, where temperatures may be up to 1800 °C, and to harness geothermal power by cycling water through their outer margins. In preparation, the US Magma Energy Program

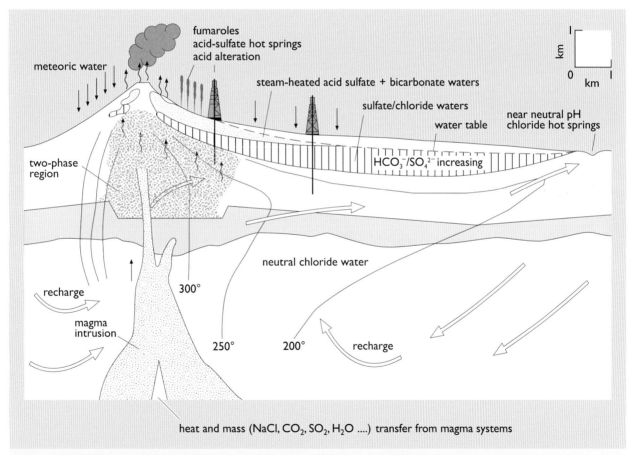

Figure 9.5 Conceptual model of a typical volcanic geothermal system in which meteoric (rain) waters percolate deep into the volcanic superstructure where they are heated by a body of magma that loses dissolved gases as it rises, and forms an intrusion (see text). The hot aqueous fluid rises and may reach the point at which water boils to form steam, producing a two-phase (steam + water) zone. The hydraulic gradient causes the geothermal fluid to migrate through any permeable rocks in the volcano flank. Here the fluids may be accessed by drilling (see drill rig symbols); the chemistry of the fluid changes during migration due to mixing with CO_2-saturated rain water

succeeded in drilling into the molten Kilauea lava lake (Hawaii) and ran successful energy extraction experiments. In any drilling operation, the drill bit is cooled and lubricated by circulating fluid (referred to as **mud** even though, usually, it is mainly water), which also lifts cuttings to the surface. In this case, as was expected, the circulating water solidified a thin shell of lava around the drill bit. The resulting tube of solid but thermally fractured rock acted as a heat exchanger, with heat being transferred to the drill hole by convecting magma. While a useful test, this is still a long step from drilling into a live magma chamber with high-pressure dissolved gases, and no further work has been undertaken in recent years.

A close encounter with magma occurred during the development of the Krafla field in northern Iceland when, in 1977, rising magma reached the depth of a borehole at 1138 m and three tonnes of magma was erupted through the hole in 20 minutes! Quite by chance, the development history of this field was dogged by a series of eruptions in 1975 and 1984, the first at Krafla for over 250 years, but progress improved once the eruptions ceased and the field now supports 60 MW$_e$ of power generation. Krafla is also the site of the Icelandic Deep Drilling Project (IDDP), the aim of which is to find and exploit supercritical fluids (>350 °C) that would transform the efficiency of power generation. The first hole, intended to be drilled to 4500 m, was suspended late in 2010 after encountering magma at 2100 m that must have chilled from >1000 °C.

Several of the world's most advanced geothermal sites (for example in northern Italy and the western USA) are located in extinct volcanic areas. Fortunately for geothermal exploitation, because rocks are such good insulators, magmatic intrusions may take millions of years to cool to ambient conditions. Such intrusions, therefore, continue to act as a focus for 'hot fluid', or hydrothermal convective cycles in permeable strata as in Figure 9.5. The nature of the resource then depends on the local conditions of pressure and temperature in the aquifer, and this determines the extraction technology and the profitability of the site.

The range of pressures and temperatures of interest to current geothermal developments lies typically between 100 and 300 °C, below the critical point at which liquid water and water vapour become indistinguishable (though conditions in parts of some fields exceed 400 °C), and in the pressure range up to about 20 megapascals (20 MPa). As noted above, however, there is increasing research interest in the possible exploitation of supercritical geothermal fluids, which would offer the prospect of very much greater conversion efficiencies.

In simple terms, high-enthalpy systems are subdivided into vapour dominated and liquid dominated, depending on the main pressure-controlling phase (i.e. steam or liquid water) in the reservoir. Vapour-dominated systems are the best and most productive geothermal resources, largely because the steam is dry (free of liquid water) and is of very high enthalpy. Where reservoir rocks are at pressures below hydrostatic, which promotes steam formation (perhaps 3–3.5 MPa at depths down to 2 km), there must be some barrier to direct vertical groundwater infiltration. The Larderello field in Italy (Figure 9.2) is of this type.

In contrast, liquid-dominated systems are at higher than hydrostatic pressures, exceeding 10 MPa at depths below 1 km (because at 1 km depth the hydrostatic pressure is about 100 bars, i.e. 10 MPa, see Box 9.3). Production of electricity from liquid-dominated systems benefits from the higher fluid pressures at depth, and water can 'flash' into steam en route to the surface. That is to say, the pressure-temperature curve in liquid-dominated systems is *below* the boiling curve for water at all depths, but when the thermal aquifer is punctured by boreholes, the reduced pressure in the well means that the rising water crosses the boiling point curve on its way to the surface. However, the steam is often wet and of lower enthalpy, which adds to the technical problems for electricity production. The famous Wairakei field in New Zealand is liquid-dominated but, typically for such systems, has developed a two-phase zone as pressures have fallen during exploitation. Fortunately, the groundwater zone has a relatively low permeability, which suppresses the tendency for natural venting of steam over most of the Wairakei area.

BOX 9.3 Pressure and depth

As noted in Chapter 5, Box 5.6, hydrostatic pressure increases by about 1 atmosphere for an increase of 10 m in depth. It follows that a geothermal aquifer 1 km thick will produce a pressure increase of 100 atmospheres (100 bar). (A pressure of 1 bar is approximately equal to 1 atmosphere.)

The SI unit for pressure is the pascal or, more appropriately for the high pressures in geothermal systems, the megapascal (MPa). One megapascal is approximately 10 atmospheres, and the 20 MPa mentioned in the text is thus 200 atmospheres.

Many geothermal aquifers also contain a steam zone and, since steam has a much lower density than water, the **vapourstatic** increase in pressure with depth (i.e. the pressure due to the weight of the column of steam) is much smaller than the hydrostatic increase. Note that it is not strictly correct to refer to water vapour as steam. Steam is condensed droplets of water (as in clouds), which is why it can be seen. Water vapour is an invisible gas. However, 'steam' is often used as a synonym for 'water vapour', though the meaning should generally be clear from the context.

The heat source in sedimentary basins

An important key to understanding many geothermal resources is the heat conduction equation:

$$q = K_T \frac{\Delta T}{z} \tag{3}$$

This is analogous to Darcy's Law, but here q is the one-dimensional vertical **heat flow** in watts per square metre (W m^{-2}). ΔT is the temperature difference across a vertical height z, and $\Delta T/z$ is thus the **thermal gradient**. The constant K_T relating these quantities is the **thermal conductivity** of the rock (in W m^{-1} K^{-1}) and is equal to the heat flow per second through an area of 1 m^2 when the thermal gradient is 1 °C per metre along the flow direction.

Values of K_T for most rock types are quite similar, in the range 2.5–3.5 W m^{-1} °K^{-1} for sandstones, limestones and most crystalline rocks. However, mudrocks (clays and shales) are the exceptions, with lower values of

1–2 W m^{-1} $^{\circ}$K^{-1}. These are also among the most impermeable rocks (Table 9.3), so mudrocks contribute two of the essential characteristics for geothermal resources: they act as impermeable cap rocks and as an insulating blanket, enhancing the geothermal gradient above aquifers in regions of otherwise normal heat flow.

So, even under conditions of average heat flow (60 mW m^{-2}), it is possible to obtain temperatures of 60 $^{\circ}$C within the top 2 km of the Earth's crust. Box 9.4 demonstrates how the differing insulating properties of rocks influence the way the temperature varies with depth. To maintain the same vertical heat flow, low-conductivity rocks require a steeper temperature gradient than a relatively good conductor, and are accordingly important in augmenting temperatures at depth.

BOX 9.4 Thermal gradient and heat flow

Consider the situation where there is a steady upward flow of heat through the top few kilometres of the Earth's crust. We can use Equation 3 to relate this flow to the temperature at any depth if we know the thermal conductivity of the rock.

If, for instance, the temperature is found to be 58 $^{\circ}$C at a depth of 2 km (2000 metres) and the surface temperature is 10 $^{\circ}$C, the temperature gradient is:

$$(58 - 10)/2000 = 0.024 \ ^{\circ}\text{C m}^{-1}$$

and if the thermal conductivity of the rock is 2.5 W m^{-1} $^{\circ}$C^{-1}, the heat flow rate is:

$$2.5 \times 0.024 = 0.060 \ \text{W m}^{-2}$$

or 60 mW m^{-2}.

Suppose, however, that this same 60 mW flows up through several layers with different thermal conductivities. Equation 3 tells us that the thermal gradient must be different in each layer, with the temperature changing most rapidly through the layer with the lowest conductivity, as in Figure 9.6 below.

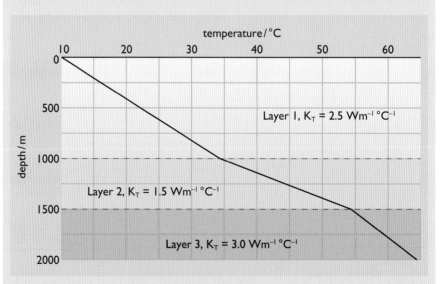

Figure 9.6 Variation of temperature with depth across three zones of differing thermal conductivity, K_T

We can check that the diagram shows the correct temperatures by using Equation 3 to calculate the temperature gradient for each layer and comparing this with the gradient read from the graph:

Layer 1

The calculated gradient is $0.060/2.5 = 0.024$ °C m^{-1}

The measured gradient is $(34.5 - 10.0)/1000 = 0.0245$ °C m^{-1}

Layer 2

The calculated gradient is $0.060/1.5 = 0.040$ °C m^{-1}

The measured gradient is $(54.5 - 34.5)/500 = 0.040$ °C m^{-1}

Layer 3

The calculated gradient is $0.060/3.0 = 0.020$ °C m^{-1}

The measured gradient is $(64.5 - 54.5)/500 = 0.020$ °C m^{-1}

Within the precision of the data therefore, the temperatures shown are consistent with a heat flow rate of 60 mW m^{-2} through each layer. Comparing this case (Figure 9.6) with the uniform rock considered above, it is obvious that the presence of the thin layer with low thermal conductivity has appreciably enhanced the temperature at a depth of 2 km.

This has led to exploration programmes aimed at locating natural waters in areas of thick sedimentary rock sequences containing mudrocks and permeable limestones or sandstones. For example, the Paris area is at the centre of a 200 km wide depression in the crystalline basement rocks. Exploration for hydrocarbon resources in the 1960s and 1970s found very little oil or gas, but was extremely successful in locating hot water between 55 °C and 70 °C at depths of 1–2 km. While low-enthalpy water resources are unsuitable for power generation (no high-pressure steam can be produced, and temperatures are too low to permit an acceptable generation efficiency), they can be of considerable benefit in meeting demands for low-grade heat (space heating, etc.). However, to be economic they must be located close to a heat load. The Paris area is ideal in this respect. Similar resources occur in some of the sedimentary basin areas of the UK, such as beneath the Yorkshire–Lincolnshire coast and in Hampshire, where the only commercial UK geothermal scheme operates in Southampton, but most are remote from suitable heat loads.

Finally, there are two extensions of the criteria discussed above which make some sedimentary basin resources more attractive:

(1) There are some basins where the background heat flow is above average, and these have led to large-scale non-electrical applications of geothermal energy in a number of countries (Table 9.2). The geological reasons for the association of high heat flow with sedimentary basins are not altogether surprising: stretching processes within the Earth's outer plate layer induce thinning that can radically raise the heat flow as well as creating a surface depression on which sedimentation occurs. Beneath the south Hungarian Plain, for example, geothermal gradients as high as 0.15 °C m^{-1} have been recorded and 120 °C water occurs at 1 km depth.

(2) In other areas, larger sedimentary thicknesses may occur. For example, high-pressure fluids at temperatures of 160–200 °C occur at 3–5 km depth in the Gulf of Mexico, southern Texas and Louisiana. Because of chemical processes occurring as a result of the depth and temperature of burial and the efficient sealing of the aquifers by impermeable rocks, pressures greatly exceed hydrostatic and 100 MPa has been recorded in local pockets of fluid. The fluids are highly saline brines with trapped gas, especially methane. These so-called **geopressured brines** are a potentially important geothermal resource for power generation, a resource that has remained untapped to date but on which there is intermittent government funding for research in the USA. The great advantage of geopressured resources is that they offer three kinds of energy: geothermal heat, 'hydraulic' energy (in water at high pressure), and the chemical energy in the large quantities of methane that are found dissolved in the fluid.

Geothermal waters

Most of the foregoing has been concerned with the source of geothermal heat. Exploitation of the heat, however, requires that the geothermal water be brought to the surface, and that brings with it a different set of problems. Water that has been in contact with rock for long periods (and geothermal waters can be thousands or even millions of years old) contains dissolved minerals. Hot water tends to be more reactive than cold water, so geothermal waters can often contain around 1% of dissolved solids. Typically, these will be carbonates, sulfates or chlorides, and dissolved silica becomes significant where waters have been in contact with rocks above 200 °C. For this reason, geothermal fluids are often called 'brines'. Dissolved gases are also common, especially at higher temperatures. Techniques are available to deal with all of these, but it is essential that they be taken into consideration at the design stage of the plant. With correct design, these contaminants can all be handled and disposed of without either operational or environmental difficulty. If they are ignored, however, or the plant designed before the water has been properly characterized, the entire system can fail within a matter of months. The various techniques that can be used are beyond the scope of this chapter, but some examples are quoted in Section 9.3.

'Hot dry rocks' or engineered geothermal systems (EGS)

Our attention now turns from sedimentary strata to the underlying crystalline 'basement'. Most rocks, especially crystalline and basement rocks, do not contain sufficient water to provide a viable geothermal resource, although far greater amounts of heat are stored in such rocks than are available in aquifers. It was thought initially that deep basement rocks would indeed be dry, and so the term **hot dry rock** (HDR) was coined around 1970 to describe the heat stored in impermeable (or poorly permeable) rock strata and the process of trying to extract that heat. More recently, it has been recognized that few if any rocks are actually dry, and there is now a general acceptance of the term 'enhanced' – or 'engineered' – 'geothermal systems' (EGS) to categorize the various projects aimed at extracting this heat.

When the permeability is too poor to allow the necessary flow of fluid, what is required is the creation of an artificial heat exchanger zone within suitably hot rocks. Because rocks are (by normal standards) poor conductors of heat, very large heat transfer surfaces (of the order of square kilometres) are required if heat is to be extracted at useful rates. This can be achieved by enhancing the natural fracture system that occurs in all such crystalline rocks. Water can then be circulated through the enhanced zone so that heat may be extracted, ideally to generate steam and, hence, electrical power. Although the technology to create suitable arrangements for reproducible heat recovery has not yet been perfected, in theory at least EGS technology could be applied over a significant proportion of the Earth's surface.

Because drilling is expensive, with costs rising exponentially with depth, only the upper 6–7 km of the Earth's crust is generally used in calculating geothermal energy potential (though some hydrocarbon and research drilling has gone as deep as 15 km). Given current technical and economic constraints on drilling depths, a minimum geothermal gradient of around $0.025\ °C\ m^{-1}$ is required if development is to be economic. With a typical thermal conductivity of $3\ W\ m^{-1}\ °C^{-1}$, this requires (from Equation 3) a heat flow of $75\ mW\ m^{-2}$, only a little above the Earth's average. In practice, however, to minimize expenditure it is customary to look for rocks with much higher heat flows (as at experimental sites in the USA, Japan and France). Granite bodies are ideal targets, because such rocks occupy large volumes of the upper crust and they crystallized from magmas that had naturally high concentrations of the chemical elements with long-lived radioactive isotopes – uranium, thorium and potassium. Here we reach a situation in which heat flow through the Earth's surface is augmented (perhaps by a factor of 2) by heat production within certain shallow crystalline rocks. If, in addition, a layer of poorly conducting sedimentary rocks overlies the granite, its 'blanketing' effect will increase the temperature gradient and make higher temperatures available at shallower depths.

9.3 Technologies for exploiting high-enthalpy steam fields

The first stage in prospecting for geothermal resources in volcanic areas involves a range of geological studies aimed at locating rocks that have been chemically altered by hot geothermal brines, and finding surface thermal manifestations, such as hot springs or mud pools. Investigations of fluid chemistry and, increasingly, the release of gases through fractured rocks allow assessment of the composition and resource potential of trapped fluids. These studies provide the first clues to the likely presence and location of exploitable resources. However, geophysical prospecting techniques, particularly resistivity surveying and other electrical methods designed to detect zones with electrically conducting fluids (i.e. brines), are probably the most effective for precise location of buried geothermal resources. Once a suitable geothermal aquifer has been located, exploration and production wells are drilled using special techniques to cope with the much higher temperatures and, in some cases, harder rock conditions than in oil and water wells. Since fluid pressures in the aquifer range up to about 10 MPa, the driller must ensure that the mud is dense enough to counteract these pressures and avoid '**blow out**', where an uncontrollable column of gas is discharged. The well is lined ('cased') with steel tubing that is cemented in place, leaving an open section or a perforated steel casing at production depths. As each string of casing has to be inserted through its predecessors as the well depth increases, the well diameter decreases with depth from perhaps 50 cm near the surface to 15 cm at production depths. A wellhead with valve gear is welded to the steel casing at ground level. This allows the well to be connected to a power plant via the network of insulated pipes that are a familiar sight in geothermal areas.

Technologies for electrical power generation depend critically on the nature of the resource – not just the fluid temperature and pressure but also its salinity and content of other gases, all of which affect plant efficiency and design. The size of any power station is determined by the economics of scale; conventional coal- or oil-fired stations are typically a few hundred megawatts per unit. A typical geothermal unit, by contrast, is usually 30–50 MW_e. This is because the amount of steam delivered by one well is usually sufficient to generate only a few MW_e, and wells are linked across the field and back to the station by pipeline. Above a certain capacity, the cost of the pipelines is such that it is cheaper to develop a separate station in another part of the field.

Given the fact that most of the costs of the electricity derived from geothermal resources are accounted for by the need to pay back the capital investment, with day-to-day operating costs being relatively minor (and insensitive to output variation), there is a great incentive to maximize the efficiency with which the relatively low-grade heat (by power generation standards) is converted into useful energy. Today there are several hundred installations operating worldwide and these include four main types, described below.

Dry steam power plant

As the name implies, this type of system (Figure 9.7(a)) is ideal for vapour-dominated resources where steam production is not contaminated with liquid. The reservoir produces superheated steam, typically at 180–225 °C and 4–8 MPa, reaching the surface at several hundred kilometres per hour and, if vented to the atmosphere, sounding like a jet engine at close proximity. Passing through the turbine, the steam expands, causing the blades and shaft to rotate and hence generating power. Temperatures up to 300–350 °C and correspondingly greater pressures are increasingly being exploited, leading to greater efficiency in electricity production.

In the simplest form of power plant, a 'back-pressure' unit, the low-pressure exhaust steam is vented directly to the atmosphere. Although such units are simple, they are also very inefficient; their main use is as temporary transportable units during the development of a new field. Once the steam supply is ensured, normal practice is then to install 'condensing' plant as shown in Figure 9.7(a). These achieve greater efficiency by condensing the exhaust steam to liquid, thus dramatically increasing the pressure drop across the turbine because liquid water occupies a volume roughly 1000 times less than the same mass of steam. Of course, the cooling towers generate waste heat in just the same way as conventional coal- and oil-power stations (or, indeed, any heat engine). At temperatures typical of geothermal fluids,

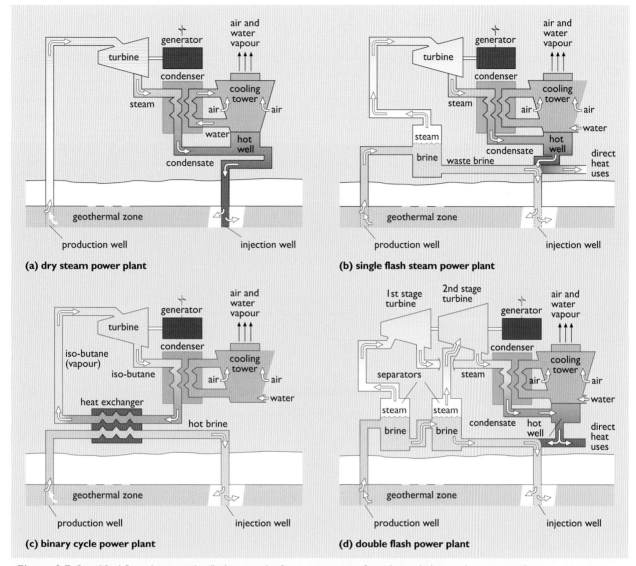

Figure 9.7 Simplified flow diagrams (a–d) showing the four main types of geothermal electrical energy production

efficiencies are low and, despite the use of high-temperature superheated steam, rarely exceed 20%. Nevertheless, whereas a 1960s plant required almost 15 kg steam per saleable kWh in optimum conditions, modern dry steam plant with higher temperature steam and better turbine designs can achieve 6.5 kg per kWh, so a 55 MW$_e$ plant requires 100 kg s^{-1} of steam.

Plant efficiency, and therefore profitability, is strongly affected by the presence in the geothermal fluid of so-called 'non-condensable' gases, such as carbon dioxide and hydrogen sulfide. When the turbine exhaust gases are cooled, achieving a suction effect on the turbine as the water condenses into liquid at around 100 °C, gases that do not similarly condense cause higher residual pressures at the back end of the turbine. Even a small percentage of such gases reduces suction efficiency and so impacts on the economics of the system; for this reason, many geothermal plants are fitted with gas ejectors. However, the ejectors themselves require either a steam supply or electric power from the turbine-generator and, consequently, reduce output. Non-condensable gases have an additional economic impact: it is no longer acceptable in most places to vent them into the atmosphere, so they must either be trapped chemically or reinjected with the waste water to avoid pollution, and both options entail additional costs.

In general, dry steam plant is the simplest and most commercially attractive. For that reason, dry steam fields were exploited early and have become disproportionately well known. In fact, only the USA and Italy have extensive dry steam resources, though Indonesia, Japan and Mexico also have a few such fields. Elsewhere, and even in these countries, liquid-dominated fields are far more common. While in some areas it is common practice to reinject the spent fluid, very little was reinjected in the largest field, the Geysers in the USA, until falling fluid pressures led to a recognition among the various private operators that the field was being over-exploited. There is now an agreed cooperative reinjection policy to make the resource more sustainable; some 70% of the mass of produced steam is now reinjected, and field production has been stabilized. (Not all of this fluid comes from the condensed steam from the turbines, because much is evaporated from the cooling towers. Instead, in a new and imaginative environmental development, treated sewage water is piped 48 km from a local community – thereby solving an unrelated local disposal problem as well as helping to maintain the geothermal reservoir.)

Single flash steam power plant

Here (Figure 9.7(b)) the geothermal fluid reaching the surface may be steam (the water having 'flashed', i.e. vaporized, within the well as pressure dropped during ascent) or hot water at high (close to reservoir) pressure. In the first case, a separator is installed simply to protect the turbine from a massive influx of water should conditions change. However, it is usually better to avoid flashing in the well because this can lead to a rapid build-up of scale deposits as minerals dissolved in the fluid come out of solution, leading to the plugging of the well. For this reason, the well is often kept under pressure to maintain fluid as liquid water. To deal with hot high-pressure water requires complex equipment designed to reduce the pressure in a controllable manner and induce flashing so that steam may be separated. Again, a conventional condensing steam turbine is at the heart of the plant, but lower steam pressures and temperatures (0.5–0.6 MPa, 155–165 °C) are common, so the plant typically requires more steam per kWh than would be required in a dry steam plant, say around 8 kg per kWh. Moreover, the bulk of the fluid produced, often up to 80%, may remain as unflashed hot brine which is then reinjected unless there are local direct use heating applications available. In general, therefore, reinjection wells must be available for fluid disposal both at single flash plants and at plants incorporating the newer types of technology described below. Increasingly, the provision of reinjection is becoming a standard requirement of many licensing schemes.

Binary cycle power plant

This type of power plant (Figure 9.7(c)) uses a secondary working fluid with a lower boiling point than water, such as pentane or butane, which is vaporized and used to drive the turbine. It is more commonly known as an organic Rankine cycle (ORC) plant (introduced earlier in Chapter 2, Box 2.4). Its main advantage is that lower-temperature resources can be developed where single flash systems have proved unsatisfactory. Moreover, chemically impure geothermal fluids can be exploited, especially if they are kept under pressure so that no flashing ever takes place. The geothermal brine is pumped at reservoir pressure through a heat exchange unit and is then reinjected; the surface loop is closed and no emissions to the environment need occur. Ideally, the thermal energy supplied is adequate to superheat the secondary fluid. (Note: a superheated fluid is a liquid above the normal boiling point, usually prevented from boiling by increasing the pressure). For geothermal fluid temperatures below about 170 °C, higher generating efficiencies are possible than in low-temperature flash steam plants. A disadvantage is that keeping the geothermal fluid under pressure and repressurizing the secondary fluid can consume some 30% of the overall power output of the system because large pumps are required. Large volumes of geothermal fluid are also involved; for example, the Mammoth geothermal plant in California uses around 700 kg s^{-1} to produce 30 MW$_e$. Nearly 250 binary cycle units are in operation today. The units are often small (5–10 MW$_e$ is typical). When larger systems are required, multiple sets are installed.

Double flash power plant

Recently, several attempts have been made to develop improved flashing techniques, particularly to avoid the high capital costs and parasitic power losses (e.g. circulating pumps for the secondary fluid) of binary plant. Double flash (Figure 9.7(d)) is ideal where geothermal fluids contain low levels of impurities and so the scaling and non-condensable gas problems that affect profitability are at a minimum. Quite simply, unflashed liquid remaining after the initial high-pressure flashing flows to a low-pressure tank where another pressure drop provides additional steam. This steam is mixed with the exhaust from the high-pressure turbine to drive a second turbine (or a second stage of the same turbine), ideally raising power output by 20–25% for only a 5% increase in plant cost. Even so, extremely large fluid volumes are required. The East Mesa plant in southern California, for example, commissioned in 1988, uses brine at 1000 kg s^{-1} from 16 wells to generate 37 MW$_e$; i.e. around five times as much fluid as for similar dry steam plant (though temperatures would be much higher in the latter case).

Future developments

As the geothermal industry continues to expand, there will be a need to develop technologies that can produce geothermal power from a variety of resources that are less ideal than dry steam. Increasing use is being made of geothermal fluids that are at either lower temperature than (but similar pressure to) those in dry steam fields, or at the same or higher temperature and much higher pressure. These are essentially liquid-dominated resources, albeit of high enthalpy, and they exist in large volumes. Inevitably, variants on the binary and double flash systems will continue to be developed; they are at the leading edge of current research. More recently, greater use has been made of the produced fluids by operating combined or hybrid cycles, using an ORC to extract further work from the main turbine exhaust or the separated water.

A number of other approaches are being developed to increase the efficiency of generation from lower temperature fluids, using either different working fluids or new power cycles, such as the Kalina cycle. The working fluid for the Kalina cycle is an ammonia-water mixture, the composition of which varies throughout the cycle. The net generating efficiency is expected to be up to 40% higher than for an ORC under the same conditions; at an inlet temperature of 130 °C the net efficiency is estimated at over 13% (58% of the theoretical Carnot efficiency – see Box 2.4). The early demonstration units (one in Iceland, one in Japan and two in Germany) are all essentially prototypes and are still showing teething problems, but the operators remain optimistic.

9.4 Technologies for direct use of geothermal energy

Some of the countries that are exploiting geothermal resources for non-electrical purposes have chosen to develop these direct use applications in areas flanking the main steam fields. Japan, New Zealand, Iceland and Italy are obvious examples, where wet steam or warm water at a range of temperatures is readily available for industrial, domestic and leisure applications. In this section, however, we leave these aside and concentrate principally on the low-temperature resources found in regions remote from plate boundaries, typically in sedimentary basins, several of which have been developed across central Europe. Drilling techniques resemble those discussed earlier, but the process is generally less hazardous since the geothermal fluid is found under much lower pressure and temperature conditions than in hot steam fields, and pumps are often required to bring the fluid to the surface at adequate flow rates. However, the hot water is usually too saline and corrosive to be allowed directly into heating systems, so once again corrosion-resistant heat exchangers are widely used. It should be pointed out that, given the cost of deep boreholes, a large market is required if sales of heat are to be large enough to pay for the development. Typical heat loads might be a vast greenhouse complex using both overhead and underground pipes, or a domestic group scheme with a combination of underfloor and radiator heaters. The dense multi-storey apartment blocks of the Paris suburbs are ideal heat loads for such local resources.

The French led the development of these low-enthalpy resources in Europe. Over the past 40 years no fewer than 55 geothermally fed group heating schemes were installed in the Paris Basin, with several more in south-western France. At the design stage, a twin production and reinjection borehole system would be planned on the basis of supplying 3–5 MW$_t$ of heat energy (25–50 l s^{-1} of water at 60–70 °C) over a lifetime of 30–50 years. The spacing of the wells must be designed to maintain high fluid pressures by reinjection while avoiding the advance of a 'cold front' (i.e. fluid at reinjection temperatures) towards the production well until capital costs are paid back, and this means that flow conditions in the aquifer need detailed study. A typical layout for a twin production well scheme, with a schematic of the heat transfer technology, is shown in Figure 9.8. Note

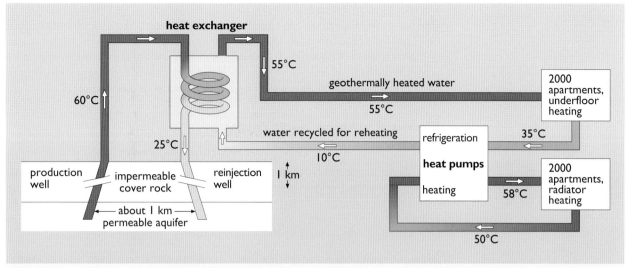

Figure 9.8 An example of a district heating scheme installed in the Paris region. This shows how the geothermal heat is exchanged to a secondary freshwater circuit. The circuit is used first to heat 2000 apartments by underfloor heating; the residual energy is then boosted using heat pumps to provide radiator heating in a further 2000 apartments. Note that the main function of the heat pumps is to lower the reinjection temperature and so extract more heat from the geothermal fluid rather than to raise the production temperature

in this example the interesting application of heat pumps to enhance the system efficiency by reducing the reinjection temperature. Heat pumps work on the same principle as refrigerators, but here produce a concentrated high temperature output. Of course, they consume electrical energy but, in the example shown in Figure 9.8, they enable the number of heated apartments to be doubled.

Although the French group heating schemes were generally a great practical success, a few suffered technical problems – mainly corrosion and scaling in the wells – and were abandoned. More seriously, their economic benefits were only marginal at times of low oil prices, increasing availability of natural gas and high interest rates and several were abandoned for financial reasons during 1989–92. Nevertheless, 34 remain in operation. At the time of writing, with rising oil and gas prices (and low interest rates) such operations look increasingly attractive again. Two new schemes had been completed by January 2011 and were scheduled to begin operation during the year – it is expected that about two new schemes per year will be developed over the next decade. Currently, they produce an annual saving of over 200 000 tonnes of oil (or equivalent in other fossil fuels) in an area which, 40 years ago, had no obvious geothermal potential. The same concept applies to the analogous UK scheme in Southampton.

In Germany, reunification had a significant effect on the way in which geothermal developments occurred, as the better (though still low temperature) resources tend to be concentrated in the eastern part of the country. Although a few schemes existed before reunification, with the freeing of capital following unification, combined with concerns about CO_2 emissions from fossil fuels, the geothermal market really began to take off. Several large-scale district heating schemes are already in operation and even more are under active development. By the end of 2009 there were 162 group heating schemes making direct use of geothermal energy in Germany, with an installed thermal power of some 250 MW_t. Several small ORC power plants are also in operation or under development in the south and south-west. There are, in addition, nearly 180 000 ground source heat pumps (see below) with a total capacity of 1860 MW_t. Substantial numbers of new schemes, both large and small, are under development.

Ground source heat pumps

The technology of the heat pump was introduced in Chapter 2, Box 2.3. It may be used for cooling, as in a refrigerator or an air conditioning plant, or for space and water heating in buildings. It is this latter use that is increasingly of interest. An electrically driven compressor can be used to raise heat from a lower temperature and deliver it at a higher temperature to a building. More heat is delivered than the electricity consumed.

Between 1995 and 2010, the total capacity of direct heat installations worldwide increased from 8664 MW_t to over 50 000 MW_t. In fact, the increase in large direct-use installations like those described above has been quite modest (for the same reasons, until recently, of high interest rates and low fossil fuel prices). Eighty percent of the increase has been accounted for by a new type of geothermal installation, the ground source heat pump (GSHP) – see Figure 9.9. In the EU Renewable Energy Directive (CEC, 2009) the energy gain is defined as 'renewable energy'.

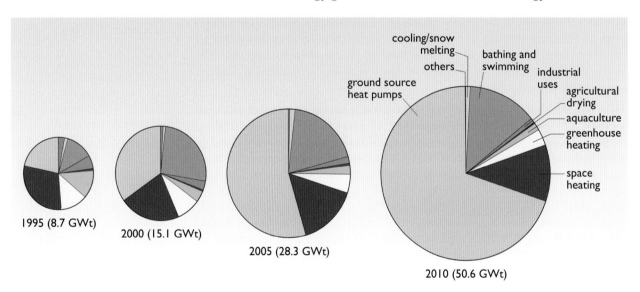

Figure 9.9 The growth worldwide of different direct geothermal applications. The area of the circles is proportional to the installed capacity

As described earlier in Chapter 2, Box 2.3, a key parameter of a heat pump is its *coefficient of performance* (COP), the ratio of the heat (or cooling energy) produced to the electricity used to drive its compressor and associated pumps. The maximum possible COP is limited by the same considerations of Carnot efficiency that limit electricity generation efficiency (see Box 2.4 and Everett et al, 2012). Heat can be raised through a small temperature difference with a high COP but, as the temperature difference increases the possible COP will decrease. Results from well-designed UK borehole GSHPs suggest that typical COPs can range from 3 to 4.5 (EEPH, 2005). Where used for space heating GSHPs may be used in conjunction with underfloor heating, which can be carried out with water at only 40 °C.

The general arrangement of a GSHP is illustrated in Figure 9.10(a). Unlike other geothermal techniques, this one relies on heat transfer by conduction from the walls of the borehole, not on the extraction of groundwater. The heat available from a well 100–150 m deep is only a few kW$_t$, but that is sufficient for a single domestic installation, and boreholes of this depth are often cheap enough to make the installation competitive with conventional heating systems. A simple loop of pipe is inserted in the well and grouted in place. A heat transfer fluid (usually water) circulates in the loop and transfers heat from the surrounding subsoil to a heat pump. More than 20 years' experience has shown that a few kilowatts can be extracted throughout the heating season; the subsoil temperature drops by a few

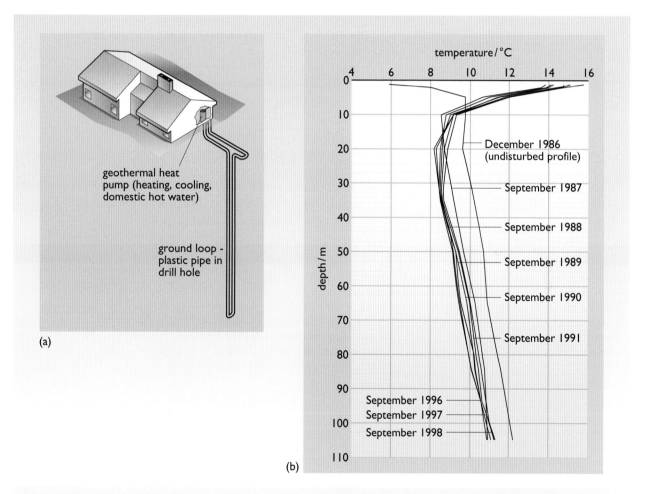

Figure 9.10 (a) The geothermal heat pump (GHP) concept, used to extract heat from warm shallow groundwater to supply a single domestic dwelling. In the winter heat is removed from the earth and delivered in a concentrated form via the heat pump. Because electricity is used, in effect, to increase the temperature of the heat, not to produce it, the GHP can deliver three to four times more energy as heat than the energy content of the electricity it consumes (b) Measured data from a Swiss GSHP installation; though old, the data remain valid. After an initial drop, the ground temperature below the solar-controlled surface zone recovers to the same value year after year. If the scheme is designed correctly, this would be a truly renewable operation

degrees but regenerates over the remainder of the year (Figure 9.10(b)). Table 9.4 shows the rate at which energy can be extracted – on a quasi-continuous basis – from typical 100–150 m deep holes.

Table 9.4 Borehole heat exchanger performance in different rock types

Rock type	Thermal conductivity /W m⁻¹ K⁻¹	Specific extraction rate/W m⁻¹	Energy yield per metre of borehole /kWh m⁻¹ a⁻¹
Hard rock	3.0	max. 80	135
Unconsolidated rock, saturated	2.0	45–50	100
Unconsolidated rock, dry	1.5	max. 30	65

Source: Rybach and Eugster, 1998

A variant on this, for new-build commercial buildings where piles carry the weight of the building down to bedrock, is for the heat exchange loops to be embedded in the piles, thereby avoiding the expense of separate boreholes. These are referred to as **energy piles**. This practice could not be recommended for friction piles (for example where a concrete building sits in a clay subsoil), however, as changes in the moisture content of the soil as a result of heat extraction could modify the behaviour of the piles.

If reversible heat pumps are used, GSHP systems can also provide cooling in summer with the added advantage of helping to recharge the underground and so increasing the sustainable heat extraction from the system. It should be emphasized that, unlike other geothermal systems, GSHPs are ideally suited for a range of scales right down to the domestic scale with a single module providing just a few kW_t.

BOX 9.5 **Solar or geothermal?**

The question is often asked whether GSHPs are truly geothermal devices, or whether they are really solar units. The answer depends on the configuration of the heat exchanger loop. All the GSHPs discussed in this chapter are 'borehole' units, with the heat exchanger located some 100 m or so below ground, and these are true geothermal systems. There is another type of GSHP, however, that uses 'ground coils', horizontal loops of piping buried just beneath the surface. These do derive their energy from the daily solar input and their performance therefore varies on a seasonal basis.

This difference in performance again has its origins in the poor thermal conductivity of soils. The thermal pulse from the daily and seasonal solar input penetrates very slowly, and for practical purposes virtually disappears at depths greater than 15–20 m (Figure 9.11). At greater depths, the temperature is controlled almost entirely by the geothermal heat flow (except where it may be influenced by local groundwater flow). If the system is not designed correcty, there is a possibility that the rate of heat extraction will exceed the recharge rate, resulting in a year-on-year reduction in ground temperature.

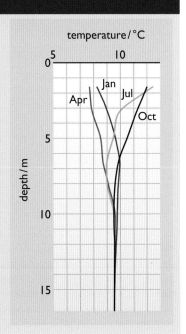

Figure 9.11 Ground temperatures typical of northern Europe (source: Sanner, 2001)

Because the presence of groundwater is not a prerequisite, this technology can be applied almost anywhere. The types of buildings that are using ground source heating and cooling in this manner range from domestic, utility or public housing, through to very large (megawatt-sized) institutional or commercial buildings exploiting multiple units. This technology can offer up to 40% reductions in CO_2 emissions against competing technologies. Better still, if the electricity to drive the heat pump is supplied from non-fossil sources, then there should be no CO_2 emissions associated with heating and cooling a building.

The concept was developed independently in the USA and Europe. While uptake seems to have been concentrated in the northern hemisphere (North America, Europe and China), the installed capacity has grown at an annual rate of around 20% over the past 15 years. It is estimated that there is now the installed equivalent of nearly 3 million 12 kW_t units (the typical size of a single dwelling installation in the USA and western Europe). In the USA, most units are sized to handle peak cooling loads, whereas most European units are focused on heating.

Large-scale arrays have also been installed in many countries to feed larger complexes where suitable supplies of deep geothermal water are not available. In the largest development to date, 4000 units – each with its own borehole – were established on a US Army base in Louisiana to provide heating and cooling. Peak electrical demand dropped by 6.7 MW_e compared with the previous installation, gas savings amounted to 2.6 TJ y^{-1} and – perhaps most telling of all – service calls dropped from 90 per day in summer to almost zero.

The UK was late in picking up this trend, but growth from the mid-1990s to 2009 was rapid. A geotechnical consulting group has been heating its offices in this way since 1996 (practising what they preach!). In 1998 a four-borehole system was fitted in the new health centre at St. Mary's, Isles of Scilly, using a 25 kW reversible heat pump to supply hot water, heating and cooling to the building. There are probably now around 12–15 000 units in operation. The situation in late 2010 was summarized as follows:

> Starting from a very low base, the level of activity is probably in the region of about 3000 – 5000 installations per year. Whilst a handful of these are larger scale open-loop systems (~500 kW – 2 MW), the majority are closed-loop systems. These range in size from 3.5 kW heating only systems in social housing, through to multi MW installations delivering heating and cooling. The main driver for this activity has been the realization that GSHPs connected to the UK grid can offer significant reductions in overall carbon emissions compared to traditional methods of heat delivery. With projected improvements in the carbon intensity of the UK electricity generation grid, GSHPs will be able to deliver even larger carbon reductions with time.
>
> (Batchelor et al., 2010)

It is worth stressing that the major driving force behind the uptake of GSHPs is the reduction in CO_2 emissions from heating and cooling systems, even though the heat pumps are electrically driven. With **coefficients of performance (COP)** – the ratio of heat output to electrical input – now in the range of 3–4.5 for borehole units in the UK (EEHP, 2005), the CO_2 savings are significant.

As other renewables start to feed into the generation mix, the CO_2 penalty of the heat pump drives may reduce even further.

Enhanced (or engineered) geothermal systems

All conventional geothermal systems (except, perhaps, GSHPs) rely on the presence of water circulating through the rock to extract heat and bring it to the surface. However, even in a good aquifer more than 90% of the heat is contained in the rock rather than in the water. Moreover, the vast majority of rocks are poorly permeable at best and the occurrence of an exploitable geothermal reservoir is a rarity. On the other hand, heat exists everywhere, and the amount of energy stored within accessible drilling depths (say, down to 7000 m, where temperatures of >200 °C would be widespread) is colossal. Cooling one cubic kilometre of rock (which is about the scale of a geothermal reservoir) by 1 °C will provide the energy equivalent of 70 000 tonnes of coal.

To put this in context, a report prepared by the Massachusetts Institute of Technology for the US Department of Energy in 2006 (MIT, 2006) calculated the heat in place at various depths in the continental USA (excluding Alaska, Hawaii and Yellowstone National Park) (Figure 9.12). The estimate was 13 million exajoules (13×10^{24} joules); by comparison, the total annual energy consumption in the USA is about 100 exajoules. Even though only a small fraction of this resource is likely to be developable, the potential of such an energy resource cannot be ignored.

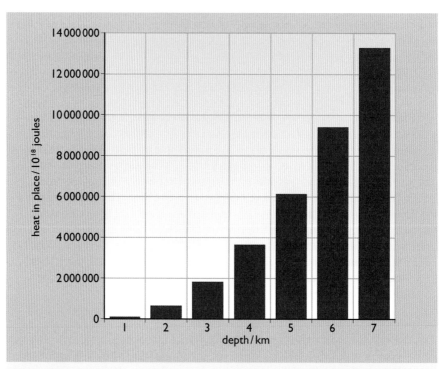

Figure 9.12 The total heat in place at various depths beneath the continental USA, excluding Alaska, Hawaii and Yellowstone (source: adapted from MIT, 2006)

This situation is not confined to the USA. All over the world, temperatures around 200 °C are accessible under a high percentage of the landmass. If this store of heat could be exploited, it would give almost every country the opportunity to generate electricity from an indigenous and (for all practical purposes) renewable resource. It is this prospect that has motivated a number of countries to spend over US$350 million over the past 40 years to find a way to exploit the resource.

The familiar concept of twin production and injection boreholes provides the basis of system designs, but here drilled into relatively hard crystalline rock and terminating several hundred metres apart. In principle, water can then be pumped down one hole, flow through the rock picking up heat, and return to the surface via the second borehole. The fundamental problem is that, as mentioned in Section 9.2, rocks are very poor conductors of heat and, to extract energy at a rate sufficient to pay back the high cost of the boreholes needed to reach these depths, very large heat transfer surfaces are needed – of the order of several square kilometres! There is now a consensus that the only practical way of achieving this figure is to work with nature, exploiting the fact that most deep rocks contain extensive networks of natural fractures. In principle, a suitable heat exchange surface can be created by opening these pre-existing fractures.

The focus has, therefore, been on learning how to stimulate and manage the fracture networks to support a useful and controllable flow of water between boreholes. The stimulation is done by using a variant of an oil industry technique known as hydro-fracturing, which consists of pumping water down the borehole at increasing pressure until fractures in the rock are opened. The progressive development of the opening fractures is followed by listening to and locating the sound of rock surfaces moving over one another. This is known as microseismic monitoring.

If the second borehole has already been drilled, it may be necessary to repeat the stimulation in that hole in order to link up with the first zone. Alternatively, the second hole may be drilled after the first stimulation to intersect the stimulated zone. A closed circuit water circulation through the fracture system is thereby generated. The trick has been in learning precisely how to control the injection conditions in each hole to ensure that water can flow through the system with a minimum of resistance. At the same time, the stimulated zones around each hole must link up in such a way that water losses are minimized, because water losses mean wasted pumping costs (and in many regions water itself is a valuable commodity). Water is circulated down the injection well, through the reservoir and up the production well to a heat exchanger and turbo-generator where the thermal energy is converted to electricity. (Lower-temperature district heating schemes are also under consideration, but the high capital costs of such an operation require an extremely large local market for the heat produced.)

Pioneering work took place at Fenton Hill in New Mexico in the 1970s and 1980s where the Los Alamos National Laboratory (LANL) developed two systems at temperatures of 200 °C and >300 °C. The Fenton Hill project proved the principle in 1979, when a 60 kW$_e$ ORC generator operated for a month on the produced water. The operating parameters were very far from those that would be needed in a commercial system, however, and a number of teams in various countries (USA, Japan, UK, France, Germany and, most recently, Switzerland) have worked cooperatively in the intervening years to understand and refine the techniques required. Notable among these activities was the full-scale (but comparatively shallow – 2000 m – and deliberately low-temperature) rock mechanics experiment at Rosemanowes Quarry in Cornwall, England, which began in 1975 and laid many of the foundations for our current understanding of the behaviour of natural fracture systems and how they might be managed.

At the beginning, it was assumed that basement rocks at depth would be devoid of fluids, so the technology was termed Hot Dry Rock (HDR). Over the years, however, it has become clear that 'Dry' is a misnomer; very few basement rocks have proved to be completely dry. Water has been found in fractured basement even at the deepest levels in the exploration boreholes in the Kola Peninsula, Russia (15 km) and in the Black Forest, Germany (>8 km). More recently, the term 'Enhanced' (or 'Engineered') 'Geothermal Systems' (EGS) has come to replace the original name, though the phrase 'hot dry rock' is too catchy to disappear easily. The overall technology has been broadly defined as 'any system in which reinjection is necessary to maintain production at commercially useful levels'. This redefinition also emphasizes the continuity which exists in the spectrum of reservoir permeabilities and geothermal technologies. A more detailed definition is given in the MIT report:

> The U.S. Department of Energy has broadly defined Enhanced (or engineered) Geothermal Systems (EGS) as engineered reservoirs that have been created to extract economical amounts of heat from low permeability and/or porosity geothermal resources. For this assessment, we have adapted this definition to include all geothermal resources that are currently not in commercial production and require stimulation or enhancement. EGS would exclude high-grade hydrothermal but include conduction dominated, low-permeability resources in sedimentary and basement formations, as well as geopressured, magma, and low-grade, unproductive hydrothermal resources. In addition, we have added coproduced hot water from oil and gas production as an unconventional EGS resource type that could be developed in the short term and possibly provide a first step to more classical EGS exploitation.
>
> (MIT, 2006, p. 1.10)

The countries principally involved in the research worked closely together throughout the 1980s and 1990s; this type of research makes great demands on both money and expertise and is aimed at what should prove to be a generally applicable technology, so it is an ideal subject for international collaboration. Following this logic, during the 1980s the various teams in the UK, France and Germany, with the support of the European Commission, agreed to pool their resources to develop a single experimental site at Soultz-sous-Forêts in the Upper Rhine Valley. The aim was to build on the results derived from the work in Cornwall, but at a site where temperatures at depth were expected to be higher. As it turned out, temperatures at depth were similar to those expected in SW England, but the work led eventually to demonstration of the practicability of the EGS concept, in this case a three-well system that has now started delivering power to the French national grid (see Box 9.6).

BOX 9.6 The EGS site at Soultz-sous-Forêts

The Soultz site, like Fenton Hill, benefits from the blanketing effect of 1000 m of sedimentary rock above the crystalline basement. The geothermal gradients through the sediments average 0.08–0.1 °C m^{-1}, falling to 0.028–0.05 °C m^{-1} in the crystalline basement beneath. Teams from France, Germany, the UK, Italy and Switzerland worked together on the site, eventually drilling four deep boreholes (>3800 m, 170 °C to 5000 m, 200 °C) as well as several shallower boreholes that are used for the microseismic monitoring system (Figure 9.13). Geologically, the site is located in the Upper Rhine Graben, where E–W tensional forces have stretched the crust and caused the granitic basement to subside. The basement is heavily fractured and even supports a small amount of natural fluid flow.

During more than 20 years of operation, the project at Soultz, coordinated and part-funded by the European Commission, became recognized as the world leader in developing EGS technology. It has been shown

that with careful control of the pressure and density of the stimulation fluid the stimulated zone can be persuaded to develop laterally so that it can be accessed by a conventional arrangement of two or more boreholes deviated in opposite directions. In late 1997, after several years of testing and hydraulic stimulation of the fracture system at 3.5 km depth, a 4-month circulation test was carried out. 25 kg s^{-1} of water was circulated on a continuous basis between wells GPK1 and GPK2. The system operated in a closed loop, with the heat produced (ca. 10 MW$_t$) being dumped via a heat exchanger. The overall loss rate was zero and no make-up water was required. In a two-well system like this, such a result was possible only because a down-hole pump was used in the production well, altering the sub-surface pressure field to ensure that losses from the injection well could be balanced by input from the natural *in situ* fluids. Tracers added to the injected fluid proved that circulation was occurring; production also

Site map

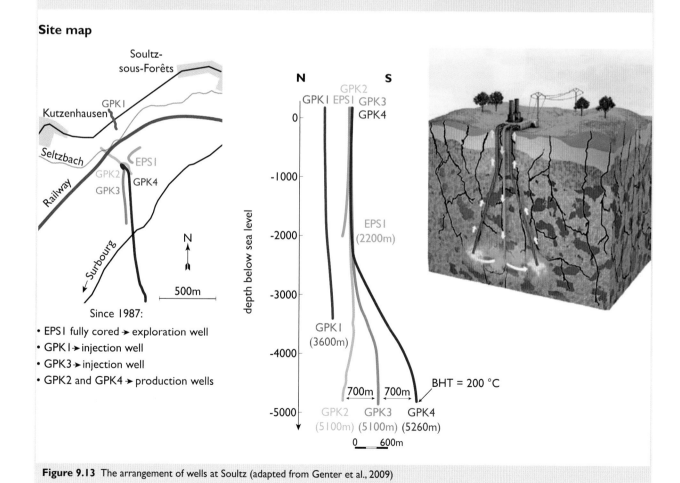

Figure 9.13 The arrangement of wells at Soultz (adapted from Genter et al., 2009)

declined very rapidly when reinjection was stopped, both observations demonstrating that this is a true EGS system under the above definition. Of equal importance was the finding that the overall system impedance was less than 0.2 MPa l^{-1} s^{-1}, closer to the targets than any previous project (target parameters for EGS developments are discussed later and are summarized in Table 9.6).

Figure 9.14 The EGS site at Soultz-sous-Forêts

Following this success, it was decided to continue to the pilot plant/proof-of-concept stage. To improve the flow distribution in the reservoir, a three-well system was designed, with two production wells flanking a single injector. GPK2 was deepened, and GPK3 and GPK4 drilled, to about 5000 m, where the bottom hole temperatures were 200 °C. The wells were deviated to give bottom hole separations of 700 m. After stimulation of the fracture zones between the wells, a 1.5 MW_e ORC turbo-generator was installed. The first power was produced in June 2008, and continuous production started in January 2011, albeit at a low initial flow rate (<35 l s^{-1}) as a precaution against induced seismicity (see Box 9.7).

After testing the system for a couple of years, the aim will be to enlarge the power plant to 6 MW_e.

As mentioned earlier, the increasing emphasis in recent years on the need to reduce CO_2 emissions and dependence on fossil fuels has caused a number of countries to increase their efforts to develop geothermal resources. Notable among these have been Germany and Australia. Germany has been involved in EGS research since the earliest days of the work at Los Alamos but, following experience gained from its work at Soultz, has begun to apply the techniques at home. Indeed, the first commercial EGS system has been completed at Landau, also in the Upper Rhine Valley about 50 km north of Soultz, to provide both heat and power to the town. It is a two-well system, with a depth of about 2700 m, producing water at 70 l s^{-1} and 160 °C. The Landau power plant was inaugurated in November 2007 and since then has been running continuously, except for some maintenance work. It is able to deliver about 3 MW_e and about 4 MW_t.

Perhaps the most surprising entrant in the geothermal stakes over the past decade has been Australia. Previously, it had been assumed that the age of the Australian crust meant that accessible temperatures would be too low to be useable, but all that changed after 2000; with Federal government support, both in terms of legislation and finance, there has been an explosion of commercial interest in geothermal prospects. By the beginning of 2011, some 56 companies held exploration licences in 418 areas (Figure 9.15). Primarily, the interest centres around the potential of EGS technology, although hot aquifers are now also being targeted in the south-west and north-east.

By 2011 it was estimated that more than AU$671 million (US$663 million) had been spent on studies, geophysical surveys, drilling, reservoir stimulation and flow tests. By 2015, investment for proof-of-concept geothermal projects is forecast to exceed AU$3227 million (US$3187 million). In a departure from practice in other countries, a significant fraction of this funding has been provided by companies and private shareholders.

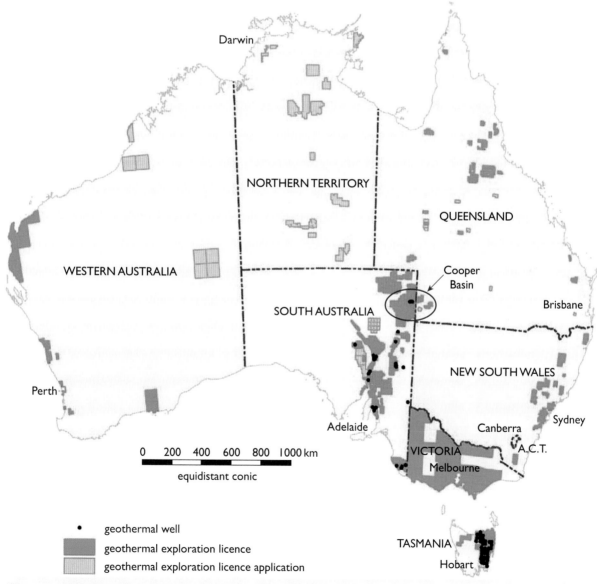

Figure 9.15 Geothermal exploration in Australia in 2011 (source: Goldstein et al., 2011)

The most advanced EGS project is that being developed by Geodynamics in the Cooper Basin. In this region Geodynamics is exploring three separate fields (Habanero, Jolokia and Savina) and has drilled a total of five wells into the granite basement to depths of 3800–5000 metres, with the aim of assessing the nature and extent of the resource present. In 2009 proof of concept was achieved at Habanero, when circulation tests confirmed sub-surface flow between Habanero #1 and #3 wells through a stimulated sub-horizontal fracture network at a depth of 4400 m. Further drilling is to be undertaken before confirming the location of a 25 MW_e commercial demonstration project (CDP), which aims to demonstrate that EGS technology is cost effective at a commercial scale. The operators expect to commission the 25 MW_e CDP geothermal power plant in 2012.

9.5 Environmental implications

The main environmental concerns associated with 'conventional' geothermal energy development are focused on those to do with site preparation, such as noise pollution during the drilling of wells, and the disposal of drilling fluids, which requires large sediment-settling lagoons. Noise is also an important factor in high-enthalpy geothermal areas during well-testing operations when steam is allowed to escape, but once a field comes into production noise levels rarely exceed those of other forms of power plant. Accidents during site development are rare, though a notable exception in 1991 was the failure of a well originally drilled in 1981 at the Zumil geothermal station on the flanks of Santiaguito volcano in Guatemala. Hundreds of tonnes of rock, mud and steam were blown into the atmosphere when the well 'blew its top', apparently because of gravitational slippage of the ground beneath the site.

Longer term effects of high-enthalpy geothermal production include ground subsidence, gaseous pollution and induced seismicity. In dry steam fields, where the reservoir pressures are relatively low and the rocks are self-supporting (as at the Geysers and Larderello), subsidence is rare. In liquid-dominated systems, however, significant reduction of the higher pressures, for example due to inadequate fluid reinjection, can induce subsidence, usually on the millimetre to centimetre scale (although maximum localized subsidence of 3 m has occurred at Wairakei as a result of early exploitation without reinjection). Reductions in reservoir pressure can also have an adverse effect on the natural manifestations (geysers, hot springs) that are a common accompaniment of high-enthalpy fields and often important to the local tourist industry. Such concerns have severely restricted the development of geothermal power generation in Japan.

Geothermal 'pollutants' are chiefly confined to the non-condensable gases: principally CO_2, with lesser amounts of hydrogen sulfide (H_2S) or sulfur dioxide (SO_2), hydrogen (H_2), methane (CH_4) and nitrogen (N_2). In the produced water there is also dissolved silica, heavy metals, sodium and potassium chlorides and sometimes carbonates, depending on the nature of the water–rock interaction at reservoir depths. Today these are almost always reinjected and this also removes the problem of dealing with waste water. Traditionally, geothermal fields have received a bad press on account of their association with the 'rotten eggs' smell of H_2S. However, this and other gaseous products of old leaking plant have now been reduced so that the environmental impact of thermal production is at a minimum. Modern plants are fitted with elaborate chemical systems to trap and destroy H_2S. Interestingly, the level of atmospheric H_2S over the Geysers field is now lower than that emitted naturally from hot springs and geysers before geothermal developments began. Nevertheless, the image of polluting geothermal systems has slowed developments at several new sites. For example, environmental legislation covering the Miravalles plant, located on the periphery of a rainforest conservation area in northern Costa Rica, delayed completion of the plant for four years. A project on Mount Apo on Mindanao Island in the Philippines was turned down by the World Bank and the Asian Development Bank on social and environmental grounds. Objectors claimed that 111 hectares of forest would be threatened, 28 rivers polluted and a national park destroyed.

The position over emissions of CO_2, an important greenhouse gas, is rather more complicated. Geothermal reservoirs often contain significant quantities of CO_2, so emissions from those power plant will also be higher in CO_2 than might otherwise have been expected. On the other hand, exploitation of the field often reduces natural emissions. Leaving aside that possible benefit, however, a survey carried out by the International Geothermal Association (IGA, 2002) shows a wide variation in CO_2 emissions from existing plant, ranging from 4 g per kWh to 740 g per kWh (though the latter figure is extreme, from a field that is naturally high in CO_2 and undoubtedly leaking large quantities long before geothermal development began). The weighted average is 122 g per kWh. Typical CO_2 emission rates from fossil-fired power stations range from about 400 g per kWh for the most up-to-date natural gas fired combined cycle plant to about 900 g per kWh for the best coal-fired stations (see Chapter 10).

Induced seismicity

The question of whether there is induced seismicity around conventional geothermal sites has been much debated, and in the case of high-enthalpy systems it must be recognized that most steam fields are located in regions already prone to natural earthquakes because of their proximity to plate boundaries. There is evidence that fluid injection lubricates fractures and increases pressures, creating small earthquakes (microseismicity), especially when reinjection is not at the same depth as the producing aquifer (mainly for reasons of fluid disposal). However, in cases where reinjection is designed to maintain reservoir pressures, seismicity is not greatly increased by geothermal production.

In conventional low-enthalpy systems, where reinjection merely maintains the natural level of reservoir pressure, induced seismicity is rare or absent.

In EGS systems, on the other hand, injection occurs at higher pressures and induced seismicity is common, at least in the project development stages (see Box 9.7). Although the vast majority of induced events are small and detectable only instrumentally, a few are large enough to be felt at the surface. This is giving rise to public concern that could severely inhibit further development, and a multinational research programme has been mounted to understand and minimize these effects.

BOX 9.7 Induced seismicity in EGS projects

Although induced seismicity has rarely been a problem in conventional geothermal developments, the question has assumed far greater significance in the context of EGS projects. The stress fields in hard crystalline rocks, which are the typical targets for EGS, are invariably anisotropic, i.e. the three principal stresses (that is, the stress field resolved along three orthogonal axes) are of different magnitudes. This means that any fracture that is not precisely aligned with the stress field will have a tendency to slip. The EGS technique of stimulating the natural fracture system exploits this property; increasing the fluid pressure within the fracture reduces the forces that are keeping it closed and allows the two surfaces to slip past one another. As fractures will always be rough, the misalignment as the fracture closes will tend to keep it propped open, and this is the basis of the stimulation. This slippage of the fracture, usually of the order of millimetres, is in fact a micro-earthquake, and locating the noise generated by these events – called 'microseismicity' – is the principal tool used to follow the progress of the stimulation. A typical stimulation operation will generate tens of thousands of micro-events, virtually all of them far below the threshold of human perception.

Figure 9.16 Schematic diagram of a fracture subjected to a non-uniform stress field

However, just occasionally an event will occur that is large enough to be detected at the surface, sometimes during stimulation but more commonly after the wells have been shut in. Typically, each of the stimulation procedures at Soultz resulted in up to five 'felt' events, all with local magnitudes (M_L) less than 3. Events of this size are very unlikely to cause material damage – they are similar to the effect of a heavy lorry passing nearby – but they can alarm the local population.

Although this may be a matter of a perceived rather than an actual problem, public perception is becoming a critical factor in EGS planning in Europe and the USA (where reinjection into the Geysers – itself an example of EGS in an already seismically active area – has resulted in enhanced seismic activity). In Landau, a succession of small seismic events – although in an area already known for low level but persistent seismic activity – has prompted the operators to reduce flow rates (and therefore output) until the causes of the increased seismicity are better understood.

Even more seriously, a proposed EGS project in Basel, Switzerland (Figure 9.17), intended to provide heat and power to the city, was forced to close when an hydraulic stimulation operation in the first borehole triggered several small events up to M_L 3.4. Although no one was hurt and material damage was minimal, pressure of public opinion forced the abandonment of the project.

Those organizations and companies who are aiming to develop EGS systems have recognized the seriousness of this problem. It may well be true that induced seismicity is very unlikely ever to pose a real threat to life or property (except perhaps in those areas with a history of significant natural seismicity), but negative public opinion could prove to be a real show-stopper. For that reason, there is a multinational research programme underway that is aiming to understand the reasons for the occurrence of these sporadic 'felt' events, and so devise methods to minimize or eliminate them.

Figure 9.17 The EGS rig at Basel

9.6 Economics and world potential

On an international scale, geothermal energy is one of the most significant 'renewable' energy resources. Its strength in this respect is that it can provide firm, predictable power on a 24-hour per day basis. Table 9.5 shows the performance of some typical high-enthalpy power plants; note the high capacity and availability factors. This high resource availability distinguishes geothermal energy from many other renewables, and results in significantly greater amounts of energy being supplied for a given installed capacity.

There was quite spectacular growth in geothermal installed capacity of approximately 14% per year following the oil embargoes of the early 1970s, at a time when conventional generating capacity grew at between 0 and

Table 9.5 Performance of typical geothermal power plant

	Italian 60 MW	Italian 20 MW	Japanese 50 MW
Year	1999	1999	1/4/97–31/3/98
Installed capacity/MW$_e$	60	20	50
Maximum load/MW$_e$	55	17	48.3
Annual produced electricity/MWh	462 845	142 248	361 651
Hours of operation of plant	8748	8483	8112
Capacity factor/%	96.1	95.5	85.5
Availability factor/%	99.9	96.8	92.6

Source: IGA, 2001

3% per year. Stabilization of oil prices brought the growth rate down to about 8% per year by the early 1990s and cheap natural gas together with liberalization of electricity markets further reduced rates to 3% during the 1990s. This trend has now reversed, driven in part by concerns about climate change and also by rising fossil fuel prices. The underlying trend is again close to 10% per year, though it has been masked in recent years by the downturn in production of the Geysers during the period of over-exploitation.

It is difficult to discuss the economics of geothermal development except in the most general terms, because the details are so location-specific. While running costs are relatively minor, geothermal projects are capital intensive and the main element of annual costs is amortization of capital. The biggest single item is drilling costs, which rise exponentially with depth. Consequently, in common with other mining operations, costs are very dependent on the quality of the resource (depth, temperature, flow rate, etc.), and hence vary greatly from country to country and from place to place. Particularly in a new area, there is also a high initial risk that the borehole will be unsuccessful in locating an exploitable resource. Moreover, the economics are also strongly dependent on country-specific fiscal and regulatory issues like interest rates, subsidies, feed-in-tariffs, carbon credits, the cost of competing fuels, etc. On the other hand, successful geothermal projects benefit from the high availability of the resource and the consequent avoided cost of back-up plant.

The best illustration of the economics of geothermal plant, therefore, is the willingness of private industry to become involved and the rate at which new plants are being commissioned. In New Zealand, for instance, where the electricity industry was privatized during the 1990s and different fuels have to compete on equal terms with hydroelectricity, a new privately owned 55 MW$_e$ plant was commissioned in 1996, a second in 1999 and several more in the period 2003–2010. Generating equipment in some of the older fields that have now been in use for nearly 50 years is also being upgraded. A recent study by the Ministry of Economic Development (Figure 9.18) showed that geothermal is currently the most competitive new-build option (Harvey et al., 2010).

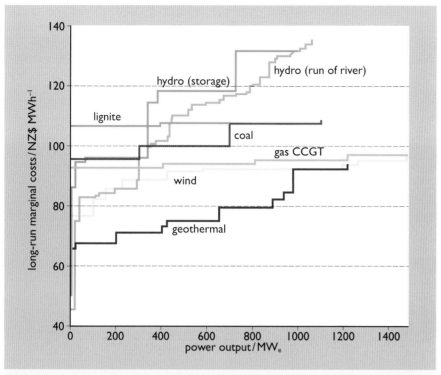

Figure 9.18 New build generation costs in New Zealand (source: Harvey et al., 2010)

Sharing of experience and R&D costs among the different operators will be a vital factor in achieving targets. The obvious economic advantages of high-enthalpy resources in providing a good return on capital have stimulated loan investments in geothermal developments by international agencies, such as the World Bank, especially in Central and South American countries. But perhaps the greatest economic gain to society in general lies in the 112 million barrels of oil a year that is already being saved (or over 200 million barrels if the electricity would otherwise have been generated by fossil fuels at 30% efficiency).

Although geothermal resources make a significant contribution in some high-enthalpy areas (e.g. 30% of energy usage in Kenya and El Salvador, 21% in the Philippines and 10% in New Zealand), the total amount of geothermal electricity produced in 2010 (some 67 TWh) accounted for only about 0.32% of global electricity consumption (Figure 9.19). Yet the long-term potential is much higher, especially in volcanically active countries, and may be realized as the technology improves. As suggested earlier, as EGS techniques improve the potential will rise dramatically, even in countries that lack high-enthalpy resources. In principle, successful development of EGS could allow most countries, even those in 'normal' areas, to generate 10% or more of their power needs from geothermal resources.

In support of this suggestion, in mid-2011 the International Energy Agency published its 'Technology Roadmap for Geothermal Heat and Power' (IEA, 2011). This roadmap foresees that by 2050 geothermal electricity generation could reach 1 400 TWh per year, i.e. around 3.5% of global electricity production. It also estimates that geothermal heat could contribute 5.8×10^{18} J annually by 2050. It qualifies the forecasts, however, by noting that 'For geothermal energy for heat and power to claim its share of the coming energy revolution, concerted action is required by scientists, industry, governments, financing institutions and the public.'

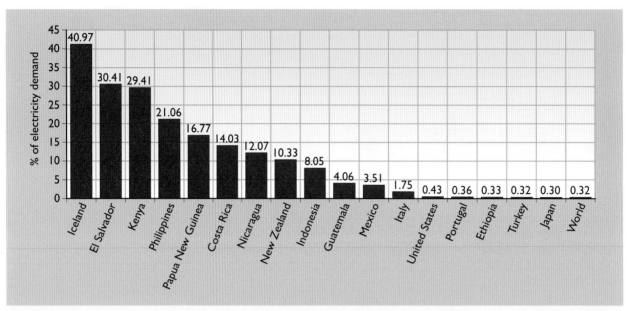

Figure 9.19 Percentage of countries' electricity demand supplied by geothermal energy in 2010

The economics of lower-grade geothermal resources are more marginal and depend on local political and economic conditions, such as the availability and price of fossil fuels, the willingness of governments to invest in new energy concepts, the degree of environmental awareness and the related tax incentives to promote 'clean' energy commercially. The awareness of the problem of greenhouse gas emissions and climate change in recent years has nevertheless provoked many governments into support of what until recently has been a neglected 'renewable' resource.

It is worth pointing out that the economics of space heating operations are also dependent on making maximum use of the geothermal resource (which is capital intensive but has low running costs). If the geothermal system is sized to meet the maximum demand of the heat load, it will lie idle for much of the year; typically, the shape of a domestic heating load duration curve throughout the year is such that some 80% of the total energy needs can be met by a system with a power output of only 40% of the peak demand. Consequently, the geothermal component of the system should be (and, in successful schemes, is) designed to meet less than half of peak demand, with the shortfall being made up by an auxiliary fossil fuel boiler.

Looking to the economic future

The economics of future EGS developments are speculative at the time of writing, and will remain so until the technology is fully demonstrated. The best estimates, derived from recent progress in reservoir development and reductions in drilling cost, and for sites with a mean temperature gradient of 35–40 °C km^{-1}, are that electricity might be produced for about €0.12–0.20 per kWh in the early pilot plant (2010 prices), reducing to half these figures for a multi-module commercial system. There is an important caveat, however: these estimates come from cost models not financial analysis, and the distinction is important.

Financial analysis can be applied to an existing operation or to a technology that is proven; all the steps in the process are known, and the costs of each step can be calculated. This information can be used to derive the break-even cost of the product in an unambiguous way. **Cost modelling**, on the other hand, is usually applied to an unproven technology – often one that is still being developed. It examines the possible costs of each step in terms of assumptions about the performance of the step itself and each preceding step. It says only that '...*if the technology performs in this way, then the cost will be...*'. The result is only as good as the initial assumptions. Cost modelling is a useful tool for setting the targets that various elements of the technology must achieve, or for establishing which aspects of the research offer the best opportunities for improvement, but it does not predict prices. This often gives rise to misconceptions, and ones that EGS research has suffered from, particularly in the UK. Used correctly, however, such analyses can be very useful.

Cost analyses of this type have been carried out for two-well systems by all the research teams involved in EGS, and resulted in general agreement on the target parameters to be achieved for a two-well EGS reservoir, aiming to produce 200 °C water for electricity generation over a 20-year reservoir life (Table 9.6).

Table 9.6 Target parameters for a two-well EGS system

Flow rate/kg s^{-1}	75–100
Effective heat exchange area/m^2	$>2 \times 10^6$
Accessible rock volume/m^3	$>2 \times 10^8$
Impedance[1]/MPa l^{-1} s^{-1}	0.1
Water losses/%	<10

[1] Strictly, impedance is not a constant, but varies with flow rate (because of pressure variation). The specified figure is the resistance to flow at the target flow rate.

Until recently, none of the projects had come close to achieving simultaneously the required flow, impedance and loss rate. In 1997–8, however, the Soultz project demonstrated a closed loop circulation at 25 l s^{-1} for 4 months with zero losses and impedance close to the necessary target. It is on the basis of extrapolating the findings from this and subsequent experiments that the previously mentioned cost estimates are derived. It is still worth repeating, however, that these figures come from cost modelling, not financial analysis. Although we can make reasonable estimates of the capital costs of EGS schemes, we still do not know for certain how to construct a good reservoir, so we can only base our estimates on what its performance and properties ought to be. Predictions of the cost of power from EGS are arguably premature until further technological developments have provided better reservoir performance data. The existing developments at Landau and Soultz, and the imminent systems in Australia and Germany will go some way towards clarifying the issues.

9.7 Geothermal potential in the United Kingdom

Sedimentary basin aquifers

As in many other countries, it was the oil crisis of the mid-1970s that spurred geothermal resource evaluation in the UK. By 1984, new maps of heat flow (Figure 9.20(a)) and of promising geothermal resource sites (Figure 9.20(b))

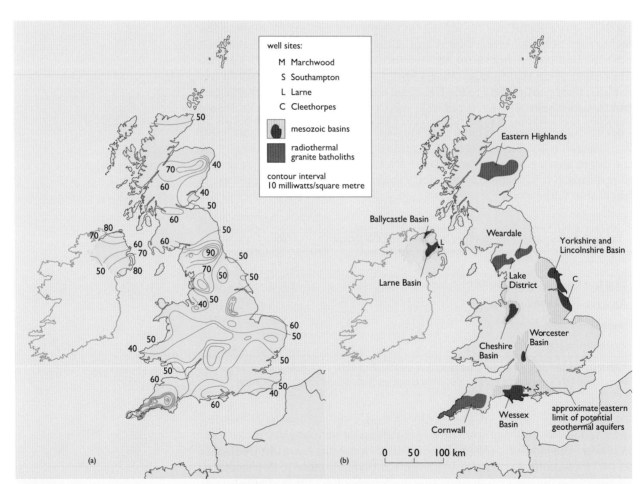

Figure 9.20 (a) Heat flow map of the UK based on all available measurements to 1984 compiled and published by the British Geological Survey (b) Distributions of radiothermal crystalline rocks (granites) and major sedimentary basins likely to contain significant geothermal aquifers in the UK

had been produced. Three radiothermal granite zones stand out with the highest heat flow values, but significant heat flow anomalies also occur over the five sedimentary basins identified, partly because these are regions of natural hot water upflow. Many shallow boreholes were drilled during this period to measure heat flow, as were four deep exploration wells that are shown in Figure 9.20(b) and Table 9.7. In each case the main aquifer is the permeable Lower Triassic Sherwood Sandstone (named after its most notable outcrops in the East Midlands).

Table 9.7 Characteristics of UK deep exploration wells

Location	Completion	Well depth (m)	Bottom hole temperature /°C	Main aquifer depth (m)	Temperature of aquifer/°C
Marchwood	Feb 1980	2609	88	1672–1686	70
Larne	July 1981	2873	91	960–1247	40
Southampton	Nov 1981	1823	77	1725–1749	76
Cleethorpes	June 1984	2092	69	1093–1490	44–55

Source: Downing and Gray (1986)

The shallower intersections with this aquifer at Larne and Cleethorpes are at a rather low temperature for geothermal exploitation but have reasonably high fluid flow rates because of the large aquifer thickness. Unfortunately, the other two wells, which intersect the aquifer at a better temperature, produce rather low flow rates because of the restricted vertical height of good aquifer rock. The yield is reduced not just because the sedimentary sequence is thinner in the Southampton area, but also because much of the Sherwood Sandstone proved to be more highly cemented and therefore less permeable. The resources are nevertheless substantial (Table 9.8).

Table 9.8 Potential UK geothermal energy resources at different temperatures

Basin	Potential resource at 40–60 °C (10^{18} J)	Potential resource at > 60 °C (10^{18} J)
East Yorks and Lincs	26.2	0.2
Wessex	2.8	1.8
Worcester	3.0	–
Cheshire	8.9	1.5
Northern Ireland	6.7	1.3
Total	**47.6**	**4.8**

Source: Downing and Gray (1986)

Assumptions behind these estimates are that the 40–60 °C resource would be exploited with the use of heat pumps, producing a reject temperature of 10 °C, whereas use of the >60 °C resource would not involve heat pumps, giving a 30 °C reject temperature. The latter resources could be doubled with heat pumps. For comparison, current UK electrical energy demand is around 350 TWh per year (1.3×10^{18} J per year, or about 30 GW$_e$ as equivalent continuous power), so we are considering here quite large resources of renewable energy, but as heat rather than electricity.

So why are geothermal aquifers not being exploited much more widely? The problem is not just one of marginal economics and geological uncertainty, but rather the mismatch between resource availability and heat load, itself a function of population density. More than half the UK resources are located in East Yorkshire and Lincolnshire, essentially rural areas lacking the concentrated populations of the Paris basin. While electricity can readily be transported over long distances from source to market, hot water is more of a problem. Transmission distances for the latter under UK conditions are likely to be restricted to just a few kilometres so the resources are likely to remain undeveloped. It is interesting to note, however, that the distance limitation arises from cost rather than heat loss; some Icelandic pipelines exceed 50 km in length with a temperature drop of only about 1 °C.

The consequence of this is that Southampton, where the geothermal borehole contributes some 2 MW_t to the city's group heating system, remains the only direct use of a deep geothermal resource yet in operation in the UK. At the end of 2010, it was reported that two UK universities had been awarded government support for exploratory boreholes aimed at extracting useful heat. Subsequently, Newcastle University had an intention to drill and test a 2000 m borehole near the city centre, but a project to be carried out by Keele University has been abandoned.

In the meantime, however, as indicated in Section 9.4, the UK has begun to adopt ground source heat pumps (GSHP) on a substantial scale. One estimate of the rate of uptake is given in Figure 9.21. It can be seen that growth was particularly rapid from 2005 to 2009; this was driven in part by government incentive schemes. What is claimed to be one of Europe's largest GSHP system was commissioned at the end of 2010 in a large shopping centre/office complex in central London. It has 13 GSHPs drawing on more

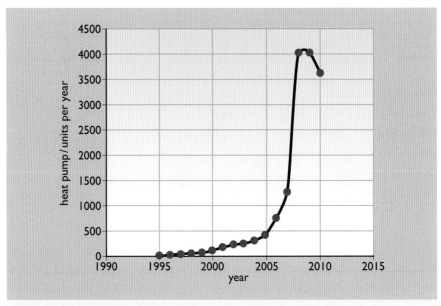

Figure 9.21 Estimated number of heat pump units installed each year in the UK (source: private communication)

than 200 energy piles and rated at 1.7 MW_t. Providing both heating and cooling, it is expected to save some 350 tonnes CO_2 per year and £300 000 in fuel costs.

However, Figure 9.21 also shows a substantial downturn since 2008, this is attributable to a number of factors such as a recession and collapse in the construction sector, temporary reduction in the oil price from an all time high and a combination of the withdrawal of one incentive scheme and delay in implementation of a new subsidy arrangement.

This is an unfortunate illustration of the fact that barriers to geothermal uptake are more often the result of non-technical and regulatory issues than of technical uncertainties.

Engineered geothermal systems

Of the three principal granite zones – in the eastern Highlands, northern England and south-west England – the latter is characterized by the highest heat flow, as shown in Figure 9.20(a). However, large areas of the more northerly granite masses are covered by low thermal conductivity sedimentary rocks and so temperatures will be higher at depth than if the granite bodies came to the surface (Figure 9.6). Nevertheless, substantial areas of Cornwall and Devon are projected in Figure 9.22 as having temperatures above 200 °C at 6 km depth and it has been estimated that the EGS resource base in south-west England alone might match the energy content of current UK coal reserves. One estimate suggests that 300–500 MW_e (about 10^{16} J y^{-1}) could be developed in Cornwall over the next 20–30 years with much more to follow later.

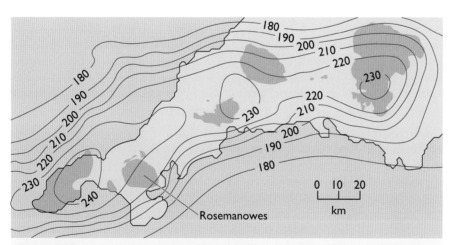

Figure 9.22 Projected temperature (°C) at 6 km depth beneath south-west England. Granite bodies that crop out at the surface are shaded

From 1976–92, Cornwall was the site of a major EGS research project that laid the foundations for the subsequent multinational European project at Soultz. Following Soultz's demonstration of the feasibility of generating power from deep fractured granite, plans are in hand for two possible EGS projects to begin in Cornwall during 2012. The first will start with a 5 km exploratory borehole on the northern edge of the Carnmenellis granite (labelled 'R' for Rosemanowes, the site of the earlier project, in Figure 9.22), with the aim of constructing a multi-well system to generate 10 MW$_e$ of power. The second will be on the St Austell granite and is designed to produce 4 MW$_e$ and several MW$_t$ of heat, primarily to supply the Eden Project (an eco-centred visitor attraction).

9.8 Legal and regulatory issues

As the exploitation of geothermal resources has moved in each country from being a technical novelty to a commercial reality, and one with serious financial implications for the developers, so it has been necessary for each country to incorporate geothermal resources into its regulatory system.

Central to this is the question of resource ownership. Not only must developers (or their banks) be confident that they have a right to the resource that they will be extracting, but there must also be provision: (a) to protect the developer's rights, and (b) to protect the rights of third parties (such as neighbouring developers or property owners). Just as every country has its own legal code, so every country with a geothermal industry has defined these rights and responsibilities according to its own background legal framework and conditions.

In the majority of cases, if not all, ownership of the resource has been reserved to the national or regional authority, which then has the power to grant licences and exact royalties under specified conditions. Typically, these conditions will impose obligations to explore and develop the resource within a defined area and within a specified period, and will include provision to protect shallow water sources and to limit the impact on neighbouring areas. For reasons of both environmental protection and reservoir management, it is now becoming common to require reinjection rather than disposal to surface. Reinjection, too, is controlled; in Italy, for example, it is not permitted to re-inject into any aquifer other than the source. In most countries a different, less onerous, set of regulations is applied to 'small' geothermal projects in order not to impose too great a burden on small-scale GSHP developments.

Unfortunately, the UK remains one of the few countries not to have established a legal basis for geothermal development. UK mining law is particularly complicated, with different types of mineral resources being subject to different ownership, etc., so there is no easy precedent to apply to geothermal resources. In the absence of an established demand, there has been no incentive to draw up new legislation. That means that there is currently no clear definition of ownership or rights, which complicates the financing of projects. Now that interest is beginning to develop, that situation will need to change. Appropriate discussions have already begun at ministerial and parliamentary level.

9.9 Summary

Over geological periods of time, the Earth's crust has accumulated a vast store of heat, principally through the decay of radioactive elements. Like any mineral deposit, this store is uneven and the highest-grade sites have been the first to be exploited. However, over more than a century of commercial exploitation technologies have been developed to allow economic exploitation of progressively lower-grade and more widespread resources.

It is a fact that exploitation of the resource at any given location entails 'mining' the heat in the short term but, given the enormous scale of the resource and the fact that exploited zones will regenerate in a comparatively short timescale, geothermal energy is ranked with the other renewable energy sources. Its major advantage is that it is continuously available irrespective of climate or daylight, while emitting few, if any, greenhouse gases.

Until recently, the principal interest in geothermal resources has focused on the availability of steam or high-temperature hot water in porous rock formations and the use of the extracted steam to generate electricity. Initially the preserve of a few countries located in volcanically-active areas, commercial use over the past 100 years has spread to more than 20 countries and should exceed 30 countries by 2015. While the impact is still minor in most of these countries, geothermal electricity is already a key component of the supply system in some.

In parallel with these developments, lower temperature waters – which have been used on a small scale throughout recorded history – are now in widespread commercial use for space heating and agriculture in more than 70 countries. Over the past 20 years or so, smaller scale systems involving heat pumps have developed even more rapidly. Exploitation at the scale of the individual dwelling has been made possible and is implemented to the extent that this application has far outstripped the deeper resources in terms of energy supplied.

To overcome the limitation that geothermal steam is available only in a few favoured locations, research has been undertaken over the past 40 years on methods of extracting the stored heat from the much greater proportion of deep hot rocks that do not naturally contain water. This work, initially termed 'hot dry rock' and now more realistically known as 'enhanced (or engineered) geothermal systems' (EGS), is now bearing fruit. The first two pilot systems are now producing electricity and several more should be commissioned over the next few years.

Successful development of EGS systems should make geothermal electricity available in most countries. Estimates are that such geothermal developments could supply around 10% of electricity demand in many countries, and some 3.5% of world demand by 2050.

References

Batchelor, A. S., Curtis, R. and Ledingham, P. (2010) 'Country update for the United Kingdom', *Proc. World Geothermal Congress 2010*, Bali, Indonesia, 25–30 April 2010, International Geothermal Association.

Bertani, R. (2010) 'Geothermal power generation in the world – 2005–2010 Update report', *Proc. World Geothermal Congress 2010*, Bali, Indonesia, 25–30 April 2010, International Geothermal Association available at http://www.geothermal-energy.org/pdf/IGAstandard/WGC/2010/0008.pdf (accessed 27 March 2012).

CEC (2009) *Directive 2009/28/EC on the Promotion of Energy from Renewable Sources*, Commission of the European Communities, [online] available from: http://eur-lex.europa.eu (accessed 20 September 2011).

Downing, R. A. and Gray, D. A. (eds) (1986) *Geothermal Energy: The Potential in the United Kingdom*. London, HMSO.

EEPH (2005) *UK Heat Pump Study*, Hard to Heat Group, Energy Efficiency Partnership for Homes, available at http://www.docstoc.com/docs/34919819/Heat-Pumps-Study (accessed 22 Feb 2012).

Everett, B., Boyle, G. A., Peake, S. and Ramage, J. (eds) (2012) *Energy Systems and Sustainability: Power for a Sustainable Future* (2nd edn), Oxford, Oxford University Press/Milton Keynes, The Open University.

Genter, A., Fritsch, D., Cuenot, N., Baumgärtner, J. and Graff, J.-J. (2009) 'Overview of the current activities of the European EGS Soultz project: from exploration to electricity production', *Proc. Thirty-Fourth Workshop on Geothermal Reservoir Engineering*, Stanford University, Stanford, California, February 9–11, 2009.

Goldstein, B., Bendall, B., Long, A. and Budd, A. (2011) 'Converting geothermal plays to projects in Australia – a national review', *Proc. Thirty-Sixth Workshop on Geothermal Reservoir Engineering*, Stanford University, Stanford, California, 31 January – 2 February 2011.

Harvey, C. C., White, B. R., Lawless, J. V. and Dunstall, M. G. (2010) '2005–2010 New Zealand Country Update', *Proc. World Geothermal Congress 2010*, Bali, Indonesia, 25–30 April 2010, International Geothermal Association.

IEA (2011) 'Technology Roadmap for Geothermal Heat and Power', OECD/IEA, Paris. Available from www.iea.org/books. http://www.iea.org/papers/2011/Geothermal_Roadmap.pdf (accessed 27 March 2012).

IGA (2001) 'Performance indicators for geothermal power plant', *IGA News*, no. 45, July–September 2001.

IGA (2002) 'Geothermal power generating plant: CO_2 emission survey', *IGA News*, no. 49, July–September 2002.

Lund, J. W., Freeston, D. H. and Boyd, T. L. (2010) 'Direct utilization of geothermal energy 2010 worldwide review, *Proc. World Geothermal Congress 2010*, Bali, Indonesia, 25–30 April 2010, International Geothermal Association.

MIT (2006) 'The future of geothermal energy: impact of EGS on the United States in the 21st century', *Report for US DoE*, Massachusetts Institute of Technology, 2006. Available at http://web.mit.edu/mitei/research/studies/documents/geothermal-energy/geothermal-energy-full.pdf (accessed 27 March 2012).

Rybach, L. and Eugster, W. (1998) 'Reliable long-term performance of BHE systems and market penetration – the Swiss success story,' *Proc. 2nd Stockton International Geothermal Conference*, March 1998, Richard Stockton College, New Jersey.

Sanner, B. (2001) 'Shallow geothermal energy', *Proc. Int. Geothermal Days 2001*, Bad Urach, Germany, 2001.

Chapter 10

Integrating renewable energy

By Bob Everett and Godfrey Boyle

10.1 Introduction

Renewable energy already makes a large contribution to world primary energy needs: in 2009 it amounted to about 13% (see Figure 1.2 in Chapter 1). Between 2005 and 2010 world fossil fuel consumption increased slowly, while the total consumption of nuclear electricity has not grown. However, various renewable energy technologies have seen high growth rates (see Figure 10.1).

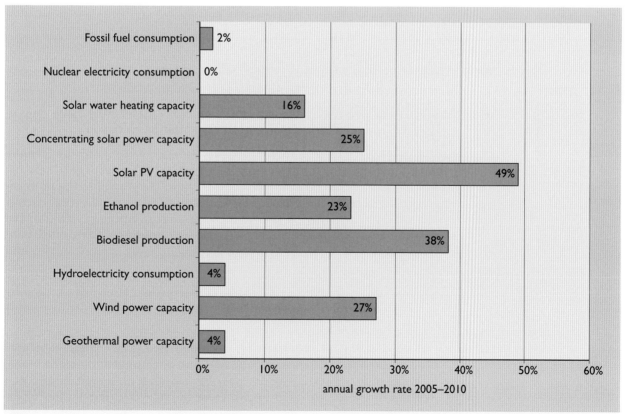

Figure 10.1 Annual growth rates of world fossil fuel, nuclear electricity and hydroelectricity consumptions plus those of renewable energy capacity and biofuels production 2005–2010 (sources: BP, 2011; REN21, 2011)

These renewable energy growth rates seem likely to continue, spurred on by concerns about climate change, limited fossil fuel reserves and the external costs of fossil and nuclear fuel use.

Chapters 2–9 reviewed each major renewable energy source in turn, giving some consideration to the contributions that each could make to future needs. Modern industrial societies demand energy in extremely large quantities and in many different forms. In order to meet these requirements, a vast world-wide network of energy supply and distribution systems has been built up. To what extent will these existing energy networks need to be modified and supplemented if renewables are to continue to make an increasing contribution? Can renewables deliver our energy, not only in significant *amounts* and at an acceptable *price*, but also in the right *form*, at the right *time*, and in the right *place*? What are the factors that currently make it difficult for renewable-energy sources to compete with conventional sources? What can society do to promote their use?

These are some of the questions addressed in this final chapter. Many of the topics discussed in earlier chapters will be revisited, and an attempt made to draw some of the threads together. Initially the situation in the UK is examined before looking further afield.

Firstly, Section 10.2 looks briefly at existing energy systems and how they currently supply our various demands. A prime reason for using renewable energy is to reduce global greenhouse gas emissions. Section 10.3 discusses the relative emissions of conventional energy and different renewable energy sources.

Next, Section 10.4 discusses the available 'resource' or 'potential' for each renewable-energy technology, and how practical and economic constraints impose limits on the contributions that each might make.

Section 10.5 looks at the current financial cost of renewable energy (particularly renewable electricity), concentrating on the UK.

Sections 10.6 and 10.7 look at renewable energy technologies in more detail, investigating the extent to which they can deliver energy in suitable forms, in the right place and at the right time. This includes the problem of integrating electricity from renewables into national electricity grids. Section 10.8 considers some possible system solutions, including the creation of a 'European Supergrid' enabling continent-wide electricity transmission, and the potential of hydrogen as an energy carrier and transport fuel.

Section 10.9 looks briefly at some of the methods that have been used by governmental and non-governmental organizations to promote renewable energy over the past few decades.

Finally, Section 10.10 describes a variety of 'energy scenarios' that have been constructed by governmental and non-governmental organizations, in order to explore the various possible patterns of energy supply and demand in the future – and in particular the roles that renewables might play in them.

10.2 The existing UK energy system

Figure 1.5 in Chapter 1 has illustrated the UK's primary and delivered (or final) energy use. Delivered energy takes five main forms:

- *liquid fuels*: about 43% in 2010 – almost entirely oil and its derivatives – petroleum (gasoline), diesel, kerosene, etc., although with an increasing percentage of ethanol and biodiesel (about 2% of liquid fuel use) (DECC, 2011a)

- *gaseous fuels*: about 34% in 2010 – mostly methane ('natural gas'), plus some 'bottled' liquefied petroleum gas (LPG – propane and butane)

- *solid fuels*: in the recent past this has been almost entirely coal and coke. In 2010 these contributed under 2%; however, with a resurgence of biomass as a fuel there are increasing contributions from fuel wood and straw, contributing about an extra 0.5%

- *electricity*: about 19% in 2010 – mostly from fossil-fuelled or nuclear power stations, with nearly 7% of this coming from hydroelectricity and other renewable sources

- *heat*: about 2% in 2010 – although only deployed to a limited extent in the UK, this involves distributing heat energy to buildings directly in the form of hot water or steam, in district heating (DH) (also known as community heating) schemes. The heat can be provided by centralized boilers (in some cases fuelled by refuse incineration) or by waste heat from power stations in combined heat and power (CHP) schemes. It can also include heat from deep geothermal sources or large scale heat pumps.

These forms of energy flow into the various distribution networks and are delivered to final consumers in the main energy-using sectors of the economy, normally categorized as domestic, services, industry and transport.

Within each of these sectors, this delivered energy is used to provide the *energy services* that we actually need.

Firstly there are services based around *heat*, i.e. warm homes, cooked food and heat for industrial processes. Heat is required in many forms – from tepid water to superheated steam – for washing, cooking, space heating and industrial processing. It can be provided either by burning fuels close to where the heat is needed, by piping heat in from a more distant CHP or DH plant, or by using electricity, either in resistance heaters or in more sophisticated devices like microwave ovens. However, a word of caution is necessary here. A warm home does not necessarily require a large supply of heat energy. The desired level of energy service can also be achieved through improved insulation and similar energy-conserving measures, such as those in the *Passivhaus* retrofit described in Chapter 2.

Secondly there are energy services based on *motive power*. A significant proportion of the energy for this is in the form of electricity for electric motors driving everything from watches and CD players through to lifts and heavy industrial machinery. Motive power is also needed, obviously, for transport (cars, trucks, buses, trains, ships and aircraft, etc.) In most cases, the form of delivered energy currently used to supply it is oil, although electricity is widely used for electric railways and in new designs of electric cars. Again a word of caution is necessary. Much of the end energy service for transportation can be described as *mobility* yet this does not necessarily require the physical transportation of people. In many cases the use of telecommunications can provide an alternative energy service to transportation.

Finally, there are *electricity-based energy services*. Electricity can be used to provide heat, motive power and lighting, each of which can also be provided by other forms of delivered energy. However, electricity is essential for those systems that simply cannot function with any other energy form – not just computers and communications systems but also more specialized electro-chemical processes, such as the manufacture of aluminium or chlorine.

Distribution

It is worth dwelling a little on the changing nature of our energy distribution systems. Most of the UK's energy demand occurs within the relatively small areas of the major cities. Many of these are built in sheltered inland valleys, deliberately to avoid the worst excesses of wind and wave energy. Many large industrial towns grew up around their fuel supplies: coal or wood. Some faded to obscurity when their fuel ran out. Others survived by importing fossil fuels.

At the end of the 19th century, the UK was a major producer and exporter of coal. Most European cities ran mainly on coal, brought by rail or by sea. It was burned when and where it was needed, or was converted into electricity or 'town gas' in local plants. The precise forms in which we make use of electricity (for example as a 230 volt AC supply) and gas (as a low-pressure piped supply) had largely been fixed by the beginning of the 20th century. Yet, at that time, oil as we know it today scarcely existed as a commercial commodity.

Today coal, oil, natural gas and electricity are distributed not just country-wide but internationally. Coal, once the major fuel in the UK, is now mainly just a fuel for power stations. Some is still locally mined, but much is imported by ship from countries such as Russia, Colombia and the USA.

The UK was a major importer of oil until the 1970s when discoveries in the North Sea made it not merely self-sufficient but a net exporter. There is an extensive internal network of pipelines conveying oil around the country. The UK's North Sea oil production peaked in 1999 and is now declining: the country is once again becoming a net oil importer.

Until the 1960s, the UK gas supplies were 'town gas', made from coal, which consisted of a mixture of carbon monoxide, methane and hydrogen. Then the large discoveries of natural gas (almost pure methane) in the North Sea revolutionized the industry. Hundreds of town gas production plants (gasworks) were closed down and a complete new infrastructure of long-distance distribution pipes was built to bring in the North Sea gas. This had a higher energy density than town gas and meant that all gas boilers and cookers had to be fitted with new burners. However, it also meant that the energy-carrying capacity of the existing pipework was doubled. This was useful since there was an immediate rise in sales of gas-fired central heating systems. Natural gas has been *the* fuel of the last quarter of the 20th century. It has a lower proportion of carbon than coal and can be burned at high efficiency in combined cycle gas turbine (CCGT) power stations. A whole new generation of these plants was constructed in the UK in the 1990s, leading to a further large increase in gas demand.

Looking beyond the UK, there is an impressive Europe-wide system of pipelines for gas distribution, stretching from gas fields in Algeria in the south to Norway in the north, from Siberia and the Caspian Sea in the east to the Republic of Ireland in the west. A major pipeline such as that shown in Figure 10.2 can cost £1 million or more per kilometre to lay.

Figure 10.2 A high-pressure gas pipeline being laid. Such pipes can carry several gigawatts of power, i.e. a similar amount to that carried by a 400 kV electricity power line

The UK's North Sea gas production peaked in 2000 and, like oil production, has been declining, with the country becoming a net gas importer after 2005. This situation is unlikely to be changed significantly by recent UK shale gas discoveries. These could amount to potential reserves of 150 billion m^3 or about two years' UK consumption (DECC, 2010a). Some natural gas is imported via pipeline from the Norwegian and Dutch areas of the North Sea, but an increasing proportion is being imported by ship as liquefied natural gas (LNG) from countries such as Algeria in North Africa and Qatar in the Persian Gulf. It is liquefied by cooling to $-162\ °C$ and can then be transported by insulated tanker and stored in large insulated storage tanks (see Figure 10.3).

Despite these large investments in infrastructure it is unclear whether natural gas will have a long-term future as a fuel in Europe as fossil fuels are phased out, as some energy scenarios suggest (see Section 10.10 below).

Turning to electricity, today it is mostly generated in large plants, usually of 200 MW or more capacity. Although much of the UK's electricity demand is in the south, it has been more convenient to site coal-fired power plants in the mining areas of Wales or the north of England. This has also avoided problems of sulfur dioxide emissions reaching urban areas. Today, most of these plants are reliant on foreign coal imported by sea. The UK's nuclear power stations are mostly sited on the coast for ease of access to cooling water. Gas-fired power stations are more likely to be sited closer to their loads, and supplied with large quantities of natural gas through the high-pressure pipelines described above. Since impurities such as sulfur are removed from the gas at source, these power stations have fewer pollution problems than their coal-fired counterparts.

(a)

(b)

Figure 10.3 (a) The LNG tanker Arctic Princess docking at the Grain LNG terminal in Kent; (b) insulated LNG tanks at the Grain terminal. Each tank in the foreground holds 190 000 m³ of LNG stored at −162 °C

The electrical output of these power stations is distributed to consumers via the National Grid, a network of high-voltage cables that covers the UK, with links to neighbouring countries (Figure 10.4). Since they are likely to be major carriers of renewably generated electricity in the future, we will look at the national grids of the UK and the Republic of Ireland in more detail later.

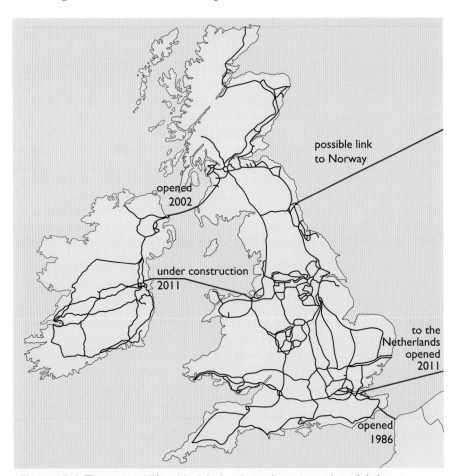

Figure 10.4 The existing UK and Irish high voltage electricity grids, with links to Norway, the Netherlands and France

Finally there are heat distribution networks. Although district heating is currently little used in the UK, in Denmark it supplies 60% of homes. Distribution pipes such as that shown in Figure 10.5 carry heat from DH and CHP plants, such as that shown at Avedøre in Denmark in Chapter 4.

10.3 How can renewable energy decrease greenhouse gas emissions?

Figure 1.7 in Chapter 1 has shown the rising global emissions of CO_2 from fossil fuel combustion. Box 4.4 in Chapter 4 described the process of CO_2 production from the combustion of a fuel such as methane. It also described how coal, a high carbon fuel, has a higher specific CO_2 emission factor than oil or methane – i.e. when burned it produces more CO_2 per unit of heat emitted.

How much CO_2 will be saved if a renewable energy source is substituted for a fossil fuel one? The answer depends critically on which fossil fuel is being replaced and whether the energy is being used for electricity generation or for heat or transport use.

Electricity generation presently accounts for about a third of global CO_2 emissions and in 2008 consumed two-thirds of world coal production. This makes electricity-generating renewables of particular interest.

Figure 10.5 District heating pipes being laid in Denmark. These have a single casing that contains both the flow and return water pipes and some insulation

Electricity-generating renewables

The production of electricity from fossil fuels involves a considerable production of waste heat. The figure for the CO_2 emitted per kilowatt-hour of electricity depends both on the specific emission factor of the fuel and on the generation efficiency. Table 10.1 shows average values for UK generation in 2010. In round numbers, the production of 1 kWh of electricity from coal produces about 0.9 kg of CO_2 and that from gas about 0.4 kg.

Table 10.1 Specific CO_2 emission factors for UK electricity in 2010 and an approximate external cost of associated carbon emissions

Fuel	Specific emission factor for electricity /g CO_2 kWh^{-1}	Assumed specific emission factor for fuel[1] /g CO_2 kWh^{-1}	Implied average generation efficiency[2] / %	External cost at an assumed carbon price of £40 t CO_2^{-1} /p kWh^{-1}
Coal	909	340	37	3.6
Oil	653	270	41	2.6
Gas	398	200	50	1.6
Average – UK generation mix	458			1.8

[1] Based on net calorific or lower heating value of fuel
[2] i.e. electricity emission factor/fuel emission factor
Sources: DECC, 2011b, AEA, 2011

The UK generation mix includes gas, coal, nuclear power and a number of other sources. Figure 10.6 shows the breakdown for 2010. Although over a quarter of generation was from coal, another quarter came from low-carbon sources such as nuclear and renewables. The overall average emission factor was under 0.5 kg CO_2 kWh^{-1}.

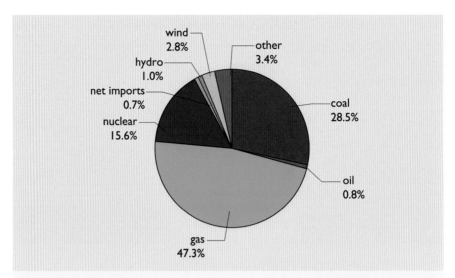

Figure 10.6 UK electricity generation fuel mix 2010. Note that 'other' includes renewable sources other than hydro and wind, and energy from waste (source: DECC, 2011c)

Carbon dioxide emissions at the power station are, of course, not the full story. There may be other greenhouse gas emissions involved, particularly those of methane. There are also indirect CO_2 and methane emissions involved in the supply of their fuel. An overall emission factor including these factors can be expressed in units of CO_2, i.e. the equivalent amount of CO_2 that would give the same amount of global warming as that resulting from the actual mix of greenhouse gases emitted. Including these extra factors would increase the values in Table 10.1 by about 15%. A full life-cycle analysis of rival technologies should also include the greenhouse gas emissions involved in the construction of the power plants and their supply infrastructure. Figure 10.7 shows the ranges of estimates of overall greenhouse gas emissions for European power plants per kilowatt-hour of electricity generated. The wide ranges for conventional fossil-fuelled plant reflects the different generation efficiencies. The lowest emission values are likely to be from CHP plant.

All electricity-generating technologies involve some emission of CO_2 and other greenhouse gases over their full life cycle. This is likely to be true even for future fossil-fuelled plants using **carbon capture and storage (CCS)**, where most of the CO_2 (possibly 90%) is separated and sent for Sequestration in deep underground aquifers – a topic that will be described further in section 10.8. However the typical emission levels for renewables are a fraction of those for conventional fossil-fuelled generation.

It is sometimes argued that the CO_2 savings from renewable energy should be compared with coal-fired generation, since this is a form of generation that needs to be removed as rapidly as possible. For a country such as China, where nearly 80% of electricity is generated using coal, this is a reasonable argument. However in most UK and European assessments the savings are likely to be compared with gas, which is the fossil fuel of choice for electricity generators.

Any economic comparison of rival technologies should include 'external costs', i.e. the cost of the damage done by the pollution emitted to the environment. Much effort has been put into trying to calculate a carbon price, a figure for the global damage done by the emission of a tonne of

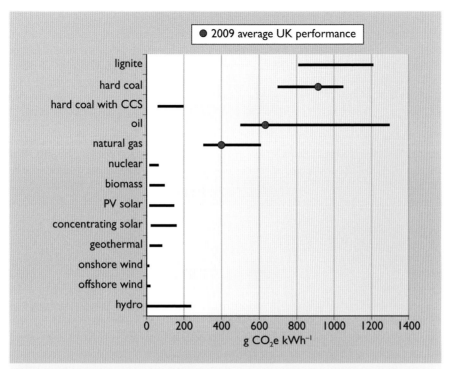

Figure 10.7 Ranges of specific greenhouse gas emissions for a variety of electricity-generation technologies in Europe: the 2010 average performance figures for UK plant are for CO_2 emissions alone. Low fossil fuel values are likely to be due to credits for CHP use (sources: EEA, 2011, DECC, 2010b)

CO_2 (or any other greenhouse gas). It is likely that this lies somewhere between £30 t CO_2^{-1} and £60 t CO_2^{-1} (see Everett et al., 2012). Table 10.1 includes values for the external costs of the electricity using a sample price of £40 t CO_2^{-1}. Given that in 2010 large UK industrial consumers paid about 5 p kWh^{-1} for their electricity the inclusion of these external costs would represent a significant increase.

Finally, greenhouse gases are not the only pollutants emitted. As described in Chapter 4, Table 4.5, coal-fired electricity may involve the emission of large amounts of sulfur dioxide, a major contributor to acid rain. Most European and many US coal-fired plants are now equipped with flue gas desulfurization (FGD) equipment to reduce this. All forms of combustion (including that of biomass) can potentially produce oxides of nitrogen (NO_x), which can also contribute to acid rain, although careful burner design can minimize their production.

Thus overall, renewable energy technologies for electricity generation have a role to play in reducing *both* global CO_2 emissions and acid rain emissions.

Heat and transport fuels

Where heat is generated from fossil fuels, the conversion efficiencies are much higher than for electricity generation, typically 80% or better for modern gas or oil-fined condensing boilers. A kilowatt-hour of useful heat from a gas-fired boiler thus involves greenhouse gas emissions of about

0.25 kg CO_2e while that from an oil-fired boiler is about 0.40 kg CO_2e. Ideally, sustainably grown biomass should not involve significant greenhouse gas emissions. However, fossil fuel energy use and greenhouse gas emissions are normally involved in fertilizer production, planting, harvesting, processing and transport. Thus heating a home by burning locally sourced wood in a well-designed stove with an efficiency of 70% is still likely to involve the emission of about 0.03 kg CO_2e per kWh in greenhouse gases (AEA, 2011).

The picture for transport fuels is more complex. Conventional fossil-fuel-based petrol and road diesel fuel (DERV) produce CO_2 when burned, but further greenhouse gas emissions (about 20%) are involved, particularly in refining. Overall it is estimated that burning a litre of 100% fossil-fuel-based petrol involves the emission of 2.7 kg CO_2e of greenhouse gases. For a litre of DERV the figure is about 3.1 kg CO_2e (AEA, 2011; BEC, 2011).

UK petrol is currently sold blended with up to 5% ethanol, which may come from a range of sources as described in Chapter 4. Ideally this should be a low-carbon fuel, but in practice there are again emissions involved in its production. The life-cycle greenhouse gas emissions from combustion of a litre of bioethanol may only be 25% of that of conventional petrol. Similarly, the combustion of a litre of biodiesel, produced from rape seed oil, is likely to involve 40% of the emissions of conventional diesel fuel. These figures are obviously highly dependent on the precise production method used for the biofuel.

10.4 How much renewable energy is available?

Let us now consider what contribution could be made by renewable energy, concentrating on the UK. It is easy to make sweeping statements about the very large natural energy flows available. These form the 'total, theoretical or available resource'. Defining how much energy a particular source or a specific technology might actually supply in practice requires looking at the various constraints – technical, social and economic – on its use. These whittle down the 'resource' or 'potential' towards something more realistic. A possible ranking of such constraints is given in Box 10.1. The 'economic potential' at the end is likely to be far smaller than any notion of a total available resource (see Figure 10.8).

BOX 10.1 Resource size terminology

Fossil fuel *resources* are stores of energy, potentially available for use once extracted from the ground. These are large, but finite. Fossil fuel *reserves* are more limited; these consist of those resources that are known *and* currently considered economic to extract within a whole range of practical, legal and environmental constraints.

In contrast, renewable energy resources are potential *flows* of energy, analogous to *the rate of consumption* of fossil fuels. It is customary to specify their potential *annual contributions*. It is important to bear in mind this distinction when comparing the potential contributions from fossil and renewable sources.

There are many definitions of the annual **resource** or **potential** of specific forms of renewable energy, differing mainly in the extent to which the definitions take into account the practical and economic limitations on its use. The following terms have all featured in recent UK energy literature.

The *Available, Theoretical or Total Resource* – the total annual energy delivered by the source; for example the total energy carried by ocean waves or the wind, or the total incident solar energy.

The *Technical Potential* (also referred to as the *Accessible Resource*) – the maximum annual energy that could be extracted from the accessible part of the Available Resource using current mature technology. Although this could change with technological advances, it may ultimately be limited by basic laws of physics (determining for instance the properties of wind turbines or the efficiency of heat engines.) It is also limited by basic accessibility constraints due to:

- practical difficulties such as the presence of roads, buildings and lakes
- institutional restrictions and the need to avoid such areas as National Parks, Sites of Special Scientific Interest (SSSIs), Areas of Outstanding Natural Beauty (AONBs), etc.

The *Practicable Potential* (or *Practical Resource*) – the technical potential, reduced by taking into account:

- constraints on using or distributing the energy – such as transportation problems, access to the electricity grid or problems of intermittent supply
- further limitations on land or technology use due to public acceptability. It may be difficult to quantify these since they may only become apparent when planning permission is sought and environmental objections are expressed.

The *Economic Potential* – the amount of the technical potential that is economically viable. Any judgement about this requires the specification of an acceptable energy cost and a discount rate that sets the cost of borrowing money for investment. (See Appendix B.)

Figure 10.8 gives a notional ranking of the different sizes of the potentials.

For example, for UK onshore wind, the theoretical resource is estimated at about 4000 TWh y^{-1}, i.e. over ten times the 2010 UK electricity demand of about 340 TWh. The technical potential is estimated to be about 660 TWh y^{-1} and the practical resource about 80 TWh y^{-1} (CCC, 2011a). Only a small amount of this resource has been developed to date; in 2010 onshore wind generated just over 7 TWh (DECC, 2011a).

These kinds of estimates, together with assumptions about economic conditions, underlie the various energy scenarios described in Section 10.10.

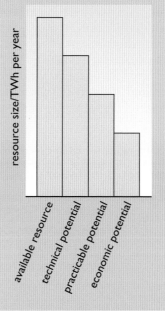

Figure 10.8 A notional ranking of renewable energy resource sizes – the 'technical', 'practicable' or 'economic' potentials may be considerably less than the 'available' resource. The precise figures will vary widely between different technologies

At a global scale the technical potential for renewable energy is extremely large. Table 10.2 gives ranges of estimates for some different electricity-generating renewables. Actual global electricity generation in 2010 from all sources, renewable and non-renewable, was just over 21 000 TWh.

Table 10.2 Global technical potential for some electricity-generating renewables

Technology	Technical potential / TWh y^{-1}
Hydroelectricity	~14 000
Wind power	24 000–160 000
Ocean energy	2 000–90 000
Geothermal	33 000–310 000

Source: IPCC, 2011

Note: 1 EJ = 278 TWh

Estimates of the technical potential for direct solar energy range between 1575 and 50 000 EJ y^{-1} (IPCC, 2011). Estimates for the global technical potential for bioenergy range to over 1500 EJ y^{-1}, i.e. 300% of current world primary energy consumption (see Chapter 4, Table 4.10). As described in Chapter 4, there are obvious possible conflicts between land use for growing food and for energy crops. A recent report by the UK Committee on Climate Change (CCC) (CCC, 2011b) warns against over-optimism about the contribution of biofuels to world energy needs, and suggests a much lower range of 10–120 EJ y^{-1}.

Renewable potential for the UK

Detailed assessments of the future UK potential for different renewable-energy technologies have been made over the past 20 years, concentrating particularly on electricity-generating sources. The rapid decarbonization of electricity supply is considered a major route to cutting UK greenhouse gas emissions by 80% by 2050. As pointed out in Box 10.1, the technical resource for onshore wind alone is about twice the current UK electricity demand of about 340 TWh y^{-1}. However, for the immediate future what are required are estimates of the practical and economic resources. Table 10.3

Table 10.3 Estimates of UK practical resource for some electricity-generating renewables

Technology	Practical Resource /TWh y^{-1}	Limitations
Solar PV	140	Intermittency. Costs, though these are likely to fall
Hydro	8	Few available sites for reservoir hydro and limited resource for run-of-river plants
Tidal range (barrage)	~40	Costs and environmental acceptability. A Severn Barrage would make up half of the resource
Tidal stream	18–200	Costs. Still at an early stage of development
Onshore wind	80	Intermittency. Public acceptability
Offshore wind	>400	Intermittency. Costs should reduce with future deployment
Wave	~40	Intermittency. Costs. Still at an early stage of development
Deep geothermal	35	A longer term possibility still not demonstrated in the UK

Source: CCC, 2011a

shows estimates of the UK practical resource for some electricity-generating renewables with some comments on their limitations.

In comparison, the UK potential for biomass may be limited. Again, there are conflicts between the need for land to grow food, wood for construction purposes, biofuels for transport and biomass for heat and electricity generation. A range of estimates has been shown in Chapter 4, Figure 4.25. In its recent report (CCC, 2011b), the UK Committee on Climate Change suggests a 2050 bioenergy potential of 400–750 PJ (about 4–8% of UK 2010 primary energy consumption). The CCC recommends that the use of bioenergy for non-CHP electricity generation should not be promoted, given the many alternatives listed above.

10.5 Is renewable energy available at an acceptable financial cost?

Some renewable-energy technologies are undoubtedly cheaper than others; and their output has to be compared with the costs of rival options. For electricity-generating renewables this includes other low-carbon technologies such as nuclear power and fossil-fuelled generation using carbon capture and storage (CCS). The following section examines the estimated costs for UK electricity-generating technologies, but is of wider applicability.

Electricity-generating renewables in the UK

It has been estimated that electricity from a new gas-fuelled CCGT plant built in 2009 would cost 6.5 p kWh^{-1} assuming a 10% discount rate on the capital repayments.

Figure 10.9 shows estimated levelized costs of a range of electricity-generating renewables (see Appendix B for a description of the term 'levelized'). These are compared with the costs of reference technologies such as conventional gas and coal generation and low-carbon alternatives such as coal-fuelled generation with CCS and nuclear power.

As mentioned earlier, any fair comparison of fossil-fuelled and low-carbon technologies should take into account a **carbon price**. The values used in the calculations for Figure 10.9 are on a sliding scale from 2009, rising in future years, but are equivalent to an average rate of about £40 per tonne of CO_2 emitted. Inclusion of a carbon price raises the overall cost of gas-fuelled electricity from 6.5 p kWh^{-1} to 8.0 p kWh^{-1}. The cost of conventional coal-fuelled electricity rises to over 10 p kWh^{-1}. The costs of coal generation with carbon capture and storage (CCS) still include a small carbon price because it is assumed that only 90% of the CO_2 can be captured. It also includes extra costs for the transport and disposal of the captured CO_2, most probably in saline aquifers deep underground or beneath the North Sea.

As described earlier, many of the other technologies involve low levels of emissions of greenhouse gases, but their carbon costs are not included in Figure 10.9.

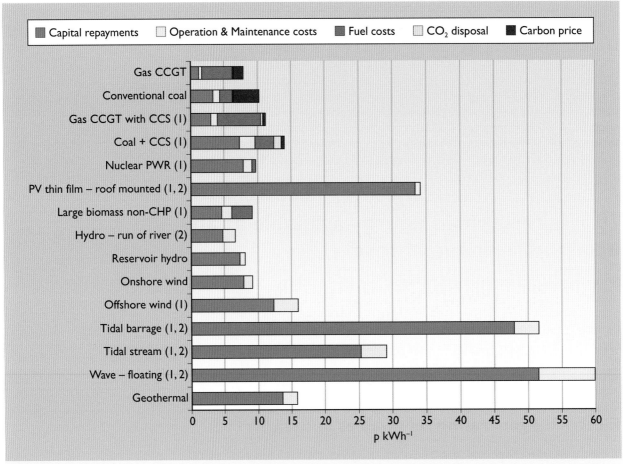

Figure 10.9 Cost comparison of conventional and low-carbon electricity-generation technologies assuming a 10% discount rate and 2009 start except where noted. Notes: (1) costed as 'First of a Kind', with potential cost reductions; (2) 2011 start and different discount rates: wave 15.5%, tidal 14.5%, hydro – run of river 7.5%, PV 7.5% (sources: Mott Macdonald, 2010, 2011)

Prices, mass production and perceived investment risk

It is obvious from Figure 10.9 that some technologies are far more expensive than others today, yet they are supported because they may become cheaper in the future.

With the exception of the biomass example, the renewable energy technologies shown are all capital-intensive. These technologies all have reasonably long lifetimes (assumed to be 18 years or more). Their fuel costs are zero and the bulk of the electricity costs consist of the repayments on the initial capital outlay. The electricity price is thus heavily influenced by the capital cost per kW of installed capacity and the discount rate applied to the finance (as explained in Appendix B, this is the 'real', inflation-corrected, interest rate charged). The importance of discount rate has been discussed in relation to hydro schemes in Chapter 5, Box 5.11, and in relation to a proposed Severn Barrage in Chapter 6. In practical projects the discount rate is likely to be determined by the perceived investment risk for private investors and the possibility of poor performance or even complete failure.

In the analysis by Mott Macdonald for Figure 10.9 some technologies, such as onshore wind, PV, and hydro have been treated as 'mature technologies'. They are considered 'low risk' investments and finance may be available at a discount rate of under 10%. Despite this, PV is still expensive because the capital costs are high, but as explained in Chapter 3, these are expected to fall in the future as production methods improve and the volume of production increases.

Offshore wind has been considered a slightly more risky investment, with finance likely to be available at a discount rate of about 12%. Again capital costs are high (typically twice those per kW of capacity of onshore wind) but are likely to fall with future volume production. Successful future deployment should also reduce the perceived investment risk, making finance available at a lower discount rate.

Then there are 'new' technologies such as tidal stream and wave power that are still only at the prototype stage. Future capital costs can only be roughly estimated and will depend on future production volumes. A 'high risk' discount rate of 14.5% has been assumed for investment for these. A 10% discount rate would reduce their electricity costs as shown in Figure 10.9 by about 20%.

Finally there are tidal barrages. Although these use conventional hydro turbine technology they are essentially 'one off' projects. In the estimates for Figure 10.9 they have been considered to be an extremely risky investment with a discount rate of 15%, resulting in an electricity price of over 50p kWh^{-1}. If this perceived risk could be reduced, then a discount rate of 10% would reduce the electricity price to about 30 p kWh^{-1} (Chapter 6, Table 6.1). It has been argued that capital-intensive projects with significant prospects of reducing national CO_2 emissions should be funded with (government) loans at a very low 'social' discount rate of 3.5%. This would dramatically reduce the electricity price to only about 11 p kWh^{-1}. Such low interest finance would, of course, also reduce the costs of rival low-carbon technologies, particularly nuclear power (see CCC, 2011a).

Heat-generating renewables in the UK

As pointed out in Chapter 1, over a third of the end-use of fuel in the UK is for low-temperature heating. Domestic space heating alone accounts for about 15% of UK primary energy consumption and domestic water heating a further 5% (DECC, 2011d). As described in Chapter 2, insulation and similar energy-conserving measures can result in dramatic reductions in space heating demand, even in existing buildings.

At present, natural gas is the prime UK heating fuel. However, it is not clear what role it will play in the future as fossil fuel use is reduced. If large amounts of electricity from low-carbon sources can be made available (particularly in winter) at an acceptable price from renewable energy and nuclear power, then heat pumps may be a preferred alternative. Biomass may also be a useful heating fuel, particularly in rural areas beyond the gas grid (subject to the availability of fuel) and solar water heating is another option. Table 10.4 shows some current levelized cost estimates for alternative domestic heating options, taking into account the capital costs of the heating system. The higher figures of the ranges are for well-insulated houses with low energy demands.

Table 10.4 Levelized costs of heat from different sources for domestic use

Heating type	Cost / p kWh^{-1}
Natural gas boiler	6.8–10.5
Air-source heat pump	10.9–16.5
Ground-source heat pump	12.8–18.6
Biomass boiler	15.3–23.7
Solar thermal water heating	26.6

Note: Based on 2011 costs, 15 year equipment life and 8% discount rate
Source: CCC, 2011a

Balancing renewables and energy efficiency

Considering the balance between investment in improved energy efficiency and investment in low-carbon energy supply is very important. Conservation and efficiency measures are usually more cost-effective than many low-carbon supply options, and they narrow the gap between energy demand and the supply of renewable energy.

The technology comparison shown in Figure 10.9 above is in terms of energy price, but this includes a notional carbon price. Another way of making a comparison is to look at the different costs of saving a tonne of CO_2 using different technologies. These can be set out as a **marginal abatement cost curve (MACC)**. Figure 10.10 shows an example of the potential savings for the UK domestic or residential sector in 2020 (CCC, 2008). It assumes that many basic measures such as improved loft insulation have already been taken up by then. The technologies are shown ranked from the lowest cost at the left to the most expensive at the right.

There is a wide range of options. Some efficiency options have a negative cost. Replacing refrigerators with more efficient models reduces electricity demand. Insulating solid walls in millions of homes can reduce the national heat demand. Both of these are likely to have a lower cost than renewable energy supply options such as photovoltaic (PV) electricity or solar water heating. This might be taken as a reason to delay the deployment of renewables until a full programme of energy-efficiency retrofits had been carried out. In practice, achieving the necessary rate of reduction in national CO_2 emissions means that all options need to be pursued simultaneously.

World-wide renewable electricity prices

The UK is a country with good wind, wave and tidal resources. However, it cannot be described as particularly sunny, nor does it have a well-developed geothermal resource. Table 10.5 shows some estimates of renewable electricity costs from a world-wide perspective. In particular this shows that in the right (sunny) locations concentrating solar

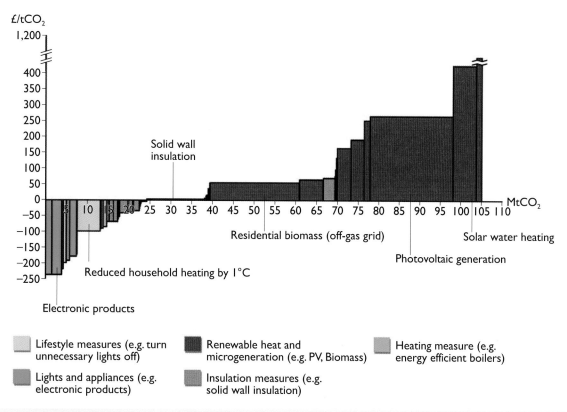

Figure 10.10 Marginal abatement cost curve (MACC) for different technologies in the UK domestic sector for 2020. UK total CO_2 emissions for 2010 were almost 500 Mt (source: CCC, 2008)

power is highly competitive with PV. Also, given the right resources, geothermal electricity generation is highly competitive with all other sources.

Table 10.5 World-wide levelized cost estimates for renewable electricity generation

Technology	Cost /US cents kWh^{-1}
Concentrating solar power	20–31
PV (residential rooftop)	23–86
PV (utility-scale fixed tilt)	15–62
Biomass co-firing	2.9–7.1
Hydro (all sizes)	2.4–15
Geothermal	4.5–13
Tidal range	23–32
Onshore wind (large turbines)	5.2–17
Offshore wind (large turbines)	12–23

Notes: 2008 data, 10% discount rate, 1 US cent ≈ 0.65 p
Source: IPCC, 2011

10.6 **Are renewable-energy supplies available *where* we want them?**

So the potential contribution of renewables to UK (and global) energy needs could be very large. Let us take our first practical constraint – can renewables supply our energy *where* we want it?

It would be helpful if renewable supplies were located at the points of maximum energy demand, the major cities. Some can be. Solar thermal or photovoltaic panels can be fitted to the roofs of buildings.

Energy crops are likely to be available in rural areas, but perhaps not those where food production is seen as more important. Fuels such as wood and forestry wastes need to be gathered and then transported from where they are grown, possibly in more remote country areas, to where they will be used, probably in towns. However, this may involve some intermediate reprocessing into wood pellets or fine powder. As has been described in Chapter 4, wood has only half the energy density of coal, and only about a third of that of fuel oil, so transporting it can be more expensive and raise issues of the transport energy and CO_2 emissions involved. This is particularly true where large amounts of fuel wood are being used for electricity generation.

Municipal solid wastes (MSW), forestry and agricultural wastes are widely used as fuels. In combined heat and power (CHP) plants they are used to make both electricity and heat. In district or community heating plants, they are just used for heat production.

In all these cases, the heat can be distributed through city-wide heating networks, as is done in Denmark. Such networks can also be used to distribute geothermal heat where it is accessible, such as in Southampton in the UK.

Biofuels with the potential to be used as petroleum substitutes, such as vegetable oils and sugars, need to be harvested and transported to processing plants, as does the fuel for their processing. All of this may involve emissions of CO_2 and other greenhouse gases. As a result, as pointed out in Section 10.3, bioethanol and biodiesel are by no means zero-carbon fuels.

Although wind power can be used at a local scale, its large-scale use in the UK (and most of north-west Europe) will involve the use of large turbines situated at prime onshore and offshore sites with high wind speeds. Similarly prime sites for wave and tidal energy are likely to be relatively remote from the main centres of demand. Integrating large-scale renewable electricity supplies into the existing distribution systems will need careful planning, investment and probably some changes in market structures.

The electricity grid in the British Isles

In the UK and Ireland, new electricity-generating renewables will have to feed into the existing National Grid structures that have been built up over the past 80 years (see Figure 10.4). In England and Wales, the National Grid developed to distribute electricity to the major cities from the large coal-fired stations close to the mines in Yorkshire and Wales (though these are now supplied by coal imported by sea) to the main load centres, which have become increasingly concentrated in the south of England. Extra grid links have been added from coastal nuclear power stations. There are also undersea links to France and the Netherlands with another under construction to Ireland scheduled for completion in 2012.

The electricity system of Scotland has been relatively self-contained, with the demand being met by a mix of coal, gas, nuclear and hydroelectric plants, yet it is linked to both that of England and Wales and that of Northern Ireland. This, in turn, is linked to that of the Republic of Ireland. There Dublin, on the east coast, forms a major load, in part supplied by power stations sited on the west coast running on imported coal.

Wave, wind and tidal power

Figure 10.11 summarizes some of the information from earlier chapters on the locations of wind, wave and tidal energy sources in the British Isles. As can be seen, these electricity-generating sources are widely distributed. Both the UK and the Irish Republic have very significant potential for renewable energy supplies. Comparison of this map with Figure 10.4 raises the question of how the existing national grids will need to evolve in order to match the new energy sources to the loads.

The best *wave* power resources are in Scotland, the Atlantic coast of Ireland and the south-west of England. Current prototype development is concentrated in Scotland.

The best *onshore wind* resources are in Scotland, Wales, Cornwall, the north and west of England and the west of Ireland. Projects in all of these areas are currently under rapid development.

The prime areas for *offshore wind* development are the shallow waters of the North Sea and Irish Sea. As shown in Figure 10.11, it would only require an area of sea 30 km by 40 km to supply 10% of the UK's electricity needs. To date, offshore wind farm development has been close inshore. However, future UK large-scale development is likely to take place further out to sea. As has been described in Chapter 7, several 'Round 3' zones have been specified for this.

The potential for *tidal barrages* is concentrated on a few large estuaries, particularly the Severn and the Solway Firth, though 'lagoon' type structures could be built in more open sea areas. The potential for *tidal stream* devices is in similar estuary locations, but there are many possibilities around prominent headlands, notably the Pentland Firth between Orkney and the Scottish mainland and the 'Race of Alderney' between Alderney in the Channel Islands and Cap la Hague on the French coast.

As Figure 10.4 shows, the UK and Irish grids mostly do not run in a concentrated 'point to point' manner. They are built as a 'mesh' and some lines are apparently duplicated. This is deliberate, to allow the system to continue operating despite the failure of any one given line. This gives flexibility of operation and potentially allows the insertion of extra sources into the existing system. The National Grid is also a key to dealing with the time variability of some renewable-energy supplies, as discussed below.

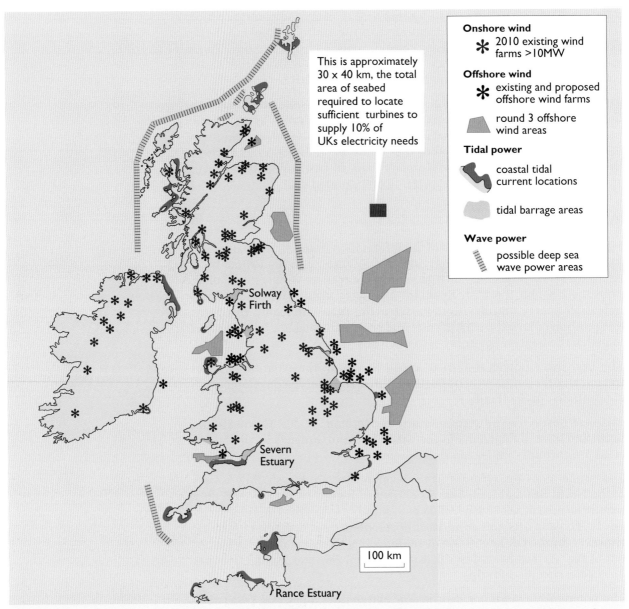

This is approximately 30 x 40 km, the total area of seabed required to locate sufficient turbines to supply 10% of UKs electricity needs

Onshore wind
✱ 2010 existing wind farms >10MW

Offshore wind
✱ existing and proposed offshore wind farms

  round 3 offshore wind areas

Tidal power
  coastal tidal current locations

  tidal barrage areas

Wave power
  possible deep sea wave power areas

Solway Firth

Severn Estuary

100 km

Rance Estuary

Figure 10.11 Geographical distributions of onshore and offshore wind farms, and possible future areas for development of offshore wind farms, wave and tidal energy projects in the British Isles (sources: DECC, 2011b, Hibernian Wind Power, 2011, Offshore Valuation Group, 2010, BERR, 2008)

A pan-European view

There are other pressures on the development of 'national' grids. The European Commission would like to develop the pan-European electricity market. From this trading perspective the 2 GW link from France to England and the 500 MW link from Scotland to Northern Ireland have been identified as 'bottlenecks', limiting international trade in electricity. Where electricity from renewable sources is concerned, the UK and Ireland have plentiful resources of wind, wave and tidal power, which could potentially be complemented by the hydro resources of Norway and Sweden, the biomass resources of central Europe and the solar resources of southern Europe and even North Africa (see Section 10.8 below).

10.7 Are renewable-energy supplies available *when* we want them?

Our demand for energy is not constant. It varies widely over the day, the week and the year. In the UK we need more energy for heating buildings in winter than in summer: the UK consumes over twice as much natural gas in a typical December as it does in a summer month.

Generally, in the 'developed' countries, there are few constraints on our demand for energy. Our electricity supply systems are organized in such a way that power is virtually certain to be available whenever we turn on a switch. Gas is always there waiting for us to turn on the cooker. Every major highway has petrol stations at regular intervals ready to serve us when we drive in.

Complex infrastructures have been put into place to enable supply to meet demand. A greater need for, say, gas will set off a whole series of pumps distributing supplies from gas-holder to gas-holder through a nationwide set of mains. Any failure of the infrastructure to deliver energy on demand – be it gas, electricity, heating oil or petrol – causes a consumer outcry.

Biofuels have many of the advantages of fossil fuels. Most of them can be easily stored and used on demand. However, most other renewable sources are variable. Wind, wave and solar energy are dependent on weather conditions. They are not completely 'firm' supplies – that is, we cannot absolutely guarantee that they will be available when we need them. Hydro power is also dependent on weather conditions but its availability is improved by the built-in storage provided by the reservoirs that form part of most hydro installations. Tidal power is variable, but entirely predictable.

This section looks at the consequences of these characteristics, first with renewable sources of heat and then with the more difficult topic of electricity.

Renewables as heat suppliers

Heat is a relatively easy topic to consider. Individual solar water heaters can be mounted on the roofs of houses or other buildings. They effectively reduce the demand for other forms of water heating and do so mainly in the summer.

As pointed out above, a wide range of biofuels and household and agricultural wastes, together with fossil fuel sources, can be used to produce heat for district or community heating networks. Such systems also allow the input of heat from solar sources, heat pumps, geothermal aquifers, or even surplus wind power. Although there have been demonstration schemes in which heat is stored 'inter-seasonally', from summer to winter, these are expensive. Most heat stores associated with community heating schemes, such as that shown in Figure 10.12, hold only sufficient supplies for a day or so. Using them frees the system designers from concerns of how the heat demand of the local buildings varies over the day, but they do need to think about how it varies over the year.

Figure 10.12 The Pimlico Accumulator in central London holds enough hot water to supply 4000 homes for a day

Municipal solid waste is not a fuel that it is desirable to store for long periods. Whatever the means used to turn it into heat and electricity – mass burning, gasification or pyrolysis – it is best to do so steadily year-round. But other biofuels can be stored and used to meet winter space heating demands.

Experience in many countries has shown that it is economically viable to insulate existing buildings to levels where their space heating demands are dramatically reduced. They can then be heated easily with relatively low-temperature district heating systems and the problem of supplying winter demand is not so difficult.

Integrating electricity from renewables

At present (2012), integrating electricity from renewable sources is a matter of considerable concern. Meeting the UK government's target for 30% of electricity supplies to come from renewable sources by 2020 will require the connection of some 30 GW of renewable electricity generation.

To understand the formidable problems involved in coping with variable inputs of such sizes it is worth describing the existing system in some detail. As will be shown, to a large extent the practical resource limitations of *where* and *when* overlap.

How the current UK electricity system works

Electricity is a much more flexible and valuable commodity than heat, but it is difficult to store. Essentially, it has to be generated immediately to suit the demand. In the UK, as in many other countries, electricity demand varies enormously over the year, and hour by hour throughout the day. Figure 10.13 shows sample national daily demand patterns for summer and winter.

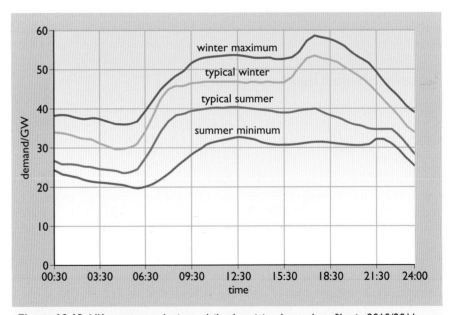

Figure 10.13 UK summer and winter daily electricity demand profiles in 2010/2011 (source: National Grid, 2011)

At night, demand is relatively low (in summer below 25 GW), but it picks up very rapidly early in the morning and flattens out during the day. In winter months there is a pronounced peak (sometimes reaching over 55 GW) in the early evening.

We can look at electricity demand on such a large aggregated scale because of the existence of the National Grid. Although individual consumers use electricity in an intermittent manner, turning on a light here and a heater there, they each usually do so at slightly differing times. This averages out to a smoothly varying electricity demand that the grid system can cope with relatively easily. This effect is known as **diversity of demand**. There are exceptions when consumer behaviour becomes synchronized, which will be discussed later.

The history of the UK National Grid's development has been described in some detail in Everett et al., 2012. When its construction was initiated in the 1930s, electricity was generated locally in hundreds of small separate power stations with low average generation efficiencies. Initially the grid linked together the most efficient power stations within regions of the UK, allowing mutual backup and coverage for peak demand. These all had to operate using alternating current (AC) at a common frequency, chosen to be 50 cycles per second (50 Hz). Hundreds of small inefficient local stations were closed down.

After experiments in 1937, it was found that it was safe to connect together all the regional systems of England, Scotland and Wales into a single national network. The output of individual stations was controlled centrally in order to optimize the overall system performance. It has largely been run as such since 1938.

At that time, many of the power stations and local distribution networks were privately owned. Although the UK industry was nationalized in 1947 and then privatized in 1989, this has not affected the national nature of the distribution system.

The grid has been repeatedly strengthened and reinforced since 1934. The main links now operate at 275 kV and 400 kV with lines capable of carrying 2 GW or more. The overall philosophy has been one of large generating plant, justified on the basis of economies of scale, centralized control, and distribution outwards and downwards in voltage to the consumer (i.e. from left to right in Figure 10.14). At the lowest voltage, the system consists of buried cables in the streets capable of supplying both 400 V, 3-phase AC and 230 V, single-phase AC (the norm for domestic consumers).

This is a picture of 'centralized generation', yet increasingly there is a trend towards **embedded generation** – smaller generators situated more locally to the loads. Solar PV panels or small CHP units may be installed in houses, schools or local leisure centres, connected to the grid at the 400/230 V level. Individual onshore wind turbines of around 1 MW output and industrial-scale CHP units are likely to be connected at the 11 000 V level. Larger offshore wind farms may have high-voltage links running ashore at 11 000 or 33 000 volts.

In practice, the grid provides enormous flexibility of operation. The demands of Oxford can be met from a power station in Yorkshire or a storage plant in Wales. Across Europe, the electricity systems operate on the common

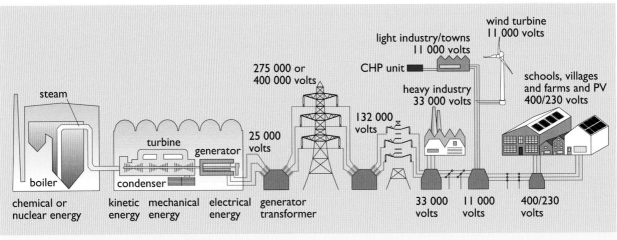

Figure 10.14 A schematic of the basic UK electricity system

frequency of 50 Hz. Because they are linked across national borders, power plants are able to share loads or export surpluses of electricity.

Matching supply and demand

The high-voltage electricity grid in England, Wales and Scotland is centrally controlled by a **system operator**, the National Grid Company. The total demand pattern is met by a mixture of generating plant, owned by different companies, which compete in a centrally controlled **power pool**, a competitive market in electricity. At any given time a computer model run by the system operator estimates what the national demand will be in the following few hours and invites bids for the supply of electricity. Power station owners reply (or rather their computers programs do), the cheapest offers are accepted and the system is adjusted to bring the appropriate stations online by remote control. The stations with the lowest running costs supply the **base load**. The lowest cost sources are likely to be renewable generators, such as wind power, which have zero fuel costs. Above wind comes nuclear, which has low fuel costs. Above them there is competition between coal- and gas-fuelled power stations. Although hydro power has zero fuel costs its high flexibility means that it is normally kept for supplying peak demands. At any time the competitive price is chosen to be sufficient to bring enough stations online to meet demand. The prices at times of peak demand can be very high indeed.

The bidding process only applies to generators above a certain size. Small generators below 1 MW are not normally constrained as to when they run, but are only paid a fixed rate for their electricity rather than a 'system price'.

The ability of different plants to provide for a changing load depends on their scale. A large coal-fired or nuclear power station may take 24 or even 36 hours to reach full output from cold. A normal CCGT station may be able to produce some power within an hour but take 8 hours to reach full output. Following a changing demand can be achieved by running stations at part load, but this reduces their overall efficiency (and increases their CO_2 emissions per kilowatt of electricity generated). It is best if they are run continuously at full output.

However, demand is not conveniently constant. There can be rapid changes in demand for several reasons:

- The daily increase in human activity in offices and factories can increase demand in the morning by 12 GW in the space of two hours.

- Simultaneously showing a popular TV programme over the whole country can produce a synchronization of behaviour of consumers. It can send a large proportion of the population rushing to use their electric kettles at the same time, during commercial breaks, leading to a **demand pickup** of over 1 GW in a matter of minutes.

- The sudden breakdown of a large power station or failure of a major transmission line may mean that generating capacity of 600 MW or more could suddenly disappear, again requiring a response in a matter of minutes.

The flexibility of hydroelectricity can go a long way to meeting these rapid changes in demand. In countries with limited hydro resources the traditional way of dealing with demand changes was to maintain 'spinning reserve' – large coal-fuelled power stations generating electricity at part load, but with sufficient steam reserves to cope with any sudden increase in demand. However, running stations in this manner reduces their efficiency and wastes fuel. So, over the years a range of other technologies has been brought into use to cope with this problem (see Box 10.2).

BOX 10.2 Matching electricity supply to short term demand fluctuations

Electricity has to be generated on demand and the voltage and frequency of the AC supply have to be held within relatively tight limits. A range of technologies is used to meet rapid and possibly unexpected increases in demand.

Pumped storage plants

These have been described in Chapter 5. At times of low demand, surplus electricity is used to pump water into high level reservoirs. At times of sudden peak demand, they can use the stored potential energy of the water to generate electricity with as little as 10 seconds' notice. The UK currently has four such plants, two in Wales and two in Scotland. Their combined peak output power is about 2.7 GW, about 5% of the UK's typical winter electricity demand. Other countries have similar schemes. Typically, such plants have an overall storage efficiency of 70–80%. Although they have a high peak power output, their water storage is only usually sufficient for a few hours operation to cover while alternative generation plant is started and connected.

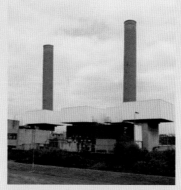

Figure 10.15 A 132 MW open cycle gas turbine (OCGT) plant in North London. It has two turbines that run on light heating oil

Open cycle gas turbine and diesel peaking plants

These can be run up to full power in half an hour or less. Small 'open cycle' gas turbines (OCGT), lacking the steam stage of the CCGT, run on natural gas or light heating oil and typically supply between 10 MW and 100 MW of electricity. Figure 10.15 shows a plant that houses two 66 MW turbines. Smaller diesel generators of around 1 MW output can be brought online in minutes. The disadvantage of both types is that they consume fossil fuel and are less thermally efficient than larger 'base-load' power stations.

Compressed air energy storage (CAES)

In a gas turbine power plant a large amount of energy is used to compress the incoming combustion air before it is burned. A CAES plant uses off-peak electricity to compress air and store it in an underground cavern. At times of peak demand, this is fed to a peaking gas turbine, reducing its gas consumption by more than 60%. The first commercial plant was a 290 MW unit built in northern Germany in 1978 using storage in a salt mine. A 110 MW plant opened in Alabama in the USA in 1991 and a 2.7 GW plant is planned for construction in Ohio. This will compress air to over 100 atmospheres pressure, storing it in an existing limestone mine nearly 700 metres under ground. A further 200 MW of capacity is planned in central Germany for construction in 2013.

Rechargeable batteries

Lead acid batteries have been used by electricity utilities for peaking power and emergency backup since the late 19th century. The largest is currently a 40 MWh system in California. Lead acid technology is limited by the number of cycles that a battery can be put through before it needs replacing.

Sodium sulfur batteries use a positive electrode of molten sulfur and a negative electrode of molten sodium. The chemicals combine to produce sodium polysulfides and electricity. The battery has to be kept at 300 °C for the reaction to take place. When the battery is recharged, the elemental sulfur and sodium are regenerated. A large number of MW scale plants have been built for electricity utilities in Japan.

Flow batteries. Unlike most conventional batteries where the key active chemicals are solid, in a flow battery, they are liquids and can be stored in tanks separately from the battery itself. This is similar to the hydrogen fuel cell (see Box 10.3), which is a 'gas battery'. Flow batteries in the MW class, based on vanadium or zinc bromide, are being developed for use in Japan.

Which is best?

Balancing the electricity grid of an entire country requires large amounts of power and appreciable amounts of stored energy. At present only pumped storage systems have power ratings of over 1 GW and the capacity to supply this for more than an hour or so, but such systems require suitable sites in mountainous regions. Peaking gas turbines can be installed almost anywhere, but consume fossil fuel in a relatively inefficient manner. CAES systems can supply 100 MW or more, but require the special geology suitable for underground high-pressure air storage.

Rechargeable batteries can potentially be installed anywhere, but at present are only available in ratings of less than 50 MW and are typically only used to supply that power for periods of an hour or less. They have the advantage of very rapid response – fractions of seconds rather than minutes. They are more likely to be used to absorb short surges and to correct control instabilities in local distribution systems. They may also act as 'starting batteries' to allow a power station to recover from a grid failure and perform what is known as a **black start**.

Including renewable energy

Where do the renewables fit into all this? Different renewable electricity sources have different characteristics, particularly as regards their variability. Let us consider the 'firm' renewables first.

Firm renewable electricity sources

Biomass and waste plants

Generation plants using municipal solid waste, waste wood or landfill gas are usually relatively small, typically in the output range 100 kW to 50 MW. As such, they are likely to be connected to the system at 11 kV or 33 kV and run fairly continuously. Many smaller plants are unmanned and run under automatic control. Biomass may also be co-fired with coal in normal large multi-megawatt power stations. In addition, new biomass plants of over 50 MW capacity are being constructed in the UK and other countries.

Apart from occasional breakdowns, their output is highly reliable and their electricity is every bit as valuable as that from larger fossil-fuelled or nuclear stations.

Geothermal plant

Although these are not used in the UK, geothermal electricity-generation plants in other countries may be of 100 MW or more capacity and their grid connection and operation is the same as that of conventional coal or nuclear plant. Since they have zero fuel costs they are likely to be run almost continuously, providing 'base load' electricity generation.

Hydroelectricity

Hydro power is perhaps the most desirable of all renewable electricity sources from the point of view of flexibility of supply. Water can be stored in reservoirs for months or even years, yet the generators can be wound up to full power and turned off again in minutes. Smaller 'run-of-river' plants are likely to operate almost continuously, subject to the availability of sufficient water.

In the UK, most plants are in the range 100 kW to 100 MW and are connected at voltages of 11 kV or above. Elsewhere in Europe, such as in Norway and Sweden, hydro power plays a major role, with the output and, in some cases, the pumped storage capacity, sold across international borders.

Although hydropower can be used as a solution to short-term variations in demand, there may be long-term year-to-year variations in its potential, depending on rainfall.

Variable renewables: the bigger picture

The remaining renewable electricity sources – solar, wind, wave and tidal power – can all make time-variable contributions to what is a time-variable demand. It is worth first considering the timing of renewable supply and demand on a European scale.

The Sun and the rotating Earth

The Sun rises in the east and sets in the west. Taking Greenwich Mean Time as a reference, dawn and the start of the working day thus arrives earlier in Germany than in the UK. Since the Earth rotates once every 24 hours, the Sun traverses 15° of longitude every hour. Berlin and Birmingham are approximately at the same latitude but displaced by 15° of longitude (see Figure 10.16). Thus the whole 'working day' in Berlin (with its associated electricity demand pattern) is likely to take place an hour earlier than in the UK.

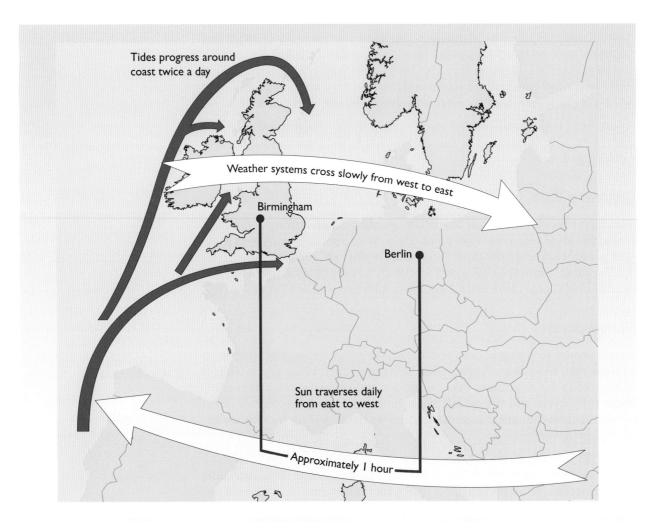

Figure 10.16 Solar, wind and tidal phenomena that influence potential electricity demand and renewable electricity generation across Europe

The Sun obviously directly affects the timing of electricity *demand* for lighting, and also any solar electricity *generation*: the peak output of a south-facing PV panel in Berlin is likely to occur 1 hour earlier than one sited in Birmingham.

Solar energy is, of course, not uniformly available throughout the day or over the year. There is more solar radiation available in summer than in winter and more in southern Europe than in northern Europe (see Figures 2.7 and 2.8 in Chapter 2).

Wind power

As described in Chapter 7, the bulk of Europe's wind resource is in the north-west. It derives from sequences of cyclones or depressions with accompanying strong winds. These weather systems move slowly from the Atlantic across Europe from west to east. They are particularly strong in autumn, winter and spring and typically about 1000 km across, taking a few days to traverse Europe. Thus on one day the wind may be blowing strongly in Ireland while Denmark is becalmed. Two days later the situation may be reversed.

The availability of stronger winds in winter means an increased wind power resource, which in part matches the increased winter electricity demand.

Tidal power

Although we normally think of tides rising and falling at one particular location, they can also be seen as progressing along the Atlantic coastline of Europe twice a day (the Mediterranean only has very small tides). The height of the tides also varies on a long two-week cycle between spring and neap tides.

Variable renewables: the details

Solar power

In the UK, most photovoltaic systems are likely to be at the kilowatt scale and connected locally at the 230 V or 400 V level. Larger arrays can have outputs rated in megawatts. These are likely to be connected to the grid at a 11 kV level.

Naturally, they only produce electricity during the day and their output will be higher in summer than in winter. Given their relative expense, they are only likely to make up a small proportion of any renewable electricity mix for the UK in the near future, although they could provide more in the longer term (see Chapter 3).

This is not the case in sunnier places such as California, Greece or Spain. Their higher levels of solar radiation mean more electricity production for a given capital investment, resulting in a lower unit solar electricity price than in the UK. Also, the electricity demand peaks not in winter but, with increasing air conditioning loads, in summer. This makes solar electricity a more desirable commodity.

The large-scale solar thermal electric plants described in Chapter 2 generate at the multi-megawatt scale and are connected to the electricity grid at high voltage. More recent plant designs have attempted to overcome the time limitations of the solar input by incorporating molten salt thermal storage to extend operation into the evening, or by combining solar and natural gas-fuelled operation.

Wind and wave power

As described in Chapter 7, modern onshore grid-connected wind turbines are likely to be rated at 2–3 MW. These are likely to be connected to the grid at the 11 kV level.

Offshore wind turbines are even larger and new designs could soon reach 10 MW. Current offshore wind farms may have total capacities of 300 MW or more and necessitate special high voltage links to the grid onshore.

Although wave power is much less developed, it is likely that individual shoreline devices will be rated at about 1 MW and offshore 'wave farms' could be 100 MW or more, with similar grid connection problems to offshore wind farms.

The output of wind and wave power generators is to a certain extent predictable, using detailed weather forecasting techniques.

Typically, a 1 MW wind turbine will produce 300–400 kW on *average over a year*. It will produce full output on a windy day but nothing on a calm one. At modest wind speeds, its output may vary considerably from minute to minute. Similarly, the output of a wave power device is dependent on the random nature of the waves. On a stormy day it will run at full power, on a flat calm one it will produce nothing, and on an intermediate day, its output will be variable.

Tidal power

As described in Chapter 6, tidal power is variable, but highly predictable.

Barrage schemes such as the proposed Severn Barrage, generating only on the ebb tide, could produce a pulse of power of up to 8 GW about six hours long every 12.4 hours (see Figure 10.17(a)). Although the peak spring output could be 8 GW, that on a neap tide a week later could only be 4 GW.

In order to accommodate this large output it would, in theory, be necessary to schedule a number of conventional power stations to shut down when the tidal plant is about to start generating; but given their thermal inertia, this would not be very fuel-efficient. It might therefore be technically better to have a large double-basin (or even triple-basin) scheme in this location with one basin kept 'high' and the other 'low' and with flexible generation and pumping between them. Although expensive, this could produce something closer to 'firm' renewable electricity.

The situation with tidal current turbines is perhaps more promising. These are likely to produce power in proportion to the cube of the speed of the water flowing through them (this is analogous to the performance of wind turbines). Their output will peak every 6.2 hours, on the incoming tide and again on the ebb tide. The output is also subject to a longer two-week cycle between spring-tide and neap-tide, the peak spring-tide output being roughly four times that of the peak neap-tide output (Carbon Trust, 2005).

The output of a single turbine might consist of a rather unpromising sequence of pulses of power three to four hours long. However, the output of two identical devices in different locations where the times of high tides differed by about 3 hours (such as Portland Bill in Dorset and the Isle of Wight) could add up to a more constant supply of electricity (see Figure 10.17(b)). In practice, it is likely that such turbines would be deployed at a large number of different locations. In total, they might produce a supply that varied little over the day, but would have a two-week variation over the spring-tide to neap-tide cycle.

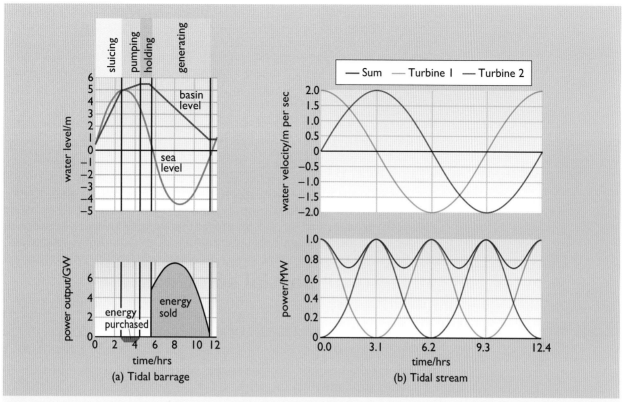

Figure 10.17 (a) A tidal power station such as the Severn Barrage would produce enormous bursts of power of up to 8 GW in magnitude and six hours long; (b) theoretically, two tidal stream turbines at two locations where the time of high tide differs by about 3 hours could in combination produce an almost continuous supply of electricity

The variability problem: a summary

In the past, electricity generation has been carried out using fossil- and nuclear-fuelled plant largely designed for continuous base-load operation. This tendency has been spurred on by the philosophy of market competition between generators to give the lowest costs. Dealing with the variability of demand or the unexpected loss of output from a large generation plant has been a secondary consideration. Part-loading of coal-fired plant and the techniques described in Box 10.2 have proved adequate.

However, the electricity grid of the future is likely to have very large inputs (tens of gigawatts) from renewable energy sources, with the variability characteristics outlined above. The associated problems, and solutions, are already apparent in countries such as Denmark and Germany where large amounts of wind power have been deployed.

Although an individual wind turbine or tidal stream turbine may have a highly variable output, if such sources are widely spaced, then just as diversity of demand adds up to a smoothly varying total demand on a national grid, so **diversity of supply** can also smooth out the local variations in output. When the wind stops blowing in Scotland, it may still be blowing in Wales. However, there is still a question of what should be done on the rare 'week with no wind power'.

A more surprising integration problem is what to do on 'the *night* with *too much* wind power'. The voltage and frequency stability of the National Grid has to be kept within tightly defined limits. Traditionally this has been ensured by the flywheel inertia and stored steam pressure of enormous 600 MW coal-fired generating plants. Yet, on a summer night with low demand and plenty of wind, it is possible that all of these would have been turned off and the system would then be *wholly powered by renewables*. This problem has already been encountered in Denmark, where in 2010 wind power provided 21% of the country's electricity. There, new offshore wind farms are being designed with equipment to maintain the frequency stability of the network and the possibility of dumping surplus wind power as heat into the country's district heating systems is being investigated.

The next section describes some technical possibilities for the integration of large amounts of renewable electricity into the electricity grid. It may, however, require a switch from a philosophy of *competition* between electricity generators to one of providing *complementary* sources of generation, with the overall goal of minimizing greenhouse gas emissions.

10.8 Some partial solutions

Future large-scale reductions in greenhouse gas emissions will require a number of tasks to be carried out that overlap in many ways:

- The decarbonization of electricity generation as described in Section 10.3
- The development of alternative transport fuels to replace fossil-fuel-based petrol and diesel
- The reduction of heat demand in the building stock
- The development of replacements for oil and natural gas as heating fuels.

At present there are three main contenders for future energy, and particularly electricity, supplies involving low greenhouse gas emissions:

- coal and natural gas-fuelled systems using carbon capture and storage (CCS)
- nuclear power
- renewable energy.

At the time of writing, the large-scale generation of electricity using CCS has yet to be demonstrated, although many of the various component technologies involved have been.

Nuclear power is a mature technology, but has problems of long-term uranium supply and public acceptability, particularly following the nuclear accident at Fukushima in Japan in 2011 and continuing concerns over links to the production of nuclear weapons.

The large-scale deployment of renewable energy is not problem-free and will require some restructuring of the existing energy (and particularly electricity) systems. Some remedies to these problems include:

- development of demand management in electricity grids
- strengthening and extension of electricity grids, both to provide short-term time diversity of supply and to tap into a wider range of supplies over the year
- development of complementary rather than competitive electricity generation
- development of hydrogen as a possible transport fuel and replacement for natural gas.

Electricity demand management

At present, in grid arrangements such as the UK's, while electricity *generation* is highly controlled there is little scope to influence electricity *demand*.

It would be convenient for fossil-fuelled, nuclear and most renewable electricity suppliers if electricity demand varied as little as possible over the day and night. This can be encouraged through the use of 'off-peak' electricity tariffs. Electricity for heating (or other) purposes is supplied at cheaper rates at night and the heat can then be stored until the next day. This allows existing power stations to continue running into the night, rather than having to build extra ones to cope with higher peak demands.

In the UK, only a relatively small amount of electricity is used for heating purposes. However, this may increase if the use of electric heat pumps is developed, particularly to replace oil heating in areas beyond the reach of the gas grid.

Taking this approach further, there are many electrical loads that are not immediately needed. One example is large-scale water pumping which, given adequate storage, could be controlled to smooth electricity demand. At times of any impending shortage of renewable electricity supply, the pumps would be turned off. At present the mechanisms for such 'load-shedding' are not well developed. However the concept of a **smart grid** involves a parallel flow of both electricity and price information that would allow increased flexibility in electricity demand and supply.

Electricity grid strengthening and enlargement

There are many reasons for strengthening national electricity grids and building international grid links. Existing electricity grids have grown up around the power stations of the past. Future large-scale renewable electricity sources are likely to be diverse and widely scattered. In the UK, connecting the considerable wind resource of Scotland to loads in the south has already required the strengthening of grid links in central Scotland and of those further south into England.

Although existing links can be upgraded by increasing the voltage and using multiple conductor wires, constructing new overhead transmission lines in areas of scenic beauty is likely to face stiff opposition. Typically, an overhead 400 kV AC line capable of carrying 3 GW costs about £1.3 million per kilometre to construct (PB Power/IET, 2012). Placing sections of the cable underground may avoid visual objections but is extremely expensive, with prices of up to £10 million per kilometre.

The renewable energy resource is not just on land but also out at sea, as shown in Figure 10.11. The large offshore 'Round 3' wind power development zones shown there are constrained by the availability of relatively shallow water so that wind turbines can be mounted on the sea bed. The development of floating wind turbines (Figure 10.18) should open up even wider possible locations for wind farms (see Offshore Valuation Group, 2010).

The connection of offshore wind farms to land necessarily requires undersea cables, but future grid links between England and Scotland could also be placed in the sea, avoiding the environmental problems of new land-based power lines (see Figure 10.19). A contract for a 2.2 MW cable through the

Figure 10.18 The Statoil 2 MW prototype floating wind turbine (here seen in 2009) potentially allows the expansion of offshore wind farms into deep water sea areas. It is secured to the sea bed by three guy wires

Irish Sea between Cheshire in England and Ayrshire in Scotland was placed in 2012. It is planned that the cable will be operational in 2016. Such links are likely to involve offshore 'supernodes', large high voltage switching and transformer stations mounted on platforms similar to oil rigs.

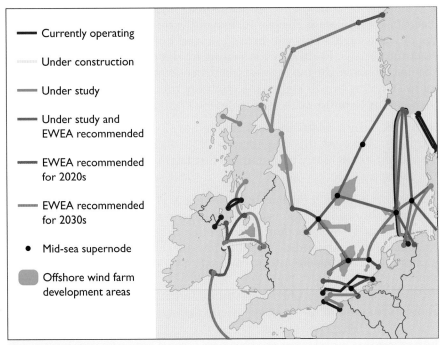

Figure 10.19 Possible future offshore and international grid links (source: Offshore Valuation Group, 2010). EWEA = European Wind Energy Association.

The existing undersea links around the British Isles have been shown in Figure 10.4. The original economic justification for the first of these was that peak electricity demands (and peak electricity prices) occur earlier in mainland Europe than in the UK. Such links are not cheap: the 1000 MW BritNed link between England and the Netherlands, 260 km long and opened in 2011, cost €600 million – over £2 million per km.

Although high voltage alternating current (AC) is normally used for land-based transmission lines, this is difficult to use for undersea cables, so the electricity is converted to high voltage direct current (HVDC) for transmission, and reconverted to AC at the opposite end. The conversion is carried out using high power semiconductor devices. Although the equipment is expensive, it is also worth using HVDC technology in new long distance land-based links over distances of more than 300 km, since the overall transmission losses can be lower than for high voltage AC.

Using grids to smooth out variability

There are many ways in which electricity grids can be used to smooth out the variability of renewable electricity supplies. For example, the movement of the tides around the coast, as shown in Figure 10.16, potentially allows the generation of electricity at different times of the day at different locations. As shown in Figure 10.17(b) the different outputs could be combined to approximate more continuous generation – but only given adequate grid capacity between these coastal sites and the ultimate customer electricity load.

Figure 10.16 has also shown how weather systems cross Europe from west to east, bringing with them alternating patterns of high and low wind speeds. The development of a chain of national and international grid links, from the far west of Ireland across the UK, Germany and Denmark to the Baltic Sea, could reduce the total variability of the output of wind farms in these countries. If the wind was not blowing in Ireland, it is probable that it would be blowing somewhere else along the line of grid connections. There is, however, a possibility that there could be no wind for a week or more at any of the wind farms, so grid expansion cannot be regarded as a complete 'solution' to the variability problem.

The UK is not alone in considering such grid strengthening. The electricity systems of Denmark, Norway, Sweden and Finland operate as a single market, the Nordpool. Denmark is a small country and has strong grid links with Sweden and Norway. This has allowed it to develop its wind power in the knowledge that it can rely on the flexible backup of over 40 GW of Swedish and Norwegian hydropower. A 580 km long 700 MW undersea cable between Norway and the north of the Netherlands was completed in 2008.

There are likely to be considerable advantages in further north-south grid strengthening within Europe, allowing northern countries to tap the considerable solar energy resource available in Spain and Italy and hydro pumped storage plants in Switzerland. This 'European Supergrid' concept is a key element in renewable energy scenarios such as that of the European Climate Foundation described later in this chapter.

Desertec

An even more ambitious proposal for electricity grid enlargement is that by the Desertec Industrial Initiative, a consortium of large and mainly German companies, which envisages strong HVDC electricity grid links between Europe and the Middle East and North African (MENA) countries (Figure 10.20). This would initially allow an interchange of solar-generated electricity from the Sahara desert, combined with wind power from the north-west African Atlantic coast, and other renewable electricity sources from central and northern Europe. In the longer term these links would be extended into the Middle East.

It is suggested that by 2050 concentrating solar power (CSP) plants with molten salt storage (see Chapter 2 Section 2.9) covering 2500 km² of desert could supply about 700 TWh y^{-1} of electricity to Europe, about 15% of its projected demand. This would require the construction of some twenty 5 GW HVDC transmission lines (Desertec, 2009).

A highly detailed study of the feasibility of a renewable energy network to provide the electricity needs of Europe and its neighbouring countries in the Middle East and North Africa can be found in Czisch, 2011.

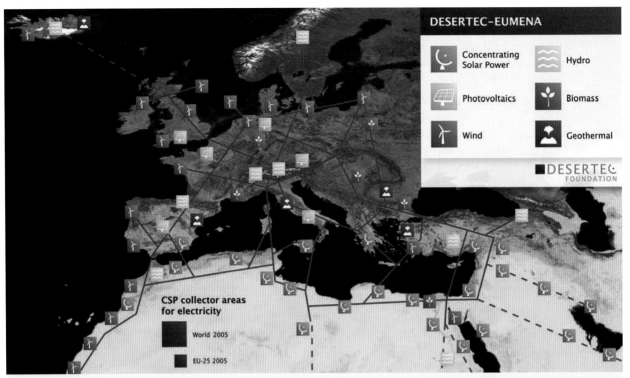

Figure 10.20 The Desertec proposal for linking renewable electricity generation between EU and Middle East and North African (MENA) countries. The brown squares at the bottom left indicate the areas of desert that would be required to supply the world's and the EU's 2005 electricity demand using concentrating solar power (CSP) technology

Complementary electricity generation

As described in the previous section, the current electricity system in the UK and many other countries is based on competition between generators. In winter, under the present market arrangements, a large coal- or gas-fuelled power plant might run continuously for months on end – as indeed it would have been designed to do. The introduction of tens of gigawatts of wind power into the UK electricity grid, and similar amounts in other European countries, changes the situation dramatically. Fossil-fuelled plants may be required to shut down at frequent intervals, possibly once a week, yet still be needed to restart at a few hours notice.

This raises two questions. The first is how 'surplus' renewable electricity might be stored or used. Box 10.3 suggests some options. The second, and more difficult, question is what kind of electricity generation plant might be *complementary* in operation to wind and other variable renewable energy sources – and available to cover 'the week with no wind'. Such plant must also be economic even though it may only be operating for 30% of the year or less. Also, ideally it should have zero CO_2 emissions.

BOX 10.3 Ways to use surplus electricity

Pumped storage

This technology has been described earlier in Box 10.2. Although it is relatively efficient at storing electricity, typically 80%, it requires large high-level reservoirs and large volumes of water. In a plant such as Cruachan in

Scotland, described in Chapter 5, Section 5.5, the storage of 1 kWh requires the movement of almost 1 tonne of water between two reservoirs differing in height by 370 m. Its total storage capacity is about 800 GWh.

Heat storage

Electricity can be used to heat water, either in domestic hot water cylinders or larger heat stores connected to district heating (DH) systems. As pointed out in Box 4.8, the specific heat capacity of water is 4.2 kJ kg^{-1} K^{-1}. Heating 1 kg of water from 10 °C to 60 °C thus requires 210 kJ. 1 kWh (3.6 MJ) of electricity thus provides enough energy to heat over 17 kg water. This low-temperature heat energy cannot be converted efficiently back into electricity because of the limitations of Carnot's Law, described in Chapter 2. However, the energy can potentially be put to good use and the required volume of storage involved is modest compared to pumped storage. If the 2500 m^3 DH store shown in Figure 10.12 were electrically heated it could absorb nearly 150 MWh of electricity.

Production of hydrogen and synthetic natural gas

If direct current electricity is passed between two electrodes immersed in water, hydrogen and oxygen can be collected at the electrodes. This process is known as **electrolysis**. Practical systems have efficiencies of 50–80%. Assuming an efficiency of 75%, it requires approximately 50 kWh of electricity to produce 1 kg of hydrogen. In principle this might be used to generate electricity in a fuel cell (see below) or even a CCGT with an efficiency of 50%, returning about 25 kWh of electricity, together with some low-temperature heat that could be used for district heating. The possible further conversion of hydrogen into methane (synthetic natural gas) is described briefly at the end of this section.

One option is the use of large combined heat and power (CHP) generation plant, producing both electricity and heat, particularly those plant using steam turbine technology, which are widely used in Denmark, fuelled by coal, natural gas and biomass. In summer, when heat demand is low, they maximize their electricity output. In winter they produce more heat, but at the cost of a reduced electricity output and generation efficiency. Although the total fuel input may remain constant, the relative proportions of heat and electricity produced can be changed quite rapidly. The inclusion of a large heat store in the system allows flexibility of operation, enabling the plant to vary its electricity output quite rapidly. Smaller CHP generation plant based on internal combustion engines and fuelled by natural gas or biogas are also capable of rapid changes in output and flexible operation.

In 2007 researchers at the University of Kassel in Germany demonstrated what they termed a 'combined power plant', an aggregation of 36 decentralized power plants based on wind, hydropower, solar and biogas energy, running together and designed to meet 1/10 000th of the total, varying, load on the German grid (Kombikraftwerk, 2007). They see this as a model for the future operation of grids as more renewable energy sources are brought into operation.

In the UK, in the short term, perhaps the cheapest large-scale complementary generation technology producing relatively low carbon electricity is the conventional gas-fuelled CCGT power station. Most modern plants are designed for continuous high-efficiency operation at full output and may require up to three hours to reach full power from a cold start. However, the temperature cycling that would be involved in repeated stop-start operation is likely to cause metal fatigue particularly in the steam boiler, leading to increased maintenance costs. Recognizing such problems, General Electric in the USA has introduced a 510 MW CCGT, the Flexefficiency 50 (GE Energy, 2010), which is designed to cope with a rapid start-up and twice as many stops and starts over its life as a normal plant, and is capable of part-load operation.

As for the economics of such plants, it can be seen from Figure 10.9 that unlike most renewable energy sources, capital costs for CCGTs only make up about 15% of the final electricity cost: the single largest item is the fuel cost. Although there is some cost penalty for leaving such a plant idle for much of the year it may well be a price worth paying.

In the longer term, such a CCGT plant might be fuelled by hydrogen produced from coal, natural gas or even biomass, coupled with the use of carbon capture and storage. It has been suggested (Starr et al., 2005; Starr, 2007) that a gasifier might be run continuously, producing hydrogen that could be stored or used elsewhere, while the CCGT operated as and when required. The effects of temperature cycling could be reduced by keeping the power plant warm using waste heat from the gasifier.

Fuel cells (described below) powered by hydrogen are also good candidates for complementary generation, since they are capable of rapid changes in output.

As yet there is little sign of nuclear power stations being designed for anything other than virtually continuous base-load operation.

Hydrogen – a fuel of the future?

Hydrogen has been widely advocated as a potential 'energy carrier' for the future. Its use as a fuel has a number of advantages:

- it can act as a store of renewable energy from season to season
- it can provide a transport fuel not dependent on the world's declining reserves of oil
- the only by-products of its combustion are water and a very small amount of nitrogen oxides, and even these emissions can be reduced to zero if fuel cells (see Box 10.4 below) are used.

Fossil-fuel-based hydrogen

The production and use of hydrogen as a fuel is well understood. Before the arrival of natural gas, 'town gas' produced from coal in the 19th century consisted mainly of a mixture of hydrogen and carbon monoxide. Hydrogen is already used in large quantities as a feedstock for the chemical industry, mainly in the manufacture of fertilizers. Currently, it is mainly produced by steam 're-forming' of natural gas (methane), a process that necessarily also produces carbon dioxide:

$$2H_2O \quad + \quad CH_4 \quad \rightarrow \quad CO_2 \quad + \quad 4H_2$$

Steam   +   Methane $\rightarrow$ Carbon dioxide   +   Hydrogen

Although technically more complex, the process of coal gasification can also be extended to produce a similar stream of CO_2 and hydrogen.

The CO_2 produced does not necessarily have to be emitted into the atmosphere. It can be 'captured', compressed and piped for long-term storage in locations such as saline aquifers deep underground. Such large-scale CO_2 disposal or **sequestration** has been carried out for over a decade by the Norwegian company Statoil in its Sleipner natural gas field (in this case the CO_2 is an unwanted contaminant of the natural gas).

More generally the separation of CO_2 produced by electricity generation or industrial processes, and its subsequent sequestration, is known as *carbon capture and storage* (CCS).

One possible vision of a future fossil-fuel-based 'hydrogen economy' involves the extensive production of hydrogen from coal or gas, combined with CCS. The hydrogen could be stored, distributed by pipeline and used to generate electricity in efficient CCGT or fuel cell power stations, where the main combustion product would simply be water vapour. Hydrogen would largely become a substitute for the existing supply of natural gas.

Although at present (early 2012) no large-scale CCS generation plants have been built, such technologies are potential competitors to renewably generated electricity.

The use of CCGTs running on hydrogen sourced from natural gas with CCS could probably produce electricity at a cost competitive with offshore wind (see Figure 10.9). On its own this technology could potentially provide 'firm' generation without any variability problems. However, as described above, it could also form the basis of a complementary generation technology to variable renewables.

Renewably based hydrogen

There also are a number of ways that renewable or 'solar' hydrogen can be produced without CO_2 by-products:

- by the electrolysis of water using electricity from non-fossil sources
- by the **thermal dissociation** of water into hydrogen and oxygen using concentrating solar collectors (probably in desert areas). To do this directly would require very high temperatures, over 2000 °C, but with more complex processes using extra chemical compounds the same result may be achievable at temperatures of under 700 °C. These processes have not yet been developed on a commercial scale
- by the gasification of biomass in a similar manner to that for coal, converting the carbon content of the biomass into CO_2. If the biomass has been sustainably grown then this CO_2 should be re-absorbed in new biomass as it is grown. The use of this technology coupled with CCS raises the interesting possibility of producing hydrogen with *negative* overall CO_2 emissions.

Other techniques under investigation include the use of photoelectrochemical cells that produce hydrogen directly from water, and artificial chemical photosynthesis.

Using hydrogen

When burned, 1 kg of hydrogen will produce 120 MJ of heat, assuming that the resulting water is released as vapour (i.e. assuming its lower calorific value). A further 20 MJ can be obtained if the water vapour can be condensed to water. Although this is nearly three times the energy per unit *mass* of petrol or diesel fuel, hydrogen has the disadvantage of being a gas, with a low energy per unit *volume* at atmospheric pressure. Hydrogen can be stored in a number of forms:

- as a gas in pressurized containers, typically at around 300 atmospheres. These containers obviously have a weight penalty

- by absorbing it into various metals, where it reacts to form a metal 'hydride': the hydrogen can be released by heating

- as a liquid, although this requires reducing its temperature to −253 °C and the use of highly insulated storage. As mentioned earlier in the chapter, natural gas (methane) is already widely shipped in liquid form, at a temperature of −162 °C.

Hydrogen can also be pumped through pipelines. Here it has a disadvantage: at atmospheric pressure, its energy density is only 10 MJ m^{-3}, about a quarter of that of natural gas. Although this would limit its use as a direct substitute in the existing heating network, a simple answer to its initial deployment might be to add a modest proportion to the existing natural gas flows, effectively reducing their overall carbon content. This would have to be limited to 15–20% hydrogen by volume before the modification of existing burners or other end-use technologies would be required.

It has been suggested that countries with plentiful supplies of renewable energy, such as Iceland, which has large untapped reserves of hydroelectricity and geothermal power, could switch to electrolytically generated hydrogen, giving up the use of fossil fuels entirely. This would include a switch to using hydrogen for road vehicles.

Starting in 2006, the European Union's HyFleet Clean Urban Transport for Europe (CUTE) project has demonstrated the use of 47 hydrogen-fuelled buses in ten cities, most of them in Europe, but also including Beijing in China and Perth in Australia (HyFleet, 2009). The buses in Berlin use internal combustion engines but the others are powered by fuel cells driving electric motors (see Figure 10.21). The principles of the fuel cell are described in Box 10.4.

Figure 10.21 A London Transport bus powered by a hydrogen fuel cell at a refuelling station in east London

BOX 10.4 **Fuel cells**

In Chapter 2 it was explained that all heat engines are inherently limited in the efficiency with which they can convert heat into motive power, and hence into electricity if the engine is driving a generator. Normally, most of the energy in the input fuel emerges as 'waste' heat (although of course this can often be harnessed and put to good use).

The fuel cell (see Figure 10.22) enables hydrogen and oxygen fuel to be converted to electricity at potentially a higher efficiency than could be achieved by burning them in a heat engine. A fuel cell is in principle a battery in which the active elements are not solids (such as the lead and lead dioxide in a car battery) or liquids (as in the flow battery plant mentioned above) but gases. Indeed, when the fuel cell was first invented by Sir William Grove in 1839 he called it a 'gas battery'.

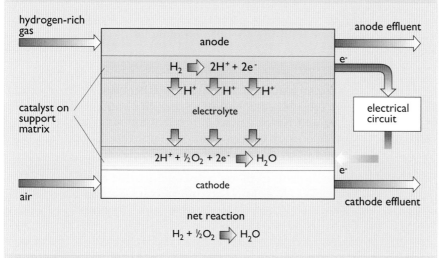

Figure 10.22 Principle of operation of a fuel cell

The principle of operation of the fuel cell is similar to electrolysis but in reverse: gases such as hydrogen and oxygen (or air) are pumped in and DC electricity is the output.

The only by-product is water and there are virtually no pollutants. There is some waste heat, but much less than in most combustion-based generation systems, and there are no moving parts. As in other types of battery, the voltage from an individual cell is low, typically about 0.7–0.8 volts, and multiple cells are connected in series to get a useful working voltage.

Despite its early invention, development work did not really start until the late 1950s. Now there is a whole range of different types including:

The Alkaline Fuel Cell (AFC). This was the first to be developed, for the US Gemini and Apollo space programmes in the 1960s. Although simple and with low manufacturing costs, AFCs are limited by the need to remove any CO_2 from the air supply to prevent contamination of the potassium hydroxide electrolyte.

The Solid Polymer Fuel Cell (SPFC). This is being developed in two forms:

- the Proton Exchange Membrane Fuel Cell (PEMFC) – a strong candidate for transport and portable power applications, this has been demonstrated in cars and buses and is available in sizes of up 250 kW

- the Direct Methanol Fuel Cell – this is another candidate being developed for transport applications. Although methanol is poisonous, it is easier to handle as a fuel than hydrogen.

The Phosphoric Acid Fuel Cell (PAFC). This is one of the most developed of the fuel cell types and is available commercially, usually as a 200 kW unit packaged with a steam reformer to allow it to run on natural gas. There has been significant PAFC development in Japan, where more than 100 plants ranging in size from 50 kW to 11 MW output are operating.

Several other fuel cell types are under development.

Typical fuel cell electricity generation efficiencies are currently in the range 40–60%. While this is better than the 25–30% that might be expected from small reciprocating engines running on natural gas or hydrogen, it is still only just competitive with the 45–60% efficiencies of large CCGT power stations.

Fuel cell developers are aiming to produce devices with costs competitive with more conventional plant and with higher efficiencies. For the moment, the strength of the fuel cell concept lies in its lack of noise, low pollution at the point of use and flexibility as a 'gas battery' since, as with other batteries, its power output can be changed very rapidly, often within fractions of a second.

For future vehicle applications there is stiff competition between fuel cells and battery electric technology. Fuel cells offer advantages of rapid refuelling compared to possible lengthy recharging times for battery-electric vehicles. Both are potentially limited by the availability of certain chemical elements: many fuel cells require 'noble metals' such as platinum as catalysts; battery-electric vehicles may be limited by global supplies of lithium for lithium-ion batteries or lanthanum for nickel metal hydride batteries.

The overall economics are also likely to depend on the relative costs of low carbon electricity for charging batteries and fuels such as hydrogen or methanol from low carbon sources for fuel cells.

The large-scale use of hydrogen would require many steps, such as those illustrated in Figure 10.23, each of which would require large capital investments. For example, surplus hydro or wind power could be electrolysed to produce hydrogen. The hydrogen would then be stored, either as high-pressure gas or as a low-temperature liquid, before being shipped to its destination in special insulated tankers such as that shown earlier in Figure 10.3(a) for liquefied natural gas. At the receiving end there would need to be further storage facilities. A 190 000 m^3 insulated tank such as that shown in Figure 10.3(b) could hold over 1.5 PJ of energy. There would also need to be provision to re-gasify the liquid hydrogen and distribute it by road tanker or pipeline. Finally, there are the individual end uses:

- in homes for cooking and heating – possibly as a mixture with natural gas
- in fuel cell power stations (of all sizes) to generate electricity and useful heat
- in road vehicles using fuel cells or hydrogen-fuelled internal combustion engines

■ and even ultimately in jet aeroplanes where the high energy content of
hydrogen per unit mass has advantages over current kerosene jet fuel.

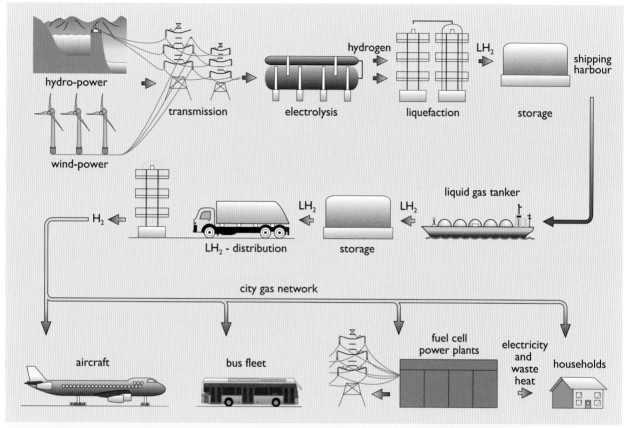

Figure 10.23 A possible future hydrogen economy. Many possible sources of renewable energy could be tapped to manufacture
hydrogen, which could be shipped to consumers and used in a variety of ways

Bringing such a vision to the current energy marketplace would require
placing a high value on hydrogen's carbon-free and pollution-free qualities,
since hydrogen would be in competition with other ways of delivering
energy services, particularly the use of electricity for surface transport and
electric heat pumps for heating.

Hydrogen would also be in competition for heating purposes with natural
gas, for which there is an extensive existing storage and distribution
infrastructure. Indeed it is possible that it may prove more economic
to convert *electrolytically generated* hydrogen and carbon dioxide into
methane (i.e. synthetic natural gas) and water, using the reverse of the
steam reforming process described above. This could be interpreted
as turning renewable energy into a fuel similar to fossil fuels, but the
intention would be to use carbon dioxide from sustainably grown biomass
sources. This uses carbon that has come from the atmosphere and will
be returned to it when the methane is burned (see Sterner et al., 2010,
and Wenzel, 2010).

10.9 **Promoting renewables**

Renewable energy can be promoted in many ways. These include:

Publicity and education

Publicity can be very effective in promoting renewables, ensuring that potential users are fully informed of the latest technologies. Once under way, some technologies, such as wind power, have become self-publicizing: photos of wind turbines now adorn the annual reports of most energy-related companies anxious to be seen to be 'green'. However, the potential of other technologies, such as some biofuels, may not be widely known.

Then there are educational initiatives at universities and educational establishments world-wide, for example this textbook and its associated courses at the Open University.

Supporting Research and Development (R&D)

Many of the experimental projects described in the previous chapters have received R&D funding from governments, the EU or the private sector. Support for R&D is very different from subsidizing the large-scale commercial deployment of new systems that may still be somewhat experimental. A modest amount of R&D investment can have significant results. In the 1980s, when a large number of small Danish companies were developing wind turbines to sell to California, the Danish Risø laboratory provided test facilities and certification procedures. These set the basis for reliable products and for the rapid expansion of the Danish turbine manufacturing industry, which is now world leading.

Setting targets

Governments can also promote renewable energy by setting targets. For example, the European Union has set a target that renewable energy must provide 20% of the EU's energy by 2020. Within this, individual EU countries have their own targets: that for the UK is 15% by 2020.

Legislation and regulation

Another important role for governments is in the specification of standards and codes for the use of new forms of renewable energy – for example the engineering recommendations that cover the connection of renewable electricity systems to the grid.

Planning procedures to obtain approval for renewable energy projects are often complex and time consuming, but these can in some cases be simplified – for example, in the UK some small renewable energy installations are considered 'permitted development', not requiring formal approval.

A more forceful approach is to include renewable energy in building codes. In the past the UK building regulations have concentrated on energy conservation. However, since 2006 they have been expressed in terms of CO_2 emission reductions and are progressively tightening with the aim that future regulations will produce 'zero-carbon' homes. These will necessarily have to incorporate a certain amount of renewable energy. In other countries, such as Greece or Israel, there are specific requirements that new housing should have solar collectors for water heating.

Financial incentives

Many countries have introduced financial incentives to encourage renewables. These can be summarized as follows.

Capital grants

These have been widely used to promote renewable energy. In the UK, they are available on a competition basis for projects including offshore wave and tidal developments and planting energy crops.

Exemption from energy taxes

In 2001, the UK government introduced a Climate Change Levy (CCL) for energy supplies to most companies (but not to domestic consumers). The current rates in Great Britain (as of April 2012) are a tax of 0.509 p kWh^{-1} on electricity, 0.177 p kWh^{-1} on gas and 1.137 p kWh^{-1} on LPG. It is roughly equivalent to a carbon tax of £10 per tonne of CO_2. Energy supplies from new renewable sources or high-quality CHP are exempt. This acts as a stimulus to the take-up of renewables and energy-efficiency improvements and should raise approximately £1 billion per year. Nuclear power is not exempt from the CCL even though it does not produce a significant amount of greenhouse gases.

The exclusion of domestic consumers avoids problems of increasing the number of households in 'fuel poverty'. In addition, companies that use large amounts of electricity such as the aluminium industry are exempt, but they do have to make a commitment to making emission savings.

Renewables obligations

These are obligations on electricity suppliers to obtain a certain proportion of their electricity from renewable sources each year (see Chapter 1, Section 1.5). How exactly they do this is left to the market. Each year this proportion is increased towards a specified target. The supply companies can pass any extra costs on to the consumers, but a ceiling has been set for these extra charges.

Renewable Energy Feed-In Tariffs (REFIT)

These specify fixed premium prices for electricity from renewable sources. This approach, used to considerable effect in Denmark, Germany and Spain, is to offer guaranteed high purchase prices for electricity generated from renewable sources – a Renewable Energy Feed-In Tariff (REFIT). The UK's REFIT scheme is known as the 'Clean Energy Cash-Back' system. Payment for electricity produced from renewables under this scheme were scaled back in 2011–12, and in 2015 the scheme is due to be replaced by a new 'Feed-in Tariff with Contracts for Differences' scheme, the details of which are at the time of writing subject to consultation.

Renewable Heat Incentive

In 2011, the UK government also announced that it would be phasing in a Renewable Heat Incentive (RHI), aimed at giving additional income per unit of energy produced to domestic and non-domestic producers of heat from renewable sources, such as solar water heaters, wood stoves, air and ground source heat pumps, etc.

10.10　Renewable energy scenarios

In this section we explore a variety of scenarios suggesting how the role of renewable energy in the energy systems of the UK, the EU and the world as a whole might develop over the next few decades. It should be stressed that such scenarios are not *predictions* of what *will* happen: they are projections of what the scenario creators consider possible, or likely to happen, given certain assumptions that they consider reasonable and the logical consequences of these assumptions. As we shall see, the assumptions of scenario creators can vary, leading to differing projections. Some scenarios assume a more-or-less linear (or in some cases, exponential) extrapolation of previous trends into the future. Other scenarios are exercises in what is sometimes called 'back-casting' – i.e. starting from a desired energy state in the future and then working out the steps that would be necessary to achieve that future state, starting from the present.

Renewable energy scenarios for the UK

How much of the UK's energy needs could renewables supply in the coming decades? Figure 1.10 in Chapter 1 illustrated how the UK's independent Committee on Climate Change (CCC), set up by the government to advise on how the UK can cut its greenhouse gas emissions, envisages the potential contribution of renewable energy to UK supplies of heat, electricity and transport energy growing in the decade from 2020 to 2030. Its analysis suggests that by 2030 renewables could be providing between 28% and 46% of UK *delivered* energy, with a renewable contribution to *electricity* supplies of 30% to 65%, renewable *heat* contributing between 35% and 50%, and a renewables contribution to *transport* energy of between 11% and 25% (CCC, 2011).

The prospects for UK renewable energy deployment on a somewhat shorter timescale, to 2020, were published in 2011 by the Department of Energy and Climate Change (DECC) in its *Renewable Energy Roadmap* (DECC, 2011e). The UK is committed under the European Union Renewable Energy directive (European Commission, 2009) to produce some 15% of its *final* energy demand from renewables by 2020, and the *Roadmap* sets out in some detail how the UK government proposes to achieve this. Table 10.6 summarizes the ranges of annual energy contributions of the eight technologies which DECC considers likely to make the most significant contributions to achieving the 15% goal, in its 'central view' of the deployment potential of each.

Looking much further ahead, to 2050, the DECC produced in 2010 a *Pathways Analysis* (DECC, 2010c) illustrating a number of different *primary* energy supply and demand 'pathways' that the UK could choose to pursue between 2010 and 2050, in order to achieve the government's goal of reducing UK greenhouse gas emissions by 80%.

Table 10.6 DECC *Renewable Energy Roadmap* central view of deployment potential of renewable energy technologies by 2020

Technology	Central range for deployment in 2020/TWh y^{-1}
Onshore wind	24–32
Offshore wind	33–58
Biomass electricity	32–50
Marine	1
Biomass heat (non-domestic)	36–50
Air-source and ground-source heat pumps (non-domestic)	16–22
Renewable transport	Up to 48
Others (including hydro, geothermal, solar and domestic heat)	14
Estimated 15% target	234

Source: DECC, 2011e

There are six of these 'illustrative' pathways, labelled A to Z, plus a Reference pathway, labelled R. Pathway A (Alpha) involves a balanced effort across all sectors – a concerted effort to reduce energy demand; an equivalent level of effort from the three main sources of low carbon energy supply – renewables, nuclear, and fossil-fuelled electricity incorporating CSS, and a concerted effort to produce and import bioenergy.

Pathway B (Beta) looks at what would be involved if it were not possible to generate electricity using CCS.

Pathway C (Gamma) looks at what would happen if no new nuclear plant were built.

Pathway D (Delta) looks at what would happen if only minimal renewable electricity capacity were built.

Pathway E (Epsilon) looks at what would happen if supplies of bioenergy were limited.

Pathway Z (Zeta) looks at what would happen if there were little behaviour change on the part of consumers and businesses.

Pathway R (the 'reference' pathway) assumes that there is little or no attempt to decarbonize and that new technologies do not materialize. Emissions targets are not met and the UK would be very vulnerable to energy security of supply shocks.

Figure 10.24 shows UK primary energy use 1960–2010, and two different projected mixes of energy supply and demand by 2050, those described in DECC pathways Alpha and Gamma. Pathway Alpha ('balanced effort') envisages renewables contributing approximately one third of UK primary energy. In the more renewables-intensive pathway Gamma the total renewable energy contribution (including waste) rises to approximately 60% by 2050.

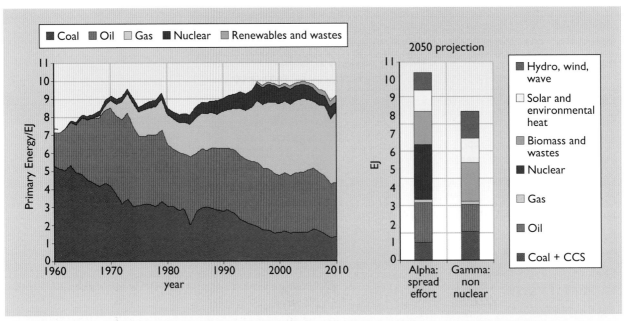

Figure 10.24 The UK primary energy supply mix 1960 to 2010, alongside two DECC projections (Alpha and Gamma) of possible energy supply mixes in 2050

A much more ambitious scenario, envisaging the UK making a transition to a zero carbon society powered (almost) entirely by renewables as early as 2030, was published by the environmental charity the Centre for Alternative Technology (CAT) in 2010 (Centre for Alternative Technology, 2010). Essentially this outcome would be achieved through two processes: 'Powering Down' – reducing energy demand without reducing quality of life; and 'Powering Up' – deploying renewable energy on a large scale and very fast.

By 2030 the CAT study suggests that the UK could have installed no less than 195 GW of offshore wind power. Due to this very high wind capacity it envisages that during very windy periods large amounts of electricity would be exported, assuming the existence of a European Supergrid. Building 195 GW of offshore wind power in less than 20 years would be an enormous challenge. But undoubtedly the UK offshore renewable energy resource, from wind, wave and tidal sources, is huge. In 2010 the *Offshore Valuation* study (Offshore Valuation Group, 2010), backed by the Crown Estate, UK and Scottish government departments and some major companies, concluded that it is equivalent in size to the UK's offshore oil and gas resources. But unlike such fossil resources, these renewable resources do not suffer from depletion and should be available for an indefinite period. The study also showed that the majority of the wind resource is located far offshore, and could be harnessed using floating turbines moored in very deep waters. The Norwegian company Statoil has developed a 2 MW floating turbine prototype, shown in Figure 10.18 above and Figure 7.43 in Chapter 7.

In 2011 another environmental charity, the World Wide Fund for Nature (WWF), published the report *Positive Energy* (WWF, 2011a) describing a range of UK electricity scenarios for 2030, based on detailed analysis by the specialist renewable energy consultancy Germanischer Lloyd Garrad Hassan (GLGH). The report demonstrated that renewables could contribute some 60–80% of the UK's electricity by 2030. The key question it aimed to answer was: Can the UK achieve a secure, sustainable and decarbonized power sector by 2030 by shifting away from polluting fossil fuels and nuclear power to an energy-efficient system built around clean and inexhaustible renewable energy?

To answer this question, GLGH created six scenarios for the UK electricity sector to 2030 (summarized in Table 10.7). The first three are 'Central' scenarios, assuming only modest attempts to reduce UK electricity demand. The first of these is a 'Core Scenario' involving 73 GW of renewable generation supplying 61% of electricity demand and in which security of

Table 10.7 Six scenarios for UK electricity supply in 2030, three 'Central' and three 'Ambitious'

	A 'Core' scenario, system security maintained primarily with gas	B 'Core' scenario, system security maintained with high levels of interconnection to Europe	C 'Stretch' scenario with high renewables and high interconnection
1 'Central scenario for electricity demand	**A1** **61% of annual electricity demand met with renewables** **Capacity mix:** • 73 GW renewables capacity • 56 GW gas capacity (18 GW requires CCS) • 3 GW interconnection capacity (already existing) Average utilization rate for gas capacity: 33%	**B1** **61% of annual electricity demand met with renewables** **Capacity mix:** • 73 GW renewables capacity • 24 GW gas capacity (17 GW requires CCS) • 35 GW interconnection capacity Average utilization rate for gas capacity: 78%	**C1** **88% of annual electricity demand met with renewables** **Capacity mix:** • 105 GW renewables capacity • 20 GW gas capacity (no CCS required) • 35 GW interconnection capacity Average utilization rate for gas capacity: 30%
2 'Ambitious' scenario for electricity demand	**A2** **62% of annual electricity demand met with renewables** **Capacity mix:** • 59 GW renewables capacity • 44 GW gas capacity (14 GW requires CCS) • 3 GW interconnection capacity (already existing) Average utilization rate for gas capacity: 33%	**B2** **62% of annual electricity demand met with renewables** **Capacity mix:** • 59 GW renewables capacity • 20 GW gas capacity (13 GW requires CCS) • 27 GW interconnection capacity Average utilization rate for gas capacity: 73%	**C2** **87% of annual electricity demand met with renewables** **Capacity mix:** • 83 GW renewables capacity • 16 GW gas capacity (no CCS required) • 27 GW interconnection capacity Average utilization rate for gas capacity: 31%

Source: WWF, 2011a

Note: Utilization rate = annual capacity factor

electricity supplies is achieved (as at present) mainly by gas-fired power stations, some of them fitted with CCS. The second 'Core Scenario' also includes 73 GW of renewables and again supplies 61% of electricity, but in this case supply security is maintained through high levels of grid interconnection to mainland Europe. The third Core Scenario is a 'Stretch' scenario including 105 GW of renewables, supplying 88% of electricity demand and also involving a high degree of interconnection to the rest of Europe.

The second set of three scenarios are 'Ambitious Scenarios for Electricity Demand', involving the implementation of more ambitious energy conservation measures. In the first of these, 59 GW of renewable generating capacity supplies 62% of electricity, but with gas generation incorporating CCS as the primary means of ensuring security of supply. In the second, 59 GW of renewables supplying 62% of electricity is again envisaged, but supply security depends mainly on interconnection to Europe. In the third scenario, a higher capacity of renewables is installed, some 83 GW supplying 87% of electricity, again with system security provided by a high level of European interconnection. As the GLGH report observes:

> In all cases, the scenarios make full provision for ambitious increases in electric vehicles (EVs) and electric heating. Energy efficiency and behavioral change lead to the reductions in demand in the ambitious demand scenarios.

> The volume of renewable capacity installed by 2020 in all scenarios is similar to that set out in the government's *Renewable Energy Roadmap* in July 2011. However, critically, the scenarios envisage installation continuing at a similar rate during the 2020s. This will avoid the risk of 'boom and bust' in the UK renewables sector [...] and mean renewables provide at least 60% of the UK's electricity by 2030.

> The amount of renewable capacity the UK can and should build is determined by economic constraints – not available resources or how fast infrastructure can be built. GLGH assumes that it is economic to supply around 60% of demand from renewables. Going beyond 60% depends on whether there's a market in other countries for the excess electricity the UK would generate at times of high renewable energy production.

> Therefore, given uncertainty over future markets, in the core scenarios (A and B) GLGH has not assumed a European market for UK renewable power despite the construction of high levels of interconnection under the B scenarios. By contrast, in the stretch scenarios (C), we assume that interconnection creates a European market for the UK's excess power, and that it becomes economic to build much more renewable capacity in the UK.
>
> (WWF, 2011a)

Renewable energy scenarios for European countries

Can renewable energies provide all, or most, of Europe's electricity? That this is possible is the conclusion of a 2010 study by the European Climate Foundation (ECF), based on analysis and scenarios constructed by the consultancy McKinsey, Imperial College London and Oxford Economics, among others (ECF, 2010). It illustrates in detail various pathways to a 'Prosperous, Low Carbon Europe', in which the renewable electricity contributions range from 40% to 100%. Figure 10.25 describes one such pathway, ECF's 80% renewable electricity scenario, with hydro, solar, wind, biomass and geothermal energy growing to provide 80% of Europe's electricity by 2050.

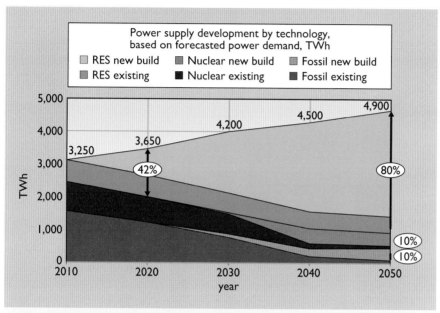

Figure 10.25 European Climate Foundation *Roadmap 2050* projection, showing how renewables could supply 80% of Europe's electricity by 2050. 'Existing' plant includes new builds up to 2010.

What contributions to energy supplies from renewables do EU countries other than the UK envisage in the coming decades? We now look at two EU nations that have taken the lead in renewable energy deployment: Denmark and Germany.

Renewable energy scenarios for Denmark

During the 1960s, Denmark's energy use expanded rapidly in an era of cheap oil (see Figure 10.26(a)). Its only indigenous energy production was a small amount of lignite mining. By 1972, it had become almost totally reliant on imported oil and was badly hit by the OPEC oil price rises in 1973 and 1979, which created considerable political embarrassment.

The situation encouraged energy researchers from Danish universities to produce, in 1983, an 'alternative energy scenario' (AE83). It outlined a programme aimed at cutting oil and coal imports to zero by 2030, to be achieved by an increase in the proportion of renewables (wind, solar and biofuels) and a sharp reduction in total energy demand. It also did not include nuclear power, which was very unpopular in Denmark at the time. The projection, shown in Figure 10.26(b), contrasted with the official 1981

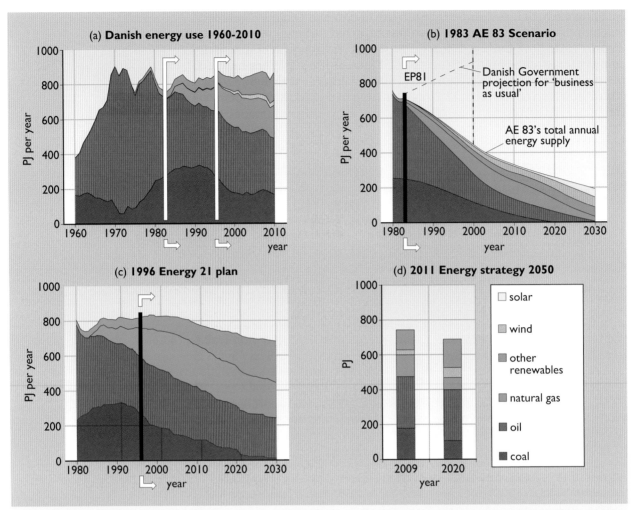

Figure 10.26 (a) Primary energy consumption in Denmark, 1960–2010; (b) Denmark's 1983 Alternative Energy scenario (AE83), and the Danish Government's 1981 'business as usual' projection (EP81); (c) 1996 Energy 21 scenario; (d) Projected energy use in 2020 in the *Energy Strategy 2050* scenario. Notes: In Figure (c) 'other renewables' includes wind. Figure (d) excludes energy use in the oil and gas industries. (sources: Norgaard and Meyer, 1989, Eurostat 1993, MEE, 1996, DEA, 2010, DEA, 2011a, Danish Government, 2011)

Danish government projection (EP81), which suggested a continued growth in energy demand (see Norgaard and Meyer, 1989).

In practice, the government solved its immediate energy security problems in several ways:

■ It pursued a policy of energy conservation, imposing high energy taxes and new regulations to encourage the insulation of buildings and promote the use of CHP. The resulting cuts in energy use for space heating were quite impressive. Between 1972 and 1985, the total area of heated building floor area increased by 30%, but the total amount of energy used to heat it *decreased* by 30% (Dal and Jensen, 2000).

■ It switched its power stations from oil-firing to coal-firing. Consequently, national reliance on oil dropped from 93% in 1972 to only 43% 20 years later in 1992.

- It developed its own oil and gas resources in the North Sea. The results started to appear in the early 1980s and by 1997 Denmark had become a net energy exporter.

- It made considerable efforts to start building up its renewable resources, particularly biomass and wind power.

Actual Danish energy consumption up to 2010 has not followed either the AE83 projection or the government EP81 projection, which confirms the point that scenarios are only pictures of what *could happen* rather than what *will happen*.

By the mid-1990s, the focus of energy policy had shifted from self-sufficiency to climate change. Denmark is a low-lying country and any future sea level rises would have serious consequences. The switch from oil to coal and natural gas meant that national CO_2 emissions had remained at their 1970s level. In 1996, the centre-left government adopted a policy of cutting these by 20% from their 1988 levels by 2005 and by 50% by 2030. This was set out in their *Energy 21* plan (MEE, 1996). Under this scenario, total primary energy use would fall only slightly and the use of coal would be replaced by increased use of natural gas and renewable energy (Figure 10.26(c)). It was suggested that by 2030 half of Denmark's electricity could come from renewable sources. In order to implement this policy, there were continued high energy taxes and support for renewable energy under a REFIT scheme.

In 2001, a new right-wing government was elected that was committed to tax reform. The new government took the attitude that the country's environmental initiative was 'ahead of schedule'. The *Energy 21* plan was dropped, orders for three offshore wind farms cancelled and many tax incentives for energy efficiency were scrapped. Despite this, the existing policies still ensured a slow increase in renewable energy use (Figure 10.26(a)).

However, issues of CO_2 emissions and energy security could not be ignored. Danish oil production reached a peak in 2004 as did its natural gas in 2006. Both have declined significantly since then, although Denmark was still a net energy exporter in 2011.

In May 2011 the Danish government produced a new 'Energy Strategy 2050' stating a long-term goal of a complete transition away from the use of fossil fuels by that date (Danish Government, 2011) (see Figure 10.26(d)). This is essentially the same goal as the 1983 AE83 scenario but with the target date delayed by 20 years. It suggests:

- a continuing reduction in overall energy demand

- a reduction in the use of coal in electricity generation, to be replaced by renewable energy, for example in the Avedøre-2 CHP power plant described in Chapter 4, Box 4.6

- a continued expansion of the use of renewable energy, particularly solid biomass and wind power. There is a target that renewables will contribute over 30% of final (i.e. delivered) energy use by 2020, i.e. about 200 PJ per year, with a 10% contribution to the transport sector. In the electricity sector it is suggested that 60% should come from renewables, 40% being from wind power.

A comparison of projected 2020 energy use with that of 2009 is shown in Figure 10.26(d). This scenario is more ambitious in terms of renewable energy deployment and less reliant on natural gas than the now 'abandoned' *Energy 21 Plan*.

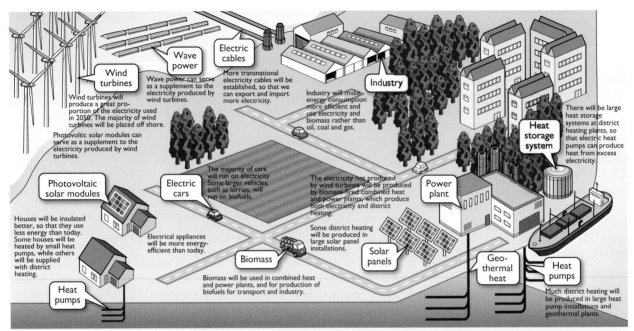

Figure 10.27 A vision of how Denmark sees it energy future in 2050, from the Danish Government's Climate Change Committee (2011) – backed up by an earlier study from the Danish Society of Engineers (2010). All fossil fuel use has been phased out, and the country runs on 100% renewable power, from a mixture of onshore and offshore wind, solar PV and thermal panels, biomass fuels (some imported), electric heat pumps, geothermal energy, CHP plants and large heat stores. International electricity links enable Denmark to export and import electricity when there are surpluses or deficits.

Overall, there are perhaps several lessons to be drawn from Danish energy policy since the 1970s:

- Primary energy use and GDP do not go hand in hand – between 1972 and 2008, Danish GDP doubled (Maddison, 2009), yet primary energy consumption has hardly changed.

- Reducing total energy demand may be difficult, particularly if the country has its own oil and gas reserves. Despite high taxes and progressive policies, Danish road transport fuel consumption increased by 80% and electricity demand has more than doubled over the same period.

- The use of CHP with district heating (DH) can be successfully promoted. By 2007 over 60% of Danish homes were fed by DH and 80% of the DH plants were fed by CHP (DEA, 2011b).

- Renewable energy use can also be promoted effectively. By 2010 it supplied 33% of the country's electricity, nearly 21% from wind power. The renewable contribution to primary energy was nearly 170 PJ per year, a far higher figure than that suggested in the AE83 scenario (although about 15% is imported).

- Denmark's high energy taxes and promotion of DH have resulted in disagreements with the European Commission in Brussels, which has had a policy of free markets and harmonization of energy prices. The Danish government successfully argued that these policies were for the protection of the environment, which is also a prime objective of European policy.

Renewable energy scenarios for Germany

In Germany, following the 2011 Fukushima nuclear accident in Japan, the government decided to adopt a revised 'Energy Concept' for the country over the coming four decades, involving a phase-out of reliance on nuclear power by 2022. According to the Federal Environment Ministry (BMU, 2011):

> The objectives are a rapid transition to the age of renewable energies and the phasing out of nuclear energy by the end of 2022. The intention is for renewable energies' share of power generation to rise from the current 17 percent of power consumption to at least 35 percent in 2020. The German government will strive to ensure this share is 50 percent by 2030, a figure that should rise to 60 percent by 2040, then 80 percent by 2050. [...] However, promoting energy efficiency will be important too if the demand for energy is to fall.

> To begin with, conventional (i.e. gas and coal-fired) power plants will still play a central role in ensuring security of supply; they are able to provide power at any time. The expansion of grids, the application of load management, the improvement of power feed-in forecasts for wind and solar energy, and the development of storage technologies will allow a power system overwhelmingly based on renewable energies to secure our supply [...] The measures planned will allow CO_2 emissions to be cut at least 80 percent by 2050.

Germany's independent advisory council on the environment (SRU) also published in 2011 a detailed study suggesting that an even more ambitious target of a 100% renewable electricity system for Germany is feasible and economic. As its report (SRU, 2011) states:

> Our scenario computations show that Germany could readily achieve a wholly renewable electricity supply that is both reliable and affordable. Providing that the relevant storage facilities and grids are implemented, the renewable energy potential in Germany and Europe would allow for the satisfaction of maximum posited electricity demand at all times throughout the year, using wind turbines, solar collectors, and other currently available technologies and despite fluctuations in the availability of renewable electricity. As the lowest cost energy resource in the run-up to 2050, wind energy, particularly from offshore wind turbines, plays a pivotal role in all of the scenarios discussed in the present report. On the other hand, the level of solar energy use in the various scenarios varies according to electricity demand and the amount of electricity that is imported. Biomass use in the scenarios involving transnational energy supply networks accounts for no more than 7% of electricity demand, largely owing to land use conflicts and the relatively high cost of this energy resource.

> [...]

In the view of the SRU, instituting a wholly renewable electricity supply in Germany by 2050 would entail economic advantages in addition to promoting climate protection, whereby the aggregate costs of such a system would be largely determined by the extent to which a network comprising other European countries is established. Inflation adjusted, a wholly renewable electricity supply using German resources only would be relatively cost intensive, ranging from 9 to 12 euro-cents per kWh, depending on demand. On the other hand, an inter-regional smaller-scale German-Danish-Norwegian or larger-scale Europe-North Africa network would provide electricity at a cost of only 6 to 7 euro-cents per kWh, including the cost of international grid and storage capacity expansion. Our rough estimates indicate that expanding the German grid would entail additional costs amounting to approximately 1 to 2 euro-cents per kWh. Over the long term, renewable electricity will prove to be less cost intensive than conventional low carbon technologies such as CCS power plants and new nuclear power plants, whose costs will rise owing respectively to limited uranium resources and storage facilities.

The eight SRU scenarios are summarized in Table 10.8 below.

Table 10.8 Eight scenarios for a wholly renewable electricity supply for Germany in 2050. In the first four scenarios, German energy demand is some 500 TWh per year (SV500). In the second four scenarios, electricity demand rises to 700 TWh per year (SV700). The four variants involve different assumptions about the extent of self-sufficiency in Germany and the strength of interconnections to neighbouring countries

Assumptions	German electricity demand in 2050: 500TWh	German electricity demand in 2050: 700TWh
Self sufficiency	Scenario 1.a: DE 100% SV-500	Scenario 1.b: DE 100% SV-700
Net self sufficiency, interchange with Denmark and Norway	Scenario 2.1.a: DE-DK-NO 100% SV-500	Scenario 2.1.b: DE-DK-NO 100% SV-700
Maximum 15% net import from Denmark and Norway	Scenario 2.2.a: DE-DK-NO 85% SV-500	Scenario 2.2.b: DE-DK-NO 85% SV-700
Maximum 15% net import from the Europe-North Africa region (EUNA)	Scenario 3.a: DE-EUNA 85% SV-500	Scenario 3.b: DE-EUNA 85% SV-700

DE: Germany; DK: Denmark; NO: Norway; EUNA: Europe and North Africa; SV: self sufficiency
SRU/SG 2011-1/Table 10-1
Source: SRU, 2011

World renewable energy scenarios

Can renewable energy provide enough energy for the world? One study demonstrating that this is possible was published in popular form in *Scientific American* in 2009 by two scientists from Stanford University in California (Jacobson and Delucchi, 2009). In the following year they backed up their arguments in two detailed papers in the refereed journal *Energy Policy* (Jacobson and Delucchi, 2010a and 2010b). Their research suggests that the world's total power demand both for electricity and other purposes, which they estimate at 11.5 – 16.9 TW by 2030, could be supplied by large numbers of wind turbines, solar power plants, water wave, hydro and geothermal installations, as shown in Table 10.9. Jacobson and Delucchi acknowledge that these numbers may seem daunting, but as they point out, installation would be spread over two decades, and since the world currently produces some 73 million cars and light trucks per year, this would suggest that the world's industrial production capacity is probably sufficient if we chose to harness it in this way.

Table 10.9 Number of Wind, Wave and Solar (WWS) power plants or devices needed to supply world final energy demand in 2030 (11.5 TW), assuming a given partitioning of the demand among plants or devices. Also shown are the footprint and spacing areas required to meet the world demand, as a percentage of the global land area, 1.446×10^8 km².

Energy technology	Rated power of one plant or device/MW	Per cent of 2030 power demand met by plant/device %	Number of plants or devices needed for world energy demand	Footprint area (% of global land area)	Spacing area (% of global land area)
Wind turbine	5	50	3.8 million	0.000033	1.17
Wave device	0.75	1	720 000	0.00026	0.013
Geothermal plant	100	4	5350	0.0013	0
Hydroelectric plant	1300	4	900[a]	0.407[a]	0
Tidal turbine	1	1	490 000	0.000098	0.0013
Roof PV system	0.003	6	1.7 billion	0.042[b]	0
Solar PV plant	300	14	40 000	0.097	0
CSP plant	300	20	49 000	0.192	0
Total		100		0.74	1.18
Total new land				0.41[c]	0.59[c]

[a] About 70% of the hydroelectric plants are already in place.
[b] The footprint area for rooftop PV does not represent an increase in land since the rooftops already exist and are not used for other purposes.
[c] Assumes 50% of the wind is over water, wave and tidal are in water. 70% of hydroelectric is already in place, and rooftop solar does not require new land.
Source: Derived from Jacobson and Delucchi, 2010a.

Another study reaching similar conclusions was published in a report by the World-Wide Fund for Nature (WWF) in its 2011 *Energy Report* (WWF, 2011b), based on analysis and scenarios by the respected Dutch energy consultancy Ecofys. It shows how, in the decades to 2050, the world could implement major energy saving measures, reducing the massive waste that is present in our current energy systems. Simultaneously we could phase in a mixture of wind, solar, geothermal, hydro and biomass energy sources to provide 95% of the world's final energy demand, as shown in Figure 10.28.

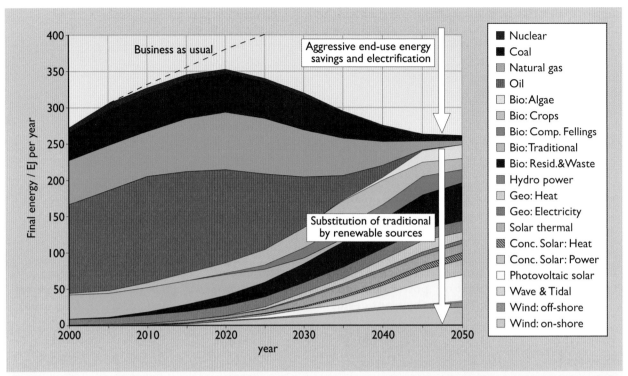

Figure 10.28 In the WWF *Energy Report* scenario, energy efficiency measures reduce final energy demand by about half by 2050, compared to what it would have been in the absence of such measures. By 2050 some 95% of the remaining energy demand is met by a combination of renewable energy sources, as shown.

A series of renewable-intensive future world energy scenarios has also been produced over the past decade by the environmental group Greenpeace, their latest version in 2010. Earlier versions were initially dismissed by sceptics as unrealistic in their projections for the growth of solar or wind power. But many of these early projections have in fact proved pessimistic, given the high growth rates of renewable energy capacity that have been achieved in recent years (see Figure 10.1).

Figure 10.29 shows the growth of world primary energy use between 1960 and 2010. It also shows projections for 2050 from Greenpeace's Advanced Energy (r)Evolution scenario (Greenpeace, 2010) and from the IEA 'Blue map' scenario (see below). As in the WWF scenario, in the Greenpeace scenario energy demand is reduced by implementing energy efficiency improvements. By 2050 in this scenario more than 80% of the world's (reduced) energy consumption would be supplied by a mixture of hydro, wind, biomass, geothermal, solar and ocean energy sources.

Greenpeace acknowledges that implementing its Advanced (r)evolution scenario would entail a high initial capital cost (US$ 17.9 trillion between 2007–2030); but because of its much lower fuel costs it would have a lower *overall lifetime* cost. It is also worth noting that these costs do not take into account the very large external social and environmental costs of conventional fossil and nuclear fuel use. Moreover, conventional economic analysis, as explained in Appendix B, uses discounting techniques that have the effect of minimizing the impact of future costs, so are biased against high capital, low fuel alternatives like renewables. Greenpeace also calculates that an extra 8 million green jobs would be created globally in its scenario.

Rather more conventional energy scenarios are published regularly by the International Energy Agency in its *Energy Technology Reports*, which outline the Agency's view of the future prospects for various energy technologies. In the 2010 report (IEA, 2010) it produced a 'Blue Map' scenario to illustrate how world CO_2 emissions could be halved by 2050, compared to current levels, in order to enable atmospheric concentrations to be kept below 450 ppm and global temperature rises to below 2 degrees Celsius – a threshold that many experts believe should be an upper limit. In the IEA's Blue Map scenario, world primary energy use continues to grow, and renewable energy by 2050 contributes almost 40% of primary energy supply (see Figure 10.29).

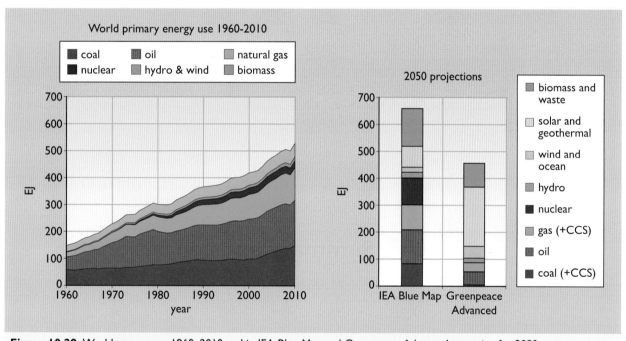

Figure 10.29 World energy use 1960–2010 and in IEA Blue Map and Greenpeace Advanced scenarios for 2050

The IEA has also published several variants on its Blue Map scenario, including a 'Blue High Renewables scenario', in which the renewable share of world *electricity* rises from 48% to 75% by 2050. In Figure 10.30 this is shown along with the electricity projection for 2050 from Greenpeace's 'Advanced Energy (r)Evolution scenario'. Both the IEA and Greenpeace expect a doubling of world electricity demand by 2050. The IEA envisages 'spread effort', including some fossil and nuclear fuels but with plenty of scope for the expansion of renewables. Greenpeace sees even more room for expansion of biomass, wind, solar and geothermal energy (but not hydro), though it rules out nuclear power.

As Figure 10.30 illustrates, in the 40 years between 1970 and 2010, world electricity generation more than tripled. In the next 40 years it seems likely

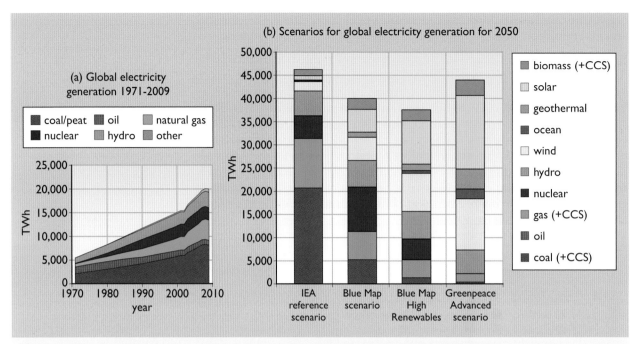

Figure 10.30 This composite chart shows: (a) World electricity generation by fuel 1971–2009. (b) Electricity projections in the IEA Reference scenario for 2050; in the IEA Blue map scenario 2050; in the IEA Blue Map High renewables scenario 2050; and in the Greenpeace Advanced Energy (r)Evolution scenario 2050

to double. Clearly, building and installing new clean electricity generation capacity and the associated networks will be a major world challenge in coming decades.

10.11 Summary and conclusions

This chapter has reviewed the conditions under which the various renewable energy sources can continue to make increasingly significant contributions to world energy needs. It began by reviewing the existing systems for distributing energy in various forms. It then looked at the potential contribution of renewables to reducing global greenhouse gas emissions. It reviewed estimates of the overall magnitude of the contribution that renewables might make to the needs of the UK, and the extent to which renewables can deliver energy close to the centres of energy demand. It then discussed ways in which the timing of output of variable renewable sources can be matched to the timing of energy demand.

Two proposals for radical changes in energy systems were described in Section 10.8: the creation of a European Supergrid, and the building of a renewable hydrogen economy.

This was followed by a brief outline of various ways in which the wider use of renewable energy could be promoted. The chapter concluded with a review of various long-term energy scenarios – for the UK, other European countries, and the world as a whole – outlining various feasible transition paths to a high-renewable future over the coming decades.

The technical, social and political challenges involved in phasing-out fossil and nuclear fuels and replacing them largely, or entirely, with renewables may seem daunting. But there is little doubt that world population, and the world economy, will continue to grow very substantially during the 21st century. An accompanying rise in global primary energy use seems extremely difficult (even if not technically impossible) to avoid. And if a substantial share of this additional energy is not to be supplied by renewables, it will have to come from fossil or nuclear fuels, with all the familiar environmental, social and resource depletion concerns that the use of these sources entails.

The extent to which renewables prove successful in increasing their share of world-wide energy supplies will depend on many factors. These include: the extent to which investment in research and development and large-scale production can bring about efficiency improvements and cost reductions; the outcome of debates about the environmental and social costs of conventional sources and the extent to which these costs are reflected in energy prices; the future patterns of world economic and population growth, and their effect on the level of demand for various forms of energy; the impact of these considerations on the priorities of governments; and the environmental and social acceptability of renewables to the public.

However, it seems difficult to avoid the overall conclusion that renewables are likely to play a greatly increased role in future energy supplies. As we have seen, the relatively conservative International Energy Agency suggests that renewables by 2050 could be supplying some 40% of world primary energy, and between 48% and 75% of world electricity demand. And if history should prove the even more optimistic WWF and Greenpeace projections to be more accurate, renewables could be supplying 90% or more of the world's energy needs by mid-century.

We hope this book will not only help to promote an improved understanding of the potential of the renewable-energy sources, but also play a part in facilitating their deployment on a world-wide basis, as countries progress towards a sustainable world economy during the 21st century.

References

AEA (2011) *2011 Guidelines to Defra / DECC's GHG Conversion Factors for Company Reporting: spreadsheet*, [online] available at http://www.defra.gov.uk (accessed 6 December 2011).

BEC (2011) *Carbon Emissions of Different Fuels*, Biomass Energy Centre, [online] available at http://www.biomassenergycentre.org.uk/portal/page?_pageid=75,163182&_dad=portal&_schema=PORTAL (accessed 2 April 2012).

BERR (2008) *Atlas of UK Marine Renewable Energy Resources*, Department of Business, Energy and Regulatory Reform, [online] available at http://www.renewables-atlas.info (accessed 12 December 2011).

BMU (2011) *Questions and answers about transforming our electricity system*, Federal Ministry for the Environment, Nature Conservation and Nuclear Safety (BMU), [online] http://www.bmu.de/english/energy_efficiency/doc/47589.php (accessed September 2011).

BP (2011) *BP Statistical Review of World Energy 2011,* London, The British Petroleum Company; available at http://www.bp.com (accessed 29 February 2012).

Carbon Trust (2005) *Variability of UK Marine Resources*, [online] available at http://www.carbontrust.co.uk (accessed 29 December 2011).

CCC (2008) *Building a low-carbon economy – the UK's contribution to tackling climate change*, Committee on Climate Change, London, available at http://www.theccc.org.uk (accessed 30 November 2011).

CCC (2011a) *The Renewable Energy Review*, Committee on Climate Change, London, available at http://www.theccc.org.uk (accessed 30 November 2011).

CCC (2011b) *Bioenergy Review*, Committee on Climate Change, London, available at http://www.theccc.org.uk (accessed 7 December 2011).

Centre for Alternative Technology (CAT) (2010) *Zero Carbon Britain 2030: A New Energy Strategy, The second report of the Zero Carbon Britain Project*, Centre for Alternative Technology, Machynlleth, Wales. Available from http://www.zerocarbonbritain.com (accessed 2 April 2012).

Czisch, G. (2011) *Scenarios for a Future Energy Supply*, Institution of Engineering and Technology, London.

Dal, P. and Jensen, H. S. (2000) *Energy Efficiency in Denmark*, Danish Energy Ministry.

Danish Government (2011) *Energy Strategy 2050*, available at http://www.denmark.dk (accessed 28 November 2011).

Danish Society of Engineers (2010) *The Danish Society of Engineers Energy Plan 2010, Summary*. Danish Society of Engineers, Copenhagen, Denmark, [online] http://ida.dk/sites/climate/introduction/Documents/Energyplan2030.pdf (accessed 17 January 2012).

DEA (2010) *Annual energy statistics: data tables*, Danish Energy Agency, available at http://www.ens.dk (accessed 25 November 2011).

DEA (2011a) *Monthly energy statistics: data tables*, Danish Energy Agency, available at http://www.ens.dk (accessed 25 November 2011).

DEA (2011b) *Danish Energy Policy 1970–2010*, Danish Energy Agency, available at http://www.ens.dk (accessed 25 November 2011).

DECC (2010a) '*The unconventional hydrocarbon resources of Britain's onshore basins – shale gas*', Department of Energy and Climate Change; available at http://og.decc.gov.uk/assets/og/bo/onshore-paper/uk-onshore-shalegas.pdf (accessed 24 April 2012).

DECC (2010b) *Digest of UK Energy Statistics (DUKES)*, Department of Energy and Climate Change, available at http://www.decc.gov.uk (accessed 5 April 2011).

DECC (2010c) *2050 Pathways Analysis,* Department of Energy and Climate Change. Available from http://www.decc.gov.uk/assets/decc/What%20 we%20do/A%20low%20carbon%20UK/2050/216-2050-pathways-analysis-report.pdf (accessed April 2 2012).

DECC (2011a) *UK Energy in Brief 2011*, Department of Energy and Climate Change, available at http://www.decc.gov.uk (accessed 2 November 2011).

DECC (2011b) *Digest of UK Energy Statistics*, Department of Energy and Climate Change, available at http://www.decc.gov.uk (accessed 30 November 2011).

DECC (2011c) *Energy Trends*, Department of Energy and Climate Change, available at http://www.decc.gov.uk (accessed 10 April 2011).

DECC (2011d) *Energy Consumption in the UK: data tables*, Department of Energy and Climate Change, available at http://www.decc.gov.uk (accessed 9 December 2011).

DECC (2011e) *UK Renewable Energy Roadmap*, Department of Energy and Climate Change. Available from http://www.decc.gov.uk/en/content/cms/ meeting_energy/renewable_ener/re_roadmap/re_roadmap.aspx (accessed 2 April 2012).

Desertec (2009), *Clean Power from Deserts: Whitebook*, Desertec Foundation, Bonn, available at http://www.desertec.org (accessed 31 December 2011).

EEA (2011) 'LCA emissions of energy technologies for energy production' [online], Copenhagen, European Environment Agency, http://www.eea. europa.eu/data-and-maps/figures/lca-emissions-of-energy-technologies (accessed 9 July 2011).

ETSU / PB Power (2002) *Concept Study – Western Offshore Transmission Grid*, available at http://www2.env.uea.ac.uk/energy/energy_links/ transmission/file15105.pdf (accessed 3 January 2012).

European Climate Foundation (2010) *Roadmap 2050: a Practical Guide to a Prosperous, Low-Carbon Europe. Volume 1, Full Report 100pp, and Appendix A: Generation.* European Climate Foundation, Berlin and Brussels. Downloadable from http://www.europeanclimate.org (accessed 5 December 2010).

European Commission (2009) *Renewable Energy Targets by 2020.* Commission of the European Communities. Available at http://ec.europa. eu/energy/renewables/targets_en.htm (accessed 2 April 2012).

Eurostat (1993) *Energy Statistical Yearbook 1992*, The Statistical Office of the European Communities.

Everett, R., Boyle, G.A., Peake, S. and Ramage, J. (eds) (2012) *Energy Systems and Sustainability: Power for a Sustainable Future* (2nd edn), Oxford, Oxford University Press/Milton Keynes, The Open University.

GE Energy (2010) *Flexefficiency 50 Combined Cycle Power Plant*. Available at http://www.ecomagination.com/portfolio/flex-efficiency (accessed 12 December 2011).

Greenpeace (2010) *Energy (r)Evolution: A Sustainable World Energy Outlook*, Greenpeace International and European Renewable Energy Council, [online] http://greenpeace.org/international/en/publications/reports/Energy-Revolution-A-Sustainable-World-Energy-Outlook/ (accessed 5 June 2012).

Hibernian Wind Power (2011), *Hibernian Wind Power Sites in Ireland; Northern Wind Power Sites in Northern Ireland*, available at http://www. hibernianwindpower.ie (accessed 2 March 2012).

HyFleet (2009) *What is HyFLEET:CUTE?* [online], available at www.global-hydrogen-bus-platform.com (accessed 9 May 2012).

IEA (2010) *Energy Technology Perspectives: Scenarios and Strategies to 2050. Part 1: Technology and The Global Energy Economy to 2050. Part 2: The Transition from Present to 2050*. International Energy Agency, Paris. Part 1: 458pp. Part 2: 252pp.

IPCC (2011) *Special report on renewable energy sources and climate change mitigation*, Intergovernmental Panel on Climate Change, available at http:// www.ipcc.ch (accessed 31 October 2011).

Jacobson, M. Z. and Delucchi, M. A. (2009) 'A path to sustainable energy by 2030', *Scientific American*, November, pp. 58–65.

Jacobson, M. Z. and Delucchi, M. A. (2010a) 'Providing all global energy with wind, water and solar power, part 1: Technologies, energy resources, quantities and areas of infrastructure, and materials', [online] available at http://www.rgo.ru/wp-content/uploads/2011/12/JDEnPolicyPt1.pdf (accessed 16 April 2012).

Jacobson, M. Z. and Delucchi, M. A. (2010b) 'Providing all global energy with wind, water and solar power, part 2: Reliability, system and transmission costs, and policies', [online] available at http://www.rgo.ru/wp-content/uploads/2011/12/JDEnPolicyPt2.pdf (accessed 16 April 2012).

Kombikraftwerk (2007) *Background Paper: The Combined Power Plant*, available at http://www.kombikraftwerk.de (accessed 4 January 2012).

Maddison, A. (2009) *Statistics on World Population, GDP and Per Capita GDP, 1–2008 AD*; available at http://www.ggdc.net/MADDISON/oriindex. htm (accessed 28 November 2011).

MEE (1996) *Energy 21*, Danish Ministry of Environment and Energy, Copenhagen.

Mott MacDonald (2010) *UK Electricity Generation Costs Update*; available at http://www.decc.gov.uk (accessed 2 March 2012).

Mott MacDonald (2011) *Costs of low-carbon generation technologies*; available at http://hmccc.s3.amazonaws.com/Renewables%20Review/ MML%20final%20report%20for%20CCC%209%20may%202011.pdf (accessed 2 March 2012).

National Grid (2011) 'National Electricity Transmission System Seven Year Statement' [online], http://www.nationalgrid.com/ (accessed 24 April 2012).

Norgaard, J. S. and Meyer, N. I. (1989) 'Planning implications of electricity conservation: the case of Denmark' in Johansson et al. (eds) *Electricity – Efficient End Use*, Lund University Press.

Offshore Valuation Group (2010) *The Offshore Valuation: a valuation of the UK's offshore renewable energy resource*, available at http://www. offshorevaluation.org (accessed 7 December 2011).

PB Power/IET (2012) Electricity Transmission Costing Study. Institution of Engineering and Technology, London, 180pp.

REN21 (2011) *Renewables 2011 Global Status Report*, Renewable Energy Policy Network for the 21st Century, REN21 Secretariat, Paris, France, 116 pp. Available from www.ren21.net (accessed December 2011).

SRU (2011) *Pathways Towards a 100% Renewable Electricity System*, Special Report, October 2011, German Advisory Council on the Environment, Berlin, Germany, 435pp. Downloadable from: http://www.solarserver. com/solar-magazine/solar-news/current/2011/kw07/sru-special-report-pathways-towards-a-100-renewable-electricity-system.html (accessed 17 January 2012).

Starr, F., Tzimas, E., Steen, M. and Peteves, S. D. (2005) *Flexibility in the production of hydrogen and electricity from fossil fuel power plants*, Proceedings International Hydrogen Energy Congress and Exhibition IHEC2005; available at http://ie.jrc.ec.europa.eu/publications/scientific_ publications/2005/P2005–096%20PUBSY%20Request%201287.pdf (accessed 5 January 2012).

Starr, F. (2007) 'Flexibility of Fossil Fuel Plant in a Renewable Energy Scenario: Possible Implications for the UK', in Boyle, G. (ed) *Renewable Electricity and the Grid: The Challenge of Variability*, Earthscan, London, pp. 121–140.

Sterner, M., Jentsch, M., Saint-Drenan, Y-M., Gerhardt, N., von Oehsen, A., Specht. M., Baumgart, F., Feigl, B., Frick, V., Stürmer, B., Zuberbühler, U. and Waldstein, G. (2010) *Renewable (power to) methane: Storing renewables by linking power and gas grids*, available at http://www.iset.uni-kassel.de (accessed 3 January 2012).

Wenzel, H. (2010) *Breaking the Biomass Bottleneck of the Fossil Free Society*, Concito report, Denmark, 34pp.

WWF (2011a) *Positive Energy: How Renewable Electricity can Transform the UK by 2030*, World Wide Fund For Nature, Godalming, UK, 64pp. Quotation on page 6. Table on page 7. Downloadable from http://assets. wwf.org.uk/downloads/positive_energy_final_designed.pdf (accessed 17 January 2012).

WWF (2011b) *The Energy Report: Part 1: 100% Renewable Energy by 2050; Part 2: The Ecofys Energy Scenario.* World Wide Fund for Nature, WWF International, Gland, Switzerland and Ecofys, Utrecht, the Netherlands. 256 pp. Downloadable from http://wwf.panda.org/what_we_do/footprint/climate_carbon_energy/energy_solutions/renewable_energy/sustainable_energy_report/ (accessed 17 January 2012).

Appendix A

Energy arithmetic – a quick reference

As explained in Chapter 1, this book has been written using standard scientific SI units; however many other units are widely in use. This appendix aims to provide a quick reference to ways of expressing very large and very small numbers, units and conversions between units. Section A1 describes the methods of specifying quantities in terms of powers of ten. Section A2 gives conversion factors between the main units used for energy and power throughout this book; it includes conversions to and from some commonly used US units. Section A3 relates some older or non-SI units to their SI equivalents. Formal definitions and more detailed accounts of the basis of the SI units can be found in *Quantities, Units and Symbols* (The Royal Society, 1975).

A1 Orders of magnitude

Discussions of energy consumption and production frequently involve very *large* numbers, and accounts of processes at the atomic level often need very *small* numbers. Two solutions to the problem of manipulating such numbers are described here: the use of a shorthand form of arithmetic and the use of prefixes.

Powers of ten

Two million is two times a million, and a million is ten times ten times ten times ten times ten times ten (that's six tens in all, being multiplied together). This can be written mathematically as:

$$2\,000\,000 = 2 \times 10 \times 10 \times 10 \times 10 \times 10 \times 10 = 2 \times 10^6.$$

The quantity 10^6 is called *ten to the power six* (or *ten to the six* for short), and the 6 is known as the **exponent**. The advantage of using this power-of-ten form is particularly obvious for very large numbers. World primary energy consumption in the year 2009, for instance, was 502 000 000 000 000 000 000 joules, and it is certainly easier to write (or say) 502×10^{18} joules than to spell out all the zeros.

The method can also be used for very small numbers, with the convention that one tenth (0.1) becomes 10^{-1}; one hundredth (0.01) becomes 10^{-2}, etc. The separation of two atoms in a typical metal, for instance, might be about 0.25 of a billionth of a metre, which is 0.000 000 000 25 m. In more compact form, it becomes 0.25×10^{-9} m.

Scientific notation is a way of writing numbers that is based on this concept, but with a more specific rule. Any number, whatever its magnitude, is written in scientific notation as a number between 1 and 10 multiplied by the appropriate power of ten. So in this form, the numbers above would be written slightly differently:

502×10^{18} becomes 5.02×10^{20}

0.25×10^{-9} becomes 2.5×10^{-10}.

In practice, styles like those on the left in each of these examples are commonly used, and a number in this form will be accepted by computers and scientific calculators – but it will be reproduced by them in either 'long decimal' or strict scientific notation, depending on the mode selected.

Prefixes

The powers of ten provide the basis for the prefixes used to indicate multiples (including sub-multiples) of units. The following table, parts of which appeared in Chapter 1 as Table 1.1, shows these in decreasing order.

Table A1 Prefixes

The sequence of the first six rows in this table may be conveniently remembered (in reverse order) with the phrase 'King Midas's Golden Touch Poisoned Everything'.

Symbol	Prefix	Multiply by	... which is
E	exa-	10^{18}	one quintillion
P	peta-	10^{15}	one quadrillion
T	tera-	10^{12}	one trillion
G	giga-	10^{9}	one billion
M	mega-	10^{6}	one million
k	kilo-	10^{3}	one thousand
h	hecto-	10^{2}	one hundred
da	deca-	10	ten
d	deci-	10^{-1}	one tenth
c	centi-	10^{-2}	one hundredth
m	milli-	10^{-3}	one thousandth
μ	micro-	10^{-6}	one millionth
n	nano-	10^{-9}	one billionth
p	pico-	10^{-12}	one trillionth

The terms 1 billion = 10^{9} and 1 trillion = 10^{12} are as used in this book. However, alternative long forms where 1 billion = 10^{12} and 1 trillion = 10^{18} may be found in older English books and in other languages.

In US usage the prefix M can be used to denote 1000 and (more commonly) MM to denote 1 million. In other circumstances US data may use M to denote 'metric', i.e. 1 Mt = 1 metric tonne. In this book we use the convention shown in Table A1.

Indian usage commonly includes the terms 1 lakh = 100 000 and 1 crore = 10 million.

A2 Units and conversions

Chapter 1 discussed some of the units used in specifying quantities of energy or power in the 'real world'. The tables in this section summarize the relationships between some of these units.

Energy

Commonly used energy units include the joule, the kilowatt-hour, the tonne of oil equivalent (toe) and tonne of coal equivalent (tce), i.e. the energy content of a tonne of an 'average' sample of these fuels. Tables A2 and A3 give the conversion factors between the most frequently used multiples of these units, on the 'household' scale (A2) and on the larger scale of national or world data (A3). The oil and coal equivalents are here expressed to two significant figures only. Since different data sources may use slightly different values and there may be confusions between lower and higher heating values (see Chapter 4, Box 4.5), you are advised always to take note of the precise conversion factors used for the energy content of fuels adopted by any source you use. Note that the factors for kWh and TWh assume a 100% conversion efficiency.

Table A2 Energy conversions at the household scale

	MJ	GJ	kWh	toe	tce
I MJ =	I	0.001	0.2778	2.4×10^{-5}	3.6×10^{-5}
I GJ =	1000	I	277.8	0.024	0.036
I kWh =	3.60	0.0036	I	8.6×10^{-5}	1.3×10^{-4}
I toe =	42 000	42	11 667	I	1.5
I tce =	28 000	28	7778	0.67	I

Table A3 Energy conversions at the national scale

	PJ	EJ	TWh	Mtoe	Mtce
I PJ =	I	0.001	0.2778	0.024	0.036
I EJ =	1000	I	277.8	24	36
I TWh =	3.60	0.0036	I	0.086	0.13
I Mtoe =	42	0.042	11.667	I	1.5
I Mtce =	28	0.028	7.778	0.67	I

Power

The kilowatt-hour is the standard unit of *energy* used for UK gas and electricity bills. *Power* is the rate at which energy is used, consumed, transferred or transformed. Its normal units are the watt and its multiple the kilowatt. A power of one kilowatt is equal to a rate of use of energy of one kilowatt-hour per hour. A power of one watt is equal to a rate of use of one joule per second.

Table A4 shows the quantities of energy per hour and per year for different constant rates in watts. Note that it can also be used to show that, for instance, 1 kWh is 3.6 MJ and 1 TWy (terawatt-year) is 31.54 EJ or 750 Mtoe.

Table A4 Power and rate of use of energy

Rate	Joules		Kilowatt-hours per year	Tonnes of oil equivalent per year	Tonnes of coal equivalent per year
	per hour	per year			
1 W	3.6 kJ	31.54 MJ	8.76	0.75×10^{-3} toe*	1.13×10^{-3} tce*
1 kW	3.6 MJ	31.54 GJ	8760	0.75 toe	1.13 tce
1 MW	3.6 GJ	31.54 TJ	8.76×10^6	750 toe	1130 tce
1 GW	3.6 TJ	31.54 PJ	8.76×10^9	0.75 Mtoe	1.13 Mtce
1 TW	3.6 PJ	31.54 EJ	8.76×10^{12}	750 Mtoe	1130 Mtce

* i.e. the energy equivalent of 0.75 kg of oil or 1.13 kg of coal

US energy units

Energy statistics in the USA are quoted in British Thermal Units (BTU). Until the 1990s UK statistics also used the BTU and a related unit, the therm.

Table A5 gives conversion factors at the 'household scale'.

Table A5 Energy conversions at the household scale (US energy units)

	MJ	Thousand BTU	kWh	Therm	toe
1 MJ =	1	0.948	0.2778	9.48×10^{-3}	2.4×10^{-5}
1000 BTU =	1.055	1	0.293	0.01	2.5×10^{-5}
1 kWh =	3.60	3.412	1	0.034	8.6×10^{-5}
1 therm =	105.5	100	29.3	1	2.5×10^{-3}
1 toe =	42 000	40 000	11 667	400	1

The power unit of the BTU h^{-1} is commonly used, even in the UK, for the ratings of domestic boilers and air conditioning plant; 1000 BTU h^{-1} = 0.293 kW.

Table A6 gives energy conversion factors at the 'national' level.

Table A6 Energy conversions at the national scale (US energy units)

	EJ	Quadrillion BTU (Quad)	TWh	Billion therms	Mtoe
1 EJ =	1	0.948	277.8	9.48	24
1 Quad =	1.055	1	293.1	10	25
1 TWh =	3.60×10^{-3}	3.41×10^{-3}	1	0.0341	0.086
1 billion therms =	0.1055	0.1	29.3	1	2.5
1 Mtoe =	0.042	0.040	11.667	0.40	1

A3 Other quantities

Table A7 gives the SI equivalents of a few other metric units and some older units that remain in common use. The final column shows the inverse relationships. For brevity, scientific notation is used for numbers greater than 10 000 or less than 0.1.

Table A7 SI equivalents

Quantity	Unit	SI equivalent	Inverse
Mass	1 oz (ounce)	$= 2.835 \times 10^{-2}\,kg$	$1\,kg = 35.27\,oz$
	1 lb (pound)	$= 0.4536\,kg$	$1\,kg = 2.205\,lb$
	1 ton ($= 2240\,lb$)	$= 1016\,kg$	$1\,kg = 0.9842 \times 10^{-3}\,ton$
	1 short ton ($= 2000\,lb$)	$= 907\,kg$	$1\,kg = 1.102 \times 10^{-3}\,short\ tons$
	1 t (tonne)	$= 1000\,kg$	$1\,kg = 10^{-3}\,t$
	1 u (unified mass unit)	$= 1.660 \times 10^{-27}\,kg$	$1\,kg = 6.024 \times 10^{26}\,u$
Length	1 in (inch)	$= 2.540 \times 10^{-2}\,m$	$1\,m = 39.37\,in$
	1 ft (foot)	$= 0.3048\,m$	$1\,m = 3.281\,ft$
	1 yd (yard)	$= 0.9144\,m$	$1\,m = 1.094\,yd$
	1 mi (mile)	$= 1.609 \times 10^{3}\,m$	$1\,m = 6.214 \times 10^{-4}\,mi$
Speed	1 km h^{-1} (kph)	$= 0.2778\,m\ s^{-1}$	$1\,m\ s^{-1} = 3.600\,kph$
	1 mi h^{-1} (mph)	$= 0.4470\,m\ s^{-1}$	$1\,m\ s^{-1} = 2.237\,mph$
	1 knot	$= 0.514\,m\ s^{-1}$	$1\,m\ s^{-1} = 1.944\,knots$
Area	1 in^2	$= 6.452 \times 10^{-4}\,m^2$	$1\,m^2 = 1550\,in^2$
	1 ft^2	$= 9.290 \times 10^{-2}\,m^2$	$1\,m^2 = 10.76\,ft^2$
	1 yd^2	$= 0.8361\,m^2$	$1\,m^2 = 1.196\,yd^2$
	1 acre	$= 4047\,m^2$	$1\,m^2 = 2.471 \times 10^{-4}\,acre$
	1 ha (hectare)	$= 10^{4}\,m^2$	$1\,m^2 = 10^{-4}\,ha$
	1 mi^2	$= 2.590 \times 10^{6}\,m^2$	$1\,m^2 = 3.861 \times 10^{-7}\,mi^2$
Volume	1 in^3	$= 1.639 \times 10^{-5}\,m^3$	$1\,m^3 = 6.102 \times 10^{4}\,in^3$
	1 ft^3	$= 2.832 \times 10^{-2}\,m^3$	$1\,m^3 = 35.31\,ft^3$
	1 yd^3	$= 0.7646\,m^3$	$1\,m^3 = 1.308\,yd^3$
	1 litre	$= 10^{-3}\,m^3$	$1\,m^3 = 1000\,litres$
	1 gallon (UK)	$= 4.546\,litres$	$1\,m^3 = 220.0\,gallons\ (UK)$
	1 gallon (US)	$= 3.785\,litres$	$1\,m^3 = 264.2\,gallons\ (US)$
	1 barrel	$= 159\,litres$	$1\,m^3 = 6.3\,barrels$
	1 acre-ft	$= 1.233 \times 10^{3}\,m^3$	$1\,m^3 = 0.811 \times 10^{-3}\,acre\text{-}ft$
	1 bushel	$= 3.637 \times 10^{-2}\,m^3$	$1\,m^3 = 27.50\,bushels$

(Continued)

Table A7 SI equivalents   *(Continued)*

Quantity	Unit	SI equivalent	Inverse
Force	1 kgf (weight of 1 kg mass)	$= 9.807\,\text{N}$	$1\,\text{N} = 0.102\,\text{kgf}$
	1 lbf (weight of 1 lb mass)	$= 4.448\,\text{N}$	$1\,\text{N} = 0.2248\,\text{lbf}$
Pressure	1 bar ($\approx$1 atmosphere)	$= 10^5\,\text{Pa (pascals)}$	$1\,\text{Pa} = 10^{-5}\,\text{bar}$
	1 kgf m^{-2}	$= 9.807\,\text{Pa}$	$1\,\text{Pa} = 0.102\,\text{kgf m}^{-2}$
	1 lbf in^{-2} (or psi)	$= 6895\,\text{Pa}$	$1\,\text{Pa} = 1.450 \times 10^{-4}\,\text{psi}$
Energy	1 barrel of oil equivalent (boe)	$= 5.7\,\text{GJ}$ (lower heating value)	$1\,\text{GJ} = 0.175\,\text{boe}$
	1 cal (calorie)	$= 4.2\,\text{J}$	$1\,\text{J} = 0.24\,\text{cal}$
	1 ft lb (foot pound)	$= 1.356\,\text{J}$	$1\,\text{J} = 0.7375\,\text{ft lb}$
	1 eV (electron-volt)	$= 1.602 \times 10^{-19}\,\text{J}$	$1\,\text{J} = 6.242 \times 10^{18}\,\text{eV}$
	1 MeV	$= 1.602 \times 10^{-13}\,\text{J}$	$1\,\text{J} = 6.242 \times 10^{12}\,\text{MeV}$
Power	1 HP (horse power)	$= 745.7\,\text{W}$	$1\,\text{W} = 1.341 \times 10^{-3}\,\text{HP}$

Reference

The Royal Society (1975) *Quantities, Units and Symbols*, London, The Royal Society.

Appendix B

Levelized costs of renewable energy

B1 Introduction

This book describes a wide range of renewable energy technologies; some, such as solar water heaters, produce only heat, while others, such as biomass combined heat and power (CHP), produce both heat and electricity. The technologies for which costs are most carefully studied are those electricity-generating ones that are in direct competition with fossil or nuclear-fuelled generating plant. This appendix aims to provide some background on how to approach the costing of electricity from such plant and explains the key terms used, although more details for each technology may be described in the individual chapters.

An ideal power plant would be available to produce electricity at any hour of the day or night throughout the year. There are, however, various constraints on many renewable energy technologies which limit the total amount of electricity that they might produce over a year.

The cost of electricity from any generating plant, whether it is fossil-fuelled, nuclear or renewably-powered has to take into account four main elements:

- fuel costs
- operation and maintenance (O&M) costs, including waste disposal
- initial capital costs
- final decommissioning costs.

In practice, for many renewable energy technologies the fuel costs are zero but there are large initial capital costs which, in some cases, can make up 80% or more of the final electricity cost. Nuclear power plants are in a similar position with low fuel costs but high capital costs as well as significant decommissioning costs. At the other extreme, generating plants fuelled by natural gas have low capital costs and high fuel costs, with the cost of the fuel making up 70% of the final electricity cost.

One accepted method of making comparisons between such different types of plant is to calculate an overall **levelized cost** for the electricity in pence per kilowatt hour over the life of the plant. This value for the electricity spreads ('levels') the costs of fuel, O&M, capital and interest repayments across a given time period, usually the life expectancy of the plant. Since inflation progressively erodes the purchasing power of money over time, such an estimate necessarily has to be expressed in money of a particular year, for example £(2010).

Levelized costs allow us to say that, for example, the cost of electricity from a land-based wind turbine is 8.26 p per kWh, while that from a wood-fuelled generation plant is 8.6p per kWh (even though the plants may operate for different numbers of hours per year and have different factors in their cost breakdowns).

The account below is very brief – greater detail on costings for different UK electricity generation technologies will be found in Mott Macdonald, 2010, and Everett et al., 2012 and particularly in Mott Macdonald, 2011 for renewable energy technologies.

B2 Basic contributions to costings

Factors affecting the amount of electricity generated

Rated capacity

This is the maximum power output of a plant; it is usually expressed in kilowatts (kW) or megawatts (MW). A combined heat and power (CHP) plant may have *two* ratings, one for electricity (MW_e) and one for heat (MW_t). The power output of a photovoltaic panel is usually expressed in terms of its peak power output under full bright sunlight (kWp).

Annual capacity factor

This is the ratio of the total electricity generated by a plant to the maximum that it might have produced if it operated continuously 24 hours a day, 365 days a year. There are many factors that can limit the annual capacity factor including:

(1) Plant availability – the proportion of the year that the plant is in full working order and not shut down for maintenance or repairs. Many mature renewable energy technologies have availabilities of 95% or more.

(2) Energy source availability – obviously the wind does not blow all the time, nor does the sun shine all the time. Technologies such as solar, wind, wave and tidal power are all quite tightly limited by the physical availability of their energy source. There are also considerations of matching electricity supply with demand (as described in Chapter 10). Some technologies, particularly large hydro plant, may be reserved for dealing with evening peaks in electricity demand rather than continuous base load generation. Other technologies such as biomass CHP plants may mainly operate in the winter when there is both a heat and an electricity demand.

Since the demand can have a significant effect on the total annual output, the capacity factor is sometimes also referred to as the *load factor*.

Typical capacity factors can range from about 10% for PV panels in cloudy countries like the UK through to 25%–40% for wind turbines and up to almost 100% for landfill gas.

The overall electricity produced per year can thus be expressed as:

electricity per year (kWh) = rating (kW) × capacity factor × 365 × 24 (hrs)

Factors affecting the total cost

Fuel costs

Although for many renewable energy technologies the 'fuel' is free, biomass technologies, in common with those using fossil fuels, have significant fuel costs. There are two 'costs' here:

- the cost of the purchased fuel
- the contribution of the fuel to the overall cost of the electricity generated.

The purchased fuel cost is likely to be expressed in £ per GJ (£ GJ^{-1}). For example, wood purchased at £60 per tonne with a calorific value of 18 GJ per tonne has a cost of £3.33 GJ^{-1} or 1.2 p kWh^{-1}. Other fuels, such as municipal solid waste (MSW) and landfill gas may be considered to have a *zero* cost, since their use is part of an overall disposal process.

The key factor in determining the contribution of this input energy cost to the final electricity cost is the *plant efficiency:*

$$\text{efficiency} = \frac{\text{energy output}}{\text{energy input}}.$$

Efficiency values may range from 25% for MSW plant up to 35% or more for modern large wood-fired steam plant.

The contribution to the final electricity cost is likely to be expressed in pence per kWh or £ per MWh (where 1 p kWh^{-1} = £10 MWh^{-1}).

$$\text{fuel cost per kwh generated} = \frac{\text{cost of energy input}}{\text{efficiency}}$$

Even though many renewable energy technologies have zero fuel costs, it should not be thought that the conversion efficiency is unimportant for them. A 1 kWpk PV panel with an efficiency of 20% will only require half the area of one with an efficiency of 10%.

Operation and maintenance (O&M) costs

These can be subdivided into two categories: 'fixed' costs which are not related to the plant output, and 'variable' costs, which are likely to be proportional to the plant output.

Fixed costs include site insurance, grid connection fees, regular safety inspections, etc. These may be expressed in terms of a certain percentage of the initial capital cost per year. Obviously access to the plant is a key issue, which is why O&M costs for offshore wind turbines are considerably higher than those for onshore ones.

Variable costs include, for example, fuel handling and ash disposal and the replacement of equipment damaged by 'wear and tear'. Variable O&M costs are usually quoted in terms of pence per kWh of output. For simplicity, the overall O&M costs may also be expressed in this way as well.

The actual figures vary widely between technologies, from 1.5 p kWh^{-1} for land-based wind up to 7 p kWh^{-1} for energy from waste plants.

Capital costs and plant life

These are often the most important factors in renewable energy plants. The capital cost is usually expressed in £ per kW of rated capacity. A large onshore wind turbine may have a capital cost of £1500 per kW of generation capacity. The cost for an offshore one may be £2500 per kW or more. This is in contrast to about £800 per kW for a competing fossil-fuelled combined cycle gas turbine (CCGT) power station.

Given these high capital costs, the expected plant life is very important. Wind turbines and PV panels may have life expectancies of over 20 years. A hydro plant or a tidal barrage might be expected to last for 60 years or more. A further factor is the construction time. A large land-based wind turbine can be erected and be operational within a year, but a large hydro plant (or a nuclear power station) may take a number of years to construct. This is time during which money is tied up without any incoming cash flow from generated electricity.

B3 Calculating costs

This section starts with a simple payback time calculation before moving on to examples of increasing complexity using levelized costing.

The case study used for these calculations is of a 1 MW (1000 kW) land-based wind turbine generator – it has a capital cost of £1500 per kW, its estimated operating and maintenance costs are £37 500 per year. It has a design life of 25 years and a capacity factor of 28%. Its construction time is very short, and there are, of course, zero fuel costs.

Payback time

For the simplest of calculations we can take the capital cost, the energy output and a competing energy price to calculate a 'payback time', the number of years taken to recover the capital outlay. This simple calculation obviously ignores O&M payments and any interest payments on the capital.

So, given a competing bulk (fossil fuel) electricity price of 6p kWh^{-1}:

> Annual electricity generation = rated capacity × capacity factor × 24 × 365
> $\quad$ = 1 × 28% × 24 × 365 = 2453 MWh
> Total capital cost = 1000 × £1500 = £1 500 000
> Value of annual electricity generation = 2453 × 1000 × 6p/100 = £147 200
> Payback time = capital cost/value of annual output
> $\quad$ = £1 500 000/£147 200
> $\quad$ = **10.2 years**

Of course during this time a further $10.2 \times £37\,500 = £382\,500$ was incurred in O&M costs which, if included, would push the payback time up further.

Simple annual levelized cost

Alternatively, a 'levelized cost' can be calculated on an annual basis. The capital can be considered to be repaid in equal annual amounts over the project lifetime (such a stream of identical payments is known as an *annuity*). Again, for simplicity the example here assumes 0% interest on the capital, so although the capital is being repaid there are no extra interest payments. The average O&M costs and fuel costs (if any) are added to these annual payments, and the average cost of the electricity is then given by:

$$\text{cost per kwh} = \frac{(\text{annual capital repayment} + \text{average annual running cost})}{\text{average annual energy output}}$$

So the calculation becomes:

Cost of capital spread over 25 years = £1 500 000/25 = £60 000 y^{-1}
Operating and maintenance costs $\qquad$ = £37 500 y^{-1}
Total annual costs $\qquad$ = £97 500 y^{-1}
Overall electricity cost = £97 500 / 2453 MWh $\qquad$ = £39.7 MWh^{-1}
$\qquad$ **= 3.97 p kWh^{-1}**.

Using discounted cash flow

In practice, these calculations are too simplistic because a pound earned or spent tomorrow is not worth the same as a pound today, for a number of reasons.

Firstly, there is our **time preference for money**. Given a choice, most people would rather have a pound today than a pound in the future. Put another way, we would need to be offered a pound plus some additional sum, say $x\%$, next year to forgo the use of one pound today.

Secondly, we have the ability to lend out money and charge interest. We can forgo the use of a pound today in order to have a pound plus an additional sum in the future. If the interest rate is high enough and the time long enough, this sum can be appreciable. For example, if we invested £100 at a 10% rate of interest per annum, we would expect to be able to withdraw £260 in 10 years' time. We can say that £100 has a **future value** of £260. Put another way, the **present value** of £260 in ten years' time at an interest rate of 10% is only £100.

We are 'discounting' future payments, saying that sums of money in the future can be expressed in terms of smaller sums today.

We can express this relationship mathematically as:

$$V_\text{p} = \frac{V_n}{(1 + r)^n}$$

where V_p is the present value of a sum of money V_n in n years time subject to a discount rate of r.

This concept leads to a technique of economic appraisal known as **discounted cash flow** (DCF) analysis. This can be carried out using standard computer spreadsheet functions.

Thirdly, **inflation** is another factor that needs to be included. This can be thought of as a 'disease of money', progressively eroding its value over time. In order to adjust for its effects, we should use a 'real' interest rate, rather than the purely monetary one. This is simple enough:

real interest rate = monetary interest rate − rate of inflation

Generally, the terms 'discount rate' and 'real interest rate' tend to be used interchangeably.

Now if a capital sum, say V_p, is to be repaid as an annuity (annual amounts of equal size running over the duration of a loan of n years) the present value of an annuity, A, given a discount rate r is:

$$V_p = \frac{A}{(1+r)} + \frac{A}{(1+r)^2} + \frac{A}{(1+r)^3} + \dots + \frac{A}{(1+r)^n}.$$

A little bit of mathematical rearrangement gives a more convenient form, in which the annuity is expressed in terms of the capital sum, discount rate and length of loan:

$$A = V_p \times r / (1 - (1+r)^{-n}$$

The ratio A/V_p is sometimes called a **capital recovery factor**.

The mathematical function for an annuity is now a standard spreadsheet function: PMT (Payment, Month, Term). However, it is not actually necessary to use a spreadsheet for very simple calculations. Before the era of computers and spreadsheets, banks used simple pre-calculated tables to work out repayments on borrowed money. This approach is still useful today. Table B1 shows the annual repayments on £1000 of borrowed capital.

Table B1 Annuitized value of capital costs (annual repayment in £ per £1000 of capital) for various discount rates and capital repayment periods

Capital repayment period / years[1]	Discount rate / %						
	0	2	5	8	10	12	15
5	200	212	231	250	264	277	298
10	100	111	130	149	163	177	199
15	67	78	96	117	131	147	171
20	50	61	80	102	117	134	160
25	40	51	71	94	110	127	155
30	33	45	65	89	106	124	152
40	25	37	58	84	102	121	151
50	20	32	55	82	101	120	150
60	17	29	53	81	100	120	150

[1] This is not necessarily equal to the total physical lifetime of the project.

Generally, a loan is agreed to be paid back over a certain number of years. However, financial institutions are often unwilling to consider loans spread over more than 25 years because of the essential uncertainty about the future, although the full working life of a hydroelectric plant may be in excess of a hundred years. Thus it is worth remembering that the capital repayment time may be shorter than the working lifetime of a scheme.

Levelized cost of electricity from a wind turbine including discounting

Using Table B1, the previous example can be revisited. Assuming a discount rate of 10% and a project lifetime of 25 years the calculation is as follows.

First, find the annuitized value of the capital cost. From Table B1 the annual repayments on £1000 over 25 years at 10% per year = £110 per year.

For a borrowed sum of £1 500 000 they will be £1500 × £110 = £165 000.

Note that this is considerably higher than the annual repayment figure of £60 000 used in the simple annual levelized cost example. (That figure can be derived using the annuitized value of £1000 over 25 years at 0% per year = £40 per year. £1500 × £40 = £60 000.)

The annual costs now are:

Cost of capital spread over 25 years	= £165 000 y^{-1}
Operating and maintenance costs	= £37 500 y^{-1}
Total annual costs	= £202 500 y^{-1}

Overall cost per kWh = £202 500/2453 MWh = £82.6 MWh^{-1}

= 8.26 p kWh^{-1}

Thus, using a discount rate of 10% and a project lifetime of 25 years, we find that the original estimate of 3.97 p kWh^{-1} is an underestimate and arrive at a much higher figure of 8.26 p kWh^{-1}.

The annuitized capital costs now make up over 80% of the total.

Levelized cost of electricity from a wood-fuelled generation plant including discounting

A useful comparison is to derive the levelized electricity cost for a generation method where the calculation includes a fuel cost and the plant efficiency.

A proposed wood-fuelled 100 MW generation plant using a fluidized bed boiler and a steam turbine has a capital cost of £2500 per kW. It is expected to run for 7500 hours per year at full power (i.e. it has a capacity factor of 7500/8760 = 85.6%) and has a design lifetime of 25 years. Its overall electrical generation efficiency is 36%. Operating and maintenance (O&M) costs have been estimated at 1.6p per kWh of electricity generated. It is assumed that wood will be available at £3.33 per GJ or 1.2 p per kWh over the lifetime of the plant.

If the station unit runs for 7500 hours per year, in each year it will produce: 7500 h × 100 MW = 750 000 MWh of electricity

The total capital cost is: £2500 × 100 000 = £250 million

First, the annuitized value of the capital cost is needed. From Table B1 the annuitized value of £1000 over 25 years at 10% per year = £110 per year.

Therefore, for a borrowed sum of £250 million the annual repayments will be: 250 000 × £110 = £27.5 million

Annuitized capital cost per kWh of electricity produced is:

27.5 × 1 000 000 × 100 / 750 000 000 = 3.67p kWh^{-1}

The fuel cost per kWh of electricity generated is:

1.2 p kWh^{-1}/ 36%	= 3.33 p kWh^{-1}
O&M costs	= 1.60 p kWh^{-1}
Total cost per kWh	= **8.60 p kWh^{-1}**

Note that in this example the fuel costs account for nearly 40% of the final electricity cost.

More complex calculations

The use of a simple annuitization table is fine if the construction time is very short, and the O&M payments and output are uniform over the life of the project. For a more accurate analysis, particularly for projects such as large-scale hydro and tidal schemes, which may take many years to build, and be subject to periodic refurbishment, a full 'Net Present Value' calculation is required.

This process involves five steps:

(1) Itemizing the capital and running costs for each year of the project life.
(2) Calculating the separate Present Values of all these annual costs using an appropriate discount rate, and summing them to give a Net Present Value (NPV – this is a standard spreadsheet function).
(3) Itemizing the electrical output for each year over the project lifetime.
(4) Calculating the value of the electricity produced. The levelized cost calculation assumes an inflation corrected cost (i.e. it is the same value, P pence per kilowatt hour, for the whole life of the plant). The monetary Net Present Value of the electricity produced will then be $P \times$ NPV (kilowatt-hours produced)
(5) Calculating the unit cost of electricity in pence per kWh as:

$$P = \frac{\text{Net Present Value of costs (pence)}}{\text{Net Present Value of output (kWh)}}.$$

Mott Macdonald, 2010 and Everett et al., 2012 contain sample calculations of this type.

Choice of discount rate

Many renewable energy technologies involve high capital costs, a characteristic also shared by nuclear power. As such, their economics are critically dependent on the cost of capital and choice of discount rate.

Past UK evaluations of the economics of renewable energy projects have used discount rates of 8% and even 15%, these being taken as the expectations of earnings in the private sector. As described in Chapter 10,

Section 10.5, in practice the 'riskiness' of the technology has to be taken into account. Mature technologies such as PV, hydro and land-based wind may be funded with lower discount rates than newer ones such as wave or tidal stream power (see Mott Macdonald, 2011).

References

Everett, B., Boyle, G. A., Peake, S. and Ramage, J. (eds) (2012) *Energy Systems and Sustainability: Power for a Sustainable Future* (2nd edn), Oxford, Oxford University Press/Milton Keynes, The Open University.

Mott MacDonald (2010) *UK Electricity Generation Costs Update* [online], Mott MacDonald; http://www.decc.gov.uk/assets/decc/statistics/projections/71-uk-electricity-generation-costs-update-.pdf (accessed 25 April 2012).

Mott MacDonald (2011) *Costs of low-carbon generation technologies*, [online], Mott MacDonald; available at http://hmccc.s3.amazonaws.com/Renewables%20Review/MML%20final%20report%20for%20CCC%209%20may%202011.pdf (accessed 25 April 2012).

Acknowledgements

Grateful acknowledgement is made to the following sources:

Chapter 1

Figures

Figure 1.4: *IEA World Energy Outlook Report* (2010), International Energy Agency; Figure 1.8: *IPCC Climate change 2007 Synthesis report*, Intergovernmental Panel on Climate Change; Figure 1.10: *Committee on Climate Change (2011) The Renewable Energy Review*, Committee on Climate Change.

Chapter 2

Figures

Figure 2.1: Arcon Solvarme A/S, Denmark; Figures 2.4, 2.6, 2.34, 2.35 and 2.41: courtesy of Bob Everett; Figure 2.16: EST (2010) *Getting Warmer: a field trial of heat pumps*, Energy Saving Trust; Figure 2.17: Energy Saving Trust; Figure 2.18: © National Energy Foundation 2011; Figure 2.24: Sunmark A/S, Denmark; Figure 2.25: Courtesy of M.G.Davies; Figure 2.27: courtesy of Derek Taylor; Figures 2.29 and 2.30: Siviour, J. B. (1977) Houses as Passive Solar Collectors, ECRM/M10710. Electricity Council Research Centre; Figures 2.31 and 2.32: Courtesy of Luwoge; Figure 2.37: Department of Energy 1987; Figure 2.38: Hockerton Housing Project; Figure 2.39: Mazra, E. (1979) *The Passive Solar Energy Book*, Rodale Press Inc; Figure 2.40: Guildhall Library; Figure 2.43: Courtesy of Biblioteca Ciudades para un futuro màs sostenible, Madrid; Figure 2.45: Koza (1983), used under a Creative Commons Attribution Licence; Figures 2.46 and 2.47: Hank Morgan/Science Photo Library; Figure 2.48: David Nunuk/Science Photo Library; Figure 2.50: www.sbp.de (Schlaich Bergermann and Partner); Figure 2.51: courtesy of Stephen Peake.

Chapter 3

Figures

Figure 3.1: Bibliothèque Nationale de France; Figure 3.3: Popperfoto/Getty Images; Figures 3.4 and 3.5: NASA; Figure 3.6: EPIA (2011) Global Market Outlook for Photoltaics to 2015 report 2011, European Photovoltaics Industry Association; Figure 3.7: IEA (2010) Trends in Photovoltaic Applications 1992–2009, Report IEA PVPS T1-19:2010, International Energy Agency; Figure 3.11: Green, M. (1982) Solar Cells, Prentice-Hall; Figure 3.14: Photo credit: Heraeus; Figure 3.15: Sharp Corporation; Figure 3.17 Spectrolab; Figure 3.18: Clean Venture 21 Corp; Figures 3.22a and 3.30: Martin Bond/ Science Photo Library; Figure 3.22b: Kaj R. Svensson/Science Photo Library; Figure 3.22c: James King-Holmes/Science Photo Library; Figure 3.23: TERI The Energy and Resources Institute; Figure 3.24: courtesy of Susan Roaf;

Chapter 4

Figures

Chapter 5

Figures

Chapter 6

Figures

Figure 6.1: Media Library, EDF/Gerard Halary; Figures 6.3, 6.31, 6.32 and 6.39: Peter Fraenkel, Marine Current Turbines Limited, www.marineturbine.com; Figure 6.7, 6.10, 6.11, 6.12, 6.14, 6.17 and 6.20: Tidal Power from the Severn Estuary, Volume 1, Energy Paper 46 1981, Dept of Energy. Crown copyright material is reproduced under Class Licence number C01W0000065 with the permission of the Controller of HMSO and the Queen's Printer for Scotland; Figure 6.8: United Kingdom Hydrographic Office (1992) Admiralty Tidal Stream Atlas: The English Channel NP250, © British Crown Copyright 2011. All rights reserved; Figure 6.9: Twidell, J.W. and Weir, A.J. (1986) Renewable Energy Sources, E. & F.N. Spon, Routledge; Figure 6.13: Report of the legislative Assembly of Western Australia, presented by Hon. Ian Thompson, November 1991, Department of Housing and Works, Australia; Figures 6.15 and 6.16: Reproduced courtesy of Mr L. Carson and Power Magazine, March 1978; Figure 6.18: © ETSU Department of Trade and Industry; Figures 6.19 and 6.21: The Severn Barrage Project: General Report, Energy paper 57, 1989, Department of Energy. Crown copyright material is reproduced under Class Licence Number C01W0000065 with the permission of the Controller of HMSO and the Queen's Printer for Scotland; Figure 6.22: Laughton, M.A. (1990) Renewable Energy Sources, Report 22, E. & F.N. Spon, Routledge; Figure 6.23: Frontier Economics 'Analysis of a Severn Barrage', Report for the NGO steering group, 2008, © Frontier Economics Ltd, London; Figure 6.25: Tidal Electric Inc, www.tidalelectric.com; Figure 6.27: DECC (2010) 'Severn Tidal Power Feasibility Study: Conclusions and Summary Report' Department of Energy and Climate Change; Figure 6.30 SeaGen Uncovered, Technical Insight on MCTs, Existing and Next Generation Devices by Frankael, P. MCT 4th International Tidal Energy Summit London Nov 2010, Marine Current Turbines Ltd; Figure 6.33: Based on an image supplied by the Engineering Business. The Patented Stingray tidal stream is being developed by the Engineering Business, who will provide project information on their website updated at regular intervals. The first Stingray generator was installed in Shetland summer 2002 with 75% DTI funding. The Stingray demonstrator is about 20m high and weighs about 40 tonnes with a nominal 150kW output in a 4 knot current, www.engb.com; Figure 6.34: Pulse Tidal Limited; Figure 6.35: Rotech Engineering; Figure 6.36: Neptune Renewable Energy Limited; Figure 6.37: Open Hydro; Figure 6.38: www.tidalstream.co.uk; Figure 6.40: The Severn Tidal Fence Group (IT Power Ltd.); Figures 6.42 and 6.43: VerdErg Renewable Energy Ltd; Figure 6.44: Hammerfest Strøm; Figure 6.45: Atlantis Resources Corporation and the Government of Novia Scotia; Figure 6.46: By courtesy of Voith Hydro; Figure 6.47: Verdant Power; Figure 6.48: Ocean Renewable Power Company.

Chapter 7

Figures

Figure 7.1: Marlec Engineering Ltd; Figure 7.2: Burroughs, W. J. et al. (1996) Weather – The Ultimate Guide to the Elements, Harper Collins; Figure 7.4:

Farndon, J. (1992) How the Earth Works, Dorling Kindersley Ltd; Figure 7.8: Needham, J. with the collaboration of Ling, W. (1965) Science and Civilisation in China, Vol 4, Part II. Cambridge University Press; Figure 7.10: Derek Taylor/Altechnica; Figure 7.11: John Mead/Science Photo Library; Figures 7.12, 7.16, 7.17a: courtesy of Bob Everett; Figure 7.13: courtesy of Vestas Wind A/S; Figure 7.14: Stewart Boyle; Figures 7.15, 7.17b and 7.38: Derek Taylor/Altechnica; Figure 7.32: Time for Action: Wind Energy in Europe (1991) European Wind Energy Association; Figure 7.34: SkyscraperPage.com; Figure 7.36: source unknown; Figure 7.37: Westmill Farm Co-operative, www.westmill.coop; Figure 7.39: GWEC (2010) The Global Wind Energy Outlook Scenarios, The Global Wind Energy Council; Figure 7.40: courtesy of Jfz. used under a Creative Commons Attribution 3.0 licence; Figure 7.41: Statoil; Figure 7.42: Musial, W. and Ram, B. (2010) Large Scale Wind Power in the United States – Assessment of Opportunities and barriers, National Laboratory for Renewable Energy; Figure 7.43b and c: Jamie Cook/Vattenfall; Figure 7.44: Atlas of UK Marine Renewable Energy Resources, Crown copyright material is reproduced under Class Licence Number C01W0000065 with permission of the Controller of HMSO and the Queen's Printer for Scotland.

Chapter 8

Figures

Figure 8.1b: courtesy of Les Duckers; Figures 8.4 and 8.32b: Martin Bond/ Science Photo Library; Figure 8.9a: adapted from Claeson, L. (1987) 'Energi fran havets vagor', Energiforkningsnamnden nr 21; Figure 8.13: Department for Business, Enterprise and Regulatory Reform (2008) Atlas of Marine Renewable Energy Resources, Crown copyright material is reproduced under Class Licence Number C01W0000065 with the permission of the Controller, Office of Public Sector Information (OPSI); Figure 8.15: adapted from Falnes, J. and Løvseth, J (1991) Energy Policy, vol 19, no 8, October 1991, Butterworth Heinemann; Figure 8.18: courtesy of Jamie Taylor, Edinburgh University Wave Power Group; Figure 8.24: Aquamarine Power; Figure 8.27: courtesy of AquaEnergy Group Ltd; Figure 8.29: OceanEnergy Ltd.; Figure 8.30: Offshore Wave Energy Limited; Figure 8.31: Oceanlinx Limited; Figure 8.33: Taylor Keogh Communications (on behalf of AWS Ocean Energy); Figure 8.34: Pelamis Wave Power.

Chapter 9

Figures

Figure 9.2a: © Atlantide Phototravel/Corbis; Figure 9.2b: © Drimi/ Dreamstime.com; Figure 9.5: reprinted from Earth Science Reviews, Vol 19, No 1, Henley and Ellis, 'Conceptual model of a typical geothermal system' © 1983 Elsevier Science; Figures 9.13 and 9.14: G.E.I.E. "Exploitation Minière de la Chaleur"; Figure 9.15: Government of South Australia; Figure 9.17: REUTERS/Stefan Wermuth; Figure 9.20: IPR/31-12 British Geological Survey © NERC, All Rights Reserved.

Chapter 10

Figures

Figure 10.2: courtesy of Transco; Figure 10.3: National Grid; Figure 10.5: courtesy of Logstor Ror; Figure 10.6: DECC Energy in Brief; Figure 10.10: Committee on Climate Change (2008) Building a low-carbon economy, Crown copyright material is reproduced under Class Licence Number C01W0000065 with the permission of the Controller, Office of Public Sector Information (OPSI); Figures 10.12 and 10.15: Bob Everett; Figure 10.13: National Grid (2010) National Grid Seven Year Statement, National Grid; Figure 10.17a: Renewable Sources of Electricity in the SWEB area: Future Prospects, SWEB and the Department of Trade and Industry; Figure 10.17b: Bodlund, B. et al. (1989) 'The challenge of choices: technology option for the Swedish electricity sector', Johansson, T. B., Bodlund, B. and Williams, R. H. eds. (1989) Electricity – Efficient end use and new technologies and their planning implications, Lund University Press; Figure 10.18: Statoil; Figure 10.19: The Offshore Valuation Group; Figure 10.20: DESERTEC Foundation, www.desertec.org; Figure 10.21: courtesy of British Petroleum; Figure 10.25: European Climate Foundation; Figure 10.27: Danish Energy Agency 2010; Figure 10.28: World Wildlife Fund (2011) The Energy Report, WWF.

Every effort has been made to contact copyright holders. If any have been inadvertently overlooked the publishers will be pleased to make the necessary arrangements at the first opportunity.

Index